Equine Genomics

Equine Genomics

Editor
BHANU P. CHOWDHARY
Texas A&M University
College Station, Texas, USA

WILEY-BLACKWELL
A John Wiley & Sons, Inc., Publication

Wiley-Blackwell is an imprint of John Wiley & Sons, formed by the merger of Wiley's global Scientific, Technical and Medical business with Blackwell Publishing.

Editorial Offices
2121 State Avenue, Ames, Iowa 50014-8300, USA
The Atrium, Southern Gate, Chichester, West Sussex, PO19 8SQ, UK
9600 Garsington Road, Oxford, OX4 2DQ, UK

For details of our global editorial offices, for customer services and for information about how to apply for permission to reuse the copyright material in this book please see our website at www.wiley.com/wiley-blackwell.

Library of Congress Cataloging-in-Publication Data

Equine genomics / editor, Bhanu P. Chowdhary, Texas A&M University, College Station, Texas, USA.
pages cm
Includes bibliographical references and index.
ISBN 978-0-8138-1563-3 (hardback : alk. paper) 1. Horses–Genetics. 2. Domestic animals–Genome mapping. I. Chowdhary, Bhanu P., editor of compilation.
SF291.E64 2013
636.1′0821–dc23

2012038413

A catalogue record for this book is available from the British Library.

Wiley also publishes its books in a variety of electronic formats. Some content that appears in print may not be available in electronic books.

Cover design by Modern Alchemy LLC

Set in 10.5/12 pt Times by Aptara® Inc., New Delhi, India
Printed and bound in Malaysia by Vivar Printing Sdn Bhd

1 2013

This is dedicated to all contributing authors for enthusiastically sharing their knowledge and discoveries and my beloved parents, Mohini and Pranveer Singh, for their love, support, and encouragement throughout my learning.

Contents

Contributors ix

Preface xi

Chapter 1 **Defining the equine genome: The nuclear genome and the mitochondrial genome** 1
Bhanu P. Chowdhary

Chapter 2 **Genetic linkage maps** 11
June Swinburne and Gabriella Lindgren

Chapter 3 **Physical and comparative maps** 49
Terje Raudsepp and Bhanu P. Chowdhary

Chapter 4 **The Y-chromosome** 73
Terje Raudsepp, Nandina Paria, and Bhanu P. Chowdhary

Chapter 5 **Unexpected structural features of the equine major histocompatibility complex** 93
Loren C. Skow and Candice L. Brinkmeyer-Langford

Chapter 6 **Assembly and analysis of the equine genome sequence** 103
Claire M. Wade

Chapter 7 **Genomic tools and resources: Development and applications of an equine SNP genotyping array** 113
Molly McCue and Jim Mickelson

Chapter 8 **Functional genomics** 125
Stephen J. Coleman, Michael J. Mienaltowski, and James N. MacLeod

Chapter 9 **Coat color genomics** 143
Samantha A. Brooks and Rebecca R. Bellone

Chapter 10	**Genomics of skin disorders** Amy E. Young, Stephen D. White, and Danika L. Bannasvch	**155**
Chapter 11	**Genomics of muscle disorders** James R. Mickelson, Stephanie J. Valberg, Carrie J. Finno, and Molly E. McCue	**171**
Chapter 12	**Genomics of skeletal disorders** Ottmar Distl	**187**
Chapter 13	**Genomics of reproduction and fertility** Terje Raudsepp, Pranab J. Das and Bhanu P. Chowdhary	**199**
Chapter 14	**Genetics of equine neurologic disease** Carrie J. Finno and Monica Aleman	**217**
Chapter 15	**Molecular genetic testing and karyotyping in the horse** M. C. T. Penedo and Terje Raudsepp	**241**
Chapter 16	**Genomics of laminitis** Jim K. Belknap	**255**
Chapter 17	**Genomics of performance** Emmeline W. Hill, Lisa M. Katz, and David E. MacHugh	**265**
Chapter 18	**Genomics of the circadian clock** Barbara A. Murphy	**285**
Chapter 19	**Mitochondrial genome: Clues about the evolution of extant equids and genomic diversity of horse breeds** Cynthia C. Steiner, Kateryna D. Makova, and Oliver A. Ryder	**311**
Index		**323**

Color plate is located between pages 216 and 217.

Contributors

Monica Aleman	University of California, Davis, California, USA
Danika L. Bannasvch	University of California, Davis, California, USA
Jim K. Belknap	College of Veterinary Medicine, The Ohio State University, Columbus, Ohio, USA
Rebecca R. Bellone	The University of Tampa, Tampa, Florida, USA
Candice L. Brinkmeyer-Langford	Texas A&M University, College Station, Texas, USA
Samantha A. Brooks	Cornell University, Ithaca, New York, USA
Bhanu P. Chowdhary	Texas A&M University, College Station, Texas, USA
Stephen J. Coleman	University of Kentucky, Lexington, Kentucky, USA
Pranab J. Das	Texas A&M University, College Station, Texas, USA
Ottmar Distl	University of Veterinary Medicine Hannover, Hanover, Germany
Carrie J. Finno	University of California, Davis, California, USA
Emmeline W. Hill	School of Agriculture, Food Science and Veterinary Medicine, University College Dublin, Belfield, Dublin, Ireland
Lisa M. Katz	School of Agriculture, Food Science and Veterinary Medicine, University College Dublin, Belfield, Dublin, Ireland
Gabriella Lindgren	Swedish University of Agricultural Sciences, Uppsala, Sweden
David E. MacHugh	School of Agriculture, Food Science and Veterinary Medicine, University College Dublin, Belfield, Dublin, Ireland
James N. MacLeod	University of Kentucky, Lexington, Kentucky, USA
Kateryna D. Makova	Department of Biology, Pennsylvania State University, University Park, Pennsylvania, USA

Molly McCue	University of Minnesota, St. Paul, Minnesota, USA
Jim Mickelson	University of Minnesota, St. Paul, Minnesota, USA
Michael J. Mienaltowski	University of Kentucky, Lexington, Kentucky, USA
Barbara A. Murphy	School of Agriculture, Food Science and Veterinary Medicine, University College Dublin, Belfield, Dublin, Ireland
Nandina Paria	Texas A&M University, College Station, Texas, USA
M. C. T. Penedo	Cornell University, Ithaca, New York, USA
Terje Raudsepp	Texas A&M University, College Station, Texas, USA
Oliver A. Ryder	Genetics Division, San Diego Zoo Institute for Conservation Research, Escondido, California, USA
Loren C. Skow	Texas A&M University, College Station, Texas, USA
Cynthia C. Steiner	Genetics Division, San Diego Zoo Institute for Conservation Research, Escondido, California, USA
June Swinburne	Animal Health Trust, Newmarket, United Kingdom
Stephanie J. Valberg	University of Minnesota, St. Paul, Minnesota, USA
Claire M. Wade	The University of Sydney, New South Wales, Australia
Stephen D. White	University of California, Davis, California, USA
Amy E. Young	University of California, Davis, California, USA

Preface

Equine Genomics largely summarizes a range of accelerated research activities conducted by equine geneticists and researchers during the past two decades. I was privileged to have been closely involved with some of the key activities and was fortunate to have witnessed the progression of several exciting ones. I can classify the two decades into two distinct eras, each of which uniquely contributed to the advancement in equine genetics. The first decade laid the foundation for a much needed knowledgebase of the equine genome, while the second built on its progress and slowly but steadily allowed us to start reaping the benefits. So, who deserves kudos for this progress? In my view, it is due to the extraordinary combined effort of several entities that enables us to read about genetics underlying about 40 equine diseases and more than 20 phenotypes in this book. Twenty years ago, the possibility of this level of achievement was unthinkable. While sitting in a researcher's chair, it would be easy for me to glorify the work of my colleagues worldwide and attribute the success to them. However, in all honesty, the credit goes much beyond this small, yet dedicated, group. The progressive community of horse owners, the scientifically demanding yet generous funding agencies – federal, state, and private – the ever approachable and helping clinicians, the increasingly open-minded breed associations, the highly supportive foundations, and the unrelenting horse enthusiasts worldwide have played a vital role in converting the "unthinkable" into "possible". Despite being a rather small community compared to researchers of other species, it is this extraordinary support, cooperation, and collaboration that allowed us to make this enviable advancement. In essence, the group has made strides, and the horse is better off with a healthier future. This book is a timely tribute to all of them.

As a group we are, and must remain to be, mindful that the real work on such discoveries has just begun. Fruits that next need to be picked will be more difficult to reach (as the lower most are already picked). Innovation, creativity, and teamwork will continue to be the mantras for our progress in the coming decades, as will be support from a more determined, vocal, and generous group of advocates who will ensure that ample funding for equine genetics research is readily available. New priority areas in equine genetics research will become more time consuming and challenging and will hence require additional funding. Moreover, many equine diseases, if not all, will be probed through the "genetics" lens. This is a reality, and the sooner the entire equine community embraces this concept, the better off we will be during coming years.

In all honesty, no project is fun unless it poses challenges. As much as challenges energize me, I soon found myself pulled in several directions after having started this project. Three things did not let me fail: the enthusiasm and support of the contributing authors, never exhausting patience of the Wiley crew Justin Jeffryes and Anna Ehler, and lastly Terje Raudsepp, who was always there to help with every "next step" along the road. Without them, this book would not have been possible.

I hope a wide spectrum of readers will find *Equine Genomics* informative and enjoyable. It is the first attempt to compile and present all research developments in this field. Despite its "scientifically inclined" tone, all horse enthusiasts will find something that they can relate to, understand, and use. Knowledge in any field is never finite. Having said that, take this book as a sincere attempt from all of us to share with you what we know in the field of *Equine Genomics*. There will always be more to report as we progress, which we will!

Bhanu P. Chowdhary, BVSc&AH, MVSc. VMD (PhD)
Professor and Associate Dean for Research and Graduate Studies

1 Defining the equine genome: The nuclear genome and the mitochondrial genome

Bhanu P. Chowdhary

A genome represents the genetic material or the core regulatory and functional machinery of any organism. In the horse, as is typically the case in most mammals, the genetic material largely resides in the nucleus of the cell, while a fractional yet functionally vital component is present in the cytoplasm. The former is packaged in the form of *chromosomes* and the latter in the form of *mitochondria*. It is difficult to assign greater significance to one over the other, hence a debate in that realm has to be left at accepting that the genes contained in each of the components have an important role to play, individually as well as in conjunction with other genes in a network or pathway, bringing wholeness to the functioning of the horse.

This chapter first provides an overview of the nuclear genome in terms of chromosome number, standard karyotype, and chromosome nomenclature, and highlights basic facts regarding equine chromosomes. The description includes a brief introduction to structural and chromosomal aberrations and their impact on overall viability of horses. Finally, a summarized overview of the current knowledge of the equine mitochondrial genome is presented. The information serves as a foundation and a reference point to all aspects of the equine genome presented and discussed in relation to mapping, genetic variation, phenotypic variations, diseases, and other such information in subsequent chapters.

Nuclear Genome of the Horse

Chromosome number, karyotype, and schematic presentation

The size of the nuclear part of the equine (*Equus caballus;* ECA) genome is around 2.7 megabase pairs (Mbps). The nuclear genome is packaged in 64 chromosomes (diploid chromosome number presented as $2n = 64$) that may be metacentric, sub-metacentric, or acrocentric. It is noteworthy that the genus *Equus* has 8 extant species and the chromosome number in these species varies from $2n = 32$ in the Hartmann's mountain zebra to 102 in the Domestic ass and the Mongolian and Transcaspian wild asses, although marked similarities in size and gene content exists between them (Chowdhary & Raudsepp, 2000). The 64 horse chromosomes are comprised of 32 pairs of homologues (one coming from sire/stallion and the other from the dam/mare), of which 31 pairs are autosomes that have an identical gene set present essentially in the same order from one end to another, and one pair is the sex chromosomes that are similar in females (two X chromosomes)

Equine Genomics, First Edition. Edited by Bhanu P. Chowdhary.

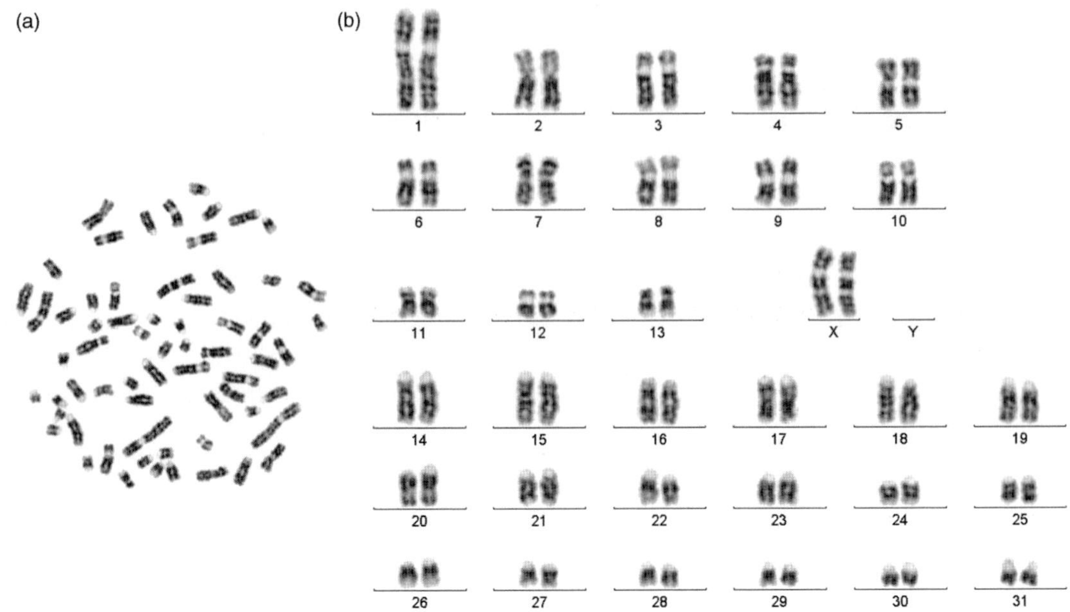

Figure 1.1 A G-banded metaphase spread (left) and a karyotype (right) of a normal female horse (2n = 64XX).

but different in the males (one X chromosome and one Y chromosome). The X-chromosome is the second-largest element among the equine chromosomes and forms about 5% of the genome, whereas the Y is one of the smallest (Chowdhary & Raudsepp, 2000).

Horse chromosomes have now been studied under the microscope for more than a century. Though Makino (1942) established that, like other mammals, horse also has an XY sex chromosome system, the correct chromosome number in horses – and the confirmation of this number by independent groups – became known much later (Rothfels et al., 1959; Moorhead et al., 1960; Makino et al., 1963). The 14 pairs of metacentric and sub-metacentric horse chromosomes, including the sex chromosomes and the 18 pairs of acrocentric autosomes, are arranged in a specific pattern (referred to as the *karyotype*; see Figure 1.1) for a quick visual assessment of the chromosomes. The arrangement has evolved over the past several decades through iterations of internationally agreed "standardizations" of which the last, and potentially the final, one was carried out by the International System for Chromosome Nomenclature of the domestic Horse (ISCNH, 1997). This standardized karyotype gave a detailed description of both G- and R-banded chromosomes along with schematic drawings (idiograms) of individual chromosome pairs showing landmarks and band numbers for each of the key banding approaches. Three size resolutions ranging from moderate to reasonably elongated chromosomes are provided to facilitate easy identification of the chromosomes. Overall, the current standard karyotype serves as a basis of "cross-talk" on individual horse chromosomes among equine cytogeneticists and researchers worldwide.

Application of different banding techniques

A variety of staining/banding techniques have been applied to analyze equine chromosomes. The ultimate goal of using these banding techniques was to exploit a range of structural and functional

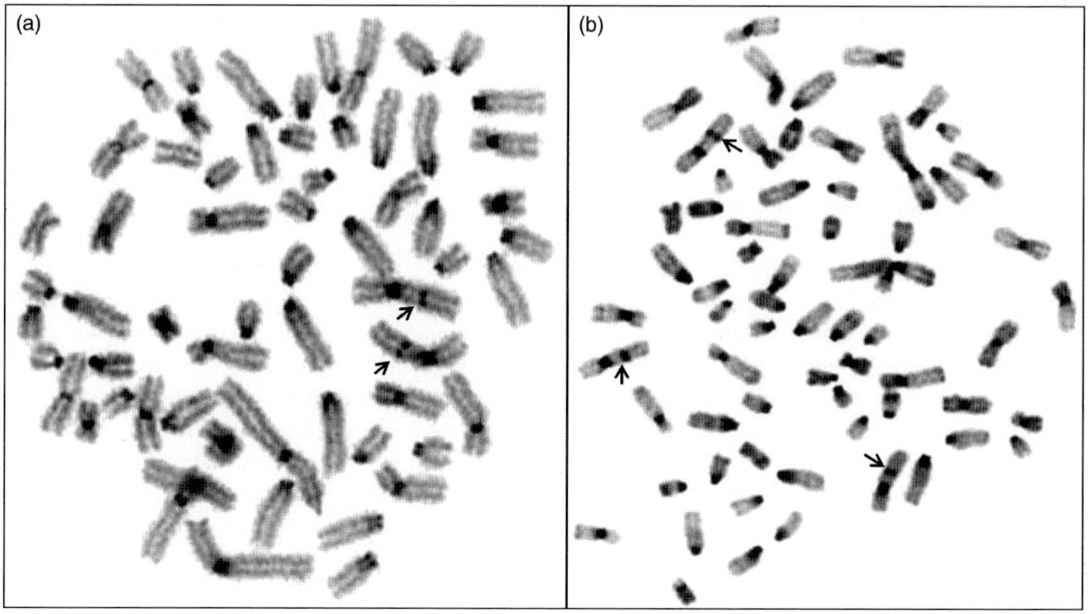

Figure 1.2 Two C-banded metaphase spreads showing distinct heterochromatic band at the centromere of most chromosomes. Additional sites (intercalary) are apparent on the long arm of the X chromosome (arrows). Left: a normal female horse (2n = 64XX); right: a female horse with X chromosome trisomy (2n = 65XXX).

features that permit unambiguous discrimination of individual pairs of homologous chromosomes. The dyes used to stain the chromosomes can be fluorescent or nonfluorescent. The most common staining technique for visualization of chromosomes is the use of Giemsa (Trujillo et al., 1962). The Q-banding (Caspersson et al., 1970) and G-banding (Seabright, 1971) techniques highlight AT-rich DNA as positive bands. The R-banding technique (positive bands are reverse to the G- and Q-positive bands) primarily highlights GC-rich DNA by fluorescent (RBG; Molteni et al., 1982) as well as heat treatment/Giemsa-staining-based approaches (RHG; e.g., Power, 1987). These bands can be enhanced by obtaining elongated chromosomes via using cell cycle synchronization and incorporation of bromodeoxyuridine (BrdU) into chromosomal DNA (e.g., Romagnano et al., 1983; Marki and Osterhoff, 1983; Power, 1987a, 1990) as done by Rønne et al. (1993). Essentially, bands are like bar codes that are unique for a chromosome pair among the pairs of homologous chromosomes in a species, which allows distinguishing the homologues from the set of chromosomes in a metaphase spread.

Some banding techniques like the C-banding specifically depict constitutive heterochromatin. The "bands" or regions are highlighted by acidic and/or alkaline and heat treatment, and are seen on almost all horse chromosomes, except ECA11. Interstitial heterochromatin is reported on ECA1pter, 12q, and Xq (most prominent), whereas the Y-chromosome is known to be largely heterochromatic (Figure 1.2). T-banding, which highlights telomeric repeat sequences present at their usual terminal location of the chromosome with no intercalary presence, is observed in horses. They were first reported using molecular cytogenetic approaches (de La Seña et al., 1995). Next, regions containing transcriptionally active ribosomal RNA genes (rDNA or nucleolus organizing regions, NORs) are visualized by NOR-banding carried out by staining chromosome preparations with silver nitrate (Goodpasture and Bloom, 1975). Molecular approaches allow identification of both active and inactive regions in a cell. Typically, ECA1, 27, 28, and 31 are considered as the NOR-bearing chromosomes. In addition to these banding techniques, attempts have also been made to produce bands on equine chromosomes specific for electron microscopy (Messier et al., 1989; Richer et al., 1989), which has even allowed the construction of a karyotype.

Chromosome aberrations – a brief overview

A vast array of reports on chromosomal abnormalities shows that equine chromosomes have been fairly extensively studied, even though the number of published reports is substantially lower than in cattle and pigs. As in other mammals, these chromosomal aberrations are classified into two major categories, namely (1) autosomal (includes autosome-sex chromosome and vice versa translocations) and (2) those involving only the sex chromosomes. Each of the categories is further classified into two subheads: *structural* and *numerical*. A large number of mosaic/chimeric karyotypes involving the sex chromosomes, as well as cases of sex-reversal, have been reported. We estimate that not more than 3,000 horses (predominantly mares) have been examined cytogenetically, resulting in the detection of about 400 cases with chromosome aberrations. This count does not include about 120 cases of sex reversal. As the majority of these cases were hand-picked, either due to phenotypic deviations or due to fertility problems, no proportionate evaluation of the frequency of chromosomal aberrations in the horse could be projected.

Rather few *structural* aberrations have been detected in horses. These include deletions (loss of part of a chromosome), translocations (both balanced and unbalanced), inversion, and duplication. Figure 1.3 shows one of the most recently discovered balanced translocations in the horse that has been analyzed using traditional and molecular techniques (Das et al., forthcoming). All *numerical* aberrations in horse are trisomies of different autosomes, all of which involve small acrocentric chromosomes (nos. 23–31). Some of the salient features of viable individuals with autosomal aberrations (both structural and numerical) are: reduced to completely impaired fertility, minor to moderate anatomical malformations primarily affecting gait or orientation, atypical poor confirmation, and moderate to notably smaller height as compared to the breed/parents average. Like in other species, production of offspring to full term by individuals bearing balanced reciprocal translocations and tandem fusions has been reported in the horse (see Power, 1991; Long, 1996; Durkin et al., forthcoming; Das et al., forthcoming); still, invariably varying degrees of reduced fertility is observed. However, foaling of an ECA26 trisomy carrier to produce a karyotypically normal colt (Bowling and Millon, 1990) is noteworthy.

The vast majority of chromosomal abnormalities reported in the horse involve the sex chromosome (predominantly the X). Of the approximately 350 abnormal karyotypes reported to date, more than 90% involve the sex chromosomes. The types of *structural* chromosome abnormalities involving the sex chromosomes vary from translocations and deletions to isochromosomes, invariably all impacting fertility. A wide range of *numerical* aberrations of the sex chromosomes are known in the horse, of which monosomy (lack) of the X chromosome (karyotype constitution 63,XO) is the most common. Few cases of pure trisomy of the X chromosome (65,XXX) have also been reported in the horse. Typically, XO and XXX mares are infertile. Next, cases of pure 65,XXY (Kubien et al., 1993) and 66,XXXY (Gluhovschi et al., 1970, 1975) have also been reported.

Finally, various categories of mosaics/chimeras involving the sex chromosomes (primarily the X) have hitherto been reported in horses. It is noteworthy that most of the reported chimeric/mosaic cases are registered as females with a wide range of clinically detectable reproductive system deviations. The animals show varying degree of virilization of the external genitalia.

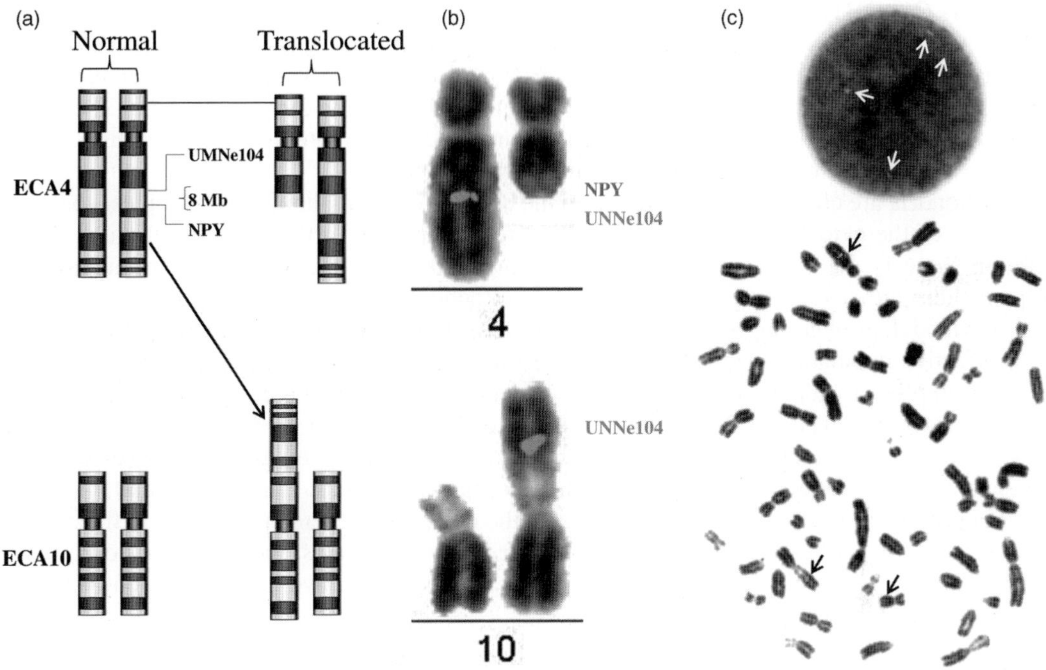

Figure 1.3 Heterozygous autosomal translocation between chromosomes 4 and 10 in a mare with a history of recurrent early embryonic loss: (a) distal half of the long arm of chromosome 4 is inverted 180° and translocated to the tip of the short arm of chromosome 10 (red arrows); (b) Fluorescence *in situ* hybridization (FISH) showing the translocation breakpoint to be in chromosome 4 between markers NPY and UMNe104; (c) Same FISH results shown in interphase (upper) and metaphase chromosomes (lower).

Sex reversal, most appropriately described as a disagreement between the karyotypic sex and the phenotypic/anatomical sex, has been described in horses. Approximately 135 such cases have to date been observed. Normally, individuals with XY sex chromosome constitution are expected to be males (64,XY). Sex reversal is a condition in which the aforementioned individuals appear more like females, with varying degrees of male-like characteristics. The terminology holds even if the karyotype indicates the animal to be a female (64,XX), but phenotype and clinical examination show the preponderance of male-like characteristics. Some of these cases and associated molecular work we carried out to analyze these cases in relation to the observed micro-deletions on the horse Y chromosome are described in Chapter 5.

Mitochondrial Genome of the Horse

Structure, function, and utility

The mitochondria (mt) are the powerhouses of cells and are responsible for more than 90% of mammalian energy production. All aerobic respiration within cells occurs in this organelle, making it one of the most critical structures for eukaryotes. There are five important protein complexes that are part of the electron transport chain that metabolizes carbohydrates and fatty chain amino acids during production of ATP. The horse mitochondrial genome is ~16,660 bp in length and includes

13 proteins (NADH1, NADH2, NADH3, NADH4L, NADH4, NADH5, NADH6, COXI, COXII, COXIII, CYTB, ATP6, and ATP8) that are part of 5 protein complexes, 22 transfer RNAs, and 2 ribosomal RNAs (12 rRNA and 16s rRNA; Xu & Arnanson 1994). The mt genome also has a highly variable control region approximately ~1,192 bp long including 1 to 29 repeats of the GTGCACCT motif often exhibiting heteroplasmy. The remaining functional proteins within the mitochondria, and those used to replicate it, are encoded in the nuclear genome.

Mitochondria are clonally replicated within the cell and occur in thousands of copies, composing up to 70% of the cytoplasm. The mitochondria are only passed to the offspring through the egg, although there is evidence for an occasional transfer through sperm. As a consequence of its biology, there are several important features of mitochondrial genome that govern its variation: (1) non-Mendelian maternal inheritance, (2) non-recombination, (3) faster evolution rate, and (4) lower effective population size. There are more variants per bp of sequence in the mitochondrial genome compared to nuclear loci because the sequence evolution rate is 5- to 15-fold higher than in the nuclear genome. The mtDNA genome may not always be useful for reconstructing population history because all mitochondrial genes are inherited as one unit. Consequently, even studies that sequence the entire mt genome only reflect the lineage of one locus. Further, the mtDNA may not reflect gene flow in the nuclear genome. Even in the case where two populations are fixed for two different mtDNA haplotypes, if dispersal is primarily through males, there may be no population structure among nuclear loci. Despite these issues, mtDNA loci have provided much insight into horse evolution and history.

Phylogenetics of Equus

The first studies that examined horse mtDNA attempted to resolve the rapid radiation of modern equids. Initially, estimates of divergence were made from restriction enzyme studies, which generated sequence divergence estimates of around 2% per million years, found that the horse was the basal in the *Equus* group, and clustered the wild asses with zebras. However, these studies failed to resolve the other relationships in this genus. The Przewalski's haplotype was grouped with the domestic horse, suggesting that it may not be a distinct species (Jansen et al., 2002). Additional studies sequenced a few gene segments, including the variable control region and 12S rRNA, and found further evidence supporting the above patterns (Oakenfull et al., 2000). In 1994, the entire mt genome of horses was sequenced. The gene content was identical to other mammals (Xu & Arnason 1994). The horse mt genome was compared to the donkey (*Equus asinus*) and exhibited an overall nucleotide difference of 6.9%, with highest divergence in the control region (11.2%), followed by the 13 protein coding genes (8.0%), 2 rRNA genes (4.1%), with the 22 tRNA genes being most conserved (3.5%) (Xu et al., 1996). The greatest amino acid divergence in coding genes was in NADH6 and ATP8 (Xu et al., 1996); both of these are proteins under positive selection across many mammals (da Fonseca et al., 2008).

Over the next decade most of the mtDNA studies used the control region to explore diversity of different breeds (Hill et al., 2002; Jansen et al., 2002). There were high levels of haplotype diversity observed – for example, 17 haplotypes among 100 thoroughbreds sampled (Hill et al., 2002). The findings have led to increased interest in determining whether these patterns reflected multiple domestication events or repeated introgression from the now extinct wild populations. Several studies examined mtDNA diversity in great detail, sequencing the control region in wider array of breeds and individuals (Jansen et al., 2002). In most breeds there were a large number of haplotypes that did not structure geographically.

Several landmark studies compared the mtDNA of archeological and subfossil horse segments with contemporary breeds. These included studies that sequenced ancient samples from Alaska, Europe, and Asia (Jansen et al., 2002; Cieslak et al., 2010; Priskin et al., 2010). In studies with large sample sizes, the mtDNA diversity of both ancient and contemporary horses was high, offering supporting evidence that the maternal lineages present in the domestic horse are the result of both high levels of ancestral variation and frequent maternal introgression from wild populations across the Eurasia (Cieslak et al., 2010). Most of these studies were based on the control region that is highly variable and has high levels of mutation rates that can mask population history (Cieslak et al., 2010; Priskin et al., 2010). The most recent studies reexamined the questions on the domestication of horses, patterns on introgression, and the number of matrilines that were involved in the formation of the domestic horse (Lippold et al., 2011; Achilli et al., 2012). These studies sequenced up to 83 entire mt genomes and supported patterns found previously (Achilli et al., 2012), indicating that there were many diverse mares founding the domestic populations, large amounts of dispersal across Eurasia, coalescence of modern haplotypes to around ~90,000–160,000 years before present (ybp), and a strong signal for population expansion around 5,000–8,000 ybp consistent with cultural evidence for domestication (Lippold et al., 2011; Achilli et al., 2012). The high level of mt genome diversity in the horse exceeds that of any other domestic animal.

New research directions

To date, the research on mt genome of horses has been limited to using these loci as markers for tracking the evolution, domestication, and population structure of horses. The future research directions on equine mtDNA are in the biology of the mitochondrion and the functional aspects of mt genomics. The mt proteins are important in the physiology of horses as they directly affect ATP production, and therefore many of these should be under selection. This suggests a mitochondrial genomic contribution to athletic performance. In addition, mt mutations have been found to cause a diverse array of disease in humans, including diplopia, ataxia, and Pearson syndrome. Mt disorder can have neurological, cardiac, respiratory, gastrointestinal, endocrinal, and ophthalmological manifestations. Considering the importance of mt genome for cellular respiration, the positive selection detected in genes such as NADH4, NADH5, and NADH6 across mammals, and the diverse array of diseases in humans caused by mt mutations, equine mt genomics is a fertile field yet to be fully explored.

Acknowledgments

Support provided by the Texas A&M CVM Link Equine Research Endowment and USDA grant # 2007-35604-17946 is gratefully acknowledged.

References

Achilli, A., Olivieri, A., Soares, P., Lancioni, H., Hooshiar Kashani, B. et al. (2012). Mitochondrial genomes from modern horses reveal the major haplogroups that underwent domestication. *Proceedings of the National Academy of Sciences of the United States of America, 109*, 2449–2454.

Bloom, S. E., & Goodpasture, C. (1976). An improved technique for selective silver staining of nucleolar organizer regions in human chromosomes. *Human Genetics*, *34*, 199–206.

Bowling, A. T., & Millon, L. V. (1990). Two autosomal trisomies in the horse: 64,XX,-26, + t(26q26q) and 65,XX, + 30. *Genome*, *33*, 679–682.

Caspersson, T., Zech, L., Johansson, C., & Modest, E. J. (1970). Identification of human chromosomes by DNA-binding fluorescent agents. *Chromosoma*, *30*, 215–227.

Chowdhary, B. P., & Raudsepp, T. (2000). Cytogenetics and physical gene maps. In A. T. Bowling & A. Ruvinsky (eds.), *The Genetics of the Horse* (pp. 171–242). Wallingford, Oxon: CABI.

Cieslak, M., Pruvost, M., Benecke, N., Hofreiter, M., Morales, A. *et al.* (2010). Origin and history of mitochondrial DNA lineages in domestic horses. *PLoS One*, *5*, e15311.

da Fonseca, R. R., Johnson, W. E., O'Brien, S. J., Ramos, M. J., & Antunes, A. (2008). The adaptive evolution of the mammalian mitochondrial genome. *BMC Genomics*, *9*, 119.

de la Seña, C., Chowdhary, B. P., & Gustavsson, I. (1995). Localization of the telomeric (TTAGGG)n sequences in chromosomes of some domestic animals by fluorescence in situ hybridization. *Hereditas*, *123*, 269–274.

Dionne, F. T., Turcotte, L., Thibault, M. C., Boulay, M. R., Skinner, J. S. *et al.* (1993). Mitochondrial DNA sequence polymorphism, VO2max, and response to endurance training. *Medical Science of Sports Exercise*, *25*, 766–774.

Gluhovschi, N., Bistriceanu, M., & Palicica, R. (1975). Les troubles de la reproduction chez les animaux domestique dus à des modifications du genome. *Medecine Veterinaire*, *44*, 155–163.

Gluhovschi, N., Bistriceanu, M., Suciu, A., & Bratu, M. (1970). A case of intersexuality in the horse with type 2A + XXXY chromosome formula. *The British Veterinary Journal*, *126*, 522–525.

Harrison, S. P., & Turrion-Gomez, J. L. (2006). Mitochondrial DNA: an important female contribution to thoroughbred racehorse performance. *Mitochondrion*, *6*, 53–63.

Hill, E. W., Bradley, D. G., Al-Barody, M., Ertugrul, O., Splan, R.K. *et al.* (2002). History and integrity of thoroughbred dam lines revealed in equine mtDNA variation. *Animal Genetics*, *33*, 287–294.

International System for Cytogenetic Nomenclature of the domestic Horse (ISCNH), Bowling, A. T., Breen, M., Chowdhary, B. P., Hirota, K. *et al.* (1997). Report of the Third International Committee for the Standardization of the domestic horse karyotype, Davis, CA, 1996. *Chromosome Research*, *5*, 433–443.

Jansen, T., Forster, P., Levine, M. A., Oelke, H., Hurles, M. *et al.* (2002). Mitochondrial DNA and the origins of the domestic horse. *Proceedings of the National Academy of Sciences of the United States of America*, *99*, 10905–10910.

Jiang, F., Miao, Y.-W., Liu, B., Shen, X., Ren, J.-F. *et al.* (2008). Mitochondrial introgressions into the nuclear genome of the domestic horse. *Zoological Research*, *29*, 577–584.

Khrapko, K. (2011). The timing of mitochondrial DNA mutations in aging. *Nature Genetics*, *43*, 726–727.

Kubien, E. M., Pozor, M. A., & Tischner, M. (1993). Clinical, cytogenetic and endocrine evaluation of a horse with a 65,XXY karyotype. *Equine Veterinary Journal*, *25*, 333–335.

Kumar, D. P., & Sangeetha, N. (2009). Mitochondrial DNA mutations and male infertility. *Indian Journal of Human Genetics*, *15*, 93–97.

Lippold, S., Matzke, N. J., Reissmann, M., & Hofreiter, M. (2011). Whole mitochondrial genome sequencing of domestic horses reveals incorporation of extensive wild horse diversity during domestication. *BMC Evolutionary Biology*, *11*, 328.

Long, S. E. (1996). Tandem 1;30 translocation: a new structural abnormality in the horse (Equus caballus). *Cytogenetics and Cell Genetics*, *72*, 162–163.

Makino, S. (1942). The chromosomes of the horse (*Equus caballus*). *Cytologia*, *13*, 26–38.

Makino, S., Sofuni, T., & Sasaki, M. S. (1963). A revised study of the chromosomes of the horse, the ass and the mule. *Proceedings of the Japanese Academy*, *39*, 176–181.

Marki, U., & Osterhoff, D. R. (1983). Horse lymphocyte cell synchronization: Improved technique for chromosome analysis. *Journal of South African Veterinary Association*, *54*, 223–224.

Messier, P. E., Drouin, R., & Richer, C. L. (1989). Electron microscopy of gold-labeled human and equine chromosomes. *Journal of Histochemistry and Cytochemistry*, *37*, 1443–1447.

Molteni, L., Cribiu, E. P., De-Giovanni-Machi, A., Beltrami, E., & Succi, G. (1982). The R-banded karyotype of the domestic horse. Paper presented at the Proceedings of the 5th European Colloquium on Cytogenetics of Domestic Animals (Milan).

Moorhead, P. S., Nowell, P. C., Mellman, W. J., Battips, D. M., & Hungerford, D. A. (1960). Chromosome preparations of leukocytes cultured from human peripheral blood. *Experimental Cell Research*, *20*, 613–616.

Oakenfull, E. A., Lim, H. N., & Ryder, O. A. (2000). A survey of equid mitochondrial DNA: Implications for the evolution, genetic diversity and conservation of Equus. *Conservation Genetics*, *1*, 341–355.

Power, M. (1987). The chromosomes of the horse:Karyotype definition, clinical applications and comparative interspecies studies. PhD Thesis, National University of Ireland, Dublin.

Power, M. M. (1990). Chromosomes of the horse. In R. A. McFeely (ed.), *Advances in veterinary science and comparative medicine, domestic animal cytogenetics* (pp. 131–167). San Diego, CA: Academic Press.

Power, M. M. (1991). The first description of a balanced reciprocal translocation [t(1q;3q)] and its clinical effects in a mare. *Equine Veterinary Journal*, *23*, 146–149.

Priskin, K., Szabo, K., Tomory, G., Bogacsi-Szabo, E., Csanyi, B. *et al.* (2010). Mitochondrial sequence variation in ancient horses from the Carpathian Basin and possible modern relatives. *Genetica*, *138*, 211–218.

Ricchetti, M., Tekaia, F., & Dujon, B. (2004). Continued colonization of the human genome by mitochondrial DNA. *PLoS Biology*, *2*, E273.

Richer, C. L., Drouin, R., & Messier, P. E. (1989). Banding of horse chromosomes for electron microscopy. Paper presented at the 6th North American Colloquium on Cytogenetics of Domestic Animals, Purdue.

Romagnano, A., Richer, C. L., & Betteridge, K. J. (1983). R-banded chromosomes of the domestic horse obtained after bromodeoxyuridine incorporation by the fluorescence-photolysis-Giemsa technique. *Journal of Dairy Science*, *66*(Suppl. 1), 251.

Rønne, M., Gyldenholm, A. O., & Storm, C. O. (1993). The RBG-banded karyotype of Equus caballus at the 525-band stage. *Hereditas*, *118*, 195–199.

Rothfels, K. H., Alexrad, A. A., Siminovitch, L., & McCulloch Parker, R. C. (1959). The origin of altered cell lines from mouse, monkey and man, as indicated by chromosome and transplantation studies. *Proceeding of Canadian Cancer Research Conference*, 3, 189–214.

Saxena, R., de Bakker, P. I., Singer, K., Mootha, V., Burtt, N. *et al.* (2006). Comprehensive association testing of common mitochondrial DNA variation in metabolic disease. *American Journal of Human Genetics*, *79*, 54–61.

Seabright, M. (1971). A rapid banding technique for human chromosomes. *Lancet*, *2*, 971–972.

Taylor, R. W., & Turnbull, D. M. (2005). Mitochondrial DNA mutations in human disease. *National Review of Genetics*, *6*, 389–402.

Trujillo, J. M., Stenius, C., Christian, L. C., & Ohno, S. (1962). Chromosomes of the horse, the donkey, and the mule. *Chromosoma*, *13*, 243–248.

Turner, C., Killoran, C., Thomas, N. S., Rosenberg, M., Chuzhanova, N. A. *et al.* (2003). Human genetic disease caused by de novo mitochondrial-nuclear DNA transfer. *Human Genetics*, *112*, 303–309.

Tyynismaa, H., & Suomalainen, A. (2009). Mouse models of mitochondrial DNA defects and their relevance for human disease. *EMBO Report*, *10*, 137–143.

Wallace, D. C. (2010). Mitochondrial DNA mutations in disease and aging. *Environtal Molecular Mutagens*, *51*, 440–450.

Xu, X., & Arnason, U. (1994). The complete mitochondrial DNA sequence of the horse, Equus caballus: Extensive heteroplasmy of the control region. *Gene*, *148*, 357–362.

Xu, X., Gullberg, A., & Arnason, U. (1996). The complete mitochondrial DNA (mtDNA) of the donkey and mtDNA comparisons among four closely related mammalian species-pairs. *Journal of Molecular Evolution*, *43*, 438–446.

2 Genetic linkage maps

June Swinburne and Gabriella Lindgren

Introduction

From the first meeting of the International Equine Gene Mapping Workshop in Lexington, Kentucky, in October 1995 until the sequencing of the horse genome in 2007 (http://www.broadinstitute.org/mammals/horse; Wade et al., 2009), the primary activity of the equine genetics academic research community was the development of integrated maps of the horse genome. Together with somatic cell hybrid, radiation hybrid, and physical/cytogenetic maps based on fluorescence in situ hybridization (FISH), the generation of genetic linkage maps was critical to achieve this goal. The primary driving force for these endeavors was to map genetic variants that underlie disease, reproduction, growth, and other interesting traits such as coat color. Knowing the genetic basis for these traits would allow for informed breeding decisions to reduce the levels of disease or select for desired characteristics. Furthermore, it would enable investigation of the population structure of horse breeds and their relationship to other equids and also contribute to the greater understanding of the evolution of the mammalian genome. Additionally, linkage maps would provide scaffolding for the assembly of the horse genome sequence. Even subsequent to the release of the horse genome sequence and the development of a single nucleotide polymorphism (SNP) microarray, the utility of the linkage maps continues. This is evidenced by recent publications about the use of genetic linkage analysis to map various congenital disorders and diseases (Mittmann et al., 2010; Swinburne et al., 2009; Zeitz et al., 2009; Lampe et al., 2009; Andersson et al., 2008). The continuing usefulness of the equine linkage map will be described later in this chapter.

Genetic Linkage Maps

Chromosomes are inherited intact from one generation to the next except where rearrangements caused by recombination events – or crossovers – occur during gamete formation. The further apart the two loci are, the more likely it is that a recombination event will take place between them. Linkage maps provide a representation of this genetic separation between loci on chromosomes – in other words, higher frequencies of recombination are represented by greater distances on the linkage map. A genetic map therefore illustrates which markers belong to the same linkage group, their relative order, and the distance between them. The distance along the map is measured in centiMorgans (cM); 1 cM is defined as a 1% probability that the two positions will be separated by

Equine Genomics, First Edition. Edited by Bhanu P. Chowdhary.

recombination in one generation. Since recombination rates vary widely across the genome, genetic distance is not directly related to physical distance; in regions of high recombination, known as recombination hot spots, the genetic distance will widen relative to the physical distance, and vice versa. On average, however, 1 cM is equal to 1 megabase (Mb).

Linkage maps therefore illustrate the likelihood of markers being inherited together, whereas physical maps provide a straightforward representation of distance in base-pairs. These maps are complementary to one another.

There are two essential components for generating linkage maps. Each of these components is described separately in the following sections. First, a large number of polymorphic markers, usually microsatellites, are required. Second, suitable reference pedigrees are necessary. The generation of a linkage map then requires the genotyping of the individuals from the pedigrees with each of the polymorphic markers. Linkage mapping software is then used to identify groups of markers that originate on the same chromosome and therefore exhibit significant linkage – hence linkage groups – by calculating logarithm of the odds (LOD) scores for each pair. LOD scores of over 3 are considered statistical evidence of linkage. A multipoint analysis is then performed on each linkage group to identify the most likely order of markers and the distance between them; these are based on the assumption that the most parsimonious order – that is, the order which necessitates the fewest crossover events – will be the most likely. Finally, each linkage group is assigned to a chromosome; this is traditionally achieved using FISH. In the generation of the linkage maps described below, each of these have utilized the software CRIMAP (Lander & Green, 1987) and MULTIMAP (Matise et al., 1994) to generate the linkage groups and perform the multipoint analysis.

Although the idea behind linkage mapping is quite simple, there are a number of pitfalls to be aware of and to accommodate for. For example, over long chromosomal distances it is possible that there will be two crossovers that could lead to recombinants being scored as non-recombinants. Therefore, recombination frequencies are not additive. Another problem is the phenomenon of interference; in positive interference a crossover has the effect of reducing the probability of a second crossover in its vicinity. Some mapping functions take interference into account whereas others do not. A number of mapping functions have been derived depending on the degree of interference assumed (Kosambi, 1944). Increasing the number of genetic markers will make the map more accurate with respect to these anomalies.

The genotyping of sufficient numbers of markers and sufficient numbers of individuals required to construct a linkage map only became possible with the advent of efficient methods for genotyping in the late 1980s–early 1990s. At this time great efforts were made to generate the first linkage maps of livestock and domesticated species such as pig, cattle, sheep, and dog based on microsatellites (Ellegren et al., 1993; Beever et al., 1994; Crawford et al., 1995; Werner et al., 1999).

Polymorphic Genetic Markers

The first essential components required for the construction of linkage maps are genetic markers that demonstrate polymorphism. Traditionally linkage maps have been constructed using microsatellite markers due to their ease of identification in the absence of genome sequence, and their increased numbers of alleles compared to SNPs. Microsatellites consist of tandem repeats where the repeat units occur immediately adjacent to one another and vary from 1-6 base pairs in length. The different alleles can be distinguished using molecular techniques such as polymerase chain reaction (PCR) followed by electrophoresis, the higher numbers of alleles providing increased power for mapping. In addition, microsatellites are abundant and widely dispersed throughout the genome.

Dinucleotide microsatellites have mainly been used in the horse; a likely explanation for this is that several research groups reported that they found it easier to isolate dinucleotide repeats than trinucleotide repeats in this species. The first horse microsatellites were identified by Ellegren et al. (1992). At present there are more than 24,000 microsatellite submissions to GenBank (http://www.ncbi.nlm.nih.gov/genbank/), which are identified using a search for ("Equus caballus"[Organism] OR horse [Organism]) AND microsatellite [All Fields]) (queried August 13, 2010). The majority of these has been submitted subsequent to the release of the genome sequence and illustrate the ease with which microsatellites can now be identified and located in the genome in an automated fashion.

Although highly polymorphic microsatellites are powerful markers for linkage mapping, these are increasingly being replaced by SNPs. The advantages of SNP arrays include fast, efficient, and highly parallel genotyping. Additionally their genotyping and analysis is more easily automated, is cheaper, and has a lower error rate. The lack of informativeness for biallelic SNPs, compared to highly polymorphic microsatellites, is offset by their dense and uniform distribution throughout the genome. Typically tens of thousands of SNPs are genotyped in a genome-wide scan, compared to several hundreds of microsatellites. SNP availability is not a limiting factor, as one SNP is found on average every 1,050 bases in the horse genome (Wade et al., 2009). However, microsatellite scans still provide an inexpensive low-resolution approach that can be performed by most basically equipped laboratories, in contrast to SNP genotyping that requires expensive equipment.

Reference Pedigrees

The construction of genetic linkage maps for animals such as the horse, where suitable reference families are difficult and expensive to generate, is challenging. The late maturity, long gestation period, and singleton pregnancies all conspire against the generation of ideal reference pedigrees; these would consist of numerous full-sibling offspring, three generations, and a high level of heterozygosity. In such situations a number of large half-sibling families have typically been used instead. These are generated by the mating of several sires to numerous dams; such family structures are widespread in production animals where prolific sires are common, and resources are not required to generate pedigrees specifically for linkage mapping. The drawbacks of half-sibling families are that the X chromosome cannot be mapped due to recombination being observed only in the male, large numbers of offspring must be genotyped to achieve sufficient power, and the pedigrees generally represent within-breed matings with a consequent lower-marker heterozygosity. Such pedigrees have been used successfully, however, to develop linkage maps in cattle (Beever et al., 1994; Ma et al., 1996), goat (Vaiman et al., 1996), and sheep (Crawford et al., 1995).

In cows, alternative approaches have also been available; it has been possible to super-ovulate cows followed by the transference of multiple embryos into synchronized recipient cows to generate large full-sibling families (Barendse et al., 1994). In the horse, however, superovulation still proves challenging (review in Scherzer et al., 2008).

Horse Genetic Linkage Maps

During the 1990s, three mapping resources were used in the horse for linkage mapping, and these are now described. The first two used large half-sibling families, and the third utilized reproductive techniques to generate full-sibling pedigrees. In addition, the available maps were merged to form a combined map.

The Uppsala map

The Uppsala map was the first low-density male autosomal linkage map of the horse genome (Lindgren et al., 1998). The reference material consisted of eight paternal half-sibling families of which four families were Icelandic horses and four were Standardbreds. These were two-generation panels with 263 offspring in total. The linkage map was generated by genotyping 140 polymorphic markers, 100 of which were arranged into 25 linkage groups on 18 different autosomes. The genetic markers used included 121 microsatellite markers, 8 protein polymorphisms, 5 restriction fragment length polymorphisms (RFLPs), 3 blood group polymorphisms, 2 PCR-RFLPs, and 1 single strand conformation polymorphism (SSCP). About one-third of the microsatellite markers had been physically mapped to chromosomes by in situ hybridization (e.g., Breen et al., 1997; Godard et al., 1997). These markers allowed twenty-two of the linkage groups to be assigned to chromosomes. The average distance between linked markers was 12.6 cM and the total map distance within linkage groups was 679 cM.

The International Horse Reference Family Panel (IHRFP) map

This linkage map was generated as an international collaborative effort and published in two stages. Phase I (Guérin et al., 1999) described the genotyping of 12 paternal half-sibling families consisting of 448 individuals, which were genotyped with 161 markers. The half-sibling families originated in the United States, Europe, and Australasia and were each comprised of 21 to 52 offspring. They represented hot-blooded, warm-blooded, draft, and pony breeds. Significant linkage was detected for 124 markers, of which 95 were unambiguously ordered with an average spacing of 14.2 cM. The markers were assembled into 29 linkage groups and 28 of these could be assigned to 26 of the 31 autosomes via either FISH or synteny mapping using a cell hybrid panel. The total map length was 936 cM.

In Phase II (Guérin et al., 2003), an additional family was added, and a further 55 individuals, and the number of markers genotyped was increased to 344. Heterozygosity in the stallions varied from 46% (a Thoroughbred family) to 66% (a Shetland pony cross family). Significant linkage was detected for 310 markers, with 257 of these unambiguously ordered with an average spacing of 10.1 cM. The markers were assembled into 34 linkage groups, which were assigned to all 31 of the autosomes. The total map length was 2262 cM.

The Newmarket map

Alternative approaches initiated by Dr. Matthew Binns were employed by Swinburne et al. (2000, 2006) to generate a reference family that would avoid the drawbacks of large half-sibling reference pedigrees. Specifically the planned family structure would avoid the need to genotype large numbers of individuals and would enable the mapping of the horse X chromosome (ECAX). The generation of such a pedigree required the use of reproductive techniques that were available locally at the Equine Fertility Unit in Newmarket, United Kingdom, led by Professor Twink Allen. The procedures

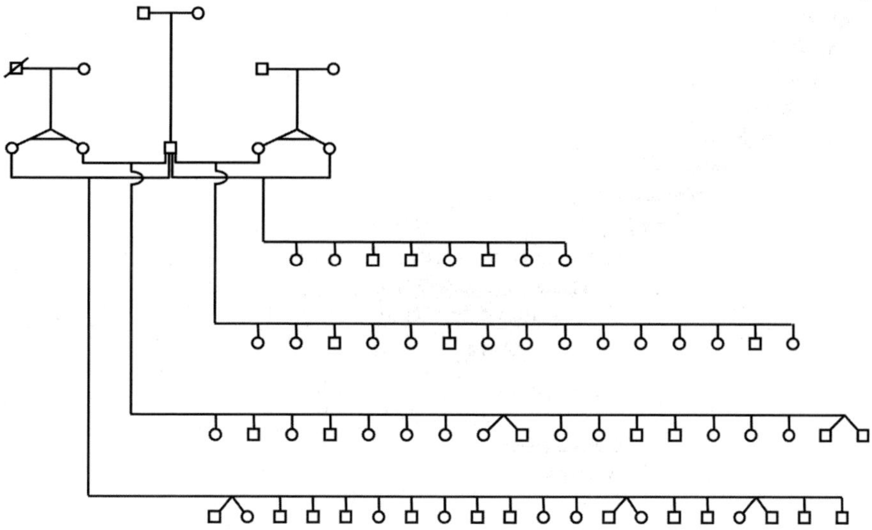

Figure 2.1 Pedigree structure of the Newmarket horse reference family used to generate the Newmarket map (Swinburne et al., 2006). Two pairs of monozygotic twin mares were covered by a single stallion to generate two families of full-sibling horse embryos. Females are depicted by circles and males are depicted by squares. All twin embryos retrieved were found to be dizygotic in origin. The breeds used in this pedigree were: Arabian, Thoroughbred, Welsh Cob, European Warmblood, and Icelandic Horse.

utilized were, first, the nonsurgical removal of equine conceptuses (Allen & Bracher 1992), and second, the generation of monozygotic twins via embryo micromanipulation (Allen & Pashen 1984). The resulting family was referred to as the Newmarket horse reference family.

The equine conceptus is unusual in its late implantation, which does not occur until 37 days post-conception. The embryo is easily recoverable via uterine lavage using videoendoscopy until this time, and will yield 3–5 mg DNA on extraction. Using this technique, in conjunction with drugs to induce estrus, up to five full-sibling embryos were obtained from each of the four mares per season. To increase the numbers of full-sibling embryos, the mares employed consisted of two pairs of monozygotic twins. Only one stallion was used on all four mares. The embryos from each pair of twins were therefore genetically full-sibling, and each full-sibling family was half-sibling with respect to the other. Over 5 seasons, 61 embryos were produced and used subsequently for genotyping. Interestingly, five pairs of dizygotic twin embryos were recovered. The structure of this pedigree is illustrated in Figure 2.1.

A third generation was also genotyped, as samples from five of the six grandparents were also available. This allowed the generation of a linkage map for ECAX as recombination in the mares could also be assayed. The founding animals represented a number of breeds, namely Arabian, Thoroughbred, Welsh Cob, European Warmblood, and Icelandic Horse, thereby maximizing the chances of heterozygosity. Only 17.1% of markers tested were homozygous across both halves of this pedigree and were therefore uninformative for mapping. Drawbacks to this family were that no phenotypes were available for the embryos, and so could not be mapped, in contrast to the half-sibling families.

The maps resulting from genotyping on this reference family have been published in two stages. The first stage (Swinburne et al., 2000) described the genotyping of all 61 F2 embryos, together with the parental and grandparental individuals, with 353 microsatellites and 6 SNPs. These were placed into 42 linkage groups of which 37 could be anchored to the physical map. The X chromosome and all autosomes except ECA28 had linkage groups assigned to them. The average spacing between the markers was 10.5 cM, and the total map length was 1780 cM.

The subsequent publication (Swinburne et al., 2006) described the genotyping of 734 microsatellites and 8 SNPs. These were assigned to 32 linkage groups, one for each of the 31 autosomes and one for ECAX. Each of these groups was assigned to a chromosome and oriented by virtue of FISH-mapped markers. The total length of this sex-averaged genetic map was 2,772 cM, with the average distance between markers being 3.7 cM. This map length is very close to the predicted length of 2,720 cM estimated from chiasma counts in horses (Scott & Long 1980). Figure 2.2 presents these 32 linkage maps as seen in Swinburne et al. (2006). In addition, the linkage maps are aligned with the contemporary RH map as described in Chowdhary et al. (2003). Alignments were also made to the human genome by *in silico* mapping, that is, comparing the unique flanking sequence of each microsatellite with the human genome sequence using BLAST; half of those markers tested identified significant and unique matches to the human genome. In a recent study (Mittmann et al., 2010), in which the order of these markers was compared with the horse genome assembly EquCab2.0 (http://www.broadinstitute.org/mammals/horse), only two discrepancies were found: marker order of TYK601 and COR020 was switched, and COR062 was incorrectly positioned on ECA19.

The International Equine Gene Mapping Workshop (IEGMW) linkage map; the merging of contemporary linkage maps

In an attempt to produce a comprehensive linkage map that would bring together linkage mapping data available at that time, the IEGMW produced a merged map (Penedo et al., 2005). A further 359 microsatellites were genotyped on the IHRFP (Guérin et al., 2003), and this was merged with the Uppsala map (Lindgren et al., 1998) and the Phase I Newmarket map (Swinburne et al., 2000), together with data from a further two half-sibling families. Since the majority of this data originated from half-sibling families, only recombination occurring in the sires was included in the analyses. In total, 766 markers were assigned to linkage groups, with only 59 markers not displaying significant linkage to another marker. The map spans a total of 3,740 cM, with an average of 6.3 cM between markers; it may be that genotyping error – which was estimated at 0.6% – has artificially inflated the map length.

The Selection of Microsatellite Mapping Panels for the Dissection of Inherited Conditions in the Horse

Once substantial linkage maps became available, efforts turned to the selection of well-spaced panels of markers for gene mapping purposes. An important contribution to this choice was the development of horse/human comparative maps to identify the likely positions of genes. Various aspects of comparative mapping are discussed in other chapters, but an important component was the use of *in silico* mapping to locate horse microsatellites with respect to the human genome (Farber & Medrano 2004; Swinburne et al., 2006; Tozaki et al., 2007). Specifically this involved the sequence comparison of unique microsatellite-flanking sequence to the human genome sequence using BLAST-like alignment tool (BLAT). Some of the comparative mapping performed during this period is illustrated in Figure 2.2.

ECA1

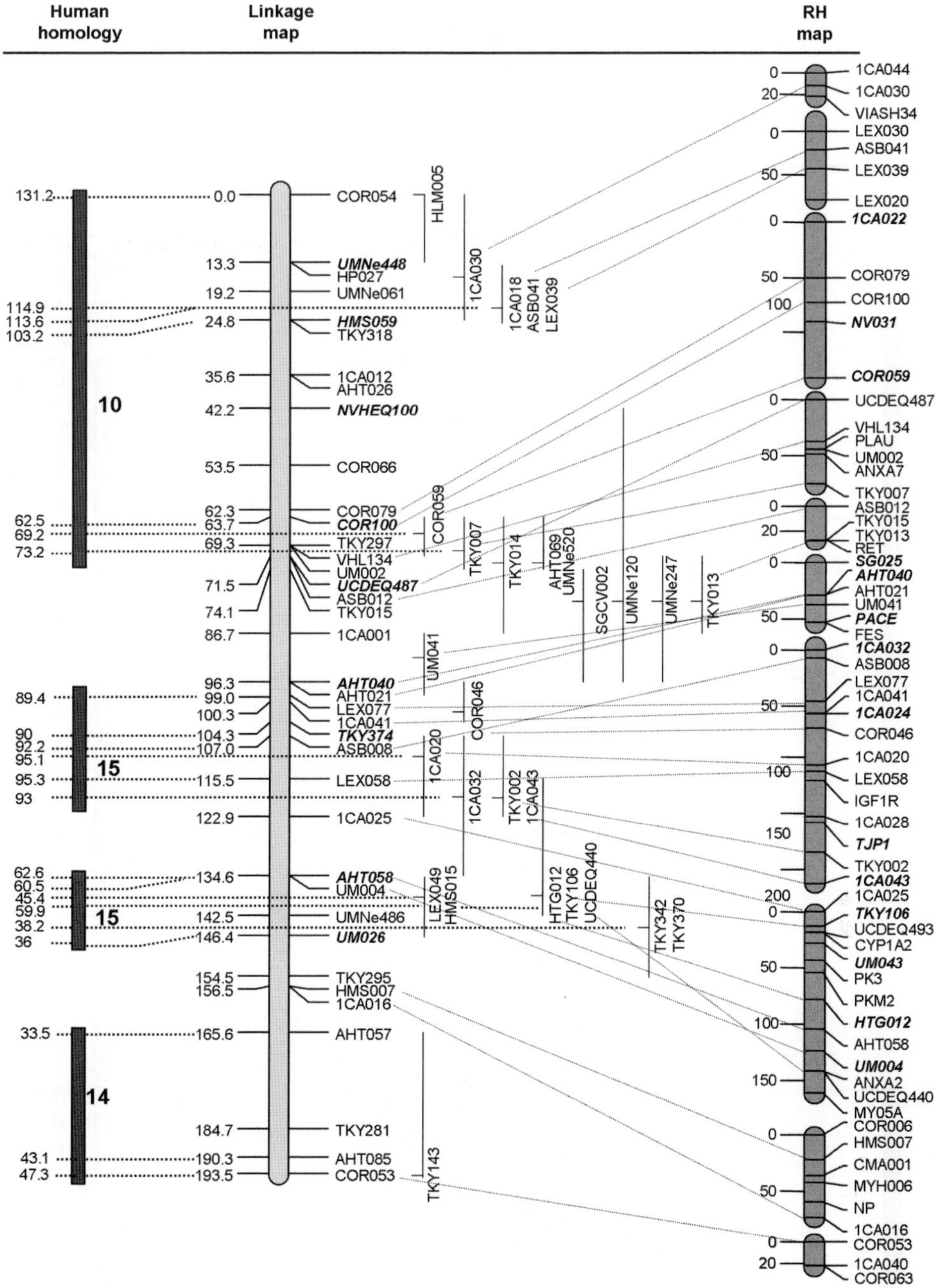

Figure 2.2 The Newmarket map (Swinburne et al., 2006). Sex-averaged genetic linkage maps of the horse autosomes and female-specific map of the X chromosome are shown as grey bars in the center of each figure. The positions of markers along the chromosome are shown in centiMorgans (cM) to the left. Framework markers are shown in bold italics. Markers that could not be ordered with a threshold of LOD > 1 are shown to the right in vertical text. The adjacent vertical line describes their probable location with their most likely position indicated by a short horizontal line. The contemporary RH map is shown to the far right (Chowdhary et al., 2003). Grey bars to the far left indicate proposed segments of conserved synteny between the horse linkage map and the human genome sequence. The identity of the human chromosome is shown adjacent to the bars. The position of match with the human chromosome is shown in megabases. The maps were orientated by reference to FISH mapped markers and genes (Chowdhary et al., 2003).

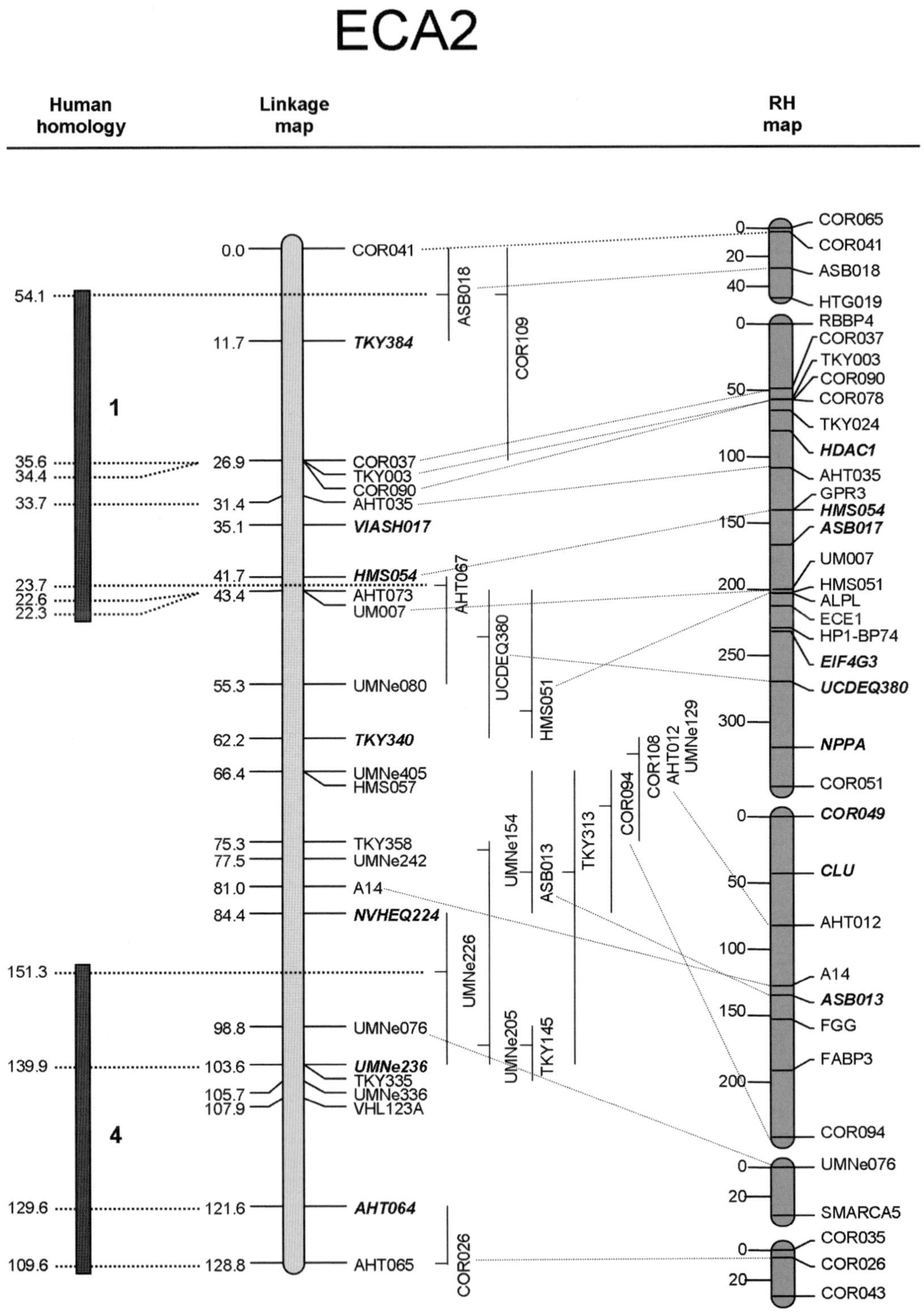

Figure 2.2 (*Continued*)

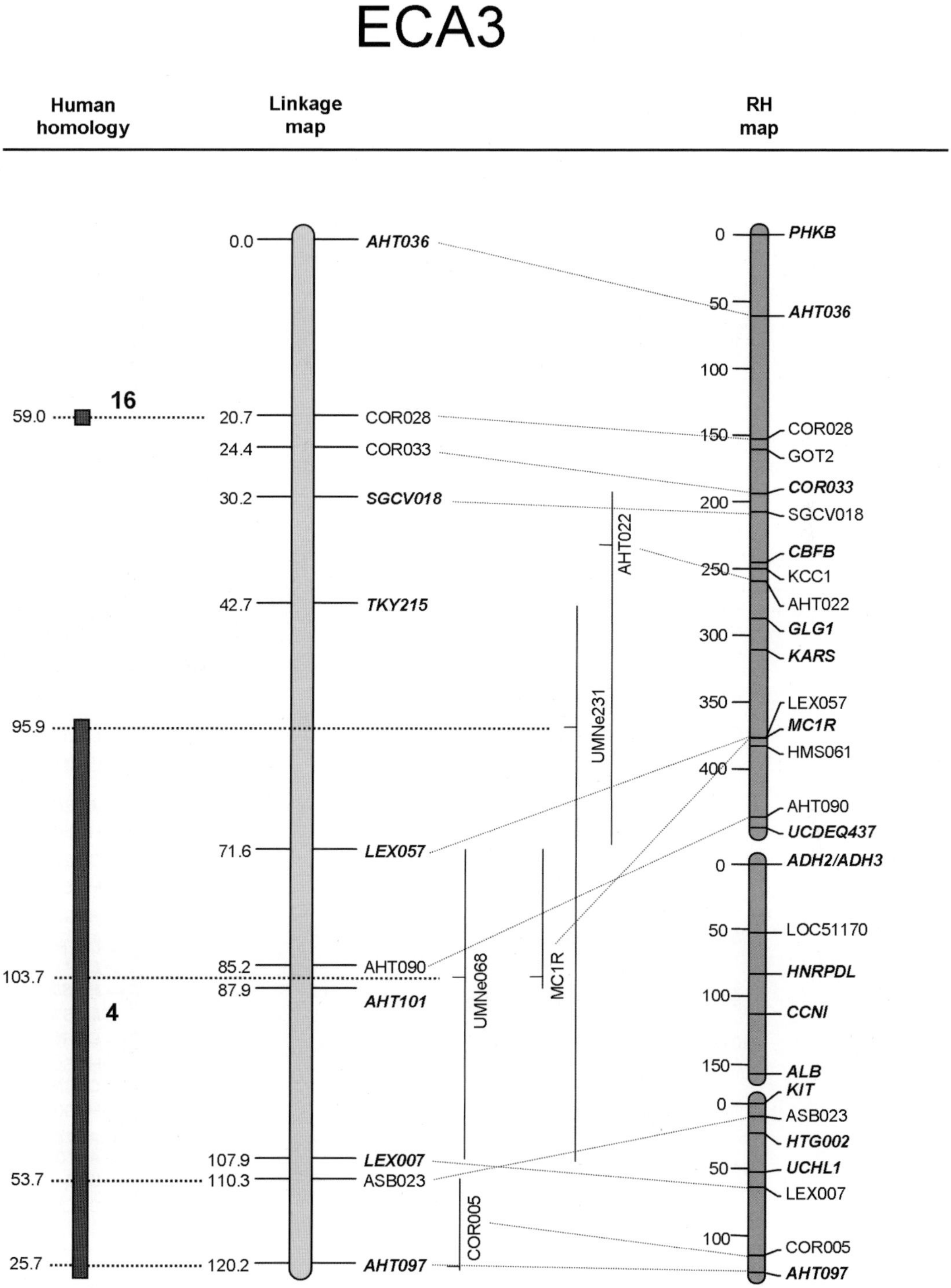

Figure 2.2 (*Continued*)

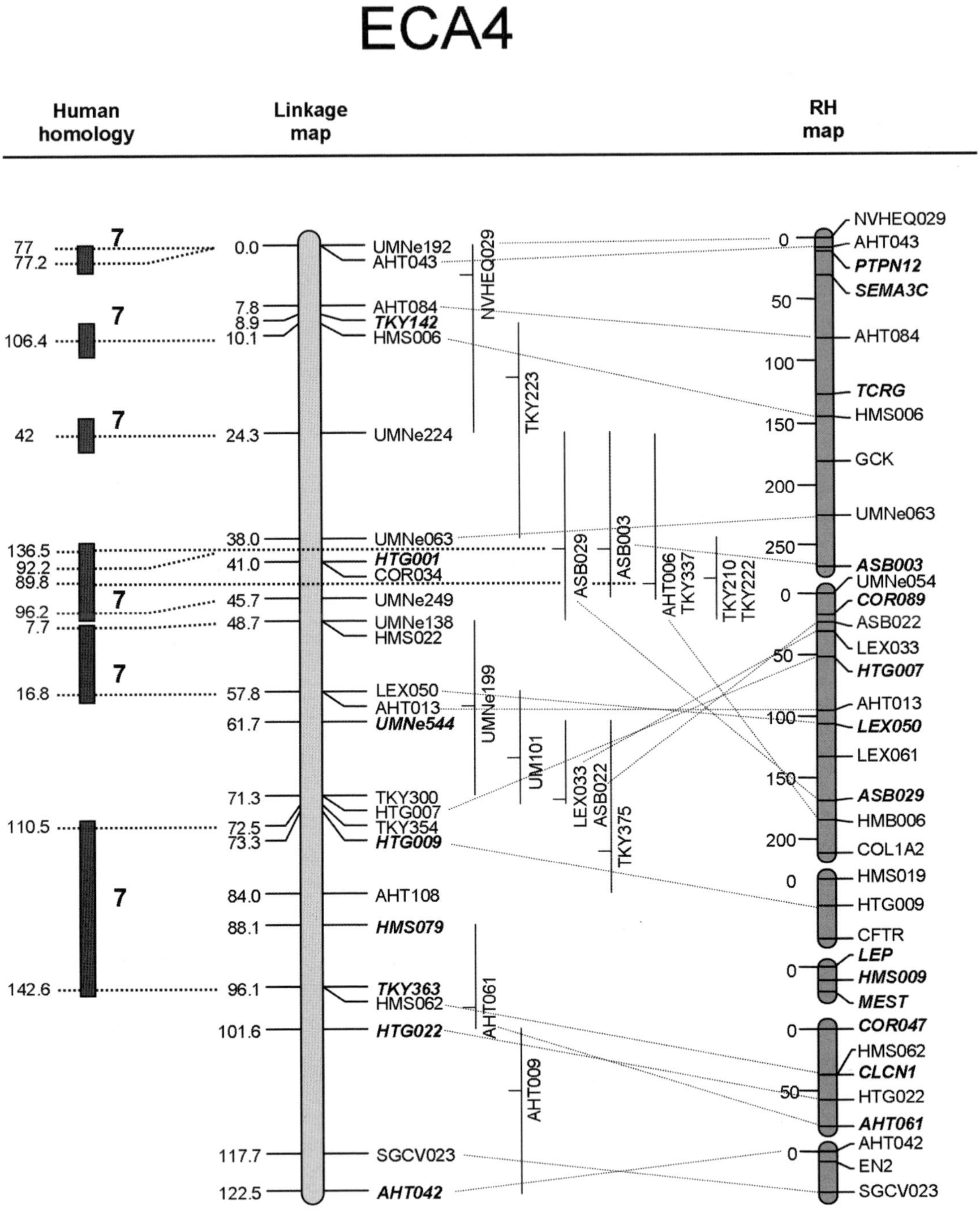

Figure 2.2 (*Continued*)

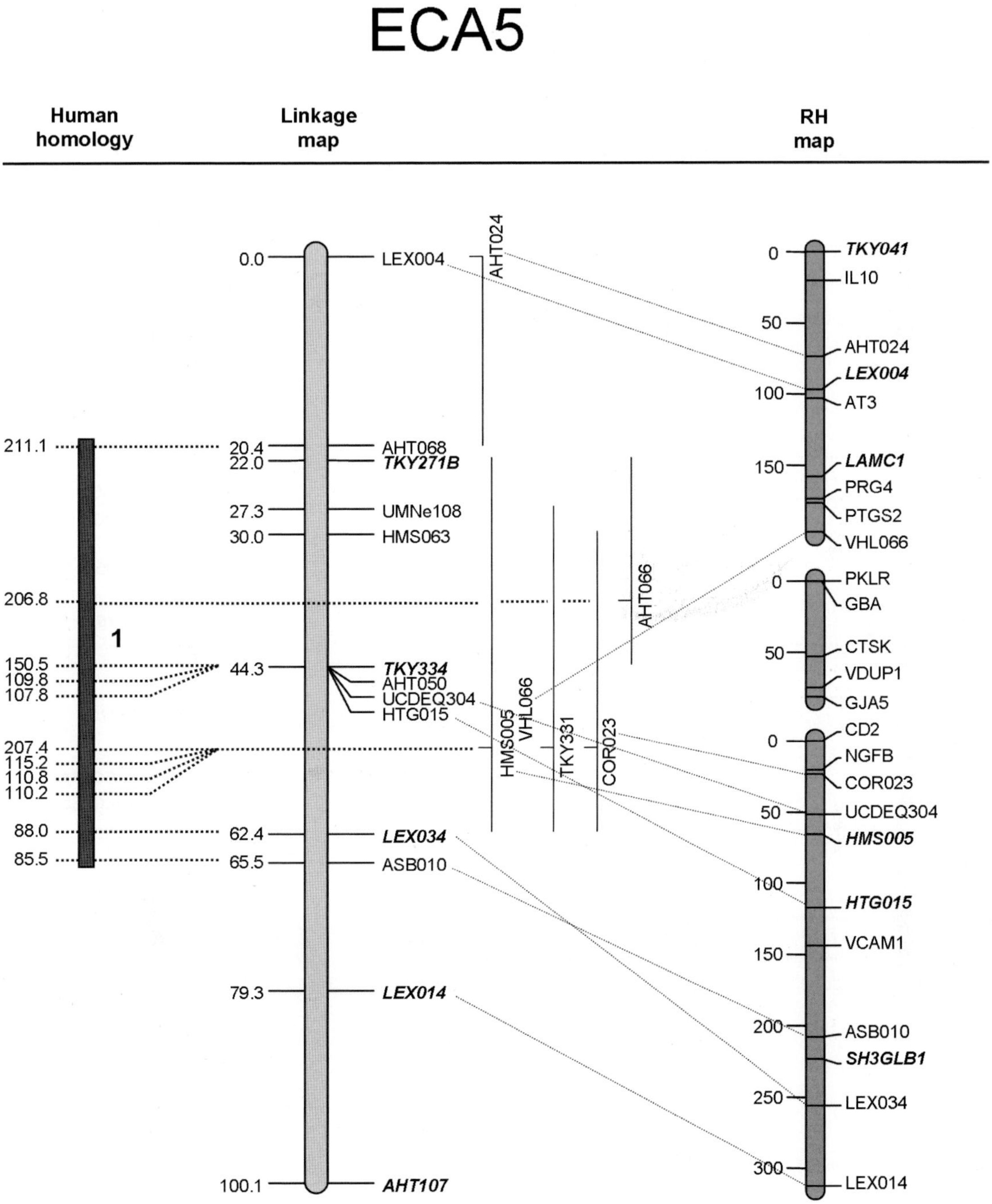

Figure 2.2 (*Continued*)

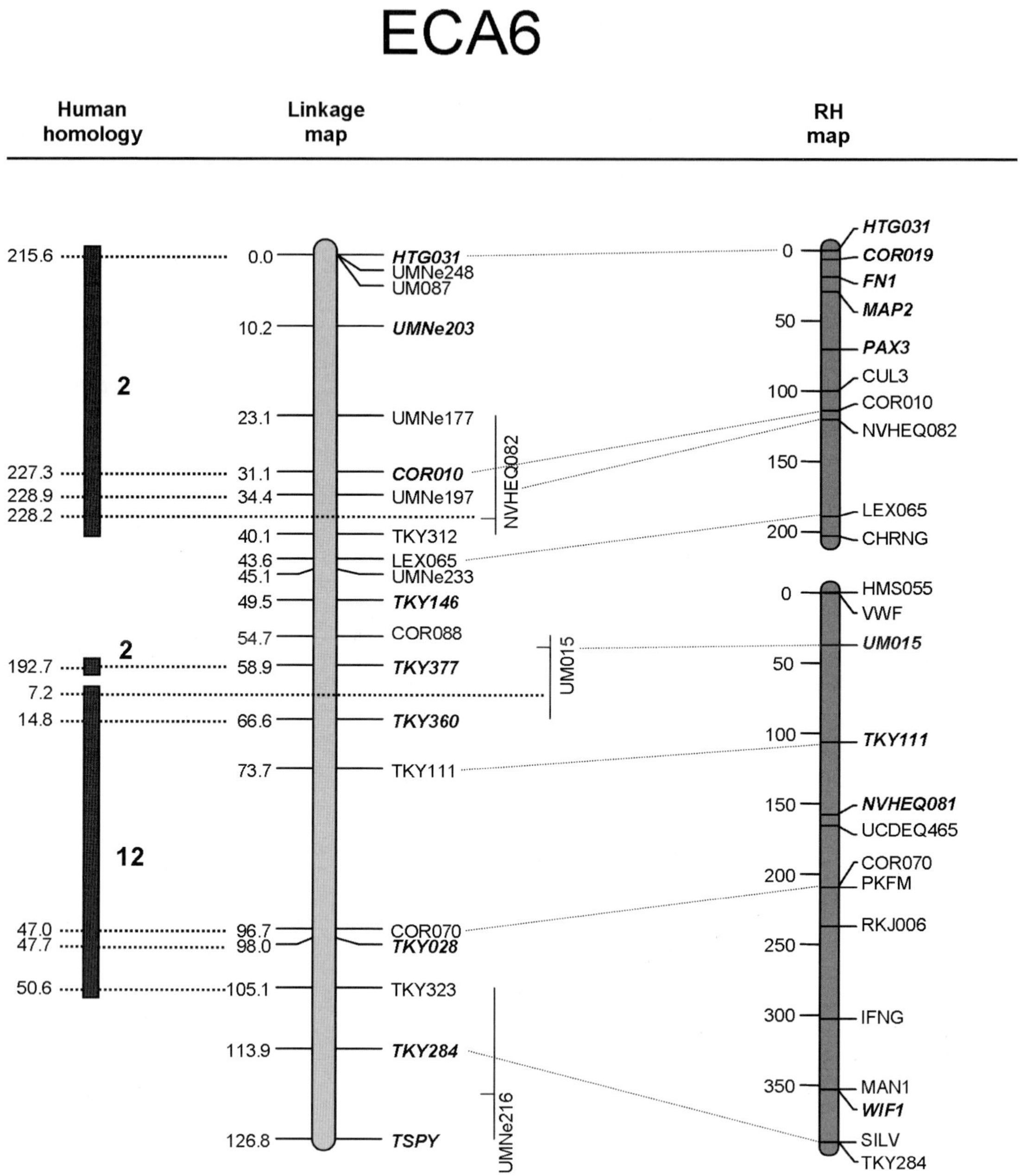

Figure 2.2 *(Continued)*

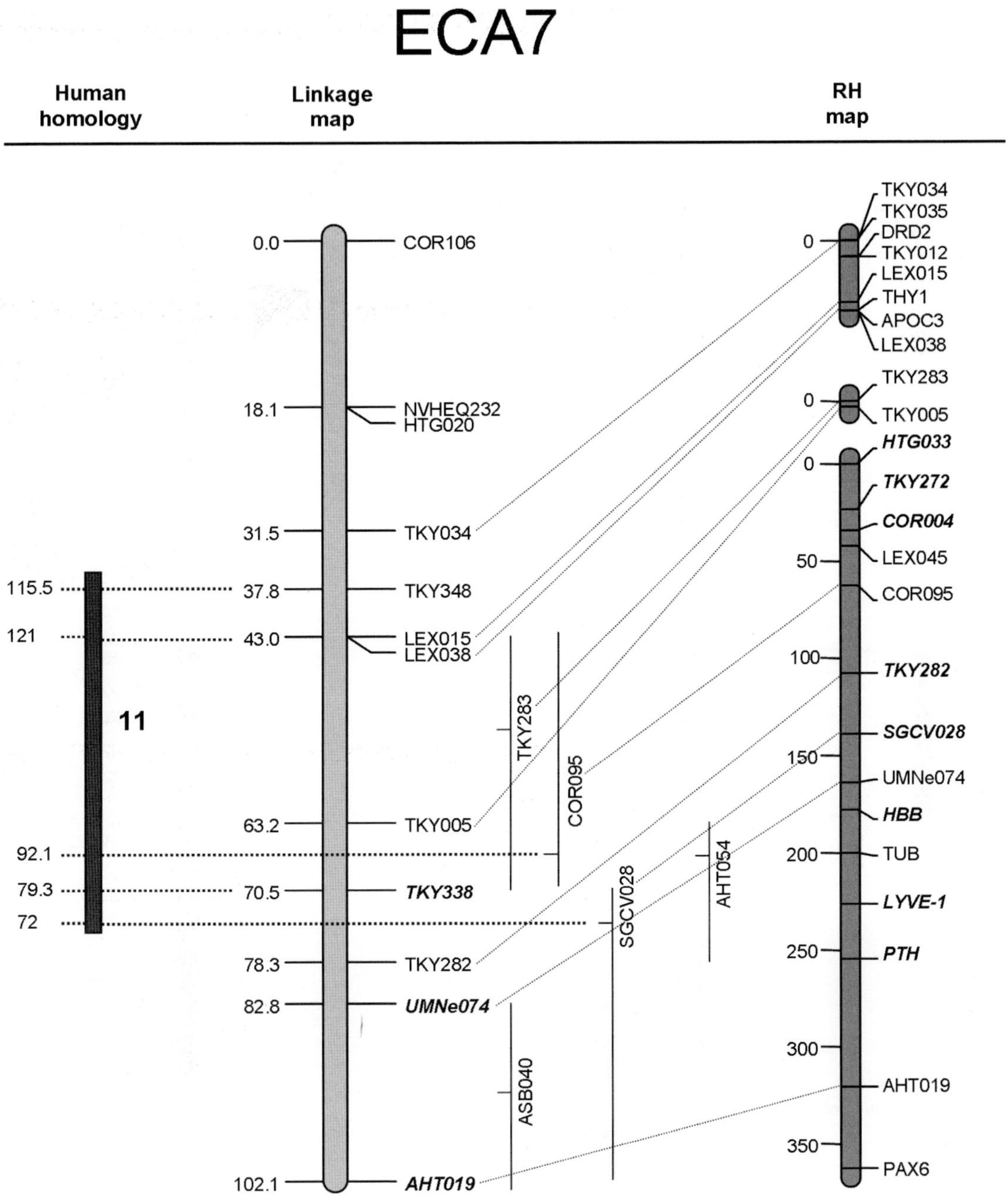

Figure 2.2 (*Continued*)

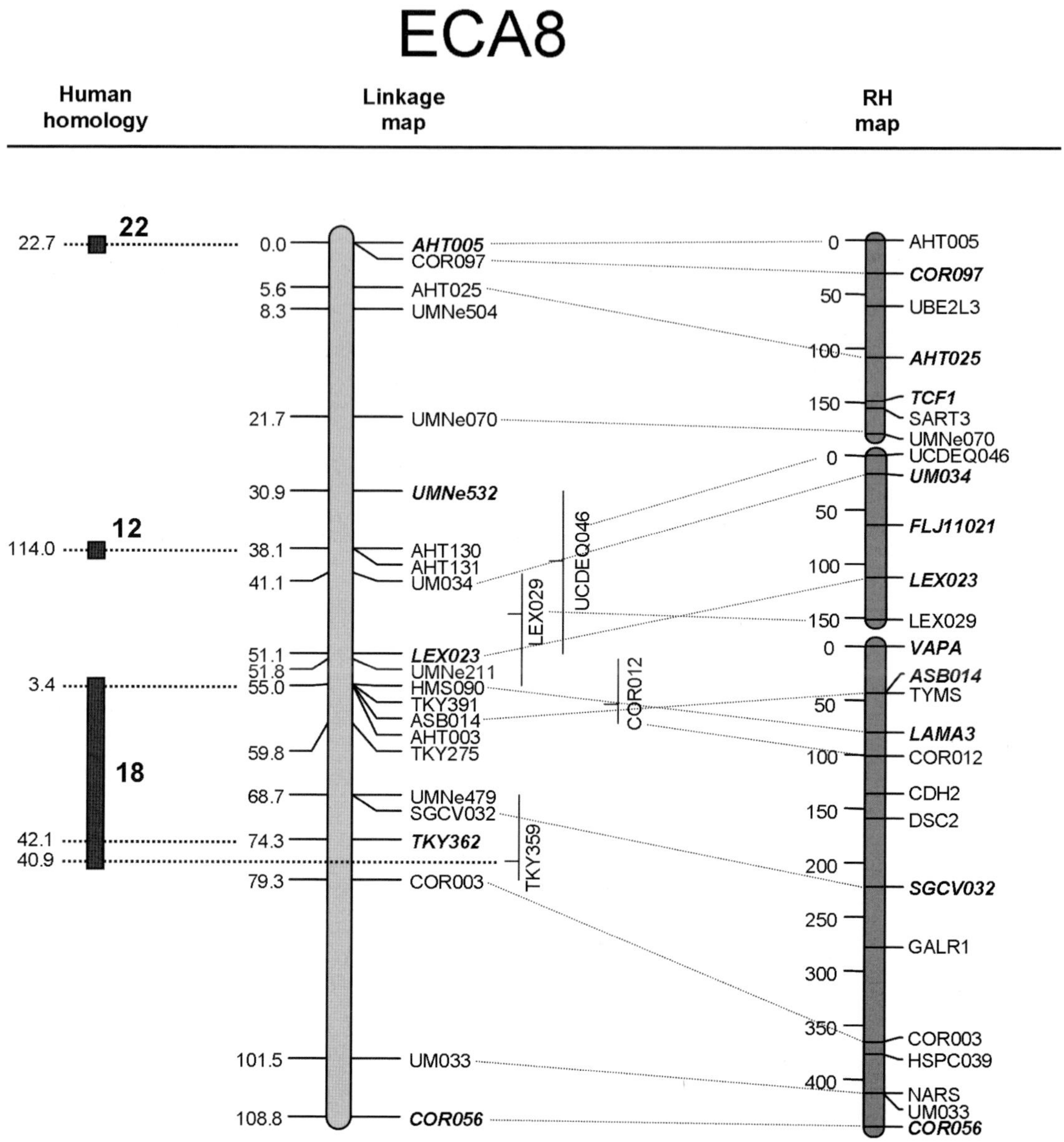

Figure 2.2 (*Continued*)

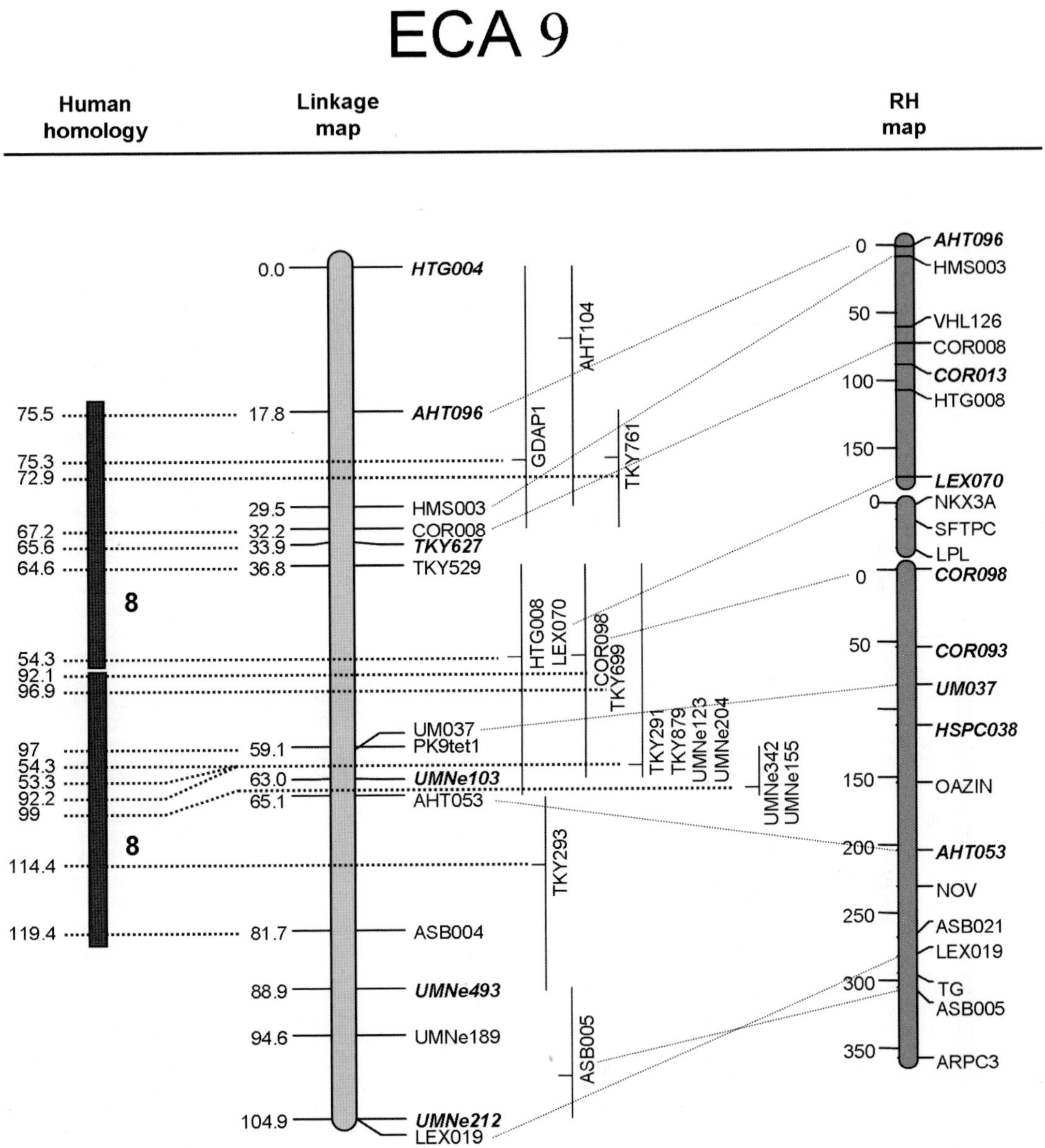

Figure 2.2 (*Continued*)

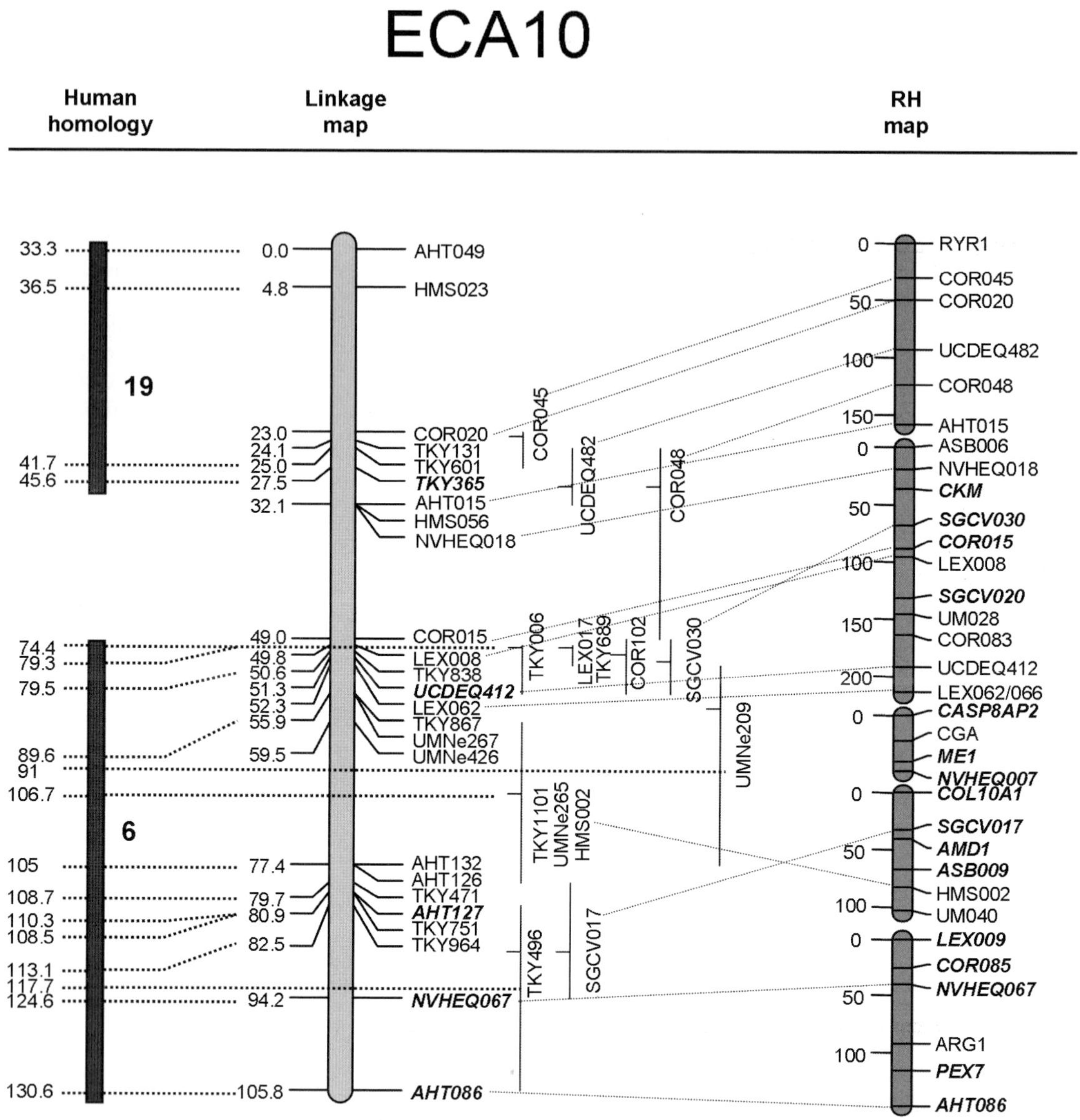

Figure 2.2 (*Continued*)

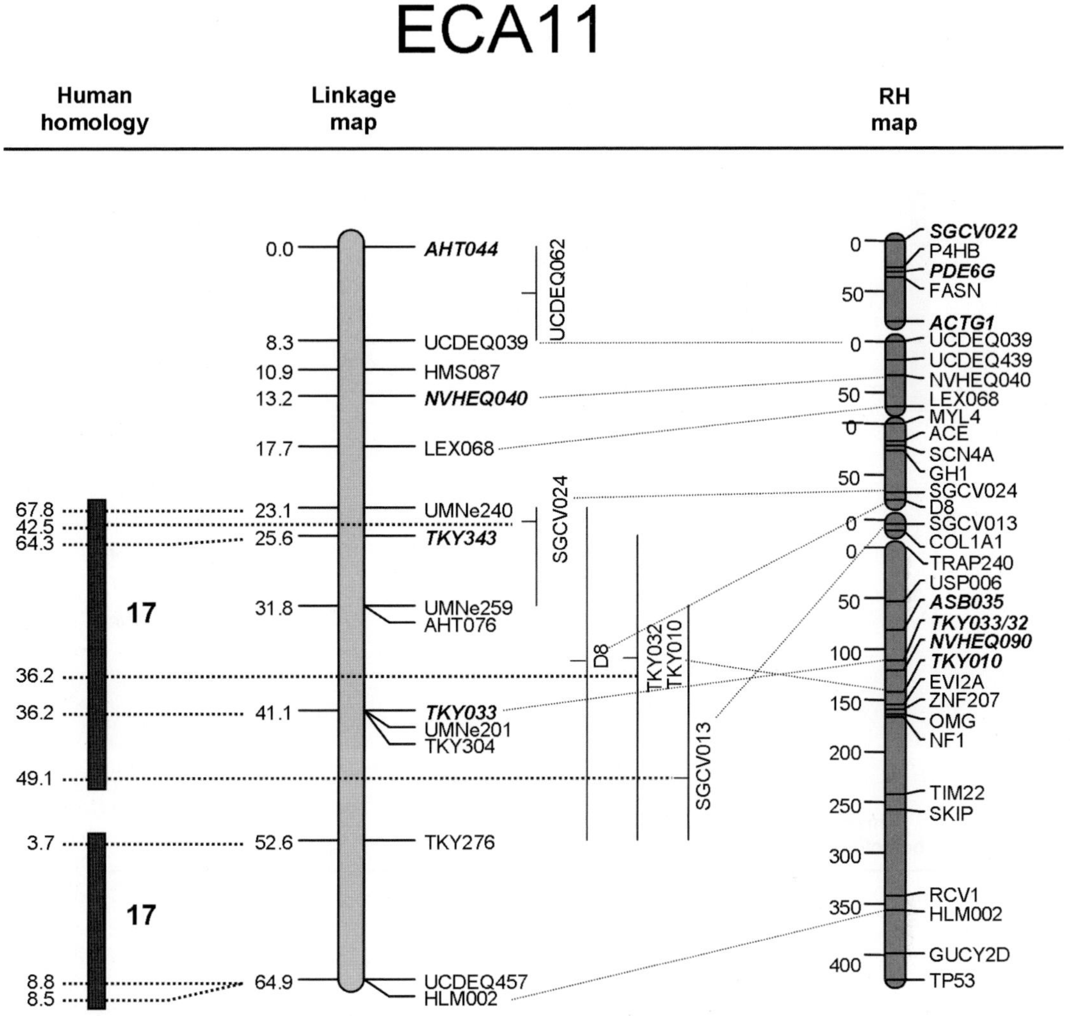

Figure 2.2 (*Continued*)

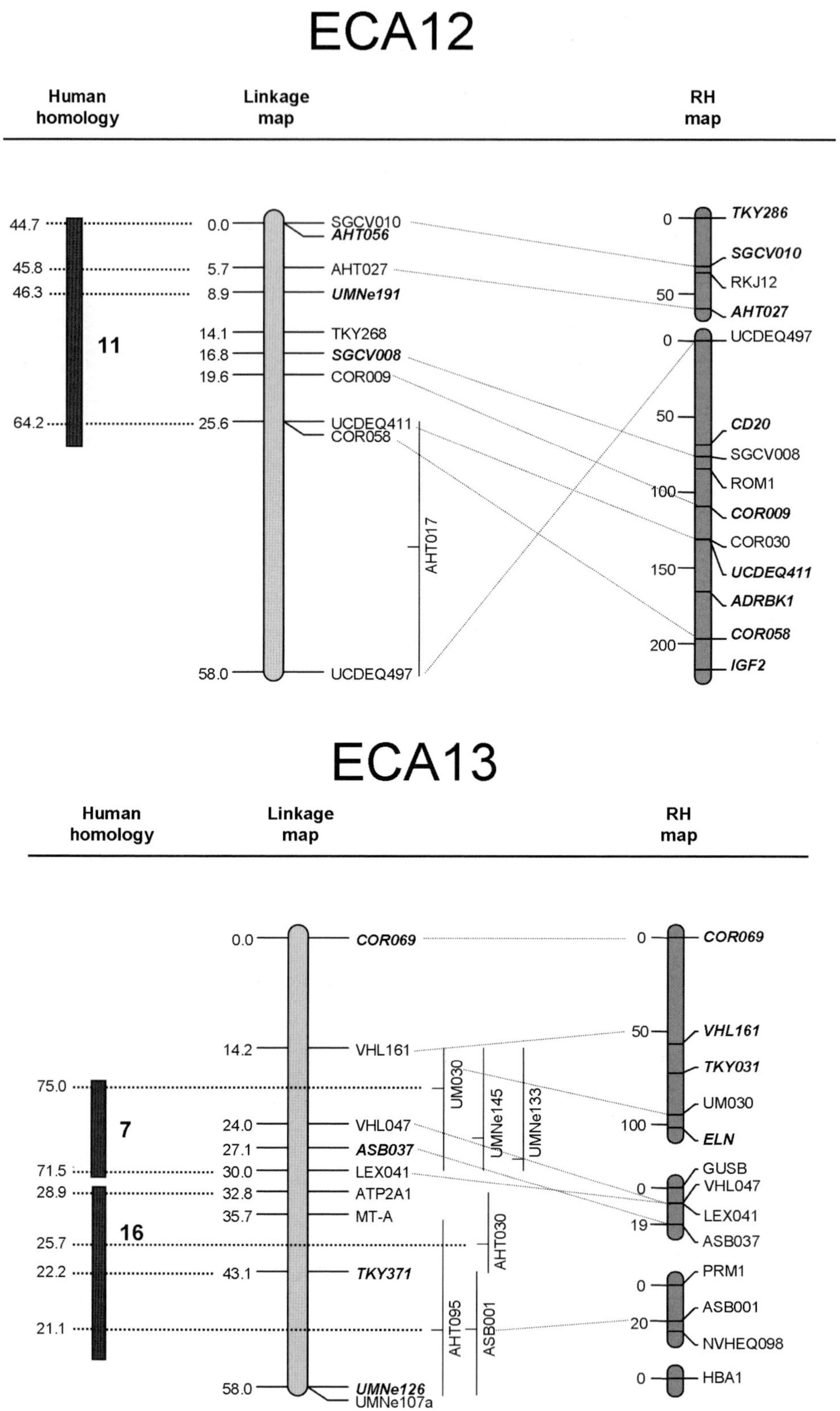

Figure 2.2 (*Continued*)

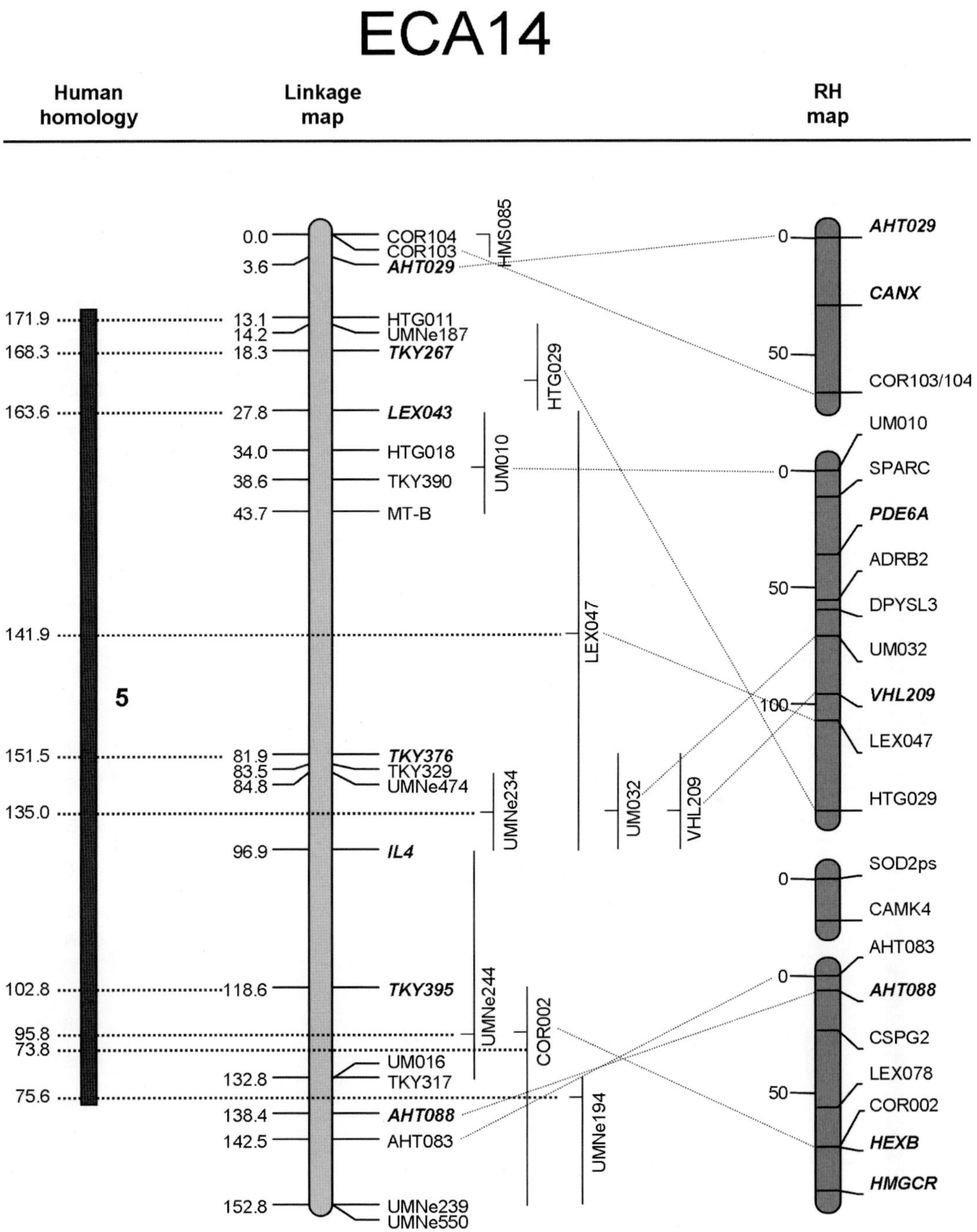

Figure 2.2 (*Continued*)

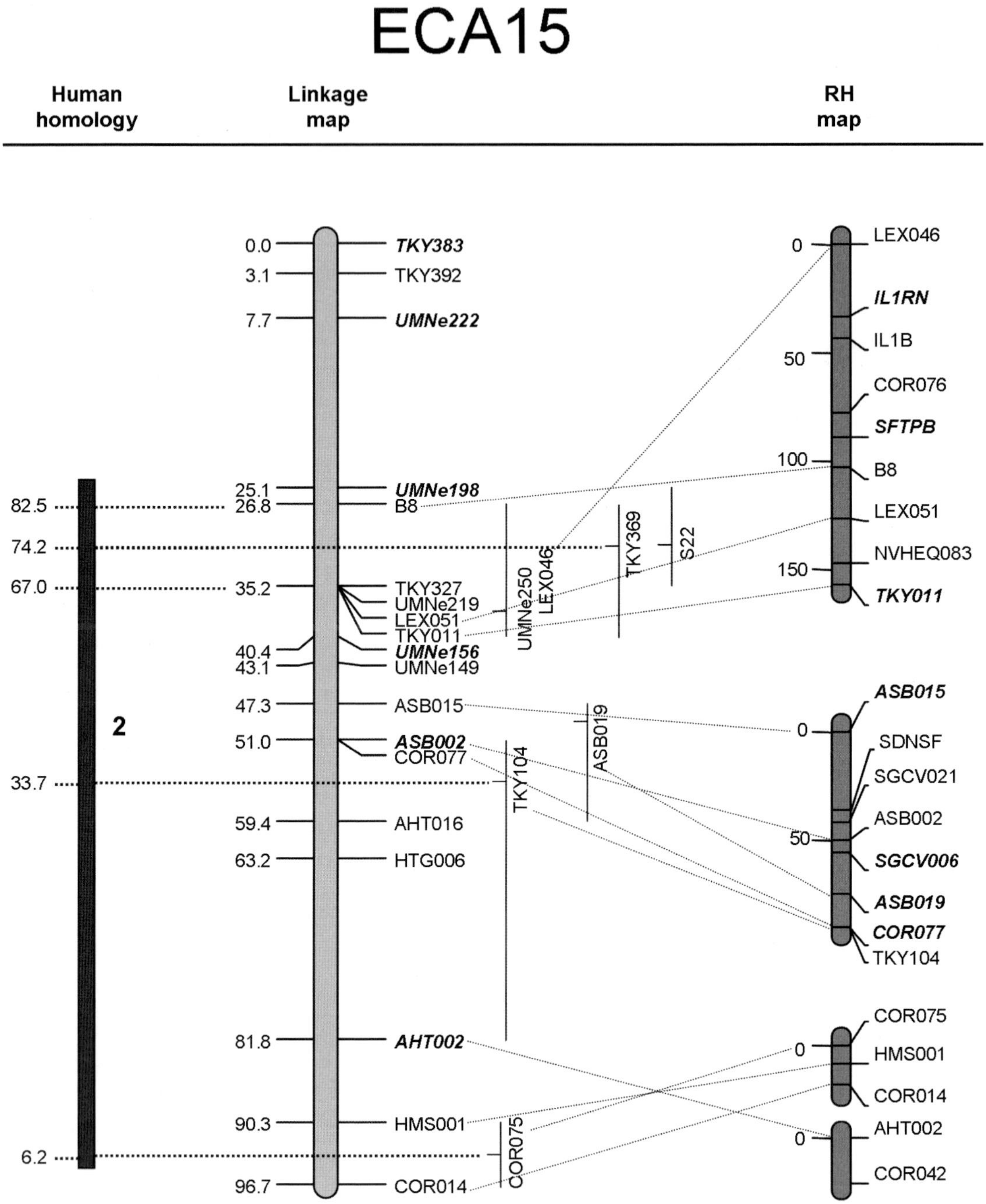

Figure 2.2 (*Continued*)

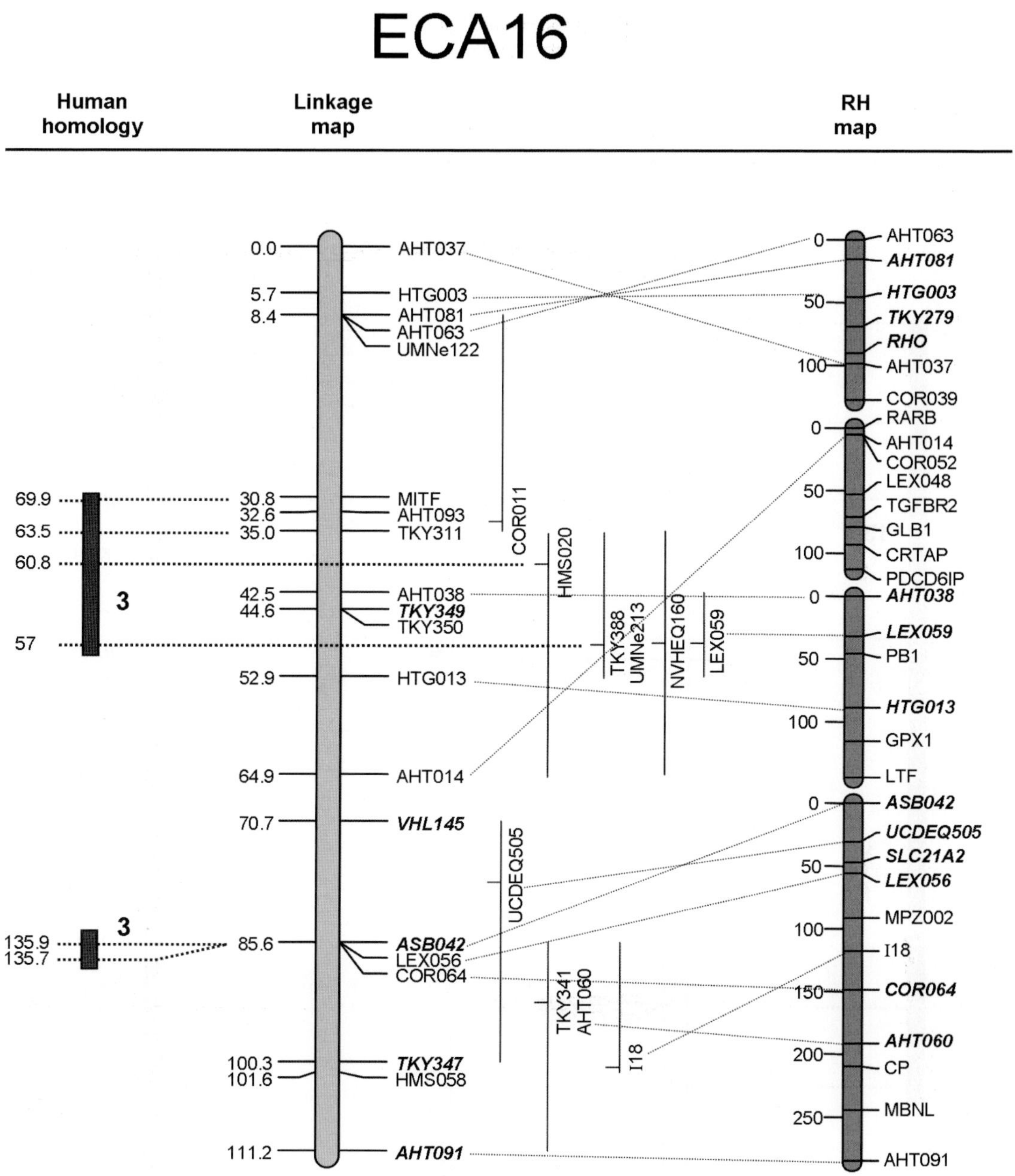

Figure 2.2 (*Continued*)

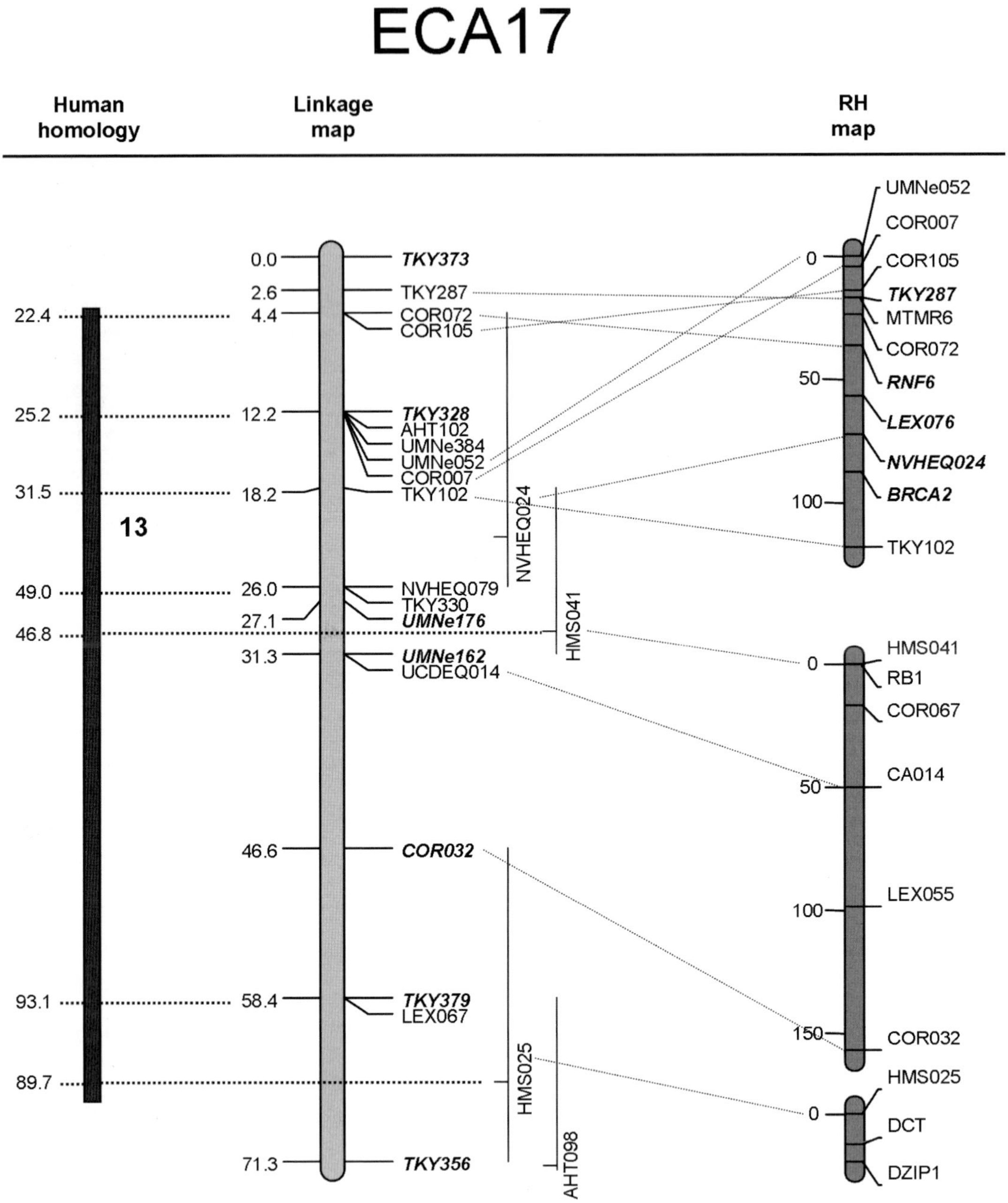

Figure 2.2 (*Continued*)

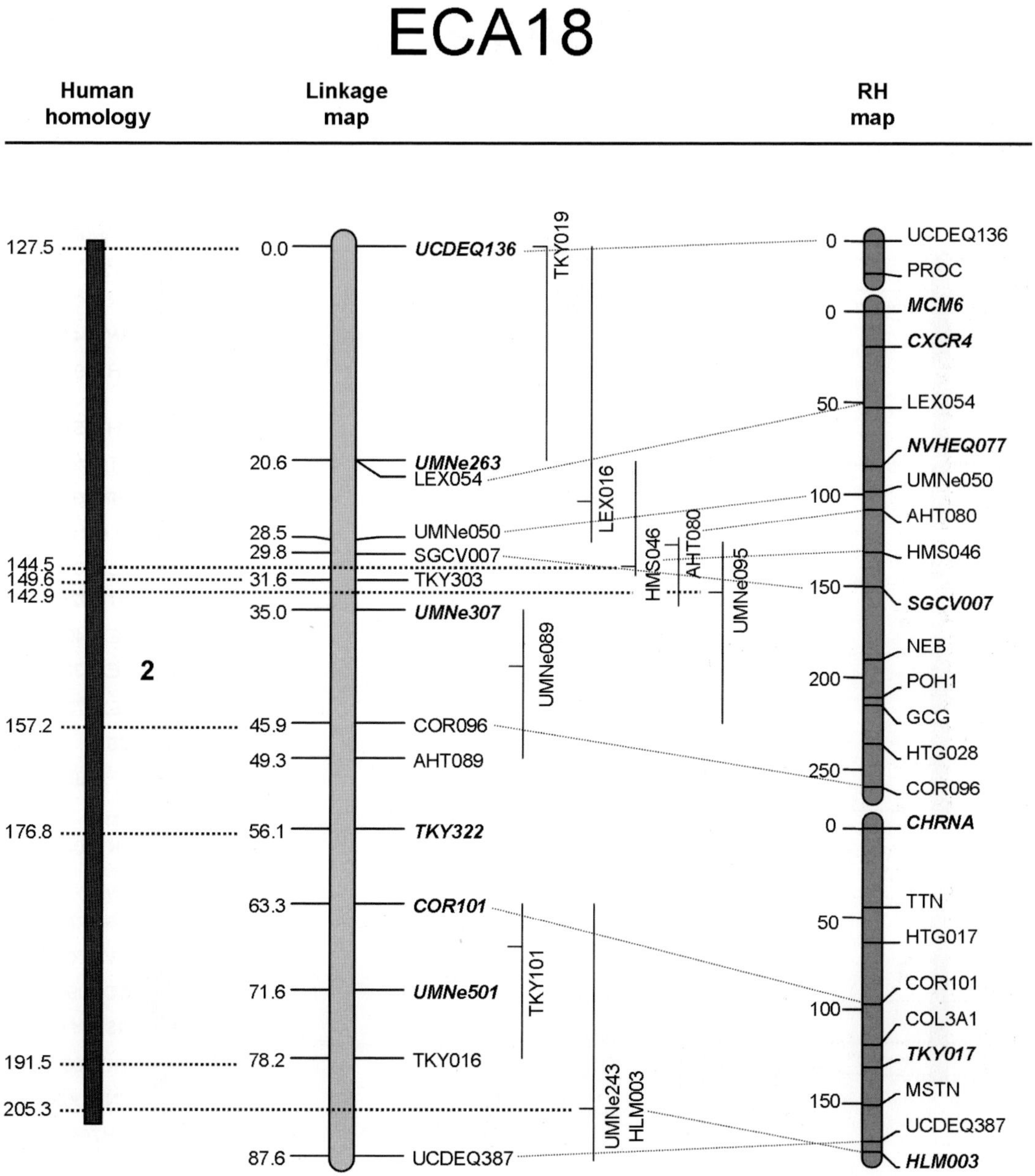

Figure 2.2 (*Continued*)

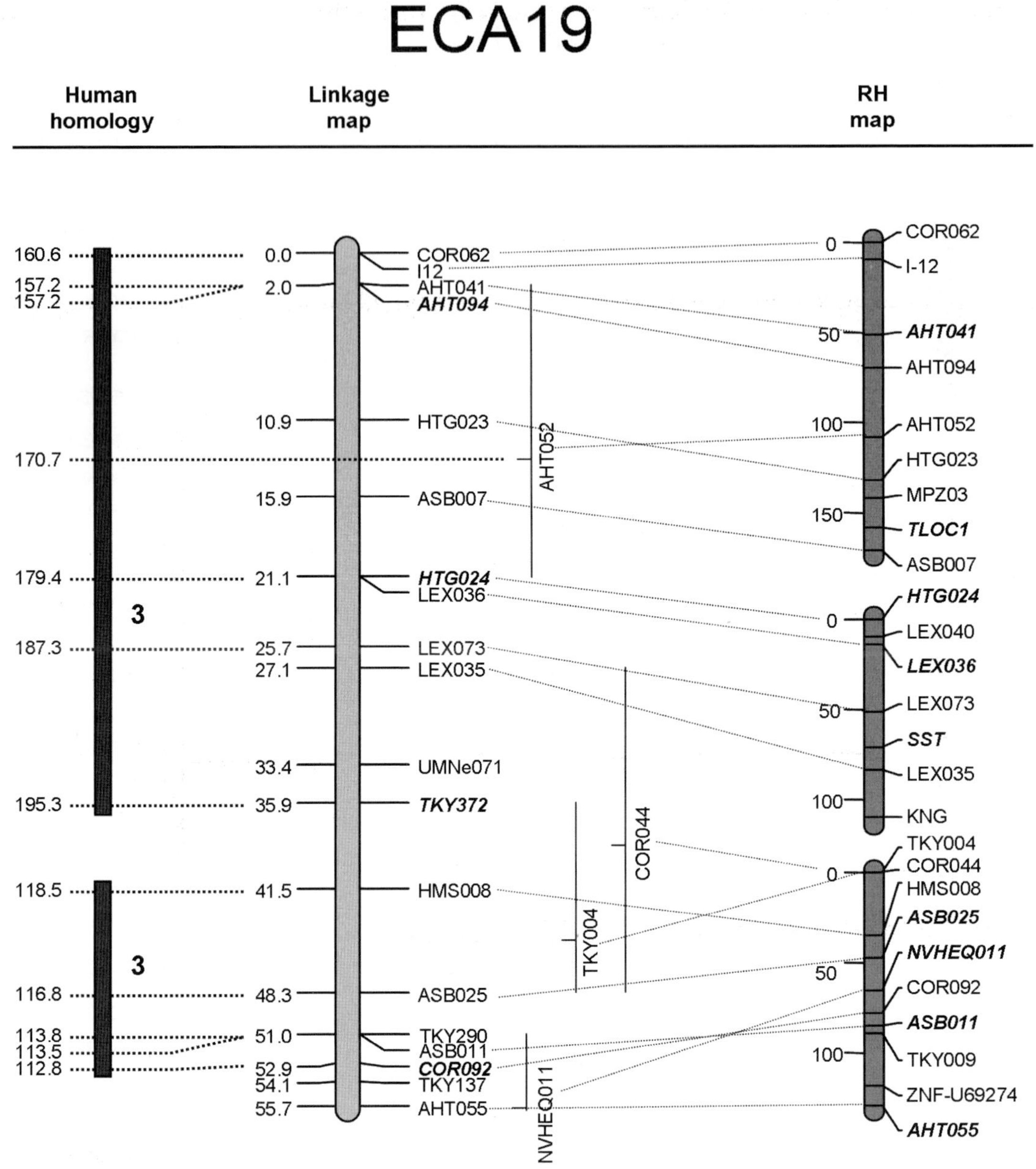

Figure 2.2 (*Continued*)

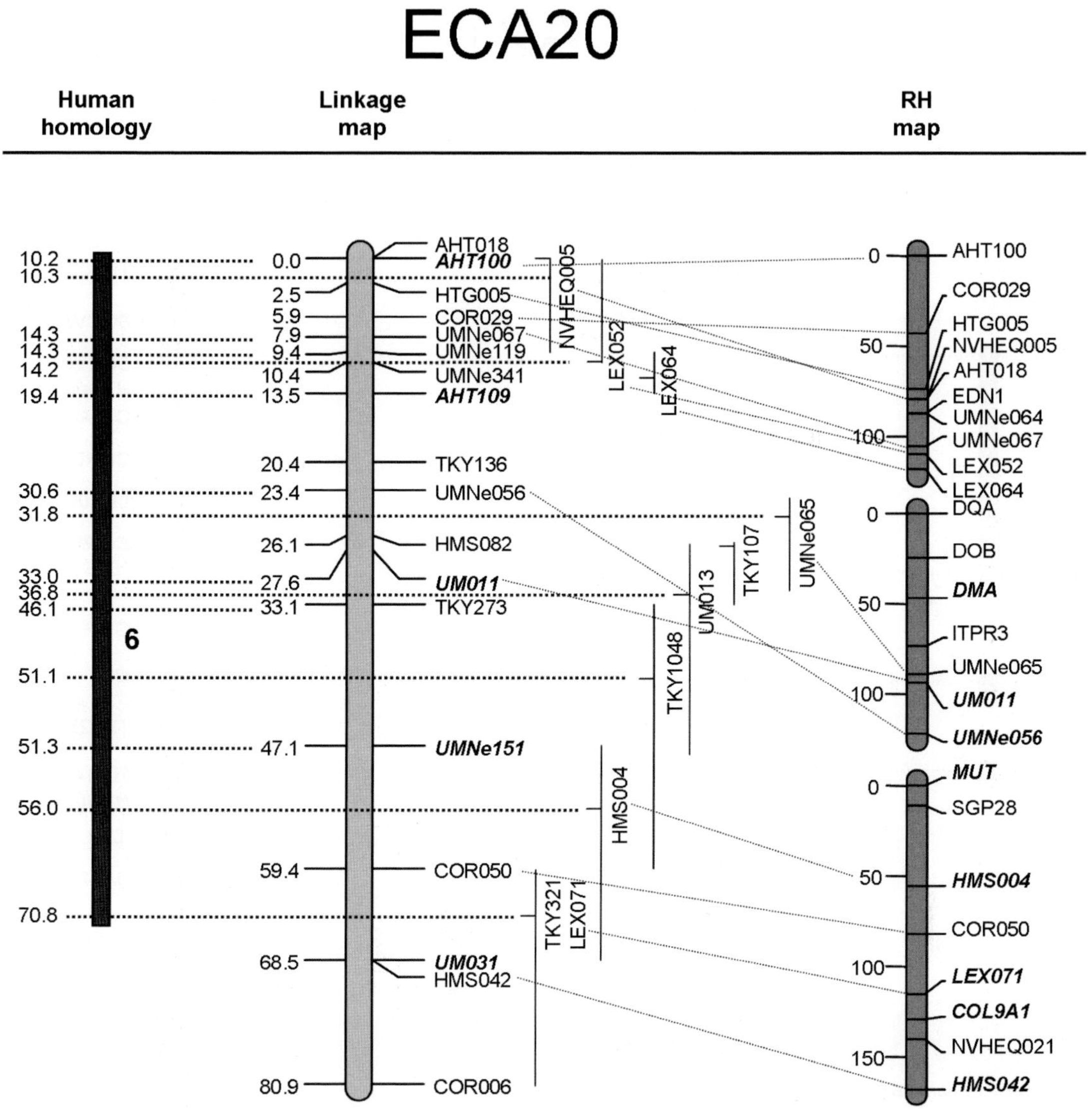

Figure 2.2 (*Continued*)

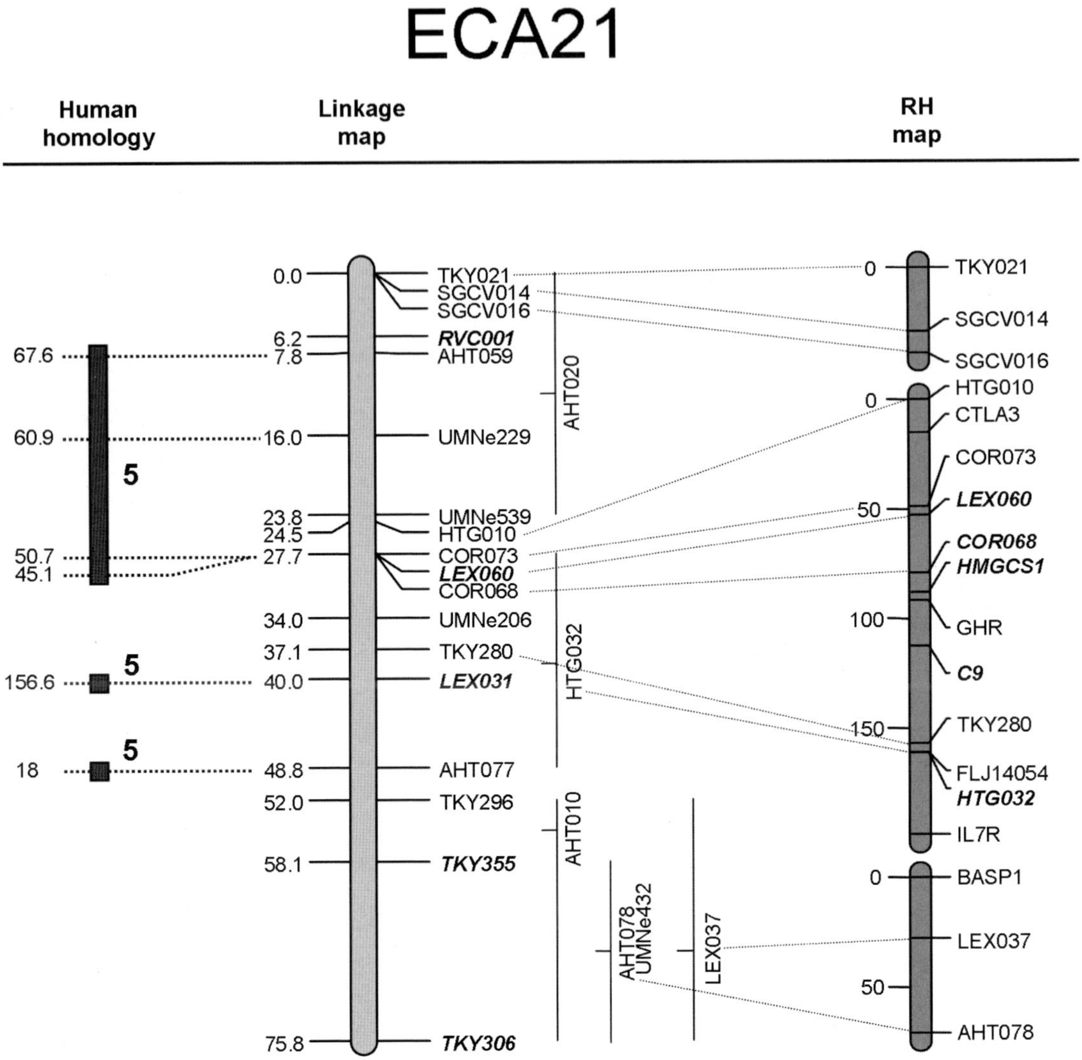

Figure 2.2 (*Continued*)

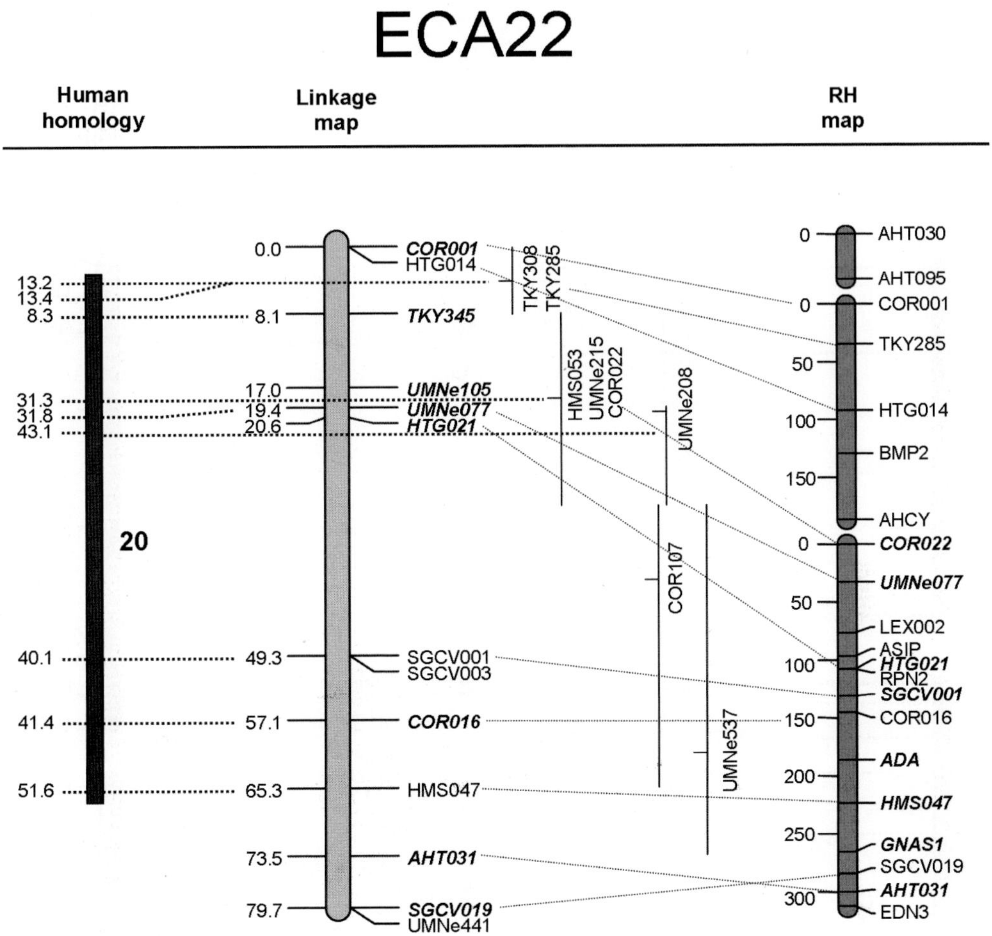

Figure 2.2 (*Continued*)

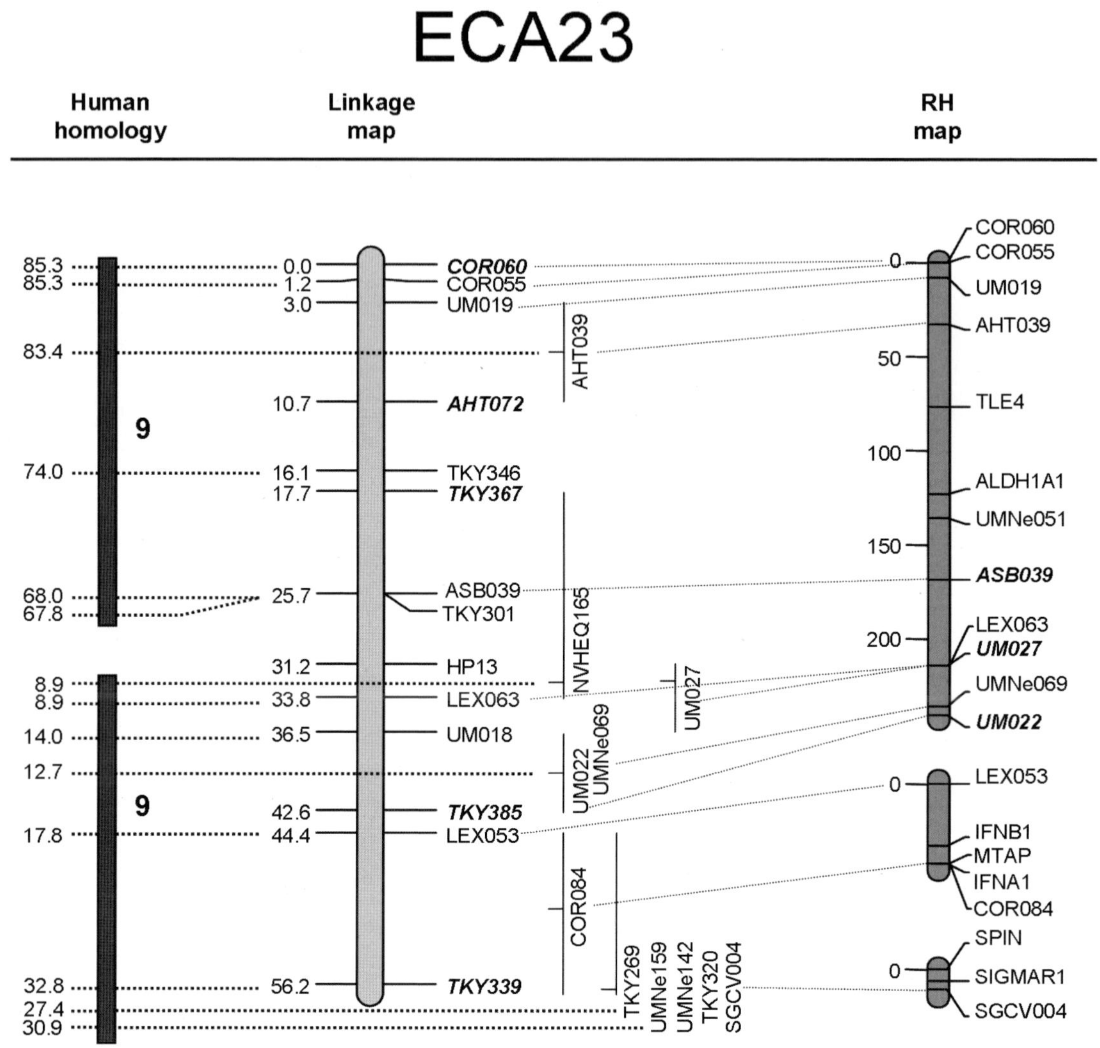

Figure 2.2 (*Continued*)

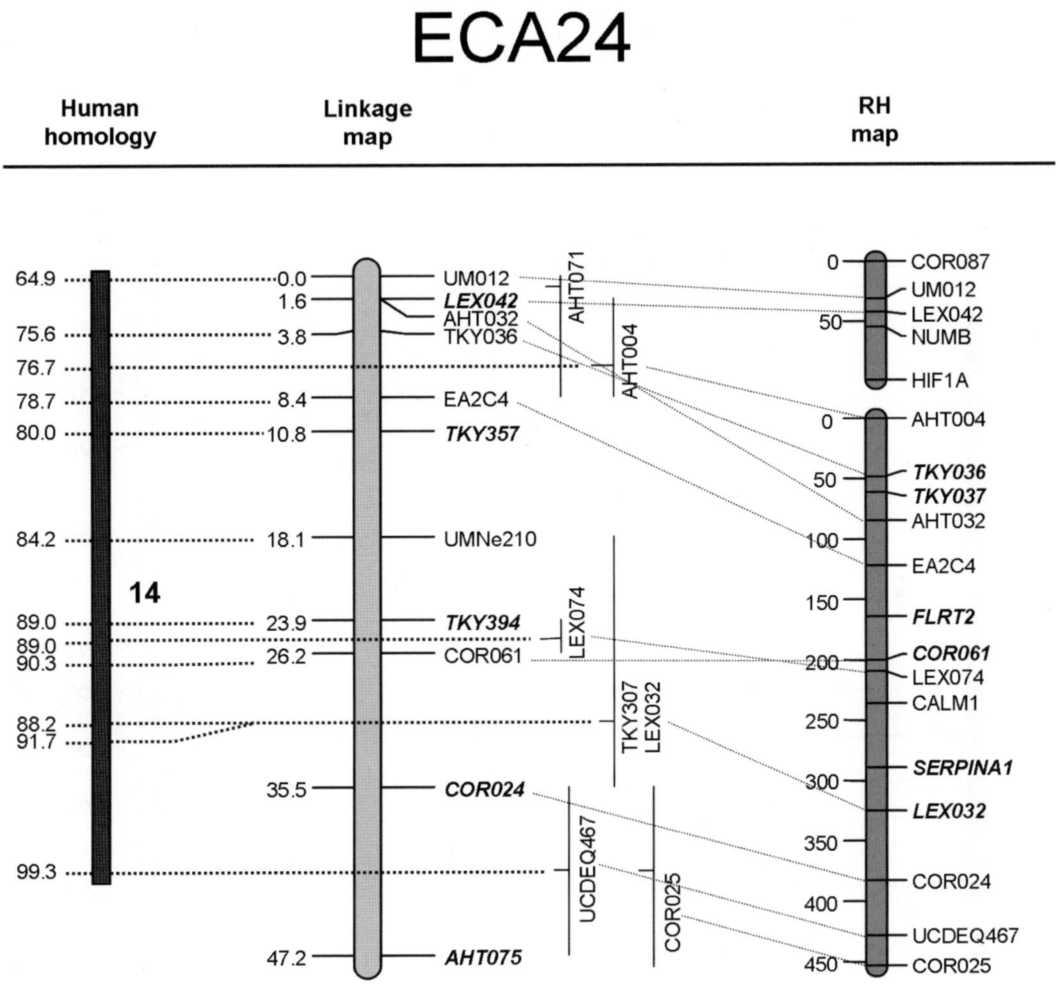

Figure 2.2 (*Continued*)

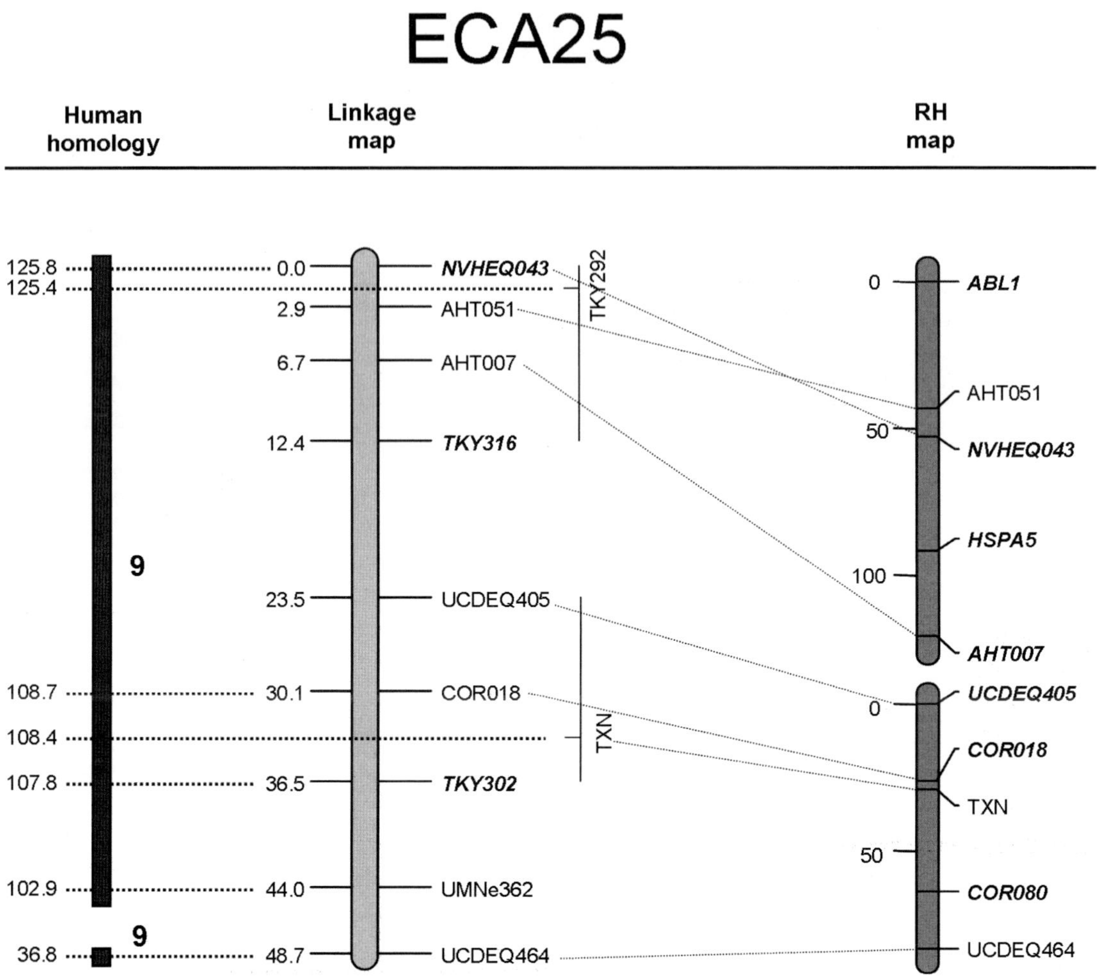

Figure 2.2 (*Continued*)

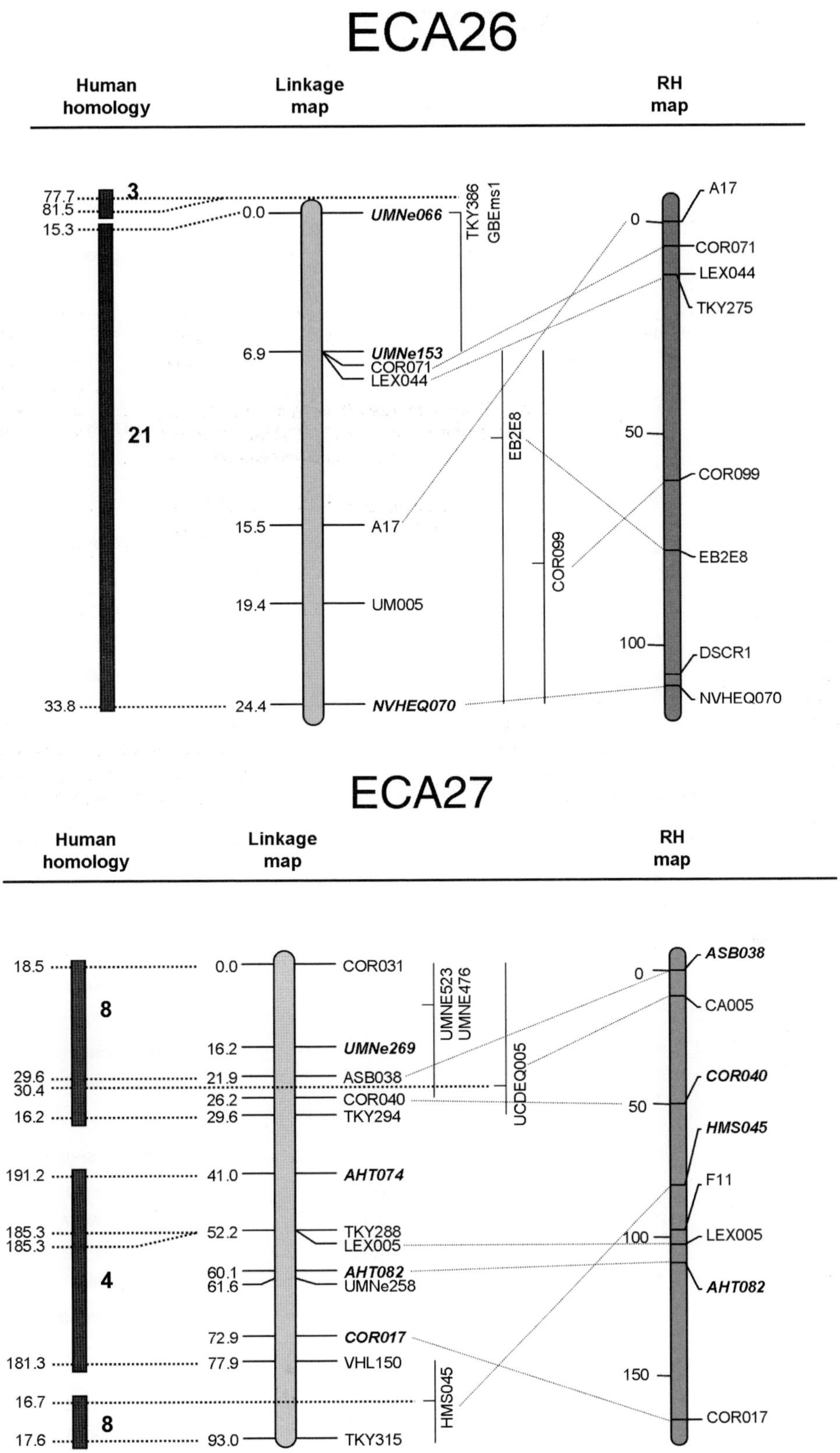

Figure 2.2 (*Continued*)

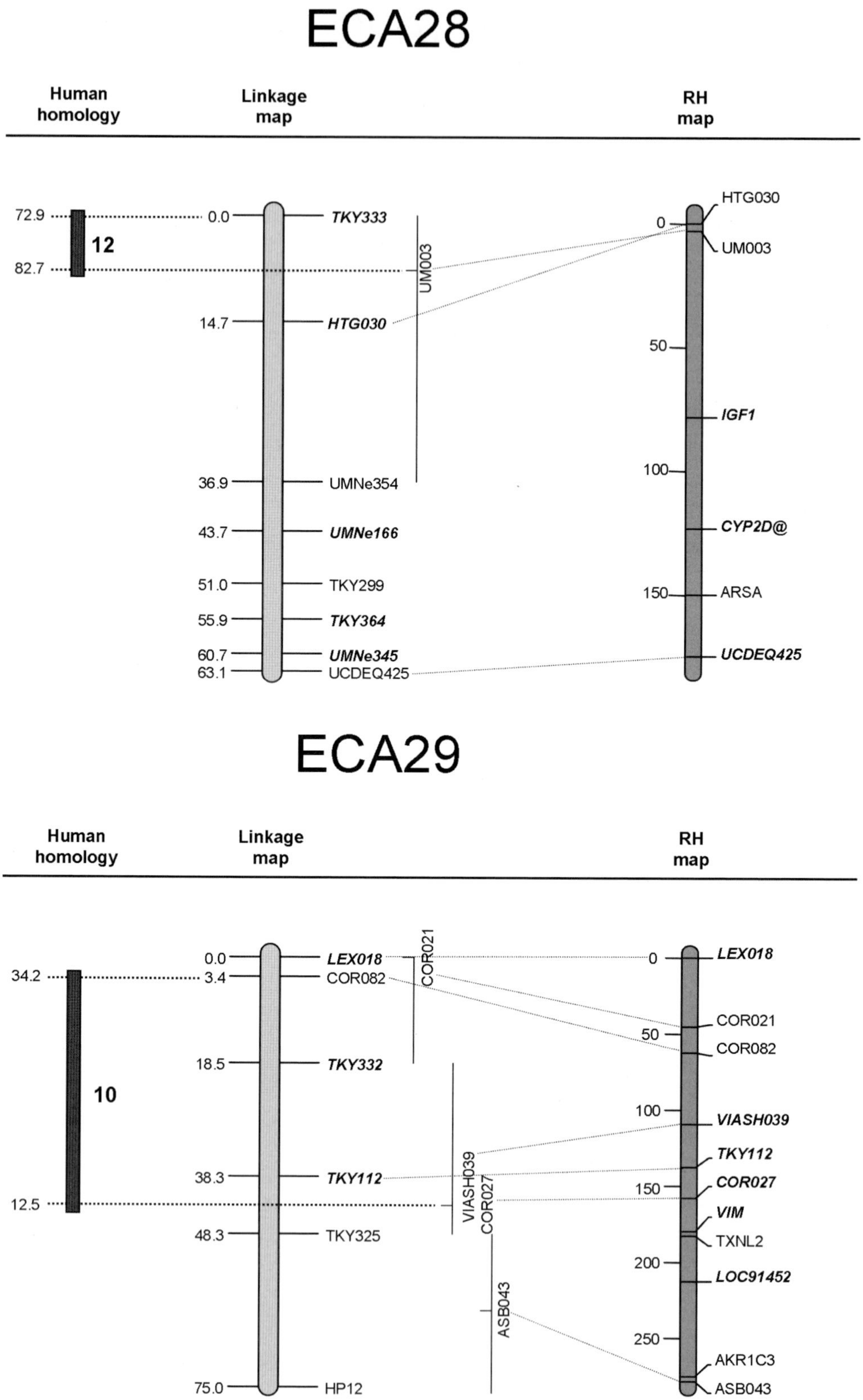

Figure 2.2 (*Continued*)

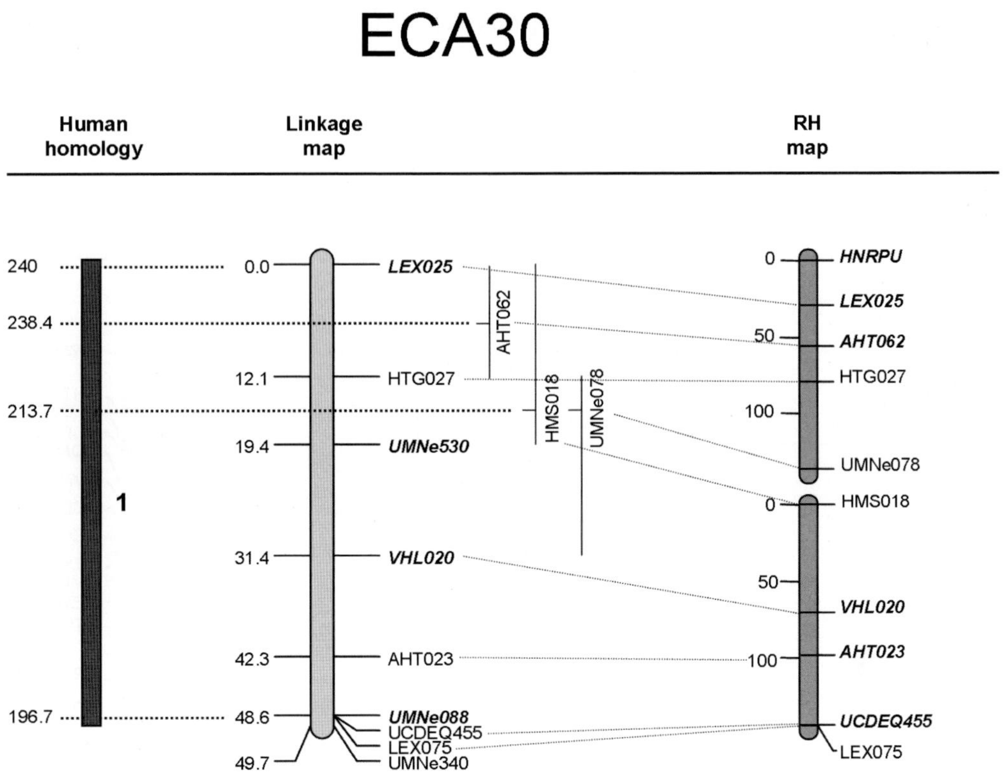

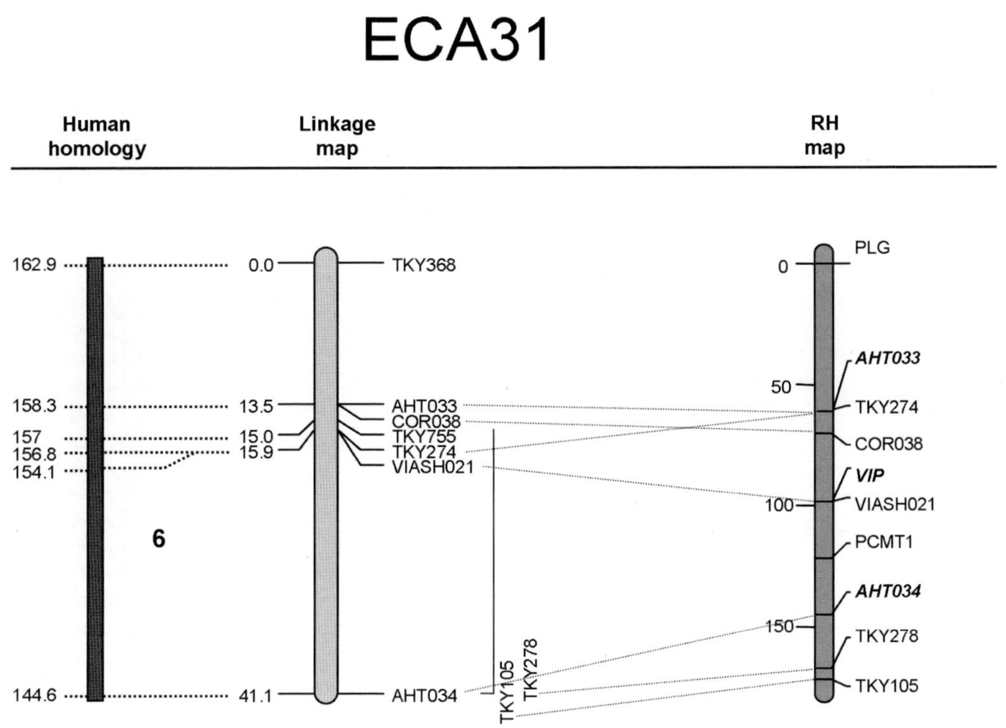

Figure 2.2 (*Continued*)

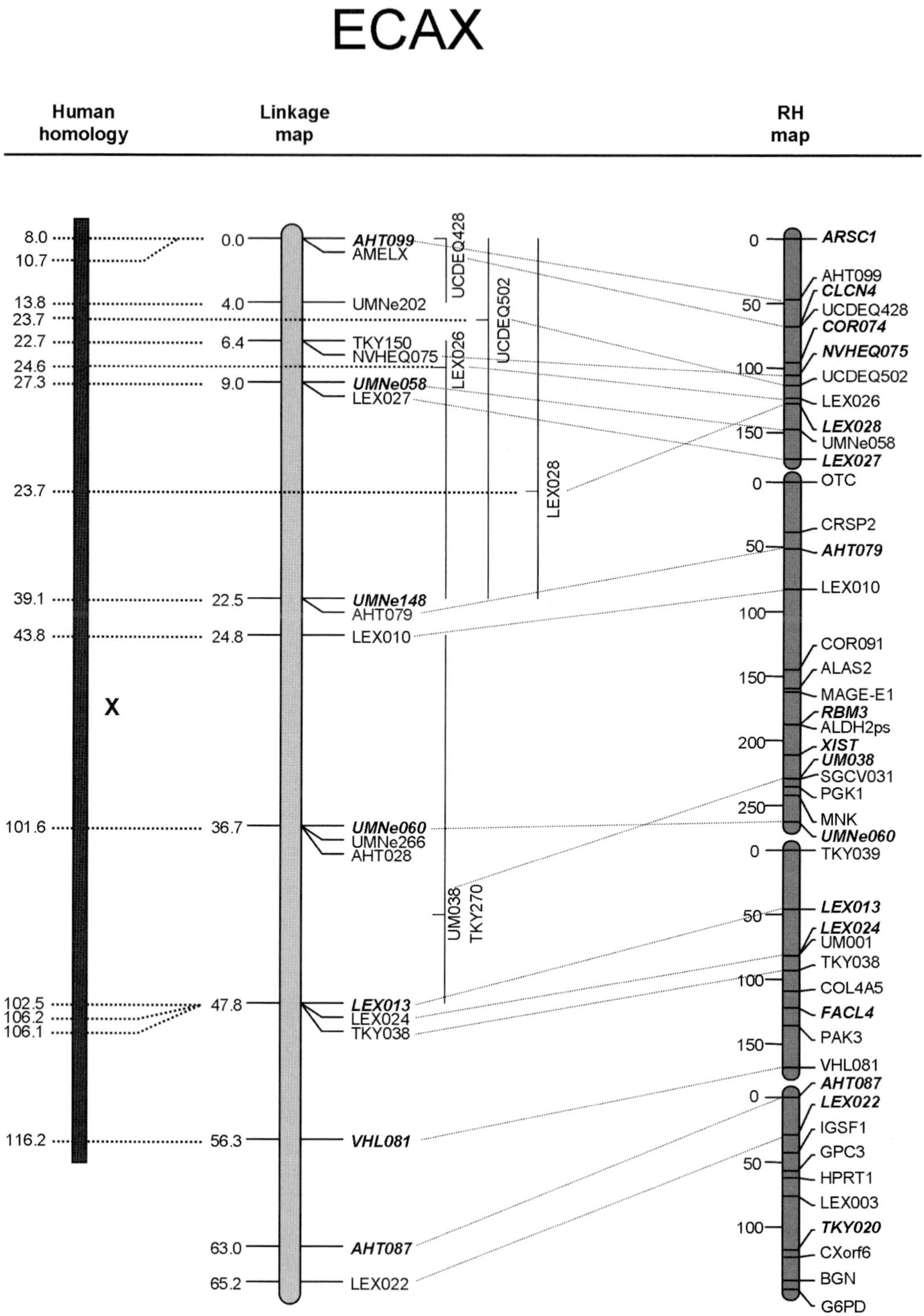

Figure 2.2 (*Continued*)

Microsatellite genome scan panels include multiplex sets of microsatellite markers that are evenly spread over the genome. Sets of markers that amplify robustly under similar PCR conditions are put together for multiplex PCR. Sets of markers that do not PCR-amplify well together are sometimes combined post-PCR before fragment separation. Microsatellite markers in the same set that have a similar fragment size are labeled with different fluorescent dyes. Panels of markers have been developed by a number of research groups; for example, a panel of 316 markers was developed by the Veterinary Genetics Laboratory, University of California, Davis (http://www.vgl.ucdavis.edu/genomic/GenomeScanPanel.pdf). More recently, with the advantage of reference to the genome sequence, Mittmann et al. (2010) identified a highly polymorphic microsatellite scan panel of 322 evenly spaced markers that covers all autosomes and ECAX.

The Continuing Usefulness of the Equine Linkage Map

Despite the recent sequencing of the horse genome and subsequent development of a SNP microarray, the utility of microsatellite panels as a first step in identifying the approximate location of a genetic variant of interest is still apparent. Further, information from the equine linkage map was used to assemble the genome sequence of this species.

For monogenic trait mapping

Genetic linkage maps can be used to identify genes or chromosome regions that regulate various phenotypic traits. Both monogenic traits, controlled by a single gene, and polygenic traits, controlled by an unknown number of genes and often environmental factors, can be studied. As described in previous sections, linkage mapping entails following the segregation of alleles in families to establish whether or not the alleles at one locus co-segregate with alleles at other loci. To be able to do this, it is necessary to determine the parental origin of each allele in the progeny. A trait can be mapped using linkage analysis by genotyping a pedigree in which this trait is segregating using a genome-wide marker panel. Markers that are inherited together (i.e., co-segregating) with the trait are genetically linked to a gene that influences that trait, and thereby indicate its approximate chromosome location. This method, known as linkage mapping, has relatively low resolution and generally results in the identification of large regions (megabases) that require further investigation by fine mapping.

The ultimate aim is to identify the gene responsible for the trait and ideally the causative mutation. To this end the critical region can be reduced in size by haplotype mapping, where more individuals of the same breed – or if possible, other breeds – are genotyped for markers in the region. Ultimately the re-sequencing of affected and control animals would be performed to search the reduced critical interval for candidate mutations. Further confirmatory studies could involve the evaluation of candidate mutations by investigating their biological function within cells.

Linkage mapping has been used to identify chromosome regions and subsequently genes for a number of disorders and phenotypic traits in horses. These include, for instance, the identification or mapping of several coat color genes (Henner et al. 2002; Locke et al., 2002; Swinburne et al., 2002; Terry et al., 2004; Brunberg et al., 2006); mapping of glycogen branching enzyme deficiency (Ward et al., 2003); and mapping of multiple congenital ocular anomalies (MCOA) (Andersson et al., 2008). These studies demonstrate the use of microsatellites for genetic mapping of monogenic traits in the horse. The mapping of monogenic traits is described in more detail in Chapters 11–18.

In assembling the genome sequence

In the same way as linkage groups were assigned to chromosomes and oriented along the chromosome arms, contiguous sequences generated during the horse genome sequencing project were anchored and oriented. This was accomplished by using the existing equine linkage maps (Guerin et al., 2003; Penedo et al., 2005; Swinburne et al., 2006) to place markers with known chromosomal locations onto the equine assembly. This procedure also ensures that the assembly is accurate. The procedure for genome assembly is described in Chapter 7.

References

Allen, W. R., & Brasher, V. (1992). Videoendoscopic evaluation of the mare's uterus. III. Findings in the pregnant mare. *Equine Veterinary Journal*, *24*, 285–291.

Allen, W. R., & Pashen, R. L. (1984). Production of monozygotic (identical) horse twins by embryo micromanipulation. *Journal of Reproductive Fertility*, *71*, 607–613.

Andersson, L. S., Juras, R., Ramsey, D. T., Eason-Butler, J., Ewart, S., Cothran, G., and Lindgren, G. 2008. Equine multiple congenital ocular anomalies maps to a 4.9 megabase interval on horse chromosome 6. *BMC Genetics*, *9*, 88.

Barendse, W., Armitage, S. M. et al. (1994). A genetic linkage map of the bovine genome. *Nature Genetics*, *6*(3), 227–235.

Beever, J. E., Da, Y., Ron, M., & Lewin, H. A. (1994). A genetic map of nine loci on bovine chromosome 2. *Mammalian Genome*, *5*(9), 542–545.

Breen, M., Lindgren, G., Binns, M. M., Norman, J., Irvin, Z., Bell, K., Sandberg, K., & Ellegren, H. (1997). Genetical and physical assignments of equine microsatellites – first integration of anchored markers in horse genome mapping. *Mammalian Genome*, *8*(4), 267–273.

Brunberg, E., Andersson, L., Cothran, G., Sandberg, K., Mikko, S., & Lindgren, G. (2006). A missense mutation in PMEL17 is associated with the Silver coat color in the horse. *BMC Genetics*, *7*, 46.

Chowdhary, B. P., Raudsepp, T., Kata, S. R. et al. (2003). The first-generation whole-genome radiation hybrid map in the horse identifies conserved segments in human and mouse genomes. *Genome Research*, *13*(4), 742–751. Erratum in: *Genome Research* (2003) 13(6a), 1258.

Crawford, A. M., Dodds, K. G. et al. (1995). An autosomal genetic linkage map of the sheep genome. *Genetics*, *140*(2), *703–724*.

Ellegren, H., Johansson, M., Chowdhary, B. P. et al. (1993). Assignment of 20 microsatellite markers to the porcine linkage map. *Genomics*, *16*(2), 431–439.

Ellegren, H., Johansson, M., Sandberg, K., & Andersson, L. (1992). Cloning of highly polymorphic microsatellites in the horse. *Animal Genetics*, *23*(2), 133–142.

Farber, C. R., & Medrano, J. F. (2004). Identification of putative homology between horse microsatellite flanking sequences and cross-species ESTs, mRNAs and genomic sequences. *Animal Genetics*, *35*(1), 28–33.

Godard, S., Vaiman, D., Oustry, A., Nocart, M., Bertaud, M. et al. (1997). Characterization, genetic and physical mapping analysis of 36 horse plasmid and cosmid-derived microsatellites. *Mammalian Genome*, *8*(10), 745–750.

Guérin, G., Bailey, E., Bernoco, D. et al. (1999). Report of the International Equine Gene Mapping Workshop: Male linkage map. *Animal Genetics*, *30*(5), 341–354.

Guérin, G., Bailey, E., Bernoco, D. et al. (2003). The second generation of the International Equine Gene Mapping Workshop half-sibling linkage map. *Animal Genetics*, *34*(3), 161–168.

Henner, J., Poncet, P. A., Guérin, G., Hagger, C., Stranzinger, G., & Rieder, S. (2002). Genetic mapping of the (G)-locus, responsible for the coat color phenotype "progressive greying with age" in horses (Equus caballus). *Mammalian Genome*, *13*(9), 535–537.

Kosambi, D. D. (1944). The estimation of map distances from recombination values. *Annals of Eugenics*, *12*, 172–175.

Lampe, V., Dierks, C., Komm, K., & Distl, O. (2009). Identification of a new quantitative trait locus on equine chromosome 18 responsible for osteochondrosis in Hanoverian warmblood horses. *Journal of Animal Science*, *87*(11), 3477–3481.

Lander, E. S., & Green, P. (1987). Construction of multilocus genetic linkage maps in humans. *Proceedings of the National Academy of Sciences*, *84*, 2363–2367.

Lindgren, G., Sandberg, K., Persson, H., Marklund, S., Breen, M. et al. (1998). A primary male autosomal linkage map of the horse genome. *Genome Research*, *8*(9), 951–966.

Locke, M. M., Penedo, M. C., Bricker, S. J., Millon, L. V., & Murray, J. (2002). Linkage of the grey coat colour locus to microsatellites on horse chromosome 25. *Animal Genetics*, *33*, 329–337.

Ma, R. Z., Beever, J. E., Da, Y. et al. (1996). A male linkage map of the cattle (*Bos taurus*) genome. *Journal of Heredity*, *87*(4), 261–271.

Matise, T. C., Perlin, M., & Chakravarti, A. (1994). Automated construction of genetic linkage maps using an expert system (MultiMap): A human genome linkage map. *Nature Genetics*, *6*(4), 384–390.

Mittmann, E. H., Lampe, V., Mömke, S., Zeitz, A., & Distl, O. (2010). Characterization of a minimal microsatellite set for whole genome scans informative in warmblood and coldblood horse breeds. *Journal of Heredity*, *101*(2), 246–250.

Penedo, M. C., Millon, L. V., Bernoco, D. et al. (2005). International Equine Gene Mapping Workshop Report: A comprehensive linkage map constructed with data from new markers and by merging four mapping resources. *Cytogenetic and Genome Research*, *111*(1), 5–15.

Scherzer, J., Fayrer-Hosken, R. A., Ray, L., Hurley, D. J., & Heusner, G. L. (2008). Advancements in large animal embryo transfer and related biotechnologies. *Reproduction in Domestic Animals*, *43*(3), 371–376.

Scott, I. S., & Long, S. E. (1980). An examination of chromosomes in the stallion (Equus caballus) during meiosis. *Cytogenetics and Cell Genetics*, *26*(1), 7–13.

Swinburne, J. E., Bogle, H., Klukowska-Rötzler, J., Drögemüller, M., Leeb, T. et al. (2009). A whole-genome scan for recurrent airway obstruction in Warmblood sport horses indicates two positional candidate regions. *Mammalian Genome*, *20*(8), 504–515.

Swinburne, J. E., Boursnell, M., Hill, G. et al. (2006). Single linkage group per chromosome genetic linkage map for the horse, based on two three-generation, full-sibling, crossbred horse reference families. *Genomics*, *87*(1), 1–29.

Swinburne, J., Gerstenberg, C. et al. (2000). First comprehensive low-density horse linkage map based on two 3-generation, full-sibling, cross-bred horse reference families. *Genomics*, *66*(2), 123–134.

Swinburne, J. E., Hopkins, A., & Binns, M. M. (2002). Assignment of the horse grey coat colour gene to ECA25 using whole genome scanning. *Animal Genetics*, *33*(5), 338–342.

Terry, R. B., Archer, S., Brooks, S., Bernoco, D., & Bailey, E. (2004). Assignment of the appaloosa coat colour gene (LP) to equine chromosome 1. *Animal Genetics*, *35*(2), 134–137.

Tozaki, T., Swinburne, J., Hirota, K., Hasegawa, T., Ishida, N., & Tobe, T. (2007). Improved resolution of the comparative horse-human map: Investigating markers with *in silico* and linkage mapping approaches. *Gene*, *392*(1–2), 181–186.

Vaiman, D., Schibler, L., Bourgeois, F., Oustry, A., Amigues, Y., & Cribiu, E. P. (1996). A genetic linkage map of the male goat genome. *Genetics*, *144*(1), 279–305.

Wade, C. M., Giulotto, E., Sigurdsson, S. et al. (2009). Genome sequence, comparative analysis, and population genetics of the domestic horse. *Science*, *326*(5954), 865–867.

Ward, T. L., Valberg, S. J., Lear, T. L. et al. (2003). Genetic mapping of GBE1 and its association with glycogen storage disease IV in American Quarter horses. *Cytogenetic and Genome Research*, *102*(1–4), 201–206.

Werner, P., Mellersh, C. S., Raducha, M. G. et al. (1999). Anchoring of canine linkage groups with chromosome-specific markers. *Mammalian Genome*, *10*(8), 814–823.

Zeitz, A., Spötter, A., Blazyczek, I., Diesterbeck, U., Ohnesorge, B. et al. (2009). Whole-genome scan for guttural pouch tympany in Arabian and German warmblood horses. *Animal Genetics*, *40*(6), 917–924.

3 Physical and comparative maps

Terje Raudsepp and Bhanu P. Chowdhary

Introduction

Physical maps represent the location and linear order of genes and markers on chromosomes, and are typically portrayed in a cytogenetic background. Thus, the construction of physical maps requires basic knowledge about the chromosomes and the karyotype of a species. The most common approaches for the construction of physical maps are in situ molecular hybridization to chromosomes leading to the development of cytogenetic maps, and genotyping markers on somatic cell and radiation hybrid panels, resulting in synteny and radiation hybrid (RH) maps. Additionally, high-resolution physical maps can be generated by constructing contigs of overlapping large insert clones or by DNA fingerprinting of entire genomic libraries. Finally, the ultimate physical map of any genomic region, chromosome, or the whole genome is the complete DNA sequence. Overall, "physical" mapping indicates that the maps reflect the actual physical organization of chromosomes, thus opposing the genetic linkage maps, which are based on meiotic recombination and are reviewed in Chapter 2.

In this chapter, we focus on cytogenetic, radiation hybrid, and comparative maps, leaving a detailed review of genome sequencing for Chapter 7. We first address equine chromosomes, standard karyotypes, and chromosome nomenclature to provide the necessary background for the construction of cytogenetic maps. This is followed by a summary of techniques and resources necessary for the development of cytogenetic maps and an overview of all *in situ* hybridization-based localizations of genetic markers in the horse genome. The importance of cytogenetic mapping in facilitating equine genome analysis by anchoring and integrating RH and linkage maps, and assisting the assembly of the whole genome (WG) sequence draft (Wade et al., 2009), is discussed. Also, a section is devoted to explaining how cytogenetic mapping has contributed to equine clinical cytogenetics.

In the second part of the chapter, we provide a brief description of somatic cell hybrid (SCH) genetics and the generation of SCH and RH panels for mapping the horse genome. The main focus in this section will be the Texas A&M University RH panel and the two major sets of gene maps generated using the panel – the first generation map comprising 730 markers (Chowdhary et al., 2003) and the second generation map comprising 4,100 markers (Raudsepp et al., 2008). The significance of high-resolution RH maps for all equine autosomes and the X chromosome and their contribution to understanding the overall organization of the equine genome is discussed.

The last part of this chapter provides a detailed overview of how the equine genome compares to human, mouse, and other mammalian and vertebrate genomes. The description includes comparisons

Equine Genomics, First Edition. Edited by Bhanu P. Chowdhary.

made on the basis of Zoo-FISH and gene mapping. The latter includes the latest data from the high-resolution RH map that provides interesting comparative information in relation to sequenced mammalian genomes.

Horse Chromosomes

Chromosome number

The first reports about horse chromosomes date back to early 1900s when it was proposed that the horse has an XO sex chromosome system (Wodsedalek, 1914) with a diploid number between 20 and 37 (Kirillow, 1912; Masui, 1919; Painter, 1924). Decades later, thanks to critical improvements in chromosome analysis methodology, these preliminary findings were revised showing that the correct diploid chromosome number of the domestic horse (*Equus caballus*, ECA) is 2n=64 (Rothfels et al., 1959). Also, like other mammalian species, the horse has the XY sex chromosome system (Makino, 1942).

Chromosome identification

A wide range of banding techniques has been used to study horse chromosomes (for review, see Chowdhary and Raudsepp, 2000). Many of these, have become methods of the past, while others like G-, R-, and DAPI banding have retained importance for chromosome identification in clinical cytogenetics (see Chapter 18) and gene mapping. *The G-banding* technique (Seabright, 1971) highlights AT-rich DNA and is still one of the most common methods for chromosome identification in cytogenetic analysis. Likewise, in fluorescence in situ hybridization (FISH)-based gene mapping, a G-banding-like pattern is produced by staining the chromosomes with a fluorescent DNA-binding dye *4′,6-diamidino-2-phenylindole (DAPI).* Alternatively, the chromosomes can be identified with *the R-banding* technique, which highlights GC-rich DNA and can be applied as fluorescent (RBG; Molteni et al., 1982) or as Giemsa staining based approach (RHG; e.g., Romagnano et al., 1987).

Beside these banding techniques, there are methods that specifically highlight certain chromosomal regions, such as constitutive heterochromatin, telomeric repeat sequences, and nucleolus organizing regions, thus providing information about the molecular organization of the chromosomes.

The C-banding method (Arrighi and Hsu, 1971) highlights centromeric heterochromatin and was in fact the first differential staining technique applied to horse chromosomes (Comings and Mattoccia, 1972). All horse autosomes, except ECA11, have darkly stained pericentromeric heterochromatic blocks (Buckland et al., 1976; Power, 1990; Ryder et al., 1978), which correspond to tandemly repeated satellite DNA sequences. Notably, molecular explanation for the absence of the C-band in ECA11 was provided only recently. Horse genome sequencing revealed that the centromere of ECA11 is evolutionarily new, has formed as a result of centromere repositioning, and has not yet acquired satellite repeats for C-banding (Carbone et al., 2006; Wade et al., 2009). As dicussed in Chapter 18, C-banding is also a useful method for quick identification of equine sex chromosomes (Buckland et al., 1976; Raudsepp et al., 2010; Ryder et al., 1978).

Classical *THA-banding* method (Dutrillaux et al., 1973), which highlights telomeric repeat sequences, has been rarely applied in the horse. Instead, the location of telomeric $(TTAGGG)_n$ repeats has been studied using molecular cytogenetic approaches (de la Sena et al., 1995). Except for their usual location, no intercalary telomeric sequences have been observed in the horse genome.

The nucleolus organizer regions (NORs) are chromosomal sites of 5.8S, 18S, and 28S ribosomal RNA (rRNA) genes and were originally detected by Ag-I technique or NOR-banding (Bloom and Goodpasture, 1976; Goodpasture and Bloom, 1975). The technique involves staining with silver nitrate and detects only the actively transcribed rRNA genes. Nowadays, FISH with rDNA probes can identify all NORs in the genome regardless of their transcriptional status. In the horse genome, NORs are located on ECA1p, 27, 28, and 31. All NORs, except the one on ECA27, were discovered using silver nitrate method (Kopp et al., 1981; Sysa et al., 1977). The cluster of rRNA genes on ECA27 was identified years later using FISH with human 18S, 5.8S, and 28S rDNA probes (Deryusheva et al., 1997). While the location and count of NORs has traditionally been a part of cytogenetic characterization of equine karyotype, assignment of rRNA genes to specific chromosomes also marks the beginning of physical gene mapping in horses.

Chromosome size

The relative length of horse chromosomes is reported only in a few studies (Hansen, 1984; Melchior and Hohn, 1976; Molteni et al., 1982). The results vary slightly between authors depending on whether G- (Melchior and Hohn, 1976), R- (Molteni et al., 1982) or Q-banding (Hansen, 1984) was used for chromosome identification. It is proposed that Q-banding followed by Giemsa staining gives the most accurate measurements (Hansen, 1984). Indeed, the relative lengths of Q-banded chromosomes as determined in 1984 are in remarkable agreement with the size of horse chromosomes according to the recent genome sequence map (Table 3.1; EcuCab2; http://uswest.ensembl.org/index.html), illustrating the quality and accuracy of the early cytogenetic studies in the horse.

The horse karyotype

Ever since the correct chromosome number in the horse was determined, the equine chromosomes have been presented in several ways. However, it was not before the correct number of metacentric and sub-metacentric (13 pairs of autosomes + the sex chromosomes) and acrocentric (18 pairs of autosomes) chromosomes was ascertained (Benirschke et al., 1965; Hsu and Benirschke, 1967) that a stable karyotype pattern started to emerge. This led to the development of the first (Ford et al., 1980) and the second (Richer et al., 1990) karyotype standards for the horse (for details, see Chowdhary and Raudsepp, 2000). Despite being useful for cytogenetic analysis, the standards fell short of an enumerated nomenclature of chromosome bands and were of limited use for gene mapping. To cater to the needs of both cytogenetics and gene mapping in correctly identifying equine chromosomes, a new high-resolution international system of chromosome nomenclature for the horse was developed (ISCNH, 1997). The new nomenclature provided a detailed description of both G- and R-banded chromosomes along with idiograms showing landmarks and band numbers for each of the banding approaches (Figure 3.1). Since then, the ISCNH 1997 has served as the internationally accepted platform for equine clinical cytogenetics and physical gene mapping.

Gene Mapping in Horses – Historical Background

Gene mapping studies in horses date back almost five decades, when *G6PD* was assigned to the X-chromosome (Mathai et al., 1966; Trujillo et al., 1965). In 1970s–1980s, significant contributions

Table 3.1 The size of horse chromosomes and a summary of physically mapped markers in the horse genome. Data is retrieved as follows: cytogenetic measurements (Hansen, 1984); molecular measurements and number of protein coding genes (EcuCab2 Ensembl: http://uswest.ensembl.org/index.html); number of loci on RH, FISH, and comparative maps (Raudsepp et al., 2008).

Horse chr.	Relative size (%), cytogenetic	Relative size, genome sequence	Size, Mb	RH mapped loci	FISH mapped loci	Compar-ative map loci	Known protein coding genes
ECA1	7.32	7.51	185.8	260	100	147	1,232
ECA2	4.67	4.88	120.8	196	55	96	802
ECA3	4.48	4.83	119.4	143	46	81	620
ECA4	4.15	4.39	108.5	196	43	98	536
ECA5	3.69	4.03	99.6	177	48	102	811
ECA6	3.64	3.42	84.7	158	43	92	769
ECA7	3.57	3.98	98.5	120	70	62	922
ECA8	3.52	3.80	94.0	120	44	74	532
ECA9	3.45	3.80	94.0	110	22	44	532
ECA10	3.37	3.39	83.9	165	51	80	685
ECA11	2.27	2.48	61.3	129	46	59	902
ECA12	1.99	1.33	33.0	55	12	27	494
ECA13	2.00	1.72	42.5	95	23	43	497
ECA14	3.92	3.80	93.9	172	48	93	508
ECA15	3.82	3.70	91.5	136	40	65	487
ECA16	3.72	3.53	87.3	135	48	66	519
ECA17	3.57	3.26	80.7	121	39	54	250
ECA18	3.43	3.33	82.5	106	21	52	302
ECA19	3.23	2.42	59.9	92	19	31	313
ECA20	3.01	2.59	64.1	134	38	53	567
ECA21	2.50	2.33	57.7	116	37	66	281
ECA22	2.47	2.02	49.9	133	52	71	403
ECA23	2.48	2.25	55.7	76	21	28	218
ECA24	2.17	1.89	46.7	59	19	21	289
ECA25	1.97	1.60	39.5	45	16	21	417
ECA26	1.83	1.69	41.8	78	16	33	154
ECA27	1.76	1.61	39.9	70	12	26	156
ECA28	1.79	1.86	46.1	57	16	29	296
ECA29	1.76	1.36	33.6	53	9	25	140
ECA30	1.81	1.21	30.0	44	9	26	127
ECA31	1.61	1.01	24.9	43	11	17	114
ECAX	5.01	5.02	124.1	222	70	122	553
ECAY	1.62	2.63	~65	23	n/a	37	n/a
TOTAL	101.6	98.67	2440.8	3,839	1,144	1,941	15,428

were made to develop equine linkage maps (Andersson and Sandberg, 1984; Sandberg, 1974; Sandberg and Andersson, 1984), but chromosome and karyotype studies in the horse had not progressed sufficiently to physically assign any of the linkage groups to specific chromosomes. With the exception of cytogenetic localization of rRNA gene clusters to autosomes in late 1970s (Sysa et al., 1977), the formal beginning of physical gene mapping in horses falls into late 1980s when the major histocompatibility complex (ELA) was assigned to ECA20 by radioactive in situ hybridization (Ansari et al., 1988; Makinen et al., 1989). The coming years showed a slow but steady progress in gene mapping. The main stimulus came in 1995 when the international equine community, under the aegis of the Dorothy Russell Havemeyer Foundation, decided to use a concerted approach to map the horse genome. This was possible because both the breeders and the horse geneticists agreed that a framework map is necessary to study the underlying genetics of numerous congenital disorders

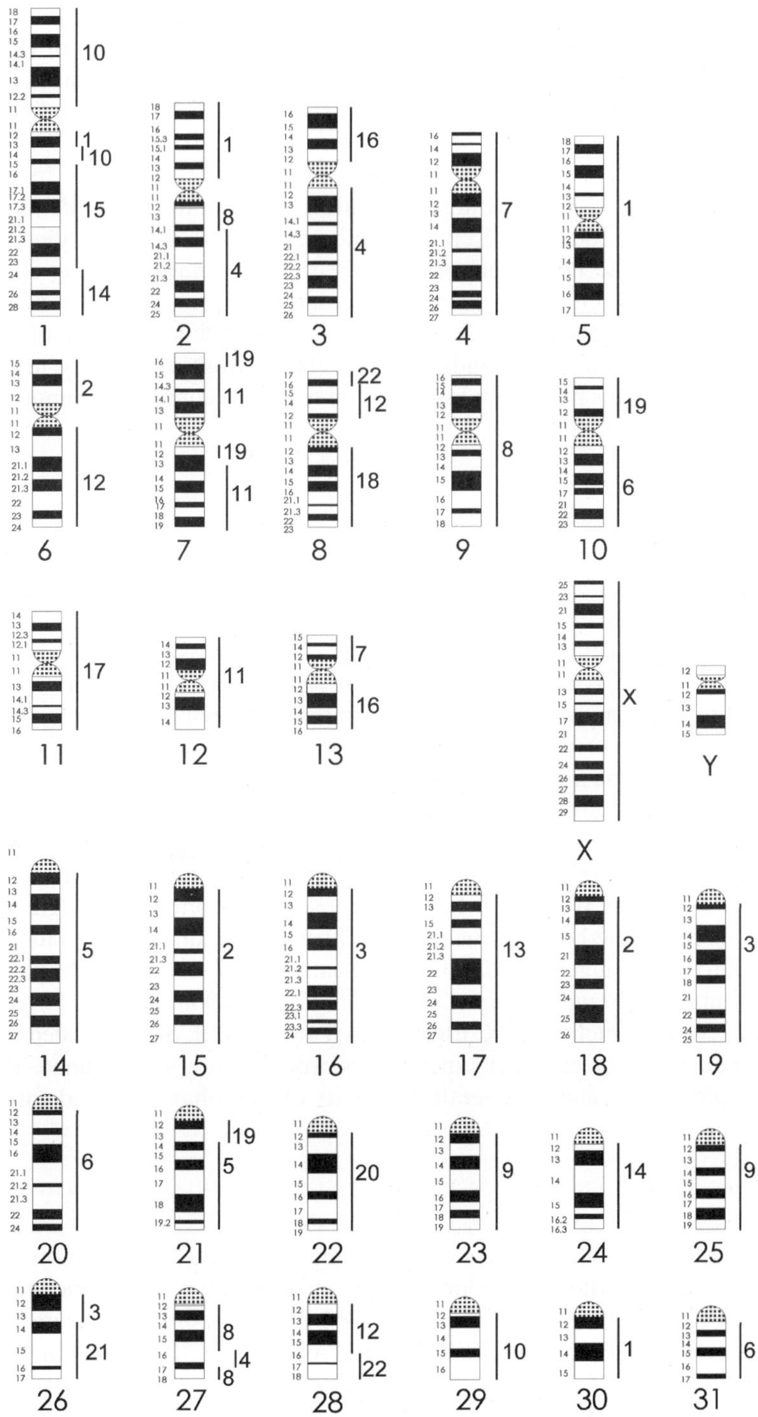

Figure 3.1 G-banded horse standard ideogram (ISCNH, 1997). Segmental homology with human chromosomes is shown to the right of each horse chromosome (Raudsepp et al., 1996; Raudsepp et al., 2008; Yang et al., 2004b).

known in horses and to find means to control them. Since then, FISH and radiation hybrid panels have occupied the central role in constructing physical maps. Development of resources, such as cDNA and bacterial artificial chromosome (BAC) libraries (for review, see Rubes et al., 2009), and identification of microsatellite and gene-specific markers (Godard et al., 1998; Godard et al., 1997; Guérin et al., 1999; Lindgren et al., 1998; Milenkovic et al., 2002; Tozaki et al., 2007) have provided tools for the construction of cytogenetic (Milenkovic et al., 2002), radiation hybrid (Chowdhary et al., 2003; Raudsepp et al., 2008) and high-resolution BAC contig maps (Brinkmeyer-Langford et al., 2008; Gustafson et al., 2003; Raudsepp and Chowdhary, 2008b). Altogether, more than 5,000 loci have been mapped in the horse genome using various physical mapping approaches (Paria, 2009; Raudsepp et al., 2008). Of these, about 900 loci integrate physical and genetic linkage maps, while about 2,000 loci are aligned with the human genome sequence and establish a comparative framework between the two genomes (Raudsepp et al., 2008). Most importantly, integrated physical, meiotic, and comparative maps have been instrumental for the assembly and annotation of the horse genome sequence (Wade et al., 2009).

Cytogenetic Map

Fluorescence in situ hybridization

Fluoresence in situ hybridization (FISH) to nuclear chromatin is the most direct approach for constructing cytogenetic maps. The method relies on Watson-Crick base pairing complementarity principle and permits the location of DNA markers in their original place, or in situ (Raudsepp and Chowdhary, 2008a). The two major components of ISH are the *probe* and the *target*. In case of cytogenetic maps, the targets can be mitotic or meiotic chromosome preparations at different stages of the cell cycle, DNA fibers (Raudsepp and Chowdhary, 2008a; Rubes et al., 2009; Speicher and Carter, 2005; Trask, 2002), or DNA tiling-arrays (Shaffer and Bejjani, 2006; Sharp, 2009). The probes vary considerably in size and origin, from a few-base-pairs-long telomeric or centromeric repeats to composite DNA sequences from the whole chromosome, chromosomal segment, or the whole genome. The probes are labeled directly with fluorochromes or indirectly with molecules that need binding by fluorochrome-conjugated antibodies to visualize the signals. Over the past 25 years, FISH methodology has improved tremendously (Dobigny and Yang, 2008; Lichter, 1997; Nederlof et al., 1989; Rubes et al., 2009; Speicher and Carter, 2005; Trask, 2002). Sophisticated imaging systems and the available number of spectrally distinct fluorochromes allow simultaneous detection of two, three or even multiple probes, and the overall sensitivity of FISH has increased about 10,000 times.

FISH mapping in the horse

Direct physical assignment of loci to horse chromosomes by in situ hybridization started more than twenty years ago using the radioactive approach. Due to the lack of horse-specific probes, human and pig genomic or cDNA clones were used. The first in situ mapped loci were the equine major histocompatibility complex (*ELA*) to ECA20q14-q22 (Ansari et al., 1988; Makinen et al., 1989) and glucosephosphate isomerase (*GPI*) to ECA10pter (Harbitz et al., 1990). The fluoresence method was for the first time used to assign hemoglobin alpha (*HBA*) to ECA13q (Oakenfull et al., 1993). From that point on, FISH has been the primary approach for the construction of cytogenetic maps in the horse.

During the early days of FISH mapping, the main limitation was the availability of suitable probes. Clones containing equine genes and microsatellites were PCR products, short cDNA sequences, or originated from relatively small-insert phage and cosmid libraries (Breen et al., 1997b; Godard et al., 1998; Godard et al., 2000; Godard et al., 1997). Sometimes, due to the lack of suitable horse probes, heterologous human or porcine BACs were used for mapping (Raudsepp et al., 1999). Despite this, in less than seven years, more than 331 DNA markers and 48 genes/ESTs were localized to equine chromosomes by FISH (for review, see Chowdhary and Raudsepp, 2000). An important breakthrough in cytogenetic mapping came with the availability of the INRA (Godard et al., 1998; Milenkovic et al., 2002), CHORI-241 (http://bacpac.chori.org/equine241.htm), and TAMU (http://hbz7.tamu.edu/homelinks/bac_est/bac.htm) large-insert genomic BAC libraries (for review, see Rubes et al., 2009). The use of BAC clones, which contain 150–200 kb fragments of the horse genome, essentially improved the quality of hybridization signals and led to rapid qualitative and quantitative improvements in FISH mapping. As a result, the number of FISH mapped markers in the horse genome started to increase rapidly. Majority of publications reported FISH assignment of a single gene or a small number of loci (e.g., (Breen et al., 1997b; Lear et al., 1998a; Lear et al., 1998b; Lear et al., 2000; Raudsepp et al., 1999; Raudsepp et al., 1997), while more comprehensive studies involved about 30–50 loci (Godard et al., 1998; Godard et al., 2000; Godard et al., 1997; Lear et al., 2001; Lindgren et al., 2001; Mariat et al., 2001; Perrocheau et al., 2005). A few studies consolidated all available map information and constructed FISH maps for all autosomes and the X chromosome (Chowdhary et al., 2003; Milenkovic et al., 2002; Perrocheau et al., 2006; Raudsepp et al., 2008). Cytogenetic mapping was also used to localize candidate genes for several equine conditions, such as recurrent airway obstruction (Jost et al., 2007; Klukowska-Rotzler et al., 2006a), genes involved in circadian clock (Murphy et al., 2007), infection (Horin et al., 2008), immunity (Musilova et al. 2005), and the development of skeletal, gastrointestinal, cardiovascular, and nervous systems (Zabek et al., 2009; Zabek et al., 2007). Also, FISH has been instrumental for validating the assembly of high-resolution BAC contig maps that were constructed for the ELA (Gustafson et al., 2003), a 3.3 Mb region on ECA21 (Brinkmeyer-Langford et al., 2008), the pseudoautosomal region (Raudsepp and Chowdhary, 2008b), and the Y chromosome (Paria, 2009; Raudsepp et al., 2004b). Likewise, FISH has been extremely useful in accurate physical alignment and orientation of synteny, genetic linkage, and RH maps (Brinkmeyer-Langford et al., 2005; Chowdhary et al., 2002b; Chowdhary et al., 2003; Goh et al., 2007; Gustafson-Seabury et al., 2005; Lee et al., 2004; Perrocheau et al., 2006; Raudsepp et al., 2008; Wagner et al., 2006), as well as for the draft assembly of the genome sequence (Lear et al., 2007; Wade et al., 2009). Presently, 1,144 loci (Raudsepp et al., 2008) have been assigned by FISH to equine chromosomes (Table 3.1). Given that the estimated size of the horse genome is 2500–2700 Mb (Wade et al., 2009), this provides an average of one cytogenetically mapped marker at every 2.2–2.4 Mb of the genome.

Multicolor and high-resolution FISH

Horse is probably the only domestic species where both multicolor and high-resolution FISH have been extensively used in physical gene mapping. Dual- or triple-color FISH on metaphase, prometaphase, and interphase chromosomes (Figure 3.2) have been pivotal in orienting and ordering closely located markers, particularly markers that are tightly linked in the RH and meiotic maps.

An essential improvement in FISH resolution is provided by fiber-FISH. Typically, DNA probes are mapped to metaphase chromosomes at a resolution of about 2–5 Mb (Raudsepp and Chowdhary, 2008a). Fiber-FISH, on the other hand, enables to distinguish probes separated by 1–2 kb and can

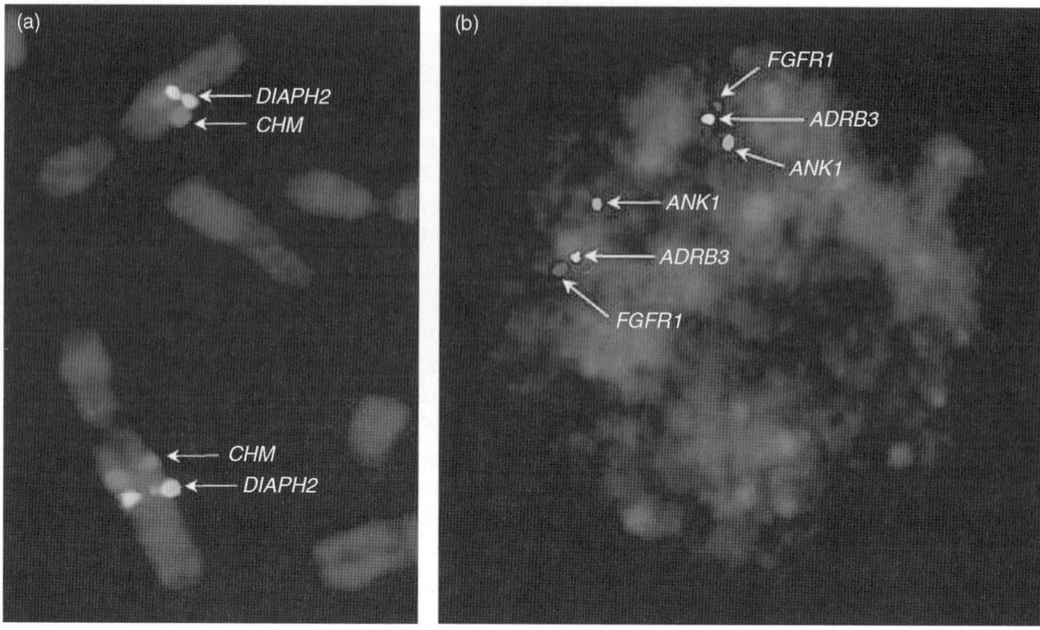

Figure 3.2 Multicolor FISH: (a) dual-color FISH to order two genes on ECAX in metaphase; (b) three-color FISH to order three closely located loci on ECA27 in interphase. (*For color details, see color plate section.*)

be used for positional cloning, determining transcriptional orientation of clones, or detection of minor chromosomal rearrangements (Laan et al., 1996). In the horse, the technique has been used for validating contigs of overlapping BAC clones in the pseudoautosomal region (Raudsepp and Chowdhary, 2008b) and the male-specific region of the horse Y chromosome (Raudsepp et al., 2004b; see also Chapter 5).

Physical mapping using flow-sorted and microdissected chromosomes

Another source of probes for cytogenetic mapping is flow-sorted and microdissected chromosomes. In flow sorting, a fluorescence activated cell sorter system (FACS) separates individual chromosomes based on their DNA content and the relative proportion of AT/GC sequences (Lebo, 1982; Lebo and Bastian, 1982). The procedure results in generating a flow karyotype, distinguishing each chromosome (Lebo, 1982). Once separated, collected, and amplified, DNA from individual chromosomes is available for various mapping purposes. In contrast to other domestic species where flow karyotypes were generated in late 1990s (see Chowdhary and Raudsepp, 2005), flow sorting of horse chromosomes had a late start. The only flow karyotype of the horse was generated in 2004 and used for comparative chromosome painting to human, donkey, and mule chromosomes (Yang et al., 2004a), and later for karyotype comparisons with other equids and Perssiodactyls (Trifonov et al., 2008). Additionally, flow-sorted ECAX has been used to detect X chromosome aneuploidies (Breen et al., 1997a). Notably, ECAY is the only flow-sorted horse chromosome that has been used for physical mapping and gene discovery (Paria, 2009; Paria et al., 2011; see Chapter 5).

A convenient alternative to flow sorting is chromosome microdissection, which allows isolation of DNA from whole chromosomes, as well as from chromosome arms, segments, or bands (Chaudhary et al., 1998; Chowdhary and Raudsepp, 2001). The technique involves manual (Guan et al., 1992; Meltzer et al., 1992) or laser-assisted (Kubickova et al., 2002) microscopic scraping of chromosomal regions of interest, followed by microdissected DNA amplification by DOP-PCR (Telenius et al., 1992), or some other whole genome amplification method (Sorensen et al., 2007). The amplified DNA is labeled and used as a painting probe for FISH experiments or as a source of DNA for cloning and mapping.

In horses, microdissected probes were first generated for all metacentric and sub-metacentric autosomes and the sex chromosomes (Raudsepp and Chowdhary, 1999), and later for all chromosomes (Bugno et al., 2009a). Additionally, arm-specific probes are available for the X chromosome (Bugno and Slota, 2007) and bi-armed autosomes (Musilova et al., 2007). These composite probes have been used for cross-species chromosome painting between horse and donkey (Raudsepp and Chowdhary, 1999) and for molecular cytogenetic analysis of various chromosomal abnormalities (Bugno and Slota, 2007; Bugno et al., 2007b, 2009a). In a few cases, chromosome-specific probes have found use for physical gene mapping. For example, microdissected ECA1 (Bowling et al., 1998) and ECA6 and ECA12 (Chowdhary et al., 1998b) were used to construct chromosome specific libraries and identify microsatellite markers. Some of these markers, namely 1CA01-44, RKJ6, and RKJ12, have been mapped to the corresponding chromosomes by FISH and/or RH analysis (Raudsepp et al., 2008).

FISH for mapping chromosome and genome rearrangements

Molecular hybridization-based methods, initially developed for physical gene mapping, have become an integral part of equine clinical cytogenetics (Durkin et al., 2010; Lear and Bailey, 2008; Rubes et al., 2009; see also Chapters 16 and 18). Molecular hybridization with chromosome-specific paints is not only an efficient method for detecting aneuploidies in chromosome preparations (Breen et al., 1997a; Bugno and Slota, 2007; Bugno et al., 2007a; Bugno et al., 2007b), but also in sperm (Bugno-Poniewierska et al., 2009) and embryos (Rambags et al., 2005). Horse BAC clones containing known genes or markers are ideal tools for detailed analysis of complex structural rearrangements (Durkin et al., 2008, 2010; Lear and Bailey, 2008; Lear et al., 2008) and for identification of chromosomes involved in autosomal trisomies (Brito et al., 2008).

Recently, as a joint effort of researchers at Texas A&M University (United States) and the University of Adelaide (Australia), and building on the available genome sequence assembly (Wade et al., 2009), a whole genome oligonucleotide tiling array was constructed for the horse (Qu et al., 2011). This next-generation research tool uses comparative genomic hybridization (CGH) as the analysis method, and is therefore also known as a CGH array. The equine tiling array comprises of 418,577 probes that are tiled into repeat-masked EquCab2 with an average resolution of 7.5 Kb. The array contains a subset of 82,354 probes representing exons of almost all 18,763 annotated equine genes. The Y chromosome is represented by 519 probes, and 26,790 probes originate from 3 Mb long sub-telomeric regions of all autosomes. This CGH array is the first comprehensive equine high-resolution resource for mapping copy number variants (CNVs) and other large-scale submicroscopic structural rearrangements (Pang et al., 2010; Shaffer and Bejjani, 2006; Sharp, 2009). It is believed that array CGH will help uncover the extent of naturally occurring structural variations in horse populations, and will also lead to the discovery of regions associated with complex traits and genetic disorders. Regardless of the specific goals, array CGH analysis of the horse genome will provide new applications to FISH although any significant copy number gain or loss needs validation by FISH on metaphase or interphase chromosomes (Liu et al., 2010).

With the whole genome sequence available (Wade et al., 2009), the practical need for FISH mapping in the horse is rapidly decreasing. For example, in 2006–2007, around a dozen publications reported FISH assignment of about 30 new loci; by 2008–2009, the number of publications had decreased to less than 10, and in 2010, there were no reports on cytogenetic mapping in the horse. Instead, equine BACs are finding use as probes for FISH mapping in the donkey (Bugno-Poniewierska et al., 2010; Bugno et al., 2009b) and other equids (Piras et al., 2009). Thus, the golden years of FISH mapping of the equine genome are irreversibly over. However, this elegant, precise, and visually attractive mapping approach will continue to carry out important tasks in equine genome analysis. FISH will be needed to complete the horse genome assembly and physically anchor the 5% of genome sequences that are currently unassigned (Wade et al., 2009). Also, FISH and equine BAC clones remain valuable tools for the detection of various genomic rearrangements, both at chromosomal and submicroscopic levels.

Somatic Cell Hybrid (SCH) Panels and Synteny Mapping

Somatic cell genetics (Faraut et al., 2009; Kao, 1983; Kucherlapati and Ruddle, 1975; Ruddle, 1981) has been the key to the development of hybrid cell panels commonly used for generating synteny maps. Over the years, very little has changed in terms of the approaches used to develop these panels in different species.

The first syntenic group identified in the horse comprised of three X-linked genes, namely *G6PD, HGPRT* and *PGK*, that were indirectly mapped by analyzing a mule x mouse hybrid cell panel (Deys, 1972). After that, four more SCH panels were constructed (Bailey et al., 1995; Lear et al., 1992; Raney et al., 1998; Shiue et al., 1999; Williams et al., 1993). Of these, only one panel was characterized using FISH with equine genomic DNA (Lear et al., 1992). By 1995, 9 syntenic groups with 25 markers were established in the horse using enzymatic analysis of lysates (Williams et al., 1993) or PCR (Bailey et al., 1995). None of the groups could be assigned to specific chromosomes. Later, the panel created at UC Davis (Shiue et al., 1999) became the main contributor to the equine synteny map. With around 450 markers analyzed on this panel, the resource was central in assigning synteny groups (comprising genes and microsatellites) to individual equine autosomes (Caetano et al., 1999a; Caetano et al., 1999b; Lindgren et al., 2001; Shiue et al., 1999) and the sex chromosomes (Shiue et al., 2000). The resource contributed also to the development of the horse-human comparative map (Caetano et al., 1999c; Chowdhary et al., 2003). The use of SCH panels and synteny mapping in the horse came to an end in mid-2000s. Part of this was attributed to depleting DNA reserves from the original panel, though the main reason was the availability of other more advanced mapping approaches.

Radiation Hybrid (RH) Panels and RH Mapping

RH mapping methodology

Radiation hybrid (RH) mapping is essentially a SCH technique with the difference that before fusion of cell lines, the cells of the species of interest (donor) are exposed to high doses of X-ray irradiation that cause fragmentation of chromosomes (Chowdhary and Raudsepp, 2005; Faraut et al., 2009). The dosage may range from as low as 3,000 rad (Gyapay et al., 1996) to as high as 50,000 rad (Lunetta et al., 1996). The higher is the radiation dose, the higher is the resolution and mapping power of the RH panel because more breaks and smaller chromosomal fragments are produced. Higher radiation doses also mean that more markers need to be genotyped in order to construct RH maps. It has been calculated that in a typical mammalian genome, an irradiation dose of 3,000 rad should allow mapping of about 3,000 markers, and about 12,000 markers can be mapped to unique positions on a 12,000 rad RH panel (Chowdhary and Raudsepp, 2005; Faraut et al., 2009). Traditionally, RH genotyping is carried out by PCR, and any type of markers, regardless of their polymorphism status, can be assigned to RH maps. The data are analyzed with dedicated software programs, of which CONCORDE (Agarwala et al., 2000) and CARTHAGENE (de Givry et al., 2005) have been recently used most.

Horse RH panels

Two horse x hamster radiation hybrid panels – a 3,000 rad (Kiguwa et al., 2000) and a 5,000 rad (Chowdhary et al., 2002a) – have been made in the horse. The 3,000 rad panel with 94 hybrid clones provided preliminary maps for horse chromosomes 1 and 10, and since then has not been used for additional analysis. Contrary to this, the 5,000 rad panel with 93 hybrid clones has been the sole resource for constructing chromosome specific and whole genome RH maps and for rapid assignment of loci of interest to chromosomes or ordering them in relation to mapped markers.

RH mapping in the horse

The 5,000 rad panel was first used to obtain low-resolution RH and comparative maps for ECA11 (Chowdhary et al., 2002b) and ECAX (Raudsepp et al., 2002). Once the two studies had demonstrated the high utility of the panel for mapping the horse genome, systematic genotyping of all available, as well as newly generated, Type I and Type II markers was initiated worldwide. These joint efforts led to the construction of the first-generation medium-density RH and comparative map of the horse genome (Chowdhary et al., 2003). The map incorporated 730 markers (258 Type I and 472 Type II) and, for the first time, integrated synteny, cytogenetic, and meiotic maps into a comprehensive consensus map for all autosomes and the X chromosome.

Since then, the 5,000 rad RH panel has been one of the primary tools for mapping the horse genome. Targeted generation of gene-specific and polymorphic markers led to the construction of high-resolution (on average 1 marker/Mb) RH and comparative maps for individual chromosomes, namely ECA17 (Lee et al., 2004), ECAX (Raudsepp et al., 2004a), ECAY (Raudsepp et al., 2004b), ECA22 (Gustafson-Seabury et al., 2005), ECA15 and ECA18 (Wagner et al., 2006), parts of ECA7, ECA10, and ECA21 (Brinkmeyer-Langford et al., 2005), and ECA14 and ECA21 (Goh et al., 2007) or chromosomal regions, specifically ECA4q (Dierks et al., 2006). Concurrently, a medium-density map of the horse genome was reported by adding 165 gene-specific markers to the existing maps (Perrocheau et al., 2006). Additionally, several research groups worldwide have used the RH panel to independently map a number of gene-specific and microsatellite markers (e.g., Berg et al., 2008; Boneker et al., 2005a; Boneker et al., 2005b; Brenig et al., 2004; Bricker et al., 2005; Klukowska-Rotzler et al., 2006a; Klukowska-Rotzler et al., 2006b; Leeb et al., 2005; Mickelson et al., 2004; Momozawa et al., 2005, 2006, 2007a; Muller et al., 2005a,b,c; Prause et al., 2007; Takahashi et al., 2004; Wagner et al., 2004a; Wagner et al., 2004b; Wagner et al., 2004c; Wittwer et al., 2005). In 2008, the concerted international mapping efforts resulted in the construction of a second-generation whole-genome RH and comparative map for the horse genome (Raudsepp

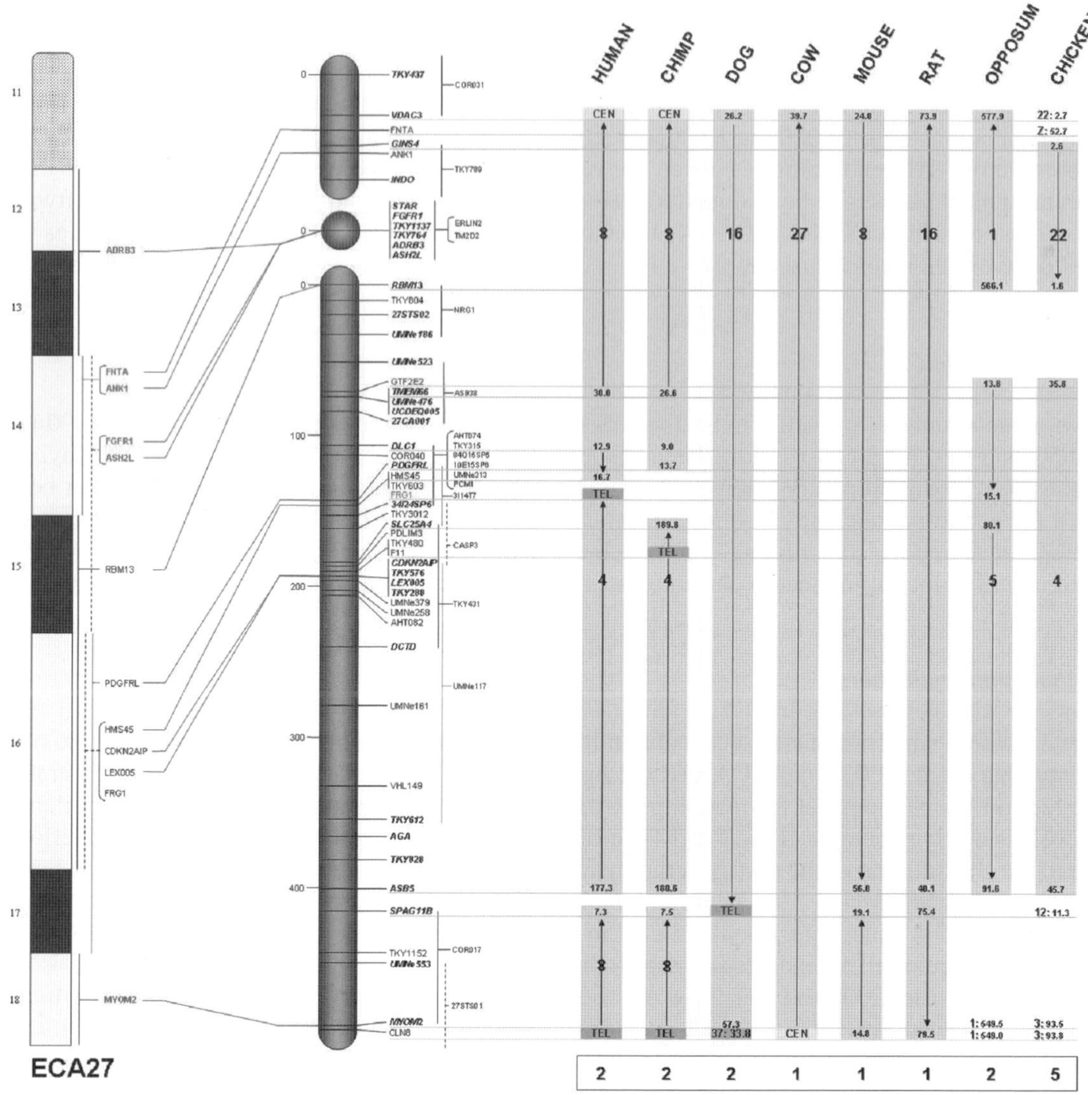

Figure 3.3 RH and comparative map for ECA27 (Raudsepp et al., 2008).

et al., 2008). This high-resolution map consists of 4,101 markers distributed over all autosomes and the X chromosome (http://content.karger.com/ProdukteDB/miscArchiv/000/151/313/Fig1.pdf). The overall resolution of the map is one marker every 720 kilobases, which represents a 6-fold improvement over the first-generation map. The map integrates available genetic linkage (907 RH mapped markers are shared with the AHT and IHRFP linkage maps; see Chapter 2) and RH mapping data into a physically ordered map for all equine chromosomes, except ECAY (Figure 3.3). The assignment and orientation of various linkage groups is strongly supported by 1,144 FISH-mapped markers. The integration of the RH and the two most recent meiotic maps (Figure 3.4) has been

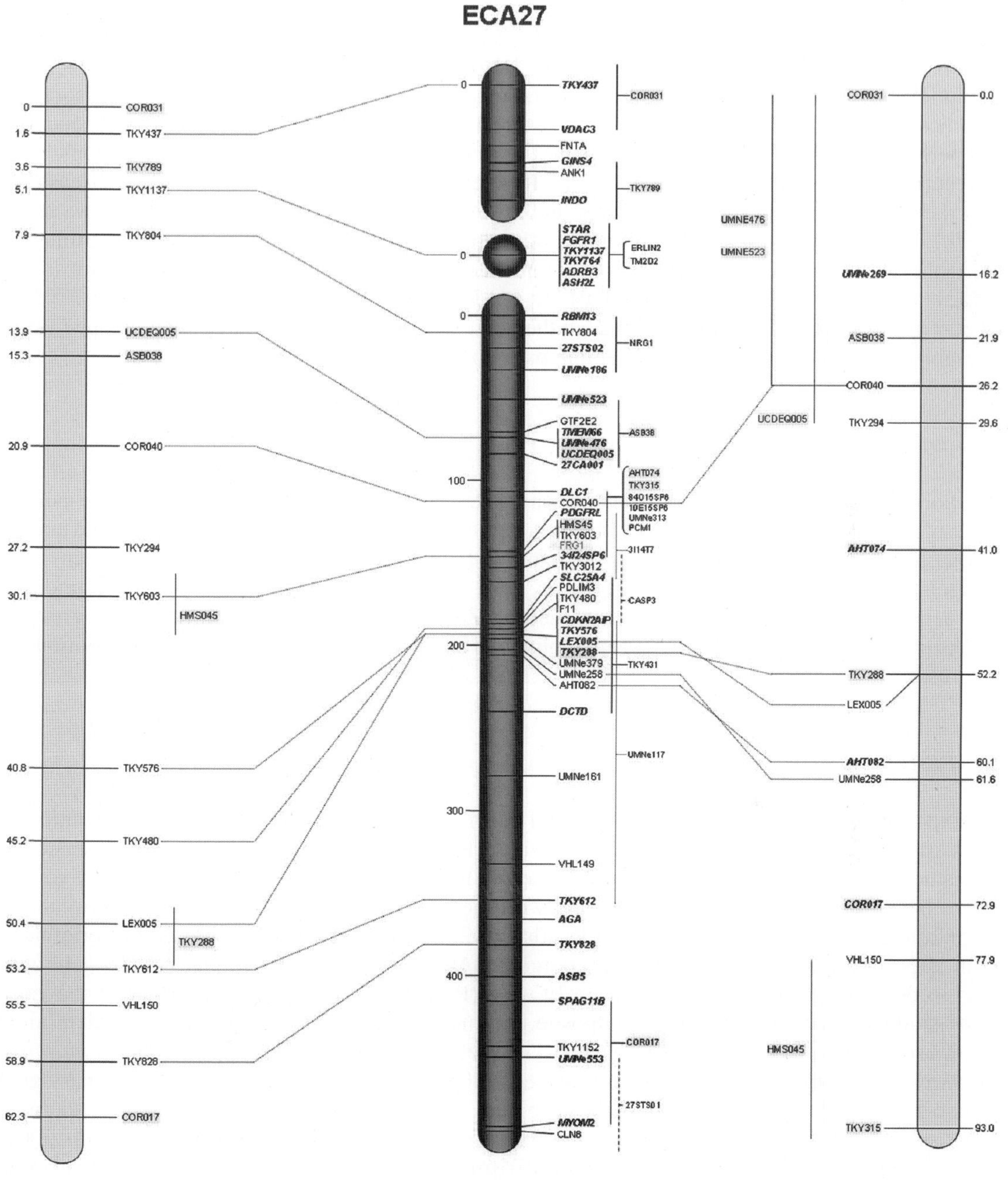

Figure 3.4 Alignment of ECA27 RH and linkage maps (Raudsepp et al., 2008).

useful in getting a comparative overview of the three maps, particularly because the linkage maps were obtained using different reference family resources, and in some instances show disagreements (http://content.karger.com/ProdukteDB/miscArchiv/000/151/313/Fig2.pdf). This high-resolution map has been a valuable tool for the identification of genes and/or markers associated with economically important traits like disease resistance or susceptibility, growth, reproduction, and performance. The RH panel has been used to map candidate genes for equine conditions, such as recurrent airway obstruction (Klukowska-Rotzler et al., 2006a), glycogen storage disease IV (Ward et al., 2003), and even temperament and anxiety (Momozawa et al., 2006; Momozawa et al., 2007b). Besides, the integrated radiation hybrid map (Raudsepp et al., 2008) has been instrumental for anchoring and orienting whole genome sequence contigs and scaffolds on horse chromosomes (Lear et al., 2007; Wade et al., 2009). Last but not least, it is noteworthy that more than 12.5% of all currently known protein coding genes in the horse genome (Table 3.1) have been mapped using a combination of RH and other physical mapping approaches.

Comparative Map

The horse genome has been compared with a variety of mammalian/vertebrate genomes using a range of approaches. The preliminary indications of segmental homologies between horse and human chromosomes were based on chromosome banding (Ronne, 1992) and early linkage, synteny, and cytogenetic maps (Chowdhary and Gustavsson, 1992). The first whole-genome segmental homology between the horse and human was determined by Zoo-FISH (Raudsepp et al., 1996). Thereafter, thanks to high-resolution RH and FISH mapping, the horse comparative map has been in continuous progress.

Zoo-FISH maps

An important breakthrough in the construction of physical chromosome maps was the invention of cross-species FISH, also referred to as Zoo-FISH (Scherthan et al., 1994). The procedure is basically the same as for regular FISH, except for some modifications of which the most critical were the use of heterologous probes with more than two times higher DNA concentration, and prolonging the hybridization time (Chowdhary and Raudsepp, 2001). In a few studies, the probes originated from large insert libraries. For example, porcine BAC clone containing human *MC1R* gene was used to determine homologous region on ECA3p (Raudsepp et al., 1999). Majority of the Zoo-FISH studies, however, have used composite chromosome-specific painting probes to determine chromosomal homologies between species. The method is, therefore, also called as *comparative chromosome painting* (Chowdhary and Raudsepp, 2001; Scherthan et al., 1994). The primary goal of Zoo-FISH was, and still is, to transfer gene map information from "map rich" species to "map poor" species and to study karyotype and chromosome evolution (Chowdhary and Raudsepp, 2001; Chowdhary et al., 1998a; Ferguson-Smith and Trifonov, 2007). With respect to the equine genome, Zoo-FISH experiments have been conducted between closely related species, namely the horse and other equids or Perissodactyls, and between distantly related species, namely the horse/equids and human (Table 3.2).

Human-horse Zoo-FISH was the first approach that delineated gross chromosomal homology between the horse and human genomes and identified 43 homologous chromosomal segments between the two species (Chaudhary et al., 1998; Chowdhary et al., 1998a; Raudsepp et al., 1996).

Table 3.2 Zoo-FISH between horse/equids and other mammalian species. Unidirectional painting is indicated by → and bidirectional by ↔

Species	References
Between distantly related mammals	
Human → horse	(Raudsepp et al., 1996)
Human → horse (selected chromosomes)	(Lear and Bailey, 1997; Rettenberger et al., 1996)
Human ↔ horse	(Yang et al., 2004a)
Human → donkey/mule	(Yang et al., 2004a)
Human → Hartmann's mountain zebra	(Richard et al., 2001)
Human↔ Burchell's zebra	(Yang et al., 2003)
Human → black rhino	(Trifonov et al., 2008)
Human →white rhino	(Trifonov et al., 2008)
Human → Malayan tapir	(Trifonov et al., 2008)
Human → Grevy's zebra	(Trifonov et al., 2008)
Between Equids	
Horse ↔ Burchell's zebra	(Yang et al., 2003)
Horse → Hartmann's mountain zebra	(Yang et al., 2003)
Horse → donkey	(Raudsepp and Chowdhary, 1999)
Horse → Przewalskii's Horse	(Yang et al., 2003)
Horse → onager	(Trifonov et al., 2008)
Grevy's zebra → onager	(Trifonov et al., 2008)
Burchell's zebra → Hartmann's mountain zebra	(Yang et al., 2003)
Between Perissodactyls	
Horse → white and black rhino	(Trifonov et al., 2008)
Horse → Baird's tapir	(Trifonov et al., 2008)
Horse → mountain tapir	(Trifonov et al., 2008)
Horse → lowland tapir	(Trifonov et al., 2008)
Horse →Malayan tapir	(Trifonov et al., 2008)
Malayan tapir → black rhino	(Trifonov et al., 2008)
Malayan tapir → Burchell's zebra	(Trifonov et al., 2008)
Malayan tapir → onager	(Trifonov et al., 2008)
Malayan tapir → Grevy's zebra	(Trifonov et al., 2008)
White rhino → Malayan tapir	(Trifonov et al., 2003; Trifonov et al., 2008)
White rhino ↔ Burchell's zebra	(Trifonov et al., 2008)
White rhino → Grevy's zebra	(Trifonov et al., 2008)

The findings provided a firsthand insight into comparative organization of horse and human genomes, but also indirectly helped relate the horse genome to other mammalian genomes. The results were shortly confirmed and refined using chromosome arm-specific paints from HSA2, 5, 6, 16, and 19 (Chaudhary et al., 1998). These landmark comparisons unambiguously laid the foundation for mapping and analyzing the equine genome using comparative approaches. The whole genome Zoo-FISH map was later verified and partially redefined by horse-human reciprocal painting (Yang et al., 2004a). The availability of next-generation human chromosome paints improved the sensitivity and resolution of Zoo-FISH and identified 17 new homology segments between the two genomes, thus taking the tally to 60 (Fig. 3.1).

In addition to the horse, Zoo-FISH homologies have been determined between human and donkey (Yang et al., 2004a) and human and Hartmann's mountain zebra (Richard et al., 2001). Also, multidirectional Zoo-FISH has been carried out between various equid and Perissodactyl species (Table 3.2). These studies have helped understand chromosome evolution within Perissodactyls and relate it to the karyotypes/genomes of other mammals.

High-resolution comparative maps

After the Zoo-FISH homology between the horse and human genomes was determined (Raudsepp et al., 1996), more detailed comparative maps started to emerge using synteny (Caetano et al., 1999a; Caetano et al., 1999b; Caetano et al., 1999c), FISH (Milenkovic et al., 2002; Perrocheau et al., 2005), and RH (Chowdhary et al., 2002b; Raudsepp et al., 2002) mapping approaches. In the first-generation whole genome RH map (Chowdhary et al., 2003), a total of 447 linearly mapped genes extended the comparison to three species: horse, human, and mouse. The first high-resolution comparative map was constructed for ECAX comparing the location of 139 genes between the horse and nine other mammalian species (Raudsepp et al., 2004a). The results revealed exceptional synteny and linkage conservation between the horse and human X chromosomes. Likewise, comparison of the organization of ECA22 and its counterparts in seven mammalian and three vertebrate species suggested that ECA22 represents an ancestral eutherian chromosome (Gustafson-Seabury et al., 2005). In contrast, multiple intra-chromosomal rearrangements were revealed while comparing the gene order between ECA17 and HSA13 (Lee et al., 2004). High-resolution RH and comparative maps were constructed also for equine counterparts of HSA2 (Wagner et al., 2006), HSA19 (Brinkmeyer-Langford et al., 2005), and HSA5 (Goh et al., 2007). The latter two studies included detailed analysis of the evolution of the homologues of HSA19 and HSA5 in 8–11 eutherian orders involving about 50 species. Until the release of the whole genome draft sequence (Wade et al., 2009), the most comprehensive comparative map for the horse was the 4,103-marker integrated physical and comparative map (Raudsepp et al., 2008). Here, the location and linear order of 1,941 equine loci were compared with 8 sequenced vertebrate genomes representing eutherian mammals (human, chimpanzee, dog, cattle, mouse, and rat), marsupials (opossum), and birds (chicken).

In addition, flanking sequences of equine microsatellite markers (Farber and Medrano, 2004; Tozaki et al., 2007) and BAC end sequences (Leeb et al., 2006) have been analyzed for possible use as comparative anchor points with genomes of other species. Also, a Comparative Mapping by Annotation and Sequence Similarity (COMPASS) strategy has been used to computationally predict the comparative location of large sets of equine expressed sequence tags (ESTs) across other mammalian species (Coleman et al., 2007).

Concluding remarks

With the draft assembly of the horse genome sequence available (Wade et al., 2009), the need for physical gene maps has been dramatically reduced, but it will not disappear completely. Lessons from humans and other sequenced species show that accurate and high-resolution integrated maps remain indispensable to validate the existing sequence assembly and to improve it (Lewin et al., 2009). Good maps are also needed for the emerging re-sequencing projects to assemble millions of short reads produced by next-generation sequencing technologies (Blow, 2008; Mardis, 2008a, 2008b). However, the increasing amount of sequence data and emergence of new genome analysis tools, such as SNP arrays, may introduce dramatic changes in the mapping strategies (Faraut et al., 2009). For example, instead of genotyping markers on the RH panel, the DNA of RH clones can be genotyped on the SNP beadchips, at once providing data for tens of thousands of markers in each RH clone. As a consequence, the construction of RH maps will need improved statistical analysis methods. Overall, the whole genome integrated maps remain to offer a well-tested resource for the analysis of genome architecture, function, and evolution, supporting the idea that "every genome sequence needs a good map" (Lewin et al., 2009).

References

Agarwala, R., Applegate, D. L., Maglott, D., Schuler, G. D., & Schäffer, A. A. (2000). A fast and scalable radiation hybrid map construction and integration strategy. *Genome Research*, *10*, 350–364.

Andersson, L., & Sandberg, K. (1984). Genetic linkage in the horse. II. Distribution of male recombination estimates and the influence of age, breed and sex on recombination frequency. *Genetics*, *106*, 109–122.

Ansari, H. A., Hediger, R., Fries, R., & Stranzinger, G. (1988). Chromosomal localization of the major histocompatibility complex of the horse (ELA) by in situ hybridization. *Immunogenetics*, *28*, 362–364.

Arrighi, F. E., & Hsu, T. C. (1971). Localization of heterochromatin in human chromosomes. *Cytogenetics*, *10*, 81–86.

Bailey, E., Graves, K. T., Cothran, E. G., Reid, R., Lear, T. L. et al. (1995). Synteny-mapping horse microsatellite markers using a heterohybridoma panel. *Animal Genetics*, *26*, 177–180.

Benirschke, K., Malouf, N., Low, R. J., & Heck, H. (1965). Chromosome complement: Differences between Equus Caballus and Equus Przewalskii, Poliakoff. *Science*, *148*, 382–383.

Berg, L. C., Mata, X., & Thomsen, P. D. (2008). Molecular characterization and chromosomal assignment of equine cartilage derived retinoic acid sensitive protein (CD-RAP)/melanoma inhibitory activity (MIA). *Gene*, *407*: 98–104.

Bloom, S. E., & Goodpasture, C. (1976). An improved technique for selective silver staining of nucleolar organizer regions in human chromosomes. *Human Genetics*, *34*: 199–206.

Blow, N. (2008). DNA sequencing: Generation next-next. *Nature Methods*, *5*, 267-–274.

Boneker, C., Kuiper, H., Wohlke, A., Drogemuller, C., Chowdhary, B. P., et al. (2005a). Assignment of the COL16A1 gene to equine chromosome 2p15.1-p15.3 by FISH and confirmation by RH mapping. *Animal Genetics*, *36*, 262–263.

Boneker, C., Muller, D., Kuiper, H., Drogemuller, C., Chowdhary, B. P. et al. (2005b). Assignment of the COL8A2 gene to equine chromosome 2p15-p16 by FISH and confirmation by RH mapping. *Animal Genetics*, *36*, 444–445.

Bowling, A. T., Millon, L. V., De Valle, A., Ponce De Leon, A. F., Carpio, C. et al. (1998). Horse chromosome 1 specific library. Paper presented at: Plant and Animal Genome VI, January 18–22, San Diego, CA.

Breen, M., Langford, C. F., Carter, N. P., Fischer, P. E., Marti, E. et al. (1997a). Detection of equine X chromosome abnormalities in equids using a horse X whole chromosome paint probe (WCPP). *Veterinary Journal*, *153*, 235–238.

Breen, M., Lindgren, G., Binns, M. M., Norman, J., Irvin, Z. et al. (1997b). Genetical and physical assignments of equine microsatellites – first integration of anchored markers in horse genome mapping. *Mammalian Genome*, *8*, 267–273.

Brenig, B., Beck, J., Hall, A. J., Broad, T. E., Chowdhary, B. P. et al. (2004). Assignment of the equine solute carrier 26A2 gene (SLC26A2) to equine chromosome 14q15–>q21 (ECA14q15–>q21) by in situ hybridization and radiation hybrid panel mapping. *Cytogenetic Genome Research*, *107*, 139.

Bricker, S. J., Brault, L. S., DelValle, A., Millon, L. V., Murray, J. D. et al. (2005). Radiation hybrid and linkage mapping of six new type I markers in the horse. *Animal Genetics*, *36*, 182–184.

Brinkmeyer-Langford, C., Raudsepp, T., Gustafson-Seabury, A., & Chowdhary, B. P. (2008). A BAC contig map over the proximal approximately 3.3 Mb region of horse chromosome 21. *Cytogenetic Genome Research*, *120*, 164–172.

Brinkmeyer-Langford, C., Raudsepp, T., Lee, E. J., Goh, G., Schaffer, A. A. et al. (2005). A high-resolution physical map of equine homologs of HSA19 shows divergent evolution compared with other mammals. *Mammalian Genome*, *16*, 631–649.

Brito, L. F. C., P. L., S., Durkin, K., Chowdhary, B. P., Turner, R. M. et al. (2008). Autosomic 27 trisomy in a standardbred colt. *Journal of Equine Veterinary Science*, 431–436.

Buckland, R. A., Fletcher, J. M., & Chandley, C. (1976). Characterization of the domestic horse (Equus caballus) karyotype using G- and C-banding techniques. *Experientia*, *32*, 1146–1149.

Bugno, M., & Slota, E. (2007). Application of arm-specific painting probes of horse X chromosome for karyotype analysis in an infertile Hutsul mare with 64,XX/65,XX + Xp karyotype: Case report. *Acta Veterinaria Hungary*, *55*, 309–314.

Bugno, M., Slota, E., & Koscielny, M. (2007a). Karyotype evaluation among young horse populations in Poland. *Schweiz Arch Tierheilkd*, *149*, 227–232.

Bugno, M., Slota, E., Pienkowska-Schelling, A., & Schelling, C. (2007b). Detection of equine X chromosome mosaicism in a mare using an equine X whole chromosome painting probe (WCPP) – a case report. *Acta Veterinaria Hungary*, *55*, 207–212.

Bugno, M., Slota, E., Pienkowska-Schelling, A., & Schelling, C. (2009a). Identification of chromosome abnormalities in the horse using a panel of chromosome-specific painting probes generated by microdissection. *Acta Veterinaria Hungary*, *57*, 369–381.

Bugno, M., Slota, E., Witarski, W., Gerber, V., & Klukowska-Roetzler, J. (2009b). Interleukin 4 receptor alpha (IL4R) and calcium-activated chloride channel 1 (CLCA1) genes map to donkey chromosome. *Hereditas*, *146*, 118–121.

Bugno-Poniewierska, M., Jablonska, Z., & Slota, E. (2009). Modification of equine sperm chromatin decondensation method to use fluorescence in situ hybridization (FISH). *Folia Histochemical Cytobiology*, *47*, 663–666.

Bugno-Poniewierska, M., Pawlina, K., Dardzinska, A., Zabek, T., Slota, E. et al. (2010). FISH mapping of six genes responsible for development of the nervous and skeletal systems on donkey (Equus asinus) chromosomes. *Hereditas*, *147*, 132–135.

Caetano, A. R., Lyons, L. A., Laughlin, T. F., O'Brien, S. J., Murray, J. D. et al. (1999a). Equine synteny mapping of comparative anchor tagged sequences (CATS) from human Chromosome 5. *Mammalian Genome*, *10*, 1082–1084.

Caetano, A. R., Pomp, D., Murray, J. D., and Bowling, A. T. (1999b). Comparative mapping of 18 equine type I genes assigned by somatic cell hybrid analysis. *Mammalian Genome*, *10*, 271–276.

Caetano, A. R., Shiue, Y. L., Lyons, L. A., O'Brien, S. J., Laughlin, T. F. et al. (1999c). A comparative gene map of the horse (Equus caballus). *Genome Research*, *9*, 1239–1249.

Carbone, L., Nergadze, S. G., Magnani, E., Misceo, D., Francesca Cardone, M. et al. (2006). Evolutionary movement of centromeres in horse, donkey, and zebra. *Genomics*, *87*, 777–782.

Chaudhary, R., Raudsepp, T., Guan, X. Y., Zhang, H., & Chowdhary, B. P. (1998). Zoo-FISH with microdissected arm specific paints for HSA2, 5, 6, 16, and 19 refines known homology with pig and horse chromosomes. *Mammalian Genome*, *9*, 44–49.

Chowdhary, B. P., & Gustavsson, I. (1992). Some aspects of equine genome analysis. Paper presented at Proceedings of an Annual Meeting of the German Society of Genetics, October 1–3, Hannover, Germany.

Chowdhary, B. P., & Raudsepp, T. (2000). Cytogenetics and physical gene maps. In A. T. Bowling & A. Ruvinsky (eds.), *The genetics of the horse* (pp. 171–242). Wallingford, Oxon: CABI.

Chowdhary, B. P., & Raudsepp, T. (2001). Chromosome painting in farm, pet and wild animal species. *Methods in Cellular Science*, *23*, 37–55.

Chowdhary, B. P., & Raudsepp, T. (2005). Mapping genomes at the chromosome level. In A. Ruvinsky & J. A. Marshall Graves (eds.), *Mammalian genomics* (pp. 23–66). Wallingford: CABI.

Chowdhary, B. P., Raudsepp, T., Fronicke, L., & Scherthan, H. (1998a). Emerging patterns of comparative genome organization in some mammalian species as revealed by Zoo-FISH. *Genome Research*, *8*, 577–589.

Chowdhary, B. P., Raudsepp, T., Honeycutt, D., Owens, E. K., Piumi, F. et al. (2002). Construction of a 5000(rad) whole-genome radiation hybrid panel in the horse and generation of a comprehensive and comparative map for ECA11. *Mammalian Genome*, *13*, 89–94.

Chowdhary, B. P., Raudsepp, T., Jeon, J.-T., and Kijas, J. (1998b). Microdissection and microcloning of chromosomes 6 and 12 in horse. Paper presented at the Proceedings of the XXVI International Conference on Animal Genetics, August 9–14, Auckland, New Zealand.

Chowdhary, B. P., Raudsepp, T., Kata, S. R., Goh, G., Millon, L. V. et al. (2003). The first-generation whole-genome radiation hybrid map in the horse identifies conserved segments in human and mouse genomes. *Genome Research*, *13*, 742–751.

Coleman, S. J., Gong, G., Gaile, D. P., Chowdhary, B. P., Bailey, E. et al. (2007). Evaluation of Compass as a comparative mapping tool for ESTs using horse radiation hybrid maps. *Animal Genetics*, *38*, 294–302.

Comings, D. E., & Mattoccia, E. (1972). DNA of mammalian and avian heterochromatin. *Experimental Cell Research*, *71*, 113–131.

de Givry, S., Bouchez, M., Chabrier, P., Milan, D., & Schiex, T. (2005). CARHTA GENE: Multipopulation integrated genetic and radiation hybrid mapping. *Bioinformatics*, *21*, 1703–1704.

de la Sena, C., Chowdhary, B. P., & Gustavsson, I. (1995). Localization of the telomeric (TTAGGG)n sequences in chromosomes of some domestic animals by fluorescence in situ hybridization. *Hereditas*, *123*, 269–274.

Deryusheva, S. E., Loginova, Y. A., Chiryaeva, O. G., Yaschak, K., & Smirnov, A. F. (1997). Distribution of ribosomal RNA genes on chromosomes of domestic horse (*Equus caballus*) revealed by fluorescence *in situ* hybridization. *Genetika*, *33*, 1281–1312.

Deys, B. F. (1972). Demonstration of X-linkage of G6PD, HGPRT, and PGK in the horse by means of mule-horse cell hybridization. PhD Thesis, University of Leiden.

Dierks, C., Momke, S., Drogemuller, C., Leeb, T., Chowdhary, B. P. et al. (2006). A high-resolution comparative radiation hybrid map of equine chromosome 4q12-q22. *Animal Genetics*, *37*, 513–.

Dobigny, G., & Yang, F. (2008). Foreword. Comparative cytogenetics in the genomics era: Cytogenomics comes of age. *Chromosome Research*, *16*, 1–4.

Durkin, K., Raudsepp, T., & Chowdhary, B. P. (2008). Precise demarcation of the breakpoint in a thoroughbred stallion carrying an ECA5 and ECA16 translocation. Paper presented at the Plant & Animal Genome XVI, January 12–17, San Diego, CA.

Durkin, K., Raudsepp, T., & Chowdhary, B. P. (2010). Cytogenetic evaluation of the stallion. In A. O. McKinnon, W. E. Vaala & D. D. Varner (eds.), *Equine Reproduction, E.L.S.* (pp. 306–320). London: Wiley-Blackwell.

Dutrillaux, B., Laurent, C., Couturier, J., & Lejeune, J. (1973). Staining of human chromosomes with acridine orange after treatment with 5 bromodeoxyuridine. *Comptes Rendus Hebdomadaires des Séances de l'Académie des Sciences Série D: Sciences Naturelles*, 276, 3179–3181.

Faraut, T., de Givry, S., Hitte, C., Lahbib-Mansais, Y., Morrison, M. et al. (2009). Contribution of radiation hybrids to genome mapping in domestic animals. *Cytogenetic and Genome Research*, *126*, 21–33.

Farber, C. R., & Medrano, J. F. (2004). Identification of putative homology between horse microsatellite flanking sequences and cross-species ESTs, mRNAs and genomic sequences. *Animal Genetics*, *35*, 28–33.

Ferguson-Smith, M. A., & Trifonov, V. (2007). Mammalian karyotype evolution. *National Review of Genetics*, *8*, 950–962.

Ford, C. E., Pollock, D. L., & Gustavsson, I. (1980). Proceedings of the First International Conference for the Standardisation of Banded Karyotypes of Domestic Animals. University of Reading Reading, England. 2nd–6th August 1976. *Hereditas, 92*, 145–162.

Godard, S., Schibler, L., Oustry, A., Cribiu, E. P., & Guerin, G. (1998). Construction of a horse BAC library and cytogenetical assignment of 20 type I and type II markers. *Mammalian Genome, 9*, 633–637.

Godard, S., Vaiman, D., Oustry, A., Nocart, M., Bertaud, M. et al. (1997). Characterization, genetic and physical mapping analysis of 36 horse plasmid and cosmid-derived microsatellites. *Mammalian Genome, 8*, 745–750.

Godard, S., Vaiman, A., Schibler, L., Mariat, D., Vaiman, D. et al. (2000). Cytogenetic localization of 44 new coding sequences in the horse. *Mammalian Genome, 11*, 1093–1097.

Goh, G., Raudsepp, T., Durkin, K., Wagner, M. L., Schaffer, A. A. et al. (2007). High-resolution gene maps of horse chromosomes 14 and 21: Additional insights into evolution and rearrangements of HSA5 homologs in mammals. *Genomics, 89*, 89–112.

Goodpasture, C., & Bloom, S. E. (1975). Visualization of nucleolar organizer regions im mammalian chromosomes using silver staining. *Chromosoma, 53*, 37–50.

Guan, X. Y., Meltzer, P. S., Cao, J., & Trent, J. M. (1992). Rapid generation of region-specific genomic clones by chromosome microdissection: Isolation of DNA from a region frequently deleted in malignant melanoma. *Genomics, 14*, 680–684.

Guérin, G., Bailey, E., Bernoco, D., Anderson, I., Antczak, D. F. et al. (1999). Report of the International Equine Gene Mapping Workshop: Male linkage map. *Animal Genetics, 30*, 341–354.

Gustafson, A. L., Tallmadge, R. L., Ramlachan, N., Miller, D., Bird, H. et al. (2003). An ordered BAC contig map of the equine major histocompatibility complex. *Cytogenetic Genome Research, 102*, 189–195.

Gustafson-Seabury, A., Raudsepp, T., Goh, G., Kata, S. R., Wagner, M. L. et al. (2005). High-resolution RH map of horse chromosome 22 reveals a putative ancestral vertebrate chromosome. *Genomics, 85*, 188–200.

Gyapay, G., Schmitt, K., Fizames, C., Jones, H., Vega-Czarny, N. et al. (1996). A radiation hybrid map of the human genome. *Human Molecular Genetics, 5*, 339–346.

Hansen, K. M. (1984). The relative length of Q-band/Giemsa stained horse chromosomes. *The 6th European Colloquium on Cytogenetics of the Domestic Animals*, 165–171.

Harbitz, I., Chowdhary, B. P., Saether, H., Hauge, J. G., & Gustavsson, I. (1990). A porcine genomic glucosephosphate isomerase probe detects a multiallelic restriction fragment length polymorphism assigned to chromosome 10pter in horse. *Hereditas, 112*, 151–156.

Horin, P., Osickova, J., Necesankova, M., Matiasovic, J., Musilova, P. et al. (2008). Single nucleotide polymorphisms of interleukin-1 beta related genes and their associations with infection in the horse. *Developments in Biology (Basel), 132*, 347–351.

Hsu, T. C., & Benirschke, K. (1967). *Mammalian chromosomes*. Berlin: Springer-Verlag.

ISCNH (1997). International system for cytogenetic nomenclature of the domestic horse. Report of the Third International Committee for the Standardization of the domestic horse karyotype, Davis, CA. *Chromosome Research, 5*, 433–443.

Jost, U., Klukowska-Rotzler, J., Dolf, G., Swinburne, J. E., Ramseyer, A. et al. (2007). A region on equine chromosome 13 is linked to recurrent airway obstruction in horses. *Equine Veterinary Journal, 39*, 236–241.

Kao, F. T. (1983). Somatic cell genetics and gene mapping. *International Review of Cytology, 85*, 109–146.

Kiguwa, S. L., Hextall, P., Smith, A. L., Critcher, R., Swinburne, J. et al. (2000). A horse whole-genome-radiation hybrid panel: chromosome 1 and 10 preliminary maps. *Mammalian Genome, 11*, 803–805.

Kirillow, S. (1912). Die Spermatogenese beim Pferde. *Archiv fur Mikroskopische Anatomie*, 79.

Klukowska-Rotzler, J., Bugno, M., Sander, P., Slota, E., Dolf, G. et al. (2006a). Chromosomal assignment of the two candidate genes (EGFR, CLCA1) for equine recurrent airway obstruction (RAO) by FISH and RH mapping. *Hereditas, 143*, 138–141.

Klukowska-Rotzler, J., Jost, U., Schelling, C., Dolf, G., Chowdhary, B. P. et al. (2006b). Characterization and RH mapping of six gene-associated equine microsatellite markers. *Animal Genetics, 37*, 305–306.

Kopp, E., Mayr, B., Czaker, R., & Schleger, W. (1981). Nucleolus organizer regions in the chromosomes of the domestic horse. *Journal of Heredity, 72*, 357–358.

Kubickova, S., Cernohorska, H., Musilova, P., & Rubes, J. (2002). The use of laser microdissection for the preparation of chromosome-specific painting probes in farm animals. *Chromosome Research, 10*, 571–577.

Kucherlapati, R. S., & Ruddle, F. H. (1975). Mammalian somatic hybrids and human gene mapping. *Annals of Internal Medicine, 83*, 553–560.

Laan, M., Isosomppi, J., Klockars, T., Peltonen, L., & Palotie, A. (1996). Utilization of FISH in positional cloning: an example on 13q22. *Genome Research, 6*, 1002–1012.

Lear, T. L., Adams, M. H., Sullivan, N. D., McDowell, K. J., & Bailey, E. (1998a). Assignment of the horse progesterone receptor (PGR) and estrogen receptor (ESR1) genes to horse chromosomes 7 and 31, respectively, by in situ hybridization. *Cytogenetic Cell Genetics, 82*, 110–111.

Lear, T. L., & Bailey, E. (1997). Localization of the U2 linkage group of horses to ECA 3 using chromosome painting. *Journal of Heredity, 88*, 162–164.

Lear, T.L., & Bailey, E. (2008). Equine clinical cytogenetics: The past and future. *Cytogenetic Genome Research, 120*, 42–49.

Lear, T. L., Brandon, R., Piumi, F., Terry, R. R., Guerin, G. et al. (2001). Mapping of 31 horse genes in BACs by FISH. *Chromosome Research, 9*, 261–262.

Lear, T. L., Coogle, L. D., and Bailey, E. (1998b). Assignment of the horse mitochondrial glutamate oxaloacetate transaminase 2 (GOT2) and v-kit Hardy-Zuckerman 4 feline sarcoma viral oncogene homolog (KIT) to horse chromosome 3 by in situ hybridization. *Cytogenetic Cell Genetics, 82*, 112–113.

Lear, T. L., Lundquist, J., Gnerre, S., Wade, C., & Lindblad-Toh, K. (2007). Anchoring supercontigs by FISH. Paper presented at Proceedings of the 7th Dorothy Russell Havemeyer International Equine Genome Mapping Workshop, August 16–19, Tahoe City, CA.

Lear, T. L., Lundquist, J., Zent, W. W., Fishback, W. D., Jr., & Clark, A. (2008). Three autosomal chromosome translocations associated with repeated early embryonic loss (REEL) in the domestic horse (Equus caballus). *Cytogenetic Genome Research, 120*, 117–122.

Lear, T. L., Piumi, F., Terry, R., Guerin, G., & Bailey, E. (2000). Horse v-fes feline sarcoma viral oncogene homologue; pyruvate kinase, muscle type 2; plasminogen; beta spectrin, non-erythrocytic 1; thymidylate synthetase; and microsatellite LEX078 map to 1q14-q15, 1q21, 31q12-q14, 15q22, 8q12-q14, and 14q27, respectively. *Chromosome Research, 8*, 361.

Lear, T. L., Trembicki, K. A., & Ennis, R. B. (1992). Identification of equine chromosomes in horse x mouse somatic cell hybrids. *Cytogenetic Cell Genetics, 61*, 58–60.

Lebo, R. V. (1982). Chromosome sorting and DNA sequence localization. *Cytometry, 3*, 145–154.

Lebo, R. V., & Bastian, A. M. (1982). Design and operation of a dual laser chromosome sorter. *Cytometry, 3*, 213–219.

Lee, E. J., Raudsepp, T., Kata, S. R., Adelson, D., Womack, J. E. et al. (2004). A 1.4-Mb interval RH map of horse chromosome 17 provides detailed comparison with human and mouse homologues. *Genomics, 83*, 203–215.

Leeb, T., Bruhn, O., Philipp, U., Kuiper, H., Regenhard, P. et al. (2005). Assignment of the equine S100A7 gene (psoriasin 1) to chromosome 5p12->p13 by fluorescence in situ hybridization and radiation hybrid mapping. *Cytogenetic Genome Research, 109*, 533.

Leeb, T., Vogl, C., Zhu, B., de Jong, P. J., Binns, M. M. et al. (2006). A human-horse comparative map based on equine BAC end sequences. *Genomics, 87*, 772–776.

Lewin, H. A., Larkin, D. M., Pontius, J., & O'Brien, S. J. (2009). Every genome sequence needs a good map. *Genome Research, 19*, 1925–1928.

Lichter, P. (1997). Multicolor FISHing: What's the catch? *Trends in Genetics, 13*, 475–479.

Lindgren, G., Breen, M., Godard, S., Bowling, A., Murray, J. et al. (2001). Mapping of 13 horse genes by fluorescence in-situ hybridization (FISH) and somatic cell hybrid analysis. *Chromosome Research, 9*, 53–59.

Lindgren, G., Sandberg, K., Persson, H., Marklund, S., Breen, M. et al. (1998). A primary male autosomal linkage map of the horse genome. *Genome Research, 8*, 951–966.

Liu, G. E., Hou, Y., Zhu, B., Cardone, M. F., Jiang, L. et al. (2010). Analysis of copy number variations among diverse cattle breeds. *Genome Research, 20*, 693–703.

Lunetta, K. L., Boehnke, M., Lange, K., & Cox, D. R. (1996). Selected locus and multiple panel models for radiation hybrid mapping. *American Journal of Human Genetics, 59*, 717–725.

Makinen, A., Chowdhary, B., Mahdy, E., Andersson, L., & Gustavsson, I. (1989). Localization of the equine major histocompatibility complex (ELA) to chromosome 20 by in situ hybridization. *Hereditas, 110*, 93–96.

Makino, S. (1942). The chromosomes of the horse (*Equus caballus*). *Cytologia, 13*, 26–38.

Mardis, E. R. (2008a). The impact of next-generation sequencing technology on genetics. *Trends in Genetics, 24*, 133–141.

Mardis, E. R. (2008b). Next-generation DNA sequencing methods. *Annual Review of Genomics & Human Genetics, 9*, 387–402.

Mariat, D., Oustry-Vaiman, A., Cribiu, E. P., Raudsepp, T., Chowdhary, B. P. et al. (2001). Isolation, characterization and FISH assignments of horse BAC clones containing type I and II markers. *Cytogenetic Cell Genetics, 92*, 144–148.

Masui, K. (1919). A spermatogenesis of domestic mammals. The spermatogenesis of the horse. *Journal of the College of Agriculture, Tokyo Imperial University*, 3.

Mathai, C. K., Ohno, S., & Beutler, E. (1966). Sex-linkage of the glucose-6-phosphate dehydrogenase gene in Equidae. *Nature, 210*, 115–116.

Melchior, I., & Hohn, H. (1976). Der Karyotyp des Pferdes (Equus caballus) dargestellt mit hilfe der G- und C-banden technik. *Giessener Beitraege zur Erbpathologie und Zuchthygiene, 6*, 179–194.

Meltzer, P. S., Guan, X. Y., Burgess, A., & Trent, J. M. (1992). Rapid generation of region specific probes by chromosome microdissection and their application. *Natural Genetics, 1*, 24–28.

Mickelson, J. R., Wagner, M. L., Goh, G., Wu, J. T., Morrison, L. Y. et al. (2004). Thirty-five new equine microsatellite loci assigned to genetic linkage and radiation hybrid maps. *Animal Genetics, 35*, 481–484.

Milenkovic, D., Oustry-Vaiman, A., Lear, T. L., Billault, A., Mariat, D. et al. (2002). Cytogenetic localization of 136 genes in the horse: comparative mapping with the human genome. *Mammalian Genome, 13*, 524–534.

Molteni, L., Cribiu, E. P., De-Giovanni-Machi, A., Beltrami, E., & Succi, G. (1982). The R-banded karyotype of the domestic horse. Paper presented at Proceedings of the 5th European Colloquium on Cytogenetics of Domestic Animals, Milan.

Momozawa, Y., Takeuchi, Y., Tozaki, T., Kikusui, T., Hasegawa, T. et al. (2005). Sequence, detection of polymorphisms and radiation hybrid mapping of the equine catechol-o-methyltransferase gene. *Animal Genetics*, *36*, 190.

Momozawa, Y., Takeuchi, Y., Tozaki, T., Kikusui, T., Hasegawa, T. et al. (2006). Polymorphism identification, RH mapping, and association analysis with the anxiety trait of the equine serotonin transporter (SLC6A4) gene. *Journal of Veterinary Medical Science*, *68*, 619–621.

Momozawa, Y., Takeuchi, Y., Tozaki, T., Kikusui, T., Hasegawa, T. et al. (2007a). SNP detection and radiation hybrid mapping in horses of nine candidate genes for temperament. *Animal Genetics*, *38*, 81–83.

Momozawa, Y., Terada, M., Sato, F., Kikusui, T., Takeuchi, Y. et al. (2007b). Assessing equine anxiety-related parameters using an isolation test in combination with a questionnaire survey. *Journal of Veterinary Medical Science*, *69*, 945–950.

Muller, D., Kuiper, H., Boneker, C., Momke, S., Drogemuller, C. et al. (2005a). Assignment of BGLAP, BMP2, CHST4, SLC1A3, SLC4A1, SLC9A5 and SLC20A1 to equine chromosomes by FISH and confirmation by RH mapping. *Animal Genetics*, *36*, 457–461.

Muller, D., Kuiper, H., Boneker, C., Momke, S., Drogemuller, C. et al. (2005b). Physical mapping of the PTHR1 gene to equine chromosome 16q21.2. *Animal Genetics*, *36*, 282–284.

Muller, D., Kuiper, H., Momke, S., Boneker, C., Drogemuller, C. et al. (2005c). Assignment of the COMP gene to equine chromosome 21q12-q14 by FISH and confirmation by RH mapping. *Animal Genetics*, *36*, 277–279.

Murphy, B. A., Lear, T. L., Adelson, D. L., & Fitzgerald, B. P. (2007). Chromosomal assignments and sequences for the equine core circadian clock genes. *Animal Genetics*, *38*, 84–85.

Musilova, P., Kubickova, S., Zrnova, E., Horin, P., Vahala, J. et al. (2007). Karyotypic relationships among Equus grevyi, Equus burchelli and domestic horse defined using horse chromosome arm-specific probes. *Chromosome Research*, *15*, 807–813.

Nederlof, P. M., Robinson, D., Abuknesha, R., Wiegant, J., Hopman, A. H. et al. (1989). Three-color fluorescence in situ hybridization for the simultaneous detection of multiple nucleic acid sequences. *Cytometry*, *10*, 20–27.

Oakenfull, E. A., Buckle, V. J., & Clegg, J. B. (1993). Localization of the horse (Equus caballus) alpha-globin gene complex to chromosome 13 by fluorescence in situ hybridization. *Cytogenetic Cell Genetics*, *62*, 136–138.

Painter, T. S. (1924). Studies in mammalian spermatogenesis. V. The chromosomes of the horse. *Journal of Experimental Zoology*, *39*, 229–248.

Pang, A. W., MacDonald, J. R., Pinto, D., Wei, J., Rafiq, M. A. et al. (2010). Towards a comprehensive structural variation map of an individual human genome. *Genome Biology*, *11*, R52.

Paria, N. (2009). Discovery of candidate genes for stallion fertility. PhD Thesis, College of Veterinary Medicine Texas A&M University.

Paria, N., Raudsepp, T., Pearks Wilkerson, A. J., O'Brien, P. C., Ferguson-Smith, M. A. et al. (2011). A gene catalogue of the euchromatic male-specific region of the horse Y chromosome: comparison with human and other mammals. *PLoS One*, *6*, e21374.

Perrocheau, M., Boutreux, V., Chadi-Taourit, S., Di Meo, G. P., Perucatti, A. et al. (2005). Equine FISH mapping of 36 genes known to locate on human chromosome ends. *Cytogenetic Genome Research*, *111*, 46–50.

Perrocheau, M., Boutreux, V., Chadi, S., Mata, X., Decaunes, P. et al. (2006). Construction of a medium-density horse gene map. *Animal Genetics*, *37*, 145–155.

Piras, F. M., Nergadze, S. G., Poletto, V., Cerutti, F., Ryder, O. A. et al. (2009). Phylogeny of horse chromosome 5q in the genus Equus and centromere repositioning. *Cytogenetic Genome Research*, *126*, 165–172.

Power, M. M. (1990). Chromosomes of the horse. In R. A. McFeely (ed.), *Advances in Veterinary Science and Comparative Medicine, Domestic Animal Cytogenetics* (pp. 131–167). San Diego, CA: Academic Press.

Prause, A., Guionaud, C. T., Klukowska-Rotzler, J., Giulotto, E., Magnani, E. et al. (2007). Chromosomal assignment of five equine HTR genes by FISH and RH mapping. *Animal Genetics*, *38*, 83–84.

Qu, Z., Chowdhary, B. P., Raudsepp, T., & Adelson, D. L. (2011). Design of an equine whole genome tiling array. Paper presented at Plant & Animal Genome XIX, January 15–19, San Diego, CA.

Rambags, B. P., Krijtenburg, P. J., Drie, H. F., Lazzari, G., Galli, C. et al. (2005). Numerical chromosomal abnormalities in equine embryos produced in vivo and in vitro. *Molecular Reproductive Development*, *72*, 77–87.

Raney, N. E., Graves, K. T., Cothran, E. G., Bailey, E., & Coogle, L. (1998). Synteny mapping of the horse using a heterohybridoma panel. Paper presented at Plant and Animal Genome VI, San Diego, CA.

Raudsepp, T., & Chowdhary, B. P. (1999). Construction of chromosome-specific paints for meta- and submetacentric autosomes and the sex chromosomes in the horse and their use to detect homologous chromosomal segments in the donkey. *Chromosome Research*, *7*, 103–114.

Raudsepp, T., & Chowdhary, B. P. (2008a). FISH for mapping single copy genes. In W. Murphy (ed.), *Phylogenomics: Methods in Molecular Biology* (pp. 310–330). New Brunswick, NJ: Humana Press.

Raudsepp, T., & Chowdhary, B. P. (2008b). The horse pseudoautosomal region (PAR): Characterization and comparison with the human, chimp and mouse PARs. *Cytogenetic Genome Research*, *121*, 102–109.

Raudsepp, T., Durkin, K., Lear, T., Das, P. J., Avila, F. et al. (2010). Molecular heterogeneity of XY sex reversal in the horse. *Animal Genetics, Special Issue on Horse Genetics and Genomics*.

Raudsepp, T., Fronicke, L., Scherthan, H., Gustavsson, I., & Chowdhary, B. P. (1996). Zoo-FISH delineates conserved chromosomal segments in horse and man. *Chromosome Research*, *4*, 218–225.

Raudsepp, T., Gustafson-Seabury, A., Durkin, K., Wagner, M. L., Goh, G. et al. (2008). A 4,103 marker integrated physical and comparative map of the horse genome. *Cytogenetic Genome Reseach*, *122*, 28–36.

Raudsepp, T., Kata, S. R., Piumi, F., Swinburne, J., Womack, J. E. et al. (2002). Conservation of gene order between horse and human X chromosomes as evidenced through radiation hybrid mapping. *Genomics*, *79*, 451–457.

Raudsepp, T., Kijas, J., Godard, S., Guerin, G., Andersson, L. et al. (1999). Comparison of horse chromosome 3 with donkey and human chromosomes by cross-species painting and heterologous FISH mapping. *Mammalian Genome*, *10*, 277–282.

Raudsepp, T., Lee, E.-J., Kata, S. R., Brinkmeyer, C., Mickelson, J. R. et al. (2004a). Exceptional conservation of horse-human gene order on X chromosome revealed by high-resolution radiation hybrid mapping. *Proceedings of the National Academy of Sciences USA*, *101*, 2386–2391.

Raudsepp, T., Otte, K., Rozell, B., & Chowdhary, B. P. (1997). FISH mapping of the IGF2 gene in horse and donkey-detection of homoeology with HSA11. *Mammalian Genome*, *8*, 569–572.

Raudsepp, T., Santani, A., Wallner, B., Kata, S. R., Ren, C. et al. (2004b). A detailed physical map of the horse Y chromosome. *Proceedings of the National Academy of Sciences USA*, *101*, 9321–9326.

Rettenberger, G., Abdo, G., & Stranzinger, G. (1996). ZOO-FISH analysis in the horse, *Equus caballus*, detects regions homologous to human chromosomes 3 and 14. *Journal of Animal Breeding and Genetics*, *113*, 145–148.

Richard, F., Messaoudi, C., Lombard, M., & Dutrillaux, B. (2001). Chromosome homologies between man and mountain zebra (Equus zebra hartmannae) and description of a new ancestral synteny involving sequences homologous to human chromosomes 4 and 8. *Cytogenetic Cell Genetics*, *93*, 291–296.

Richer, C. L., Power, M. M., Klunder, L. R., McFeely, R. A., & Kent, M. G. (1990). Standard karyotype of the domestic horse (Equus caballus). Committee for standardized karyotype of Equus caballus. The Second International Conference for Standardization of Domestic Animal Karyotypes, INRA, Jouy-en Josas, France, May 22–26, 1989. *Hereditas*, 112, 289–293.

Romagnano, A., Drouin, R., & Richer, C. L. (1987). Idiograms of horse chromosomes at prometaphase, early metaphase, and midmetaphase after R-banding by BrdU incorporation followed by the fluorochrome-photolysis-Giemsa technique. *Genome*, *29*, 674–679.

Ronne, M. (1992). Putative fragile sites in the horse karyotype. *Hereditas*, *117*, 127–136.

Rothfels, K. H., Alexrad, A. A., Siminovitch, L., & McCulloch Parker, R. C. (1959). The origin of altered cell lines from mouse, monkey and man, as indicated by chromosome and transplantation studies. *Proceeding of the Canadian Cancer Research Conference*, *3*, 189–214.

Rubes, J., Pinton, A., Bonnet-Garnier, A., Fillon, V., Musilova, P. et al. (2009). Fluorescence in situ hybridization applied to domestic animal cytogenetics. *Cytogenetic and Genome Research*, *126*, 34–48.

Ruddle, F. H. (1981). A new era in mammalian gene mapping: Somatic cell genetics and recombinant DNA methodologies. *Nature*, *294*, 115–120.

Ryder, O. A., Epel, N. C., & Benirschke, K. (1978). Chromosome banding studies of the Equidae. *Cytogenetic Cell Genetics*, *20*, 332–350.

Sandberg, K. (1974). Linkage between the K blood group locus and the 6-PGD locus in horses. *Animal Blood Groups Biochemical Genetics*, *5*, 137–141.

Sandberg, K., & Andersson, L. (1984). Genetic linkage in the horse. I. Linkage relationships among 15 blood marker loci. *Hereditas*, *100*, 199–208.

Scherthan, H., Cremer, T., Arnason, U., Weier, H. U., Lima-de-Faria, A. et al. (1994). Comparative chromosome painting discloses homologous segments in distantly related mammals. *National Genetics*, *6*, 342–347.

Seabright, M. (1971). A rapid banding technique for human chromosomes. *Lancet*, *2*, 971–972.

Shaffer, L. G., & Bejjani, B. A. (2006). Medical applications of array CGH and the transformation of clinical cytogenetics. *Cytogenetic Genome Research*, *115*, 303–309.

Sharp, A. J. (2009). Emerging themes and new challenges in defining the role of structural variation in human disease. *Human Mutations*, *30*, 135-–144.

Shiue, Y. L., Bickel, L. A., Caetano, A. R., Millon, L. V., Clark, R. S. et al. (1999). A synteny map of the horse genome comprised of 240 microsatellite and RAPD markers. *Animal Genetics*, *30*, 1–9.

Shiue, Y. L., Millon, L. V., Skow, L. C., Honeycutt, D., Murray, J. D. et al. (2000). Synteny and regional marker order assignment of 26 type I and microsatellite markers to the horse X- and Y-chromosomes. *Chromosome Research*, *8*, 45–55.

Speicher, M. R., & Carter, N. P. (2005). The new cytogenetics: Blurring the boundaries with molecular biology. *National Review of Genetics*, *6*, 782–792.

Sysa, P. S., Hohn, H., & Schmid, I. (1977). Nucleolus organizers in chromosomes od some domestic animals. *Annales de Genetique et Selection Animale*, *9*, 540–541.

Takahashi, T., Yawata, M., Raudsepp, T., Lear, T. L., Chowdhary, B. P. et al. (2004). Natural killer cell receptors in the horse: evidence for the existence of multiple transcribed LY49 genes. *European Journal of Immunology*, *34*, 773–784.

Telenius, H., Pelmear, A. H., Tunnacliffe, A., Carter, N. P., Behmel, A. et al. (1992). Cytogenetic analysis by chromosome painting using DOP-PCR amplified flow-sorted chromosomes. *Genes Chromosomes Cancer*, *4*, 257–263.

Tozaki, T., Swinburne, J., Hirota, K., Hasegawa, T., Ishida, N. et al. (2007). Improved resolution of the comparative horse-human map: Investigating markers with in silico and linkage mapping approaches. *Gene*, *392*, 181–186.

Trask, B. J. (2002). Human cytogenetics: 46 chromosomes, 46 years and counting. *National Review of Genetics*, *3*, 769–778.

Trifonov, V., Yang, F., Ferguson-Smith, M. A., & Robinson, T. J. (2003). Cross-species chromosome painting in the Perissodactyla: Delimitation of homologous regions in Burchell's zebra (Equus Burchellii) and the white (Ceratotherium Simum) and black rhinoceros (Diceros Bicornis). *Cytogenetic Genome Reseach*, *103*, 104–110.

Trifonov, V. A., Stanyon, R., Nesterenko, A. I., Fu, B., Perelman, P. L. et al. (2008). Multidirectional cross-species painting illuminates the history of karyotypic evolution in Perissodactyla. *Chromosome Research*, *16*, 89–107.

Trujillo, J. M., Walden, B., O'Neil, P., & Anstall, H. B. (1965). Sex-linkage of glucose-6-phosphate dehydrogenase in the horse and donkey. *Science*, *148*, 1603–1604.

Wade, C., Giulotto, E., Sigurdsson, S., Zoli, M., Gnerre, S. et al. (2009). Genome sequence, comparative analysis and population genetics of the domestic horse (*Equus caballus*). *Science*, *326*, 865–867.

Wagner, M. L., Goh, G., Wu, J. T., Morrison, L. Y., Alexander, L. J. et al. (2004a). Sixty-seven new equine microsatellite loci assigned to the equine radiation hybrid map. *Animal Genetics*, *35*, 484–486.

Wagner, M. L., Goh, G., Wu, J. T., Raudsepp, T., Morrison, L. Y. et al. (2004b). Radiation hybrid mapping of 63 previously unreported equine microsatellite loci. *Animal Genetics*, *35*, 159–162.

Wagner, M. L., Goh, G., Wu, J. T., Raudsepp, T., Morrison, L. Y. et al. (2004c). Radiation hybrid mapping of 75 previously unreported equine microsatellite loci. *Animal Genetics*, *35*, 68–71.

Wagner, M. L., Raudsepp, T., Goh, G., Agarwala, R., Schäffer, A. A. et al. (2006). A 1.3-Mb interval map of equine homologs of HSA2. *Cytogenetic Genome Research*, *112*, 227–234.

Ward, T. L., Valberg, S. J., Lear, T. L., Guerin, G., Milenkovic, D. et al. (2003). Genetic mapping of GBE1 and its association with glycogen storage disease IV in American Quarter horses. *Cytogenetic Genome Research*, *102*, 201–206.

Williams, H., Richards, C. M., Konfortov, B. A., Miller, J. R., & Tucker, E. M. (1993). Synteny mapping in the horse using horse-mouse heterohybridomas. *Animal Genetics*, *24*, 257–260.

Wittwer, C., Chowdhary, B. P., & Distl, O. (2005). Radiation hybrid mapping of equine CDK2, DGKA, DNAJC14, MMP19, CTSL and GAS1. *Animal Genetics*, *36*, 536–537.

Wodsedalek, J. E. (1914). Spermatogenesis of the horse with special reference to the accessory chromosome and the chromatoid body. *The Biological Bulletin*, *27*, 295–325.

Yang, F., Fu, B., O'Brien, P. C., Nie, W., Ryder, O. A. et al. (2004a). Refined genome-wide comparative map of the domestic horse, donkey and human based on cross-species chromosome painting: insight into the occasional fertility of mules. *Chromosome Research*, *12*, 65–76.

Yang, F., Fu, B., O'Brien, P. C. M., Nie, W., Ryder, O. A. et al. (2004b). Refined genome-wide comparative map of the domestic horse, donkey and human based on cross-species chromosome painting: insight into the occasional fertility of mules. *Chromosome Research*, *12*, 65–76.

Yang, F., Fu, B., O'Brien, P. C. M., Robinson, T. J., Ryder, O. A. et al. (2003). Karyotypic relationships of horses and zebras: results of cross-species chromosome painting. *Cytogenetic Genome Research*, *102*, 235–243.

Zabek, T., Bugno, M., Klukowska-Rotzler, J., Gerber, V., & Slota, E. (2009). Chromosomal assignment of equine genes involved in the development of skeletal, gastrointestinal, cardiovascular and nervous system. *Hereditas*, *146*, 177–179.

Zabek, T., Bugno, M., Klukowska-Rotzler, J., Uhlmann, B., Gerber, V. et al. (2007). Chromosomal assignment of five equine genes responsible for the development of the skeletal and nervous systems. *Animal Genetics*, *38*, 425–426.

4 The Y-Chromosome

Terje Raudsepp, Nandina Paria, and Bhanu P. Chowdhary

Introduction

The Y chromosome is one of the most structurally, functionally, and evolutionarily distinct regions in the mammalian genome. It is believed that mammalian sex chromosomes, the Y and the X, originate from a pair of autosomes and were homologs more than 300 million years ago (Ohno, 1967; Lahn & Page, 1999; Graves et al., 2002; Graves, 2006; Marshall Graves, 2008; Graves, 2010). Since then, the Y chromosome has acquired male-sex-determining and other male-beneficial functions and has undergone structural rearrangements that suppressed crossing over with the X chromosome. This led to gradual loss of the ancestral gene set and accumulation of repetitive sequences. As a result, the contemporary Y chromosome is constitutively haploid, male specific, and one of the gene-poorest and perhaps the smallest elements in the mammalian genome (Graves, 2006; Marshall Graves, 2008; Graves, 2010).

It was thought for a long time that due to the unusual evolutionary history, the Y chromosome has lost most of the functional genes, preserving only the ones involved in male sex determination. Consequently, very limited concerted efforts were directed to study the organization and gene content of the Y chromosome in the majority of the domestic species, including the horse. The chromosome was inappropriately ignored until as late as mid-1980s as a "barren wasteland" or "junk" (Sinclair et al., 1990; Lahn & Page, 1997; Quintana-Murci & Fellous, 2001). However, systematic discovery, mapping, and functional analysis of Y-linked genes in human (Lahn & Page, 1999; Skaletsky et al., 2003; Kirsch et al., 2005) and mouse (McElreavey & Mitchell, 2002; Toure et al., 2004a; Toure et al., 2004b; Toure et al., 2005; Ellis et al., 2007; Grzmil et al., 2007), and complete sequencing of the male-specific region of Y (MSY) in human (Skaletsky et al., 2003) and chimpanzee (Hughes et al., 2010), have dramatically changed these views. The studies clearly show that the mammalian Y chromosome carries a rich repertoire of functional genes, several of which play a role in spermatogenesis and male fertility. Therefore, Y chromosome studies have been initiated in several species, including the horse, with the aim to identify all Y-linked genes, generate a detailed physical map for the chromosome, and sequence it.

The Y chromosome exhibits unique structural and functional features that are very different compared to other chromosomes. Because of this, the techniques used to study the molecular organization of the Y are slightly different and sometimes more complicated than those used for the autosomes and the X chromosome. Typically, the Y chromosome is not included in mammalian

Equine Genomics, First Edition. Edited by Bhanu P. Chowdhary.

whole genome sequencing projects and is studied separately. Likewise, the horse whole genome draft sequence (Wade et al., 2009) does not include the Y chromosome.

Following human (Skaletsky et al., 2003), mouse (Ellis et al., 2007), and cat (Murphy et al., 2006; Pearks Wilkerson et al., 2008), horse is among the few domestic species where the Y chromosome is relatively well studied. This chapter describes the cytogenetic and molecular organization of the horse (*Equus caballus*, ECA) Y chromosome (ECAY), including a detailed map and gene catalog of the pseudoautosomal region (PAR), which is an integral part of ECAY. The chapter also elaborates on the advances made to close in on Y-linked male fertility genes and the overall role of the Y chromosome in equine sex and reproduction related (SRR) disorders. Finally, a brief overview is provided summarizing the current knowledge about the Y chromosome in other equids and Perissodactyls. Possibility to use horse Y chromosome sequence polymorphism for population studies is discussed.

Cytogenetics

The Y chromosome is one of the smallest chromosomes in the horse karyotype (2n=64), constituting about 1.6–1.9% of the total haploid genome (Hansen, 1984). For comparison, its evolutionary counterpart or gametologue, the X chromosome, makes up of about 5% of the genome and is the third-largest chromosome (Hansen, 1984). The G- and C-banding patterns of ECAY were described already in the 1970s (Hageltorn & Gustavsson, 1974; Buckland et al., 1976; Cribiu & De Giovanni, 1978; Ford et al., 1980), followed by R-banding (Romagnano et al., 1987; Ronne et al., 1993) a decade later. Initially, it was proposed that the horse Y chromosome is morphologically acrocentric (Buckland et al., 1976) or telocentric (Ford et al., 1980; Hansen, 1984). However, the most recent equine chromosome nomenclature defines ECAY as sub-metacentric with a minute short arm (ISCNH, 1997). The G-banding pattern of ECAY is simple with just one distinct G-positive band at the distal end of Yq (Figure 4.1a), which corresponds to the euchromatic portion of the chromosome. Substantial part (~75%) of the remaining chromosome is heterochromatic and stains light with the G-banding method. The R-banding pattern of ECAY is not very distinct and does not show a typical reverse banding to G-bands (Romagnano et al., 1987; Power, 1988; ISCNH, 1997) (Figure 4.1b). The C-banding method that highlights heterochromatin stains most of ECAY dark (Figure 4.1c), and is the preferred method for quick and unequivocal identification of ECAY in cytogenetic analysis.

The horse Y chromosome varies considerably (more than twice as compared to the smallest known size) in length among normal individuals (Hansen, 1984; Power, 1988; our unpublished observations). This polymorphism is primarily attributed to the various amount of Y heterochromatin, though earlier studies also suggest variation in the euchromatic portion of the chromosome (Power, 1988). Size polymorphism of the horse Y chromosome is not considered to be breed dependent or related to any clinical or behavioral abnormalities (Hansen, 1984; Power, 1988).

Compared to autosomes and the X chromosome, relatively few numerical or structural cytogenetic abnormalities have been reported for ECAY. This might be because rearrangements in the gene-poor Y chromosome have milder effect on the phenotype, as a result of which the affected individuals escape attention and are not subjected to cytogenetic analysis. Also, the lack of molecular tools for the study of ECAY has considerably hindered the detection of fine-scale cytogenetic aberrations. Majority of abnormalities involving the Y chromosome are mixoploid, mosaic, or chimeric conditions where different cell populations, such as 63,X/64,XY (Hughes & Trommershausen-Smith, 1977; Trommershausen-Smith et al., 1979; Dunn et al., 1981),

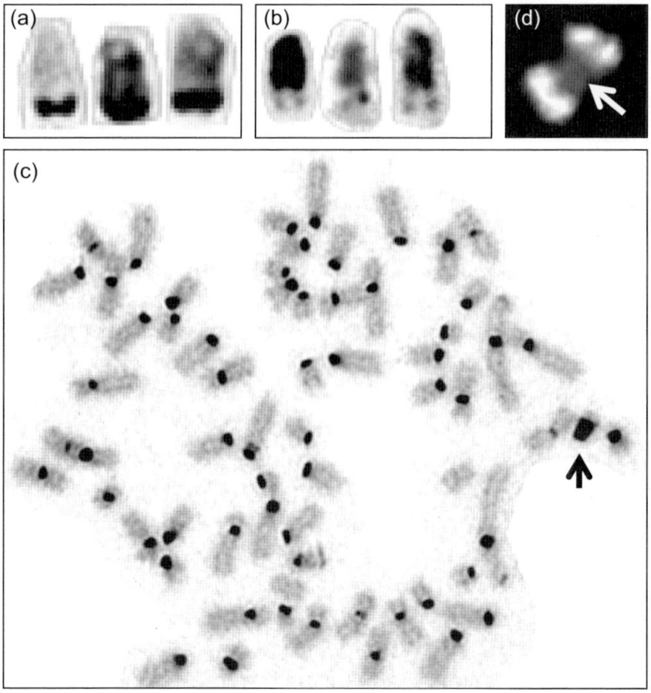

Figure 4.1 Horse Y chromosome cytogenetics: (a) GTG-banded and (b) R-banded Y chromosomes from three different cells (ISCNH 1997); (c) C-banded male metaphase spread showing (arrow) prominent staining of ECAY heterochromatin; and (d) Isochromosome Y: Terminal bright FISH signals show the location of MSY and the PAR, the central region of dull fluorescence corresponds to the heterochromatin, and the arrow shows the location of the centromere.

64,XX/64,XY (McIlwraith et al., 1976; Dunn et al., 1981), 64,XX/96,XXY (Power & Leadon, 1990), or even 63,XO/64,XX/65,XXX/65,XXY/66,XXXY/66,XXYY (Klunder et al., 1990) are present in the same individual. Most of these conditions are associated with abnormal sex determination and/or sexual development (for review, see Villagomez et al., 2009). However, the XX/XY blood chimerism that has been found in equine heterosexual co-twins does not affect fertility (Miyake et al., 1982; Bugno et al., 2007; Juras R. et al., 2010). This is contrasts the freemartin condition in cattle where blood-chimaeric female co-twins are typically sterile (Padula, 2005). Additionally, there are two reports about infertile stallions having XXY sex chromosomes – an equine counterpart to human Kleinfelter syndrome (Kubien et al., 1993; Makinen et al., 2000).

The only true Y chromosome structural rearrangements, reported so far, are large-scale ECAY deletions in 64,XY sex reversal mares (Raudsepp et al., 2010a), which are discussed in detail later in this chapter, and the presence of Y isochromosome (Herzog et al., 1989; our unpublished data). The latter is a Y chromosome that has lost the short arm and replaced it with an identical copy of the long arm. The resulting isochromosome i(Y) is metacentric with both arms corresponding to Yq (Figure 4.1d). Since the long arm of ECAY carries the MSY and the PAR, the cells with i(Y) are genetically similar to those with Y chromosome disomy (YY). The two individuals carrying isochromosome Y had mosaic karyotypes 63,XO/64,Xi(Y), and both animals were phenotypically abnormal. The horse described by Herzog et al. (1989) was a pseudo-hermaphrodite, and the most recent case (our unpublished data) was an intersex with ambiguous external genitalia. It is, therefore, not clear whether haploinsufficiency for Xp genes in XO cells or the overdose of MSY and PAR genes in Xi(Y) cells is the cause of abnormal sex determination and differentiation in these animals.

Molecular Probes

An important breakthrough in horse Y chromosome research came with the availability of new genomics tools and resources, such as large-insert genomic bacterial artificial chromosome (BAC) libraries (for review, see Rubes et al., 2009), chromosome flow sorting and microdissection technology, and various genomic DNA amplification systems (Telenius et al., 1992; Sorensen et al., 2007). This allowed isolation of ECAY sequences and generation of Y-specific molecular probes. The probes have been used for molecular hybridizations (fluorescence in situ hybridization, or FISH) in clinical cytogenetics, Y chromosome mapping, sequencing, gene discovery, and comparative Y chromosome studies between the horse and other equids/ Perssodactyls.

The first ECAY-specific whole-chromosome painting probes were generated by chromosome microdissection (Raudsepp and Chowdhary, 1999; Kubickova et al., 2002) and used for quick identification of ECAY in cytogenetic preparations (Kubickova et al., 2002), for sperm FISH to detect ECAY aneuploidy (Bugno-Poniewierska et al., 2009), and for horse-donkey Zoo-FISH (Raudsepp and Chowdhary, 1999). Typically, only 10–20 copies of ECAY are microdissected and amplified by DOP-PCR (Telenius et al., 1992). Due to the largely heterochromatic nature of ECAY and repeat-biased amplification of DOP-PCR, microdissected probes tend to be enriched with repetitive sequences, while euchromatic sequences remain underrepresented. Therefore, microdissected Y-chromosome probes are good for FISH but not suitable for cloning and sequencing of the male-specific or pseudoautosomal regions of the chromosome.

An alternative method for the generation of Y-chromosome-specific probes is chromosome flow sorting, which allows isolation of thousands of Y chromosomes in one run. Compared to other domestic species, such as pig and dog, where flow karyotypes were available already in mid-1990s (Langford et al., 1993; Yerle et al., 1993; Langford et al., 1996), flow sorting of horse chromosomes had a late start. According to our best knowledge, the only horse flow karyotype, including the Y chromosome, was generated in 2004 and used for comparative chromosome painting on human, donkey, and mule chromosomes (Yang et al., 2004), and later for karyotype comparisons with other equids and Perssiodactyls (Trifonov et al., 2008). Most importantly, flow-sorted horse Y chromosome (provided by M. A. Ferguson-Smith, Cambridge Resource Centre for Comparative Genomics) has been instrumental for the discovery of Y-linked genes and for generating a comprehensive map for horse MSY (Raudsepp et al., 2008; Paria, 2009; Paria et al., 2011) as discussed later in this chapter.

Horse Y chromosome research has essentially benefitted from the fact that all three available equine BAC libraries, namely CHORI-241 (http://bacpac.chori.org/equine241.htm), INRA (Milenkovic et al., 2002) and TAMU (http://hbz7.tamu.edu/homelinks/bac_est/bac.htm), (see Chapter 3 in this volume), have been constructed from male individuals. Genomic clones containing ECAY sequences have been used for the discovery of Y-linked genes and microsatellites, and as "building bricks" for BAC contig maps of the male specific (Raudsepp et al., 2008; Paria, 2009) and the pseudoautosomal regions (Raudsepp & Chowdhary, 2008). Large-insert clones containing ECAY sequences are also the best probes for FISH analysis either for mapping (Raudsepp et al., 2004; Raudsepp et al., 2008), identification of multicopy sequences (Raudsepp et al., 2004; Raudsepp et al., 2008; Paria, 2009; Paria et al., 2011), or for the detection of large-scale ECAY deletions (Raudsepp et al., 2010a).

Genes

The genes present in the Y chromosome fall into two broad categories: the pseudoautosomal genes, which are shared between the X and the Y chromosome and are discussed later in this chapter, and the male-specific genes that are present only in the Y chromosome.

The study of male-specific genes in horses started about a decade ago, though significant progress in this research has been made only during the past few years. The first gene loci were assigned to ECAY indirectly by PCR showing that primers for *TSPY* (Manz et al., 1998), *ZFY* (Senese et al., 1999), and *SRY* (Hasegawa et al., 1999) amplify exclusively from male genomic DNA and have no amplification in females. Amelogenin primers amplified two distinct products corresponding to *AMELY* and *AMELX* in males and only the *AMELX* product in females (Hasegawa et al., 2000), thus assigning *AMELY* to the Y chromosome. A few years later, *SRY*, *ZFY*, and *STS* were synteny mapped to the horse Y chromosome by somatic cell hybrid analysis (Shiue et al., 2000). The location of *SRY* on *Yq13* and *ZFY* on *Yq15* was further refined using FISH (Hirota et al., 2001) (Fig. 4.2a). Systematic discovery of ECAY genes started in 2004 when the location and linear order of eight MSY and one PAR gene (Figure 4.2b) was determined using a combination of radiation hybrid analysis, FISH, and BAC contig mapping (Raudsepp et al., 2004). Since then, the ECAY gene catalog has significantly expanded. Recent cDNA selection experiments (Del Mastro & Lovett, 1997) from horse testis, combined with sequencing Y-specific BACs and BAC ends, resulted in the discovery, functional analysis, and mapping of 37 MSY genes (Table 4.1) (Paria, 2009; Paria et al., 2011). Together with the 19 PAR genes (Raudsepp & Chowdhary, 2008), there are currently 56 genes and ESTs assigned to the horse Y chromosome (Figure 4.2c). This makes the ECAY gene catalog one of the most comprehensive Y-specific gene collections in eutherian mammals, standing next to those of human (Skaletsky et al., 2003), chimpanzee (Hughes et al., 2010), and mouse (Yamauchi et al., 2009).

According to the origin of sequences, the horse MSY genes fall into three categories: (1) X-degenerate, (2) ampliconic, and (3) acquired (Table 4.1). The X-degenerate genes are typically single-copy sequences with an ancestral homologue or a gametologue, on the X chromosome. Examples of such gene pairs are *AMELY-AMELX*, *ZFY-ZFX*, or *USP9Y-USP9X*, and the horse Y chromosome carries 20 X-degenerate genes. A characteristic to X-degenerate genes is ubiquitous expression, so their transcripts are found in most of the body tissues.

The most unique and distinctive feature of MSY is the presence of ampliconic or multicopy genes. Many of those are Y-borne (Murphy et al., 2006; Paria, 2009; Paria et al., 2011), but some have moved to MSY from autosomes or the X chromosome. Amplification of these sequences is the result of gene conversion and/or duplication, which are typical functional features of the non-recombining Y chromosome (Skaletsky et al., 2003). Compelling evidence from human and mouse studies suggest that amplification of these sequences has occurred because they play a critical role in male fertility (Skaletsky et al., 2003; Ellis et al., 2007; Cocquet et al., 2009). Indeed, the majority of the 15 equine Y-linked ampliconic genes are expressed exclusively or predominantly in testis. It is also noteworthy that 10 of these genes are novel and have not been found yet in any other mammalian species. Finally, demarcation between the ampliconic, X-degenerate, and acquired genes is not always distinct. For example, four X-degenerate genes, namely *CUL4BY*, *RBMY*, *TSPY*, and *UBE1Y*, and an acquired gene *ZNF33bY* have become multicopy in the course of evolution.

The Y chromosome is also the target of gene acquisition by transposition from other genomics locations. The horse MSY has acquired five such genes from various autosomes and one from the mitochondrial genome. All are expressed ubiquitously, except *ZNF33bY*, which is testis specific (Paria, 2009; Paria et al., 2011). The horse MSY gene sequences have been studied for the presence

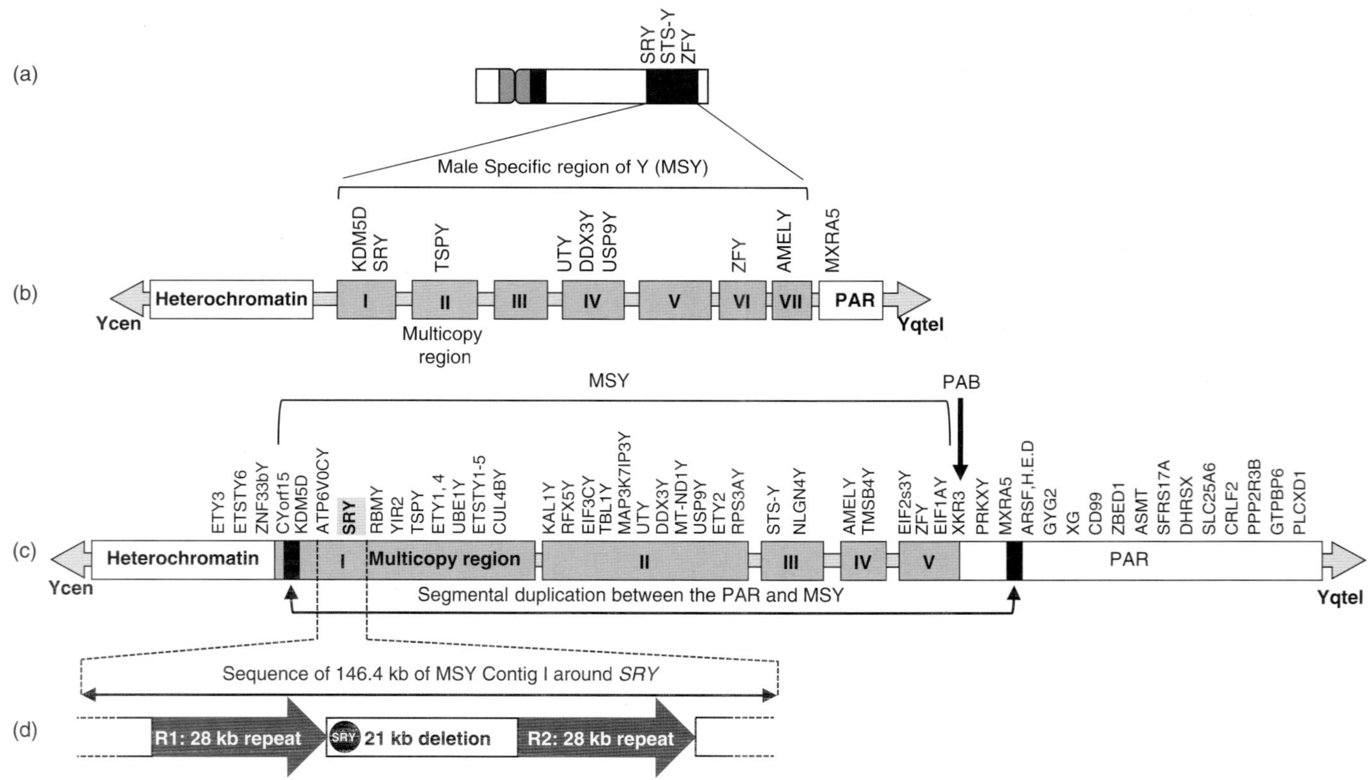

Figure 4.2 Evolution of the horse Y chromosome gene map: (a) ECAY GTG-banded ideogram (ISCNH, 1997) showing the location of the first three synteny (Shiue et al., 2000) and FISH (Hirota et al., 2001) mapped genes; (b) ECAY gene map from 2004 (Raudsepp et al., 2004). Roman numerals mark the seven BAC contigs; (c) current status of ECAY gene map comprising of 37 MSY (Paria, 2009) and 19 PAR (Raudsepp and Chowdhary, 2008) genes. Black rectangles in the PAR and MSY Contig I indicate segmental duplication between the two regions; (d) detailed sequence map of a 146.4 kb region around the *SRY* locus. Two 100% identical 28 kb directional repeats (black block arrows) flank a 21 kb region that contains the *SRY* gene. This region is deleted in XY SRY-negative mares.

Table 4.1 Gene catalog of the horse Y chromosome

Gene symbol	Gene name	Gene category	Copy number	Expression	ORFs	Presence in other species
AMELY	Amelogenin, Y-linked	X-degenerate	single copy	no expression	none	human, chimpanzee, cat, pig, cattle
ATP6V0CY	ATPase, H+ transporting, lysosomal 16kDa, V0 subunit c	acquired	single copy	ubiquitous	none	Not found
CUL4BY	Cullin 4 B Y	X-degenerate	multicopy	intermediate	present	cat, dog
CYorf15	Chromosome Y open reading frame 15	X-degenerate	single copy	ubiquitous	present	human, chimpanzee SC, cat
DDX3Y	Dead box Y	X-degenerate	single copy	ubiquitous	present	human, chimpanzee, mouse, cat, rat, pig, cattle
EIF1AY	Translation initiation factor 1A Y	X-degenerate	single copy	ubiquitous	none	human, chimpanzee, cat
EIF2s3Y	Eukaryotic translation initiation factor 2, subunit 3 gamma, Y- linked	X-degenerate	single copy	ubiquitous	none	mouse, cat, pig
EIF3CY	Eukaryotic translation initiation factor 3, subunit C on Y	acquired	single copy	ubiquitous	present	n/a
ETSTY1	Equus Testis-specific transcript Y1	ampliconic	multicopy	testis-specific	none	not found
ETSTY2	Equus Testis-specific transcript Y2	ampliconic	multicopy	testis-specific	none	not found
ETSTY3	Equus Testis-specific transcript Y3	ampliconic	multicopy	testis-specific	none	not found
ETSTY4	Equus Testis-specific transcript Y4	ampliconic	multicopy	testis-specific	none	not found
ETSTY5	Equus Testis-specific transcript Y5	ampliconic	multicopy	testis-specific	none	not found
ETSTY6	Equus Testis-specific transcript Y6	ampliconic	multicopy	testis-specific	none	not found
ETY1	Equus transcript Y1	ampliconic	multicopy	intermediate	none	not found
ETY2	Equus transcript Y2	ampliconic	single copy	ubiquitous	none	not found
ETY3	Equus transcript Y3	ampliconic	multicopy	no expression	none	not found
ETY4	Equus transcript Y4	ampliconic	multicopy	ubiquitous	none	not found
KAL1Y	Kallmann Syndrome 1 on Y	X-degenerate	single copy	no expression	none	human pseudogene, pig, cattle PAR
KDM5D (SMCY)	Jumonji, AT rich interactive domain 1D	X-degenerate	single copy	ubiquitous	present	human, chimpanzee, mouse, cat, dog, cattle, pig
MAP3K7IP3Y	Mitogen-activated protein kinase 7 interacting protein 3 on Y	X-degenerate	single copy	ubiquitous	none	n/a
MT-ND1Y	Mitochondrially encoded NADH dehydrogenase 1 on Y	acquired	single copy	ubiquitous	none	n/a
NLGN4Y	Neuroligin 4 isoform Y	X-degenerate	single copy	intermediate	none	human, mouse, chimpanzee, cattle PAR
RBMY	RNA-binding motif Y	X-degenerate	multicopy	testis-specific	none	human, mouse
RFX5Y	Regulatory factor X 5 on Y	acquired	single copy	intermediate	none	n/a

(continued)

Table 4.1 (*Continued*)

Gene symbol	Gene name	Gene category	Copy number	Expression	ORFs	Presence in other species
RPS3AY	Ribosomal Protein S3A	X-degenerate	single copy	ubiquitous	present	n/a
SRY	Sex determining region Y	X-degenerate	single copy	intermediate	present	human, mouse, chimpanzee, cattle, pig, cat, rat
STS-Y	Steroid sulfatase (microsomal), isozyme S on Y	X-degenerate	single copy	n/a	none	human, chimpanzee pseudogene, mouse, cattle, pig, dog PAR
TBL1Y	Transducin (beta)-like 1, Y-linked	X-degenerate	single copy	n/a	none	human, chimpanzee, cattle PAR
TMSB4Y	Thymosin (beta) 4 Y	X-degenerate	single copy	ubiquitous	none	human, chimpanzee
TSPY	Testis-specific Protein Y	X-degenerate	multicopy	testis-specific	present	human, chimpanzee, rat, cattle, goat, pig, mouse pseudogene
UBE1Y	Ubiquitin activating enzyme Y	X-degenerate	multicopy	testis-specific	present	mouse, rat, pig, cat
USP9Y	Ubiquitin-specific protease 9 Y	X-degenerate	single copy	ubiquitous	present	human, mouse, chimpanzee, cat, rat, pig
UTY	Ubiquitously transcribed tetratricopeptide repeat gene, Y linked	X-degenerate	single copy	ubiquitous	none	human, mouse, chimpanzee, cat, pig
YIR2	Inverted repeat 2 Y	ampliconic	multicopy	intermediate	none	human
ZFY	Zinc finger Y	X-degenerate	single copy	ubiquitous	none	human, cat, chimpanzee, rat, pig, mouse
ZNF33bY	Zinc Finger protein 33b on Y	acquired	multicopy	testis-specific	none	n/a

of open reading frames and protein-coding ability, and full-length cDNA sequences have been generated for *TSPY* and a few novel testis-specific transcripts (Table 4.1).

Comparison of ECAY gene catalog with other mammalian species shows that only the X-degenerate genes can serve as comparative loci, whereas the majority of ampliconic and acquired sequences are horse specific (Table 4.1).

Maps

As described in the previous section, the first ECAY maps showed Y-linkage of a few genes without defining their precise location or relative order. Thereafter, the first Y-specific microsatellite sequences were isolated using representation difference analysis (RDA; Wallner et al., 2004). These "seed" sequences were instrumental for initiating systematic analysis of the horse Y chromosome leading to the construction of the first detailed physical map of MSY (Raudsepp et al., 2004). The first-generation composite map (Figure 4.2b) (Raudsepp et al., 2004) combined genotyping on the 5,000 rad radiation hybrid (RH) panel (Chowdhary et al., 2002), BAC contig mapping, BAC fingerprinting, and FISH analysis. The use of four different mapping approaches was necessary because of the unusual and unique features of ECAY of which the most outstanding is the presence

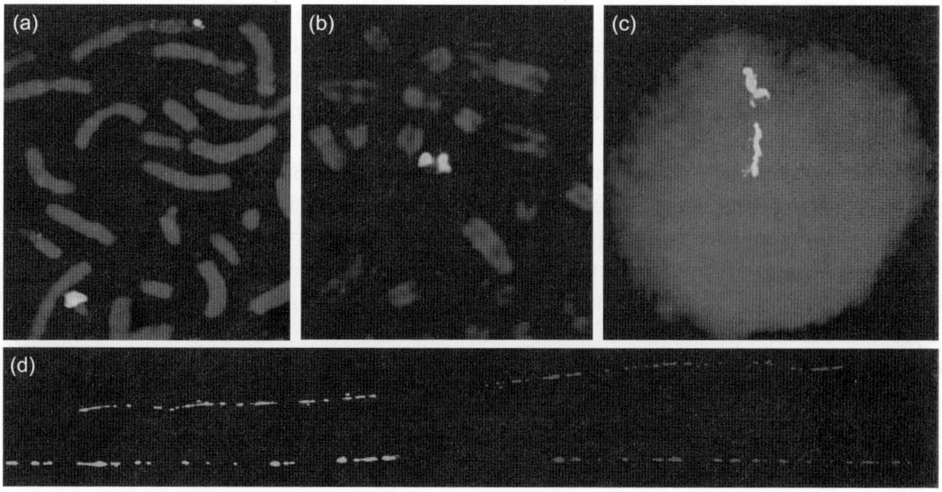

Figure 4.3 Examples of metaphase (a, b), interphase (c), and fiber FISH (d) to study the horse Y chromosome using (a) a pool of ECAY and PAR BAC clones; (b) BAC clone containing *RBMY, SRY, TSPY*, and *YIR2*; (c) BAC clone containing novel multicopy genes *ETY4* and *ETSTY5*; and (d) BAC clones containing *KDM5D* (green) and *SRY* (red). (*For color details, see color plate section.*)

of multiple copies of gene-specific and non-genic sequences. This sets limitations to PCR-based mapping approaches such as RH genotyping and sequence tagged site (STS) content analysis. Both methods must be supported by metaphase- (Figures 4.3a, 4.3b), interphase- (Figure 4.3c), and fiber FISH (Figure 4.3d) in order to anchor markers to specific regions, identify multicopy sequences, and validate physical distances and overlaps between the BAC clones on the map.

The first-generation ECAY map (Figures 4.2a, 4.2b) shows that the euchromatic region of the chromosome lies in the distal one-quarter of the long arm, where the PAR is located terminally. The rest of the chromosome is predominantly heterochromatic, which is in agreement with cytogenetic studies (Figure 4.1c). Analysis of the 5,000 rad RH panel provided a baseline map spanning 88 centirays with 8 genes and 15 STS markers. Isolation of BAC clones for markers mapped by RH, end sequencing of the BACs, STS development from the end sequences, and bidirectional chromosome walking yielded 109 markers (100 STS and 9 genes) contained in 73 BACs. STS content mapping grouped the BACs into seven physically ordered contigs (Figure 4.2b) that were verified by high-resolution FISH and BAC fingerprinting. It was estimated that the BAC contigs in this map provided ~20–25% coverage of the euchromatic region of the chromosome.

The first-generation composite map served as an important foundation for expanded studies aimed at developing a minimum tiling path of BACs over the MSY that could finally be used for obtaining complete sequence of the region. The second-generation map of the horse MSY was completed recently (Figure 4.2c) (Raudsepp et al., 2008; Paria, 2009) and comprises of 197 BAC clones that are arranged into a physical map over the male-specific region of ECAY. The five contigs of overlapping BAC clones contain 267 linearly ordered markers of which 230 are STSs and 37 are specific genes or ESTs. Proximally, the map extends into Y chromosome heterochromatin, while the most distal clones span the pseudoautosomal boundary (PAB) and join MSY with the PAR (Raudsepp and Chowdhary, 2008). At the distal end of Contig 1 there is a region containing multiple copies of known genes (*CUL4BY*, *RBMY*, *TSPY*, *UBE1Y*), several novel equine-specific expressed sequence tags (ESTs), and non-genic sequences. Notably, the horse *SRY*, despite being a single-copy gene

(Paria et al., 2011), is embedded in the multicopy region (Figure 4.2c,d). Unique to the horse Y chromosome is also the presence of a few functional genes in the heterochromatin. The size of MSY, including the four gaps that collectively span about 4–5 Mb, is approximately 16 Mb. Given that the heterochromatic portion counts for about two-thirds of the chromosome, the horse Y chromosome contains about 60 Mb DNA.

While the map gives the most comprehensive overview about the organization of ECAY, the contig of overlapping BAC clones serves as an ideal template for complete sequencing of MSY in the horse. The minimum tiling path (MTP) of MSY comprises of 54 BACs (our unpublished data). However, because of the unusual organization of the Y chromosome, its complete sequencing needs higher redundancy of clones than is necessary for autosomal or X-linked regions (Skaletsky et al., 2003). Currently, more than 100 MSY BAC clones have been selected for next-generation sequencing at Texas A&M University.

The Pseudoautosomal Region (PAR)

The pseudoautosomal region is a region of true sequence homology between the Y and X chromosomes (Skaletsky et al., 2003; Galtier, 2004; Ross et al., 2005; Blaschke & Rappold, 2006). The PAR is essential for sex chromosome pairing and segregation in male meiosis (Ellis & Goodfellow, 1989; Blaschke & Rappold, 2006; Flaquer et al., 2008), and is critical for spermatogenesis and male fertility. Furthermore, recent studies with human embryonic stem cells suggest that the PAR genes might have an important role in early development (Urbach & Benvenisty, 2009). Despite these outstanding biological roles, the physical domain of the PAR, which lies between the terminal end of the sex chromosomes and the pseudoautosomal boundary (PAB), has been determined in only a few mammalian species, including the horse (Raudsepp & Chowdhary, 2008; Raudsepp et al., 2011).

Synaptonemal complex analysis of meiotic prophase chromosomes in stallion spermatocytes shows that at diakinesis, the Y chromosome is associated with the short arm of the X chromosome (Scott & Long, 1980; Safronova & Pimenova, 1988; Power et al., 1992). The exact location of the PAR in ECAY, however, remained unclear (von Nett et al., 1996) until the first pseudoautosomal loci were physically mapped to the terminal regions in ECAYq and Xp (Raudsepp & Chowdhary, 2008).

Using the foundation provided by the first-generation map of the Y chromosome (Raudsepp et al., 2004), a high-resolution BAC contig map has been constructed for the entire horse PAR, extending also into the X- and Y-specific regions (Raudsepp & Chowdhary, 2008). The map consists of 71 BAC clones that are arranged into two contigs, shows linear order of 19 genes (Figure 4.2c) and 110 STS markers, and, for the first time, demarcates the pseudoautosomal boundary (PAB) on both sex chromosomes. The size of equine PAR is about 1.8 Mb. Notably, the draft assembly of the horse genome (Wade et al., 2009) contains only proximal part of the PAR, up to *SFRS17A* locus (Figure 4.2c), thus missing about 800 kb most distal sequences and six PAR genes.

The physical domain of the PAR is proximally determined by the PAB – a border across which sequence homology between the sex chromosomes reduces and recombination stops (Blaschke & Rappold, 2006; Mangs & Morris, 2007). Equine PAB is located between *PRKXY* and *EIF1AY* loci on the Y (Figure 4.2c), and between *PRKXY* and *NLGN4X* on the X chromosome. More recently, the complete sequence of the equine PAB was obtained (Raudsepp et al., 2010b). While sequences of the X and Y chromosomes are 98–100% similar in the PAR, this homology disappears after the PAB because a massive LINE L1 element starts immediately proximal to the PAB on the X chromosome. Interestingly, the equine PAB spans a gene (an ortholog of human autosomal *XKR3*)

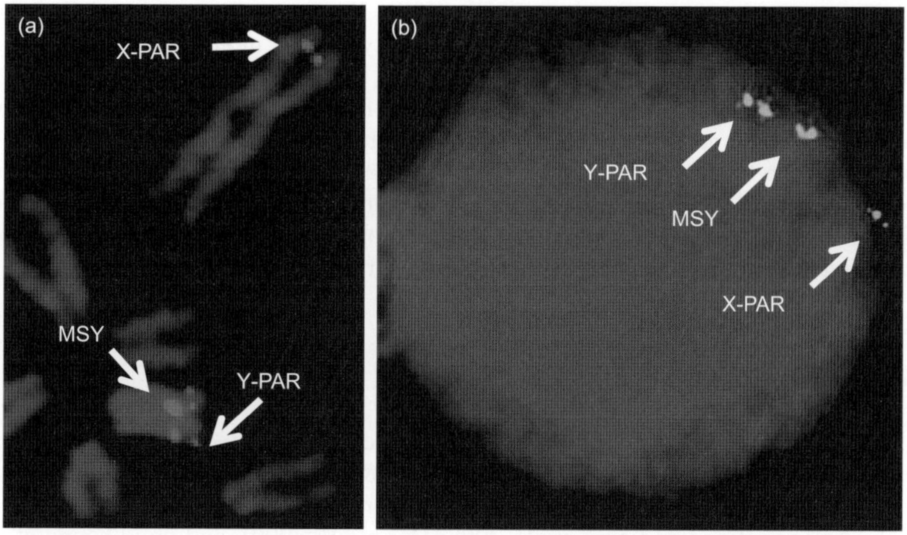

Figure 4.4 FISH on metaphase (a) and interphase (b) chromosomes showing the duplication between the proximal part of MSY and the PAR. (*For color details, see color plate section.*)

such that part of the gene is in the PAR and part in the Y-specific region. The gene is truncated and probably nonfunctional on the X chromosome (our unpublished data).

The most unusual discovery regarding the equine PAR is the presence of a ~200 kb duplication between the middle region of the PAR and the proximal part of MSY (Raudsepp & Chowdhary, 2008) (Figure 4.2c). In situ hybridizations with BAC clones from the two regions consistently show three signals on male metaphase chromosomes: one each on the terminal ends of ECAXp and ECAYq (i.e., the PAR) and one on the most proximal part of the MSY (Figure 4.4). This might be the result of duplications combined with unequal crossover events between the sex chromosomes during evolution. Interestingly, similar duplications between MSY and the PAR were detected in two other equids – donkey and Hartmann's mountain zebra (our unpublished data) – suggesting that these evolutionary rearrangements happened before the divergence of extant equids. The likely impact of the observed duplication on pairing between the sex chromosomes during meiosis and the possible effect of the presence of three copies of functional sequences in males are intriguing.

Disorders

Male infertility

The idea that selective pressures will favor the accumulation of male benefit genes, those that improve male fertility, on the Y chromosome was first suggested in 1931 by Sir Ronald Fisher (Fisher, 1931). The validity of these ideas was proved 45 years later in humans (Tiepolo and Zuffardi, 1976), and another 10 years later in mice (Burgoyne et al., 1986), when correlation between male infertility and deletions in the Y chromosome was clearly established. Since then, many more Y chromosome microdeletions and mutations have been described and associated with a diverse spectrum of defective spermatogenic phenotypes in human and mouse (Ellis & Affara,

2006; McElreavey et al., 2006; Ellis et al., 2007). While Y chromosome microdeletions are used as molecular markers for diagnostic tests in human fertility clinics, mouse Y chromosome genes are extensively studied to understand their function.

Recent studies of MSY genes in horses (Paria, 2009; Paria et al., 2011), leading to the discovery of multicopy and testis-specific transcripts, strongly suggest the likely involvement of MSY genes in regulating stallion fertility. Two pilot studies – one showing statistically significant differential expression of selected MSY genes in the testis of normal and subfertile stallions, and another showing overexpression of *TSPY* in the testis of a stallion with morphologically abnormal sperm – provide the first experimental support to these ideas (our unpublished data).

The PAR and sex chromosome aneuploidies

To date, the PAR has been studied and the PAB defined in only five mammalian species: horse (Raudsepp & Chowdhary, 2008), human/chimp (Ellis & Goodfellow, 1989; Hughes et al., 2005; Ross et al., 2005; Kuroki et al., 2006), mouse (Palmer et al., 1997; Gianfrancesco et al., 2001; Perry et al., 2001), cattle (Van Laere et al., 2008; Das et al., 2009), and dog (Young et al., 2008). Remarkably, all five PARs are different in size and gene content. For example, the 1.8 Mb equine PAR is ~0.9 Mb smaller than human PAR1 but contains orthologs from an additional 1 Mb region beyond the human PAR1 (Ross et al., 2005; Raudsepp & Chowdhary, 2008).

Given the proposed role of PAR genes in early embryonic development (Urbach & Benvenisty, 2009), the size and gene content of the PAR might be decisive in determining the viability and the severity of phenotypic effects of sex chromosome aneuploidies in different species. Indeed, the Molecular Cytogenetic Lab at Texas A&M University has recorded during the past 10 years chromosome analysis data for 210 female horses, 62 dogs, 22 pigs, and 62 cattle – all referred to the lab because of various reproductive problems. Remarkably, 20% of the infertile mares have X chromosome monosomy, while no sex chromosome aneuploidies were detected among infertile cows, sows, or bitches (Raudsepp et al., 2010b; Raudsepp et al., 2011). These observations are in line with other cytogenetic studies in domestic species (Villagomez et al., 2009) and support the idea that viable XO individuals are present in species with small PARs, such as horses (Raudsepp & Chowdhary, 2008), humans (Ross et al., 2005), and mice (Perry et al., 2001) but are a rare or absent in species where the PAR is substantially larger like cattle/ruminants (Van Laere et al., 2008; Das et al., 2009), dogs (Young et al., 2008), and pigs (our unpublished data).

Sex reversal syndrome

Male-to-female sex reversal in horses is a developmental disorder in which phenotypic females or female-like animals are genetically males with 64,XY karyotype. The condition has been found in several breeds (Kent et al., 1986; Kent et al., 1988a; Makinen et al., 1999; Bugno et al., 2003; Switonski et al., 2005; Raudsepp et al., 2010a) and is the second-most common sex chromosome abnormality in horses after X monosomy (Lear & Bailey, 2008; Villagomez et al., 2009). Phenotypes of the XY sex reversal horses can vary in a broad range, from a very feminine to a greatly masculinized mare (Kent et al., 1988b). While most equine male-to-female sex reversal cases are sporadic, familiar patterns of inheritance have been observed in certain Arabian sire lines (Kieffer, 1976; Bowling et al., 1987; Kent et al., 1988b).

Despite the relatively frequent occurrence of XY mares among animals with chromosome abnormalities, the molecular causes of the condition are yet poorly understood. This is partially because only limited molecular information has been available for the horse Y chromosome (ECAY). A test using PCR with *SRY* primers has been the only molecular method to study the XY mares since 1995 (Pailhoux et al., 1995). Two molecularly distinct types of XY females have been identified: mares with no *SRY* gene (Pailhoux et al., 1995; Abe et al., 1999; Makinen et al., 1999; Bugno et al., 2003) and *SRY*-positive mares (Switonski et al., 2005).

Recently, using markers from the second generation ECAY (Raudsepp et al., 2008; Paria, 2009) and PAR (Raudsepp & Chowdhary, 2008) maps, the first comprehensive molecular study of the Y chromosome in XY mares was carried out (Raudsepp et al., 2010a). The study involved 18 64,XY female horses of which 13 were *SRY*-negative and 5 *SRY*-positive. It was shown that the XY sex reversal condition in horses has two molecularly distinct forms: (1) a Y-linked form, which is characterized by massive Y chromosome deletions; and (2) a non-Y-linked form where the Y chromosome of affected females is molecularly the same as in normal males. The Y-linked form is molecularly heterogeneous and can involve deletions as small as $\sim$ 21 kb or deletions of the entire euchromatic region. Regardless of the size, all deletions include the *SRY* gene. Sequence analysis of selected ECAY BAC clones show that *SRY* is located between two 100% identical directional repeats (Figure 4.2d). As proposed recently by Lange et al. (2009), the likely mechanism for *SRY* deletion is inter-chromatid recombination between the repeats resulting in the deletion of the region between them.

It has also caught attention that *SRY*-negative sex reversal condition occurs in some species, such as horse, human, and mouse, but is extremely rare or not found at all in others like sheep, pig, and dog (Raudsepp et al., 2010a). These differences might be related to the diversity in the organization of the Y chromosome in different species (Skaletsky et al., 2003; Raudsepp et al., 2004; Murphy et al., 2006; Hughes et al., 2010). For example, if *SRY* is a single-copy gene, like it is in horses (Raudsepp et al., 2008; Paria, 2009; Paria et al., 2011), any Y chromosome rearrangement that causes the loss of *SRY* will result in an *SRY*-negative condition. In contrast, in species with multiple copies of *SRY*, like cat (Pearks Wilkerson et al., 2008), rabbit (Geraldes & Ferrand, 2006) or rat (Turner et al., 2007), deletion of one *SRY* copy leaves other copies intact to carry out their function. It can also be that *SRY* location in relation to inverted and directional repeats is different across species, thus facilitating *SRY* deletions in some species but not in others.

Detailed molecular analysis of the Y chromosomes of *SRY*-positive females (Switonski et al., 2005; Raudsepp et al., 2010a) has not revealed any differences between the Y chromosomes of sex-reversed females and normal males. Very recently, using array comparative genomic hybridization, a large genomic deletion in ECA29 was found in some, but not all, *SRY*-positive females, suggesting that the condition is molecularly heterogeneous (Raudsepp et al., 2012). Thus, the molecular causes of *SRY*-positive 64,XY condition in the horse remain largely undefined.

Polymorphism and Population Studies

The male-specific, constitutively haploid, and non-recombining MSY is an ideal target for developing polymorphic markers for the study of evolutionary events within species, populations, and breeds. Human Y chromosome haplotypes have been instrumental in tracking the genetic history of human populations worldwide (Kayser, 2010; Stoneking & Delfin, 2010) but also for unveiling domestication patterns, genetic origins of breeds, and genetic diversity in pigs (Cliffe et al., 2010), cattle (Bollongino et al., 2008; Ginja et al., 2010), and sheep (Meadows et al., 2004).

With similar intentions, microsatellite markers and single nucleotide polymorphisms (SNPs) have been searched from the horse Y chromosome. The first six ECAY microsatellites were genotyped in 49 male horses of 32 different breeds and in 9 other equid species (Wallner et al., 2004). Surprisingly, only a single haplotype was found in the domestic horse, whereas the same markers showed considerable variation in the genus *Equus*. Simultaneously with the development of the first (Raudsepp et al., 2004) and second (Raudsepp et al., 2008; Paria, 2009) generation ECAY maps, more than 60 new microsatellite markers have been identified (our unpublished data). Disappointingly, genotyping these markers on more than 100 males from ~50 horse breeds has not revealed any variation. Similarly, no success has been achieved by searching MSY sequence variations at single nucleotide level. No polymorphic nucleotide sites were found by direct sequencing of five MSY regions with a total length of 1,887 base-pairs (bp) in 21 stallions from 14 breeds (Wallner et al., 2003), or by screening 14,300 bp of intronic regions for SNPs in another 52 male horses from 15 breeds (Lindgren et al., 2004). More recently, taking advantage of the BAC contig map of the entire ECAY (Paria, 2009), a systematic search for SNPs in MSY and the PAR was conducted (Oom do Mar et al., 2008). From BAC end sequences, 54 STSs were generated collectively covering ~30,000 bp. The STSs were sequenced in 39 male horses and ponies representing 11 diverse breeds. Similarly to previous studies, no polymorphic sites were found.

While all efforts have failed to detect MSY polymorphism in the majority of European, Asian, and North American horse breeds, a single polymorphic MSY marker was recently detected in indigenous Chinese horses (Ling et al., 2010). Furthermore, recent studies using DNA from eight ancient wild horses and one 2,800-year-old domesticated horse found considerable level of ancestral diversity in MSY sequences (Lippold et al., 2011). The loss of this diversity in modern horse breeds has been attributed to a strong sex bias in horse domestication during which only a limited number of patrilines contributed to the genetic structure of modern horse breeds (Lindgren et al., 2004).

Y chromosome in Other Equids and Perissodactyls

Information about the Y chromosome in other equids and Perissodactyls is limited to cytogenetic analysis and to a few studies where ECAY markers have been used to analyze their Y chromosome in other equids.

Cytogenetics

The size, morphology, and C-banding pattern of the Y chromosome are very similar between all extant equids: horses, asses, onagers, and zebras (Ryder et al., 1978; Houck et al., 1998; Raudsepp et al., 2000; O'Brien, 2006). In a majority of the species it is a small sub-metacentric chromosome, except Grevy's zebra where the Y is metacentric (Musilova et al., 2007), indicating a possible inversion during the evolution of this equid lineage. In tapirs the Y chromosome is medium or small acrocentric (Houck et al., 2000) and in white rhinoceros it is a sub-metacentric, albeit with a distinct C-banding pattern (Houck et al., 1994).

Cross-species molecular hybridizations using microdissected or flow-sorted ECAY as a probe show considerable degree of sequence similarity between horse and donkey (Raudsepp & Chowdhary, 1999), horse and onager (Trifonov et al., 2008), and horse and Hartmann's zebra (Yang et al., 2003) Y chromosomes. Notably, ECAY repeat sequences are not strictly Y-specific in the donkey and hybridize to the majority of autosomal centromeres (Raudsepp & Chowdhary, 1999).

Genes and mapping

Donkey (*Equus asinus*, EAS) is probably the only equid where horse Y chromosome markers have been used for mapping and gene expression analysis (Raudsepp et al., 2008; Paria, 2009; Paria et al., 2011). According to these limited data, ECAY and EASY have similar DNA sequence, gene content, gene expression profiles, and share the duplication between MSY and the PAR (described earlier in this chapter).

Polymorphism

While the search for Y chromosome sequence polymorphism among horse breeds has been disappointing, ECAY markers show considerable individual variation in almost all other equid species, including the Przewalski's Horse, and the only other domesticated equid, the donkey (Wallner et al., 2003; Lindgren et al., 2004; Wallner et al., 2004; Oom do Mar et al., 2008). This indicates that low level of Y chromosome sequence variability is not a general feature among equids. It is also noteworthy that while recent genome sequence analysis did not recognize the domestic horse and the Przewalski's Horse as discrete species (Lau et al., 2009; Wade et al., 2009), Y chromosome data indicate clear divergence between the two equids (Wallner et al., 2003; Lindgren et al., 2004).

Concluding Remarks

Steady progress in horse Y chromosome mapping, gene discovery, and functional analysis have established an important foundation for the study of Y-linked component of stallion fertility and for complete sequencing of this chromosome in horses, thus adding the missing component to the genome sequence data. Information about the molecular structure and function of ECAY presents also the "Perissodactyl contribution" to our knowledge about mammalian sex chromosomes.

Acknowledgments

Horse Y chromosome research has been supported by the LINK Endowment, Texas AgriLife Research, Grayson-Jockey Club Research Foundation, American Quarterhorse Research Foundation, and U.S. Department of Agriculture grant # 2007-35604-17946.

References

Abe, S. et al. (1999). Deletion of the Sry region on the Y chromosome detected in a case of equine gonadal hypoplasia (XY female) with abnormal hormonal profiles. *Equine Veterinary Journal*, *31*, 336–338.

Blaschke, R. J., & Rappold, G. (2006). The pseudoautosomal regions, SHOX and disease. *Current Opinions in Genetic Development*, *16*, 233–239.

Bollongino, R. et al. (2008). Y-SNPs do not indicate hybridisation between European aurochs and domestic cattle. *PLoS One*, *3*, e3418.

Bowling, A. T. et al. (1987). An update of chromosomal abnormalities in mares. *Journal of Reproductive Fertility Supplement*, *35*, 149–155.

Buckland, R. A. et al. (1976). Characterization of the domestic horse (Equus caballus) karyotype using G- and C-banding techniques. *Experientia, 32*, 1146–1149.

Bugno, M. et al. (2003). A sporadic case of the sex-reversed mare (64,XY; SRY-negative): Molecular and cytogenetic studies of the Y chromosome. *Theriogenology, 59*, 1597–1603.

Bugno, M. et al. (2007). Karyotype evaluation among young horse populations in Poland. *Schweiz Arch Tierheilkd, 149*, 227–232.

Bugno-Poniewierska, M. et al. (2009). Modification of equine sperm chromatin decondensation method to use fluorescence in situ hybridization (FISH). *Folia of Histochemical Cytobiology, 47*, 663–666.

Burgoyne, P. S. et al. (1986). Spermatogenic failure in male mice lacking H-Y antigen. *Nature, 320*, 170–172.

Chowdhary, B. P. et al. (2002). Construction of a 5000_{rad} whole-genome radiation hybrid panel in the horse and generation of a comprehensive and comparative map for ECA11. *Mammalian Genome, 13*, 89–94.

Cocquet, J. et al. (2009). The multicopy gene Sly represses the sex chromosomes in the male mouse germline after meiosis. *PLoS Biology, 7*, e1000244.

Cribiu, E. P., & De Giovanni, A. (1978). Le caryotype du cheval domestique (Equus caballus), de l'ane (Equus asinus) et du mullet par le méthode de bande C. *Annales de Génétique et Sélection Animale, 10*, 161–170.

Das, P. J. et al. (2009). Characterization of the bovine pseudoautosomal region and comparison with sheep, goat, and other mammalian pseudoautosomal regions. *Cytogenetic Genome Research, 126*, 139–147.

Del Mastro, R. G., & Lovett, M. (1997). Isolation of coding sequences from genomic regions using direct selection. *Methods in Molecular Biology, 68*, 183–199.

Dunn, H. O. et al. (1981). Two equine true hermaphrodites with 64,XX/64,XY and 63,XO/64,XY chimerism. *Cornell Veterinary, 71*, 123–135.

Ellis, N., & Goodfellow, P. N. (1989). The mammalian pseudoautosomal region. *Trends in Genetics, 5*, 406–410.

Ellis, P. J., & Affara, N. A. (2006). Spermatogenesis and sex chromosome gene content: An evolutionary perspective. *Human Fertility, 9*, 1–7.

Ellis, P. J. et al. (2007). Bidirectional transcription of a novel chimeric gene mapping to mouse chromosome Yq. *BMC Evolutionary Biology, 7*, 171.

Fisher, R. A. (1931). The evolution of dominance. *Biology Review, 6*, 345–368.

Flaquer, A. et al. (2008). The human pseudoautosomal regions: A review for genetic epidemiologists. *European Journal of Human Genetics, 16*, 771–779.

Ford, C. E. et al. (1980). Proceedings of the First International Conference for the Standardisation of Banded Karyotypes of Domestic Animals. University of Reading Reading, England, August 2–6, 1976. *Hereditas, 92*, 145–162.

Galtier, N. (2004). Recombination, GC-content and the human pseudoautosomal boundary paradox. *Trends in Genetics, 20*, 347–349.

Geraldes, A., & Ferrand, N. (2006). A 7-bp insertion in the 3′ untranslated region suggests the duplication and concerted evolution of the rabbit SRY gene. *Genetics & Selective Evolution, 38*, 313–320.

Gianfrancesco, F. et al. (2001). Differential divergence of three human pseudoautosomal genes and their mouse homologs: implications for sex chromosome evolution. *Genome Research, 11*, 2095–2100.

Ginja, C. et al. (2010). Origins and genetic diversity of New World Creole cattle: inferences from mitochondrial and Y chromosome polymorphisms. *Animal Genetics, 41*, 128–141.

Graves, J. A. (2006). Sex chromosome specialization and degeneration in mammals. *Cell, 124*, 901–914.

Graves, J. A. (2010). Review: Sex chromosome evolution and the expression of sex-specific genes in the placenta. *Placenta, 31*(Suppl), S27–S32.

Graves, J. A. (2002). Evolution of the human X – a smart and sexy chromosome that controls speciation and development. *Cytogenetic Genome Research, 99*, 141–145.

Grzmil, P. et al. (2007). The influence of the deletion on the long arm of the Y chromosome on sperm motility in mice. *Theriogenology, 67*, 760–766.

Hageltorn, M., & Gustavsson, I. (1974). The application of new staining techniques in the identification of individual chromosome pairs in domestic animals. *1st World Congress on Genetics Applied to Livestock Production, October 7–11, Madrid, Spain* (pp. 203–211).

Hansen, K. M. (1984). The relative length of Q-band/Giemsa stained horse chromosomes. *The 6th European Colloquium on Cytogenetics of the Domestic Animals* (pp. 165–171).

Hasegawa, T. et al. (1999). Linear SRY transcript in equine testis. *Journal of Veterinary Medical Science, 61*, 97–100.

Hasegawa, T. et al. (2000). Sex determination by simultaneous amplification of equine SRY and amelogenin genes. *Journal of Veterinary Medical Science, 62*, 1109–1110.

Herzog, A. et al. (1989). Gonosomenmosaik beim maskulinen Pseudohermaphroditismus des Pferdes. *Tierärztliche Praxis, 17*, 171–175.

Hirota, K. et al. (2001). FISH assignment of two equine BAC clones containing SRY and ZFY. *Animal Genetics, 32*, 326–327.

Houck, M. L. et al. (1994). Diploid chromosome number and chromosomal variation in the white rhinoceros (Ceratotherium simum). *Journal of Heredity, 85*, 30–34.

Houck, M. L. et al. (1998). Chromosomal rearrangements in a Somali wild ass pedigree, Equus africanus somaliensis (Perissodactyla, Equidae). *Cytogenetic Cell Genetics, 80*, 117–122.

Houck, M. L. et al. (2000). Comparative cytogenetics of tapirs, genus tapirus (Perissodactyla, tapiridae). *Cytogenetic Cell Genetics, 89*, 110–115.

Hughes, J. F. et al. (2005). Conservation of Y-linked genes during human evolution revealed by comparative sequencing in chimpanzee. *Nature, 437*, 100–103.

Hughes, J. F. et al. (2010). Chimpanzee and human Y chromosomes are remarkably divergent in structure and gene content. *Nature, 463*, 536–539.

Hughes, J. P., & Trommershausen-Smith, A. (1977). Infertility in the horse associated with chromosomal abnormalities. *Australian Veterinary Journal, 53*, 253–257.

ISCNH, Bowling, A. T., Breen, M., Chowdhary, B. P., Hirota, K. et al. (1997). Report of the Third International Committee for the Standardization of the domestic horse karyotype, Davis, CA, 1996. *Chromosome Research, 5*, 433–443.

Juras R. et al. (2010). XX/XY blood lymphocyte chimerism in heterosexual dizygotic American Bashkir Curly horse twins. *Journal of Equine Veterinary Science, 30*, 575–580.

Kayser, M. (2010). The human genetic history of Oceania: Near and remote views of dispersal. *Current Biology, 20*, R194–R201.

Kent, M. G. et al. (1988a). Concentration of serum testosterone in XY sex reversed horses. *Journal of Endocrinological Investigation, 11*, 609–613.

Kent, M. G. et al. (1988b). XY sex reversal syndrome in the mare: Clinical and behavioral studies, H-Y phenotype. *Human Genetics, 79*, 321–328.

Kent, M. G. et al. (1986). XY sex-reversal syndrome in the domestic horse. *Cytogenetic Cell Genetics, 42*, 8–18.

Kieffer, N. M. (1976). Male pseudohermaphroditism of the testicular feminizing type in a horse. *Equine Veterinary Journal, 8*, 38–41.

Kirsch, S. et al. (2005). Interchromosomal segmental duplications of the pericentromeric region on the human Y chromosome. *Genome Research, 15*, 195–204.

Klunder, L. R. et al. (1990). Six separate sex chromosome anomalies in an Arabian mare. *Equine Veterinary Journal, 22*, 218–220.

Kubickova, S. et al. (2002). The use of laser microdissection for the preparation of chromosome-specific painting probes in farm animals. *Chromosome Research, 10*, 571–577.

Kubien, E. M. et al. (1993). Clinical, cytogenetic and endocrine evaluation of a horse with a 65,XXY karyotype. *Equine Veterinary Journal, 25*, 333–335.

Kuroki, Y. et al. (2006). Comparative analysis of chimpanzee and human Y chromosomes unveils complex evolutionary pathway. *Natural Genetics, 38*, 158–167.

Lahn, B. T., & Page, D. C. (1997). Functional coherence of the human Y chromosome. *Science, 278*, 675–680.

Lahn, B. T., & Page, D. C. (1999). Four evolutionary strata on the human X chromosome. *Science, 286*, 964–967.

Lange, J. et al. (2009). Isodicentric Y chromosomes and sex disorders as byproducts of homologous recombination that maintains palindromes. *Cell, 138*, 855–869.

Langford, C. F. et al. (1993). Preparation of chromosome-specific paints and complete assignment of chromosomes in the pig flow karyotype. *Animal Genetics, 24*, 261–267.

Langford, C. F. et al. (1996). Chromosome-specific paints from a high-resolution flow karyotype of the dog. *Chromosome Research, 4*, 115–123.

Lau, A. N. et al. (2009). Horse domestication and conservation genetics of Przewalski's horse inferred from sex chromosomal and autosomal sequences. *Molecular Biology Evolution, 26*, 199–208.

Lear, T. L., & Bailey, E. (2008). Equine clinical cytogenetics: The past and future. *Cytogenetic Genome Research, 120*, 42–49.

Lindgren, G. et al. (2004). Limited number of patrilines in horse domestication. *Natural Genetics, 36*, 335–336.

Ling, Y. et al. (2010). Identification of Y chromosome genetic variations in Chinese indigenous horse breeds. *The Journal of Heredity, 101*, 639–643.

Lippold, S. et al. (2011). Discovery of lost diversity of paternal horse lineages using ancient DNA. *Nature Communications, 2*, 450.

Makinen, A. et al. (1999). Infertility in two mares with XY and XXX sex chromosomes. *Equine Veterinary Journal, 31*, 346–349.

Makinen, A. et al. (2000). Two sterile stallions with XXY-syndrome. *Equine Veterinary Journal, 32*, 358–360.

Mangs, A., & Morris, B. J. (2007). The human pseudoautosomal region (PAR): Origin, function and future. *Current Genomics, 8*, 129–136.

Manz, E. et al. (1998). Identification of an equine Y chromosome specific gene locus (eTSPY) with potential in preimplantation sex diagnosis. *Theriogenology*, *49*, 364.

Marshall Graves, J. A. (2008). Weird animal genomes and the evolution of vertebrate sex and sex chromosomes. *Annual Review of Genetics*, *42*, 565–586.

McElreavey, K., & Mitchell, M. J. (2002). The sexually transmitted genome. *Trends in Genetics*, *18*, 13–14.

McElreavey, K. et al. (2006). Y chromosome variants and male reproductive function. *International Journal of Andrology*, *29*, 298–303.

McIlwraith, C. W. et al. (1976). An equine cryptorchid with testicular and ovarian tissues. *Equine Veterinary Journal*, *8*, 156–160.

Meadows, J. R. et al. (2004). Nucleotide diversity on the ovine Y chromosome. *Animal Genetics*, *35*, 379–385.

Milenkovic, D. et al. (2002). Cytogenetic localization of 136 genes in the horse: Comparative mapping with the human genome. *Mammalian Genome*, *13*, 524–534.

Miyake, Y. I. et al. (1982). Chromosomal analysis and blood type examination of multiple births in equine. *Japanese Journal of Veterinary Research*, *30*, 11–18.

Murphy, W. J. et al. (2006). Novel gene acquisition on carnivore Y chromosomes. *PLoS Genetics*, *2*, e43.

Musilova, P. et al. (2007). Karyotypic relationships among Equus grevyi, Equus burchelli and domestic horse defined using horse chromosome arm-specific probes. *Chromosome Research*, *15*, 807–813.

O'Brien, S. J., Menninger, J. C., & Nash, W. G. (2006). *Atlas of mammalian chromosomes*. Hoboken, NJ: Wiley & Sons.

Ohno, S. (1967). *Sex chromosomes and sex linked genes*. New York: Springer-Verlag.

Oom do Mar, M. et al. (2008). Single Nucleotide Polymorphism (SNP) discovery from the horse Y chromosome (ECAY). *Proceedings of the XXI Conference of the International Society of Animal Genetics, July 20–24, Amsterdam, The Netherlands*, 2156.

Padula, A. M. (2005). The freemartin syndrome: an update. *Animal Reproductive Science*, *87*, 93–109.

Pailhoux, E. et al. (1995). Molecular analysis of an XY mare with gonadal dysgenesis. *Hereditas*, *122*, 109–112.

Palmer, S. et al. (1997). A gene spans the pseudoautosomal boundary in mice. *Proceedings of the National Academy of Sciences USA*, *94*, 12030–12035.

Paria, N. (2009). *Discovery of candidate genes for stallion fertility*. College Station: College of Veterinary Medicine, Texas A&M University.

Paria, N. et al. (2011). A gene catalogue of the euchromatic male-specific region of the horse Y chromosome: Comparison with human and other mammals. *PLoS One*, *6*, e21374.

Pearks Wilkerson, A. J. et al. (2008). Gene discovery and comparative analysis of X-degenerate genes from the domestic cat Y chromosome. *Genomics*, *92*, 329–338.

Perry, J. et al. (2001). A short pseudoautosomal region in laboratory mice. *Genome Research*, *11*, 1826–1832.

Power, M. M. (1988). Y chromosome length variation and its significance in the horse. *Journal of Heredity*, *79*, 311–313.

Power, M. M. et al. (1992). Synaptonemal complex analysis of an autosomal trisomy in the horse. *Cytogenetic Cell Genetics*, *61*, 202–207.

Power, M. M., & Leadon, D. P. (1990). Diploid-triploid chimaerism (64, XX/96,XXY) in an intersex foal. *Equine Veterinary Journal*, *22*, 211–214.

Quintana-Murci, L., & Fellous, M. (2001). The human Y chromosome: The biological role of a "functional wasteland." *Journal of Biomedical Biotechnology*, *1*, 18–24.

Raudsepp, T., & Chowdhary, B. P. (1999). Construction of chromosome-specific paints for meta- and submetacentric autosomes and the sex chromosomes in the horse and their use to detect homologous chromosomal segments in the donkey. *Chromosome Research*, *7*, 103–114.

Raudsepp, T., & Chowdhary, B. P. (2008). The horse pseudoautosomal region (PAR): Characterization and comparison with the human, chimp and mouse PARs. *Cytogenetic Genome Research*, *121*, 102–109.

Raudsepp, T. et al. (2000). Cytogenetics of donkey chromosomes: Nomenclature proposal based on GTG-banded chromosomes and depiction of NORs and telomeric sites. *Chromosome Research*, *8*, 659–670.

Raudsepp, T. et al. (2004). A detailed physical map of the horse Y chromosome. *Proceedings of the National Academy of Sciences USA*, *101*, 9321–9326.

Raudsepp, T. et al. (2008). Sequencing-ready physical map of the horse Y chromosome. *XXXI Conference of the International Society for Animal Genetics (ISAG), July 20–24, Amsterdam, The Netherlands*, P2107.

Raudsepp, T. et al. (2010a). Molecular heterogeneity of XY sex reversal in the horse. *Animal Genetics Special Issue on Horse Genetics and Genomics*. Animal Genetics 41 Suppl 2:41–52.

Raudsepp, T. et al. (2010b). The role of Y chromosome and PAR genes in male and female fertility in horses. *Plant & Animal Genome XVIII, January 9–13, 2010, San Diego, CA, USA*, W549.

Raudsepp, T. et al. (2012). Genomic copy number variation in horses with disorders of sexual development. *Plant & Animal Genome XX, January 14–18, San Diego, CA.*

Romagnano, A. et al. (1987). Idiograms of horse chromosomes at prometaphase, early metaphase, and midmetaphase after R-banding by BrdU incorporation followed by the fluorochrome-photolysis-Giemsa technique. *Genome*, *29*, 674–679.

Ronne, M. et al. (1993). The RBG-banded karyotype of Equus caballus at the 525-band stage. *Hereditas*, *118*, 195–199.

Ross, M. T. et al. (2005). The DNA sequence of the human X chromosome. *Nature*, *434*, 325–337.

Rubes, J. et al. (2009). Fluorescence in situ hybridization applied to domestic animal cytogenetics. *Cytogenetic Genome Research*, *126*, 34–48.

Ryder, O. A. et al. (1978). Chromosome banding studies of the Equidae. *Cytogenetic Cell Genetics*, *20*, 332–350.

Safronova, L. D., & Pimenova, T. I. (1988). Karyotypes of cattle (Bos taurus) and horses (Equus caballus) on the basis of synaptonemal complexes. *Genetika*, *24*, 708–714.

Scott, I. S., & Long, S. E. (1980). An examination of chromosomes in the stallion (Equus caballus) during meiosis. *Cytogenetic Cell Genetics*, *26*, 7–13.

Senese, C. et al. (1999). A HaeIII PCR-RFLP in the ZFY/ZFX genes of horses. *Animal Genetics*, *30*, 390–391.

Shiue, Y. L. et al. (2000). Synteny and regional marker order assignment of 26 type I and microsatellite markers to the horse X- and Y-chromosomes. *Chromosome Research*, *8*, 45–55.

Sinclair, A. H. et al. (1990). A gene from the human sex-determining region encodes a protein with homology to a conserved DNA-binding motif. *Nature*, *346*, 240–244.

Skaletsky, H. et al. (2003). The male-specific region of the human Y chromosome is a mosaic of discrete sequence classes. *Nature*, *423*, 825–837.

Sorensen, K. M. et al. (2007). Whole genome amplification on DNA from filter paper blood spot samples: An evaluation of selected systems. *Genetic Testing*, *11*, 65–71.

Stoneking, M., & Delfin, F. (2010). The human genetic history of East Asia: Weaving a complex tapestry. *Current Biology*, *20*, R188–R193.

Switonski, M. et al. (2005). Sex reversal syndrome (64,XY; SRY-positive) in a mare demonstrating masculine behaviour. *Journal of Animal Breeding Genetics*, *122*(Suppl 1), 60–63.

Telenius, H. et al. (1992). Cytogenetic analysis by chromosome painting using DOP-PCR amplified flow-sorted chromosomes. *Genes Chromosomes Cancer*, *4*, 257–263.

Tiepolo, L., & Zuffardi, O. (1976). Localization of factors controlling spermatogenesis in the nonfluorescent portion of the human Y chromosome long arm. *Human Genetics*, *34*, 119–124.

Toure, A. et al. (2004a). A protein encoded by a member of the multicopy Ssty gene family located on the long arm of the mouse Y chromosome is expressed during sperm development. *Genomics*, *83*, 140–147.

Toure, A. et al. (2004b). A new deletion of the mouse Y chromosome long arm associated with the loss of Ssty expression, abnormal sperm development and sterility. *Genetics*, *166*, 901–912.

Toure, A. et al. (2005). Identification of novel Y chromosome encoded transcripts by testis transcriptome analysis of mice with deletions of the Y chromosome long arm. *Genome Biology*, *6*, R102.

Trifonov, V. A. et al. (2008). Multidirectional cross-species painting illuminates the history of karyotypic evolution in Perissodactyla. *Chromosome Research*, *16*, 89–107.

Trommershausen-Smith, A. et al. (1979). Cytogenetic and clinical findings in mares with gonadal dysgenesis. *Journal of Reproductive Fertility* (Suppl), 271–276.

Turner, M. E. et al. (2007). Genomic and expression analysis of multiple Sry loci from a single Rattus norvegicus Y chromosome. *BMC Genetics*, *8*, 11.

Urbach, A., & Benvenisty, N. (2009). Studying early lethality of 45,XO (Turner's syndrome) embryos using human embryonic stem cells. *PLoS One*, *4*, e4175.

Van Laere, A. S. et al. (2008). Characterization of the bovine pseudoautosomal boundary: Documenting the evolutionary history of mammalian sex chromosomes. *Genome Research*, *18*, 1884–1895.

Villagomez, D. A. et al. (2009). Classical and molecular cytogenetics of disorders of sex development in domestic animals. *Cytogenet Genome Research*, *126*, 110–131.

von Nett, P. et al. (1996). Neue Erkenntnisse uber die Chromosomenpaarung in der Meiose. *Deutsch Tierarztl Wschr*, *103*, 369–448.

Wade, C. M. et al. (2009). Genome sequence, comparative analysis, and population genetics of the domestic horse. *Science*, *326*, 865–867.

Wallner, B. et al. (2003). Fixed nucleotide differences on the Y chromosome indicate clear divergence between Equus przewalskii and Equus caballus. *Animal Genetics*, *34*, 453–456.

Wallner, B. et al. (2004). Isolation of Y chromosome-specific microsatellites in the horse and cross-species amplification in the genus Equus. *Journal of Heredity*, *95*, 158–164.

Yamauchi, Y. et al. (2009). Live offspring from mice lacking the Y chromosome long arm gene complement. *Biology of Reproduction*, *81*, 353–361.

Yang, F. et al. (2003). Karyotypic relationships of horses and zebras: Results of cross-species chromosome painting. *Cytogenetic Genome Research*, *102*, 235–243.

Yang, F. et al. (2004). Refined genome-wide comparative map of the domestic horse, donkey and human based on cross-species chromosome painting: Insight into the occasional fertility of mules. *Chromosome Research*, *12*, 65–76.

Yerle, M. et al. (1993). Accurate characterization of porcine bivariate flow karyotype by PCR and fluorescence in situ hybridization. *Genomics*, *16*, 97–103.

Young, A. C. et al. (2008). Tackling the characterization of canine chromosomal breakpoints with an integrated in-situ/in-silico approach: The canine PAR and PAB. *Chromosome Research*, *16*, 1193–1202.

5 Unexpected structural features of the equine major histocompatibility complex

Loren C. Skow and Candice L. Brinkmeyer-Langford

The major histocompatibility complex (MHC) is arguably one of the most interesting and biologically puzzling regions of the vertebrate genome. On the one hand, the MHC contains the most polymorphic genes in the vertebrate genome, but allelic differences in these genes predate speciation to generate trans-species polymorphisms (Klein, 1987). Sequence analysis of the definitive antigen-presenting genes provide strong evidence for balancing selection in most populations (reviewed by Hedrick, 1999), but some populations apparently thrive with an essentially monomorphic MHC (McGuire et al., 1985; Nizitec et al., 1988). The vast amount of sequencing data now available for many different vertebrate species is providing a detailed evolutionary perspective to MHC diversity and promises insights into the functional significance of such diversification. The recent completion of the equine genome project significantly contributes to the comparative genomics of the MHC as the horse is the only representative of the order *Perissodactyla* for which a whole genomic sequence (WGS) is presently available. This review discusses the organizational features of the equine leucocyte antigen (ELA) complex as revealed by the assembly of WGS data with special emphasis on those features that appear to be unique to the *Perissodactyl* MHCs.

Organization and Gene Content of the Model MHC

To consider the biological significance of organizational diversity in the MHC, one must first understand the general properties of this region of the vertebrate genome. The consensus MHC is one of the most gene-dense regions of vertebrate genomes, containing more than 230 translatable genes distributed over approximately 3 Mb of DNA. These genes appear to be syntenic in all vertebrate taxa except the bony fishes (*Osteoichthys*), where MHC homologous genes are distributed over four different chromosomes (Sato et al., 2000). Since genes in the MHCs of cartilaginous fishes are also syntenic (Ohta et al., 2000), the disrupted organization in the MHCs of bony fishes is thought to be a derived state related to ancestral tetraploidy. Many of the translated genes in the MHC encode proteins of immunological function, the definitive genes of which encode cell surface receptors that present antigenic polypeptides to CD8+ cytotoxic T lymphocytes (MHC class I receptors) or CD4+ helper T lymphocytes (MHC class II receptors). Proteins encoded at class I and class II loci participate in the initial stages of the acquired immune response and play pivotal roles in self:non-self discrimination of polypeptides. Not surprisingly, genetic variations in the MHC are associated with predispositions to numerous disease and parasite susceptibilities, autoimmune disorders, and

Equine Genomics, First Edition. Edited by Bhanu P. Chowdhary.

other immune phenotypes (see review by Trowsdale, 2011). Genetic variation in the human HLA has been associated with more than 200 different diseases (de Bakker et al., 2006).

Based on the "mammalian model," the typical organizational structure of the MHC consists of two regions of antigen presenting genes, designated class I and class II, that bracket a third recognizable region of highly conserved genes termed the class III region. Class III genes are functionally heterogeneous with many functions that are not apparently related to immune response. If one defines the boundaries of the consensus MHC by the gamma-aminobutyric acid receptor 1 (GABBR1) locus at the terminus of the class I region and by the antigenic peptide transporter binding protein, tapasin (TAPBP) locus at the terminus of the class II region, most MHCs comprise approximately 3.6 Mb of DNA and contain about 230 translatable genes. Recently it has become appropriate to also consider the MHC extended regions when investigating MHC-related phenotypes (Walsh et al., 2003; Lie & Thorsby, 2005).

Some MHC Genes Are Highly Polymorphic

The "classical" class I and the class II receptors are the most polymorphic proteins yet to be identified, with more than 6,000 alleles demonstrated for the human HLA class I genes (Tait et al., 2011). Most of the nucleotide substitutions that give rise to the protein polymorphisms are concentrated in codons that line the antigen binding domains of both class I and class II receptors. Analysis of the ratios of non-synonymous: synonymous nucleotide substitutions in the exons of both classical (Ia) class I and the class II genes reveals that the codons involved in antigen presentation are subject to balancing selection (ns:s > 1) whereas nucleotide substitutions in the other regions of the cds demonstrate ns:s < 1, as expected for genes under purifying selection (Harris & Myer, 2006). Most MHCs appear to contain 15–20 class I genes, of which only three or four meet the characteristics of classical antigen presenting genes. The remaining class I genes are designated nonclassical (Ib) and show tissue-specific expression, have low levels of polymorphism and no evidence for overdominance. An ns:s < 1 across the antigen-binding domains of these genes suggest they are under purifying selection. The functional role(s) of nonclassical class I genes remain to be fully defined, but presumably these genes encode receptors with high epitope specificity or have acquired functions other than antigen presentation in the acquired immune response.

The Molecular Map of the Equine Leucocyte Antigen Complex

The original genetic characterizations of ELA were based largely on serological phenotypes, using polyclonal antisera raised during pregnancy (Lazary et al., 1988). This approach identified 19 serologic specificities in ELA. It is apparent that the early serological methods primarily identified genetic variation in the ELA class I region as pregnancy-induced antibodies in horses are developed against paternal class I receptors expressed on trophoblasts of the chorionic girdle (Donaldson et al., 1990). A more complete development of the genetic map of ELA, and of the horse genome in general, proceeded rapidly beginning in the 1990s, in large part due to the new molecular biology technologies and the spinoffs from the human genome project. A singularly significant event in horse genomics occurred in 1995 with the establishment of the First International Equine Gene Mapping Workshop held in Lexington, Kentucky, under the aegis of the Dorothy Russell Havemeyer Foundation (Bailey,

2010). This international consortium has continued to meet in alternate years, supported by The Havemeyer Foundation, and has produced a rich collection of genomics tools to produce a deep and detailed integrated genetic map of the horse genome sufficient to support a genome sequencing project. These tools include gene-mapping populations (Guerin et al., 1999, 2003; Swinburne et al., 2000; Perrocheau et al., 2006), microsatellite marker panels (Shiue et al., 1999), BAC libraries and radiation hybrid libraries (Chowdhary et al., 2002, 2003), and integrated molecular cytogenetic maps (Raudsepp et al., 2008). In 2006, the Broad Institute initiated sequencing of the genome of a Thoroughbred mare, Twilight, and published the assembled genome sequence in 2009 (Wade et al., 2009). A 54K equine SNP chip panel was subsequently designed (McCue, 2012) for whole genome association studies, and the genomics community can anticipate a wealth of additional sequence data from other breeds (i.e., Doan et al., 2012) in the near future that will reveal much about the evolution of this species.

Based on the Eca version 2.2 assembly of the equine genome and homologies with HLA and HLA extended sequences, the ELA complex spans approximately 6 Mb of DNA (Ecab 20: 27,403 526..33,865,081) near the center of chromosome 20. Annotation of this region predicted approximately more than 230 genes and identified 4,392 SNPs. A comparison of sequence-based maps of HLA (version 36) and ELA (version 2.2) is presented in Figure 5.1.

In many ways, the alignment of ELA sequence with HLA demonstrated a great deal of organizational similarity and conserved synteny as expected (Gustafson et al., 2003). Sequence alignments also confirmed earlier mapping studies that found ELA class I sequences distributed in three clusters, including two class I loci within the class II region (Gustafson et al., 2003; Tallmadge et al., 2005). However, noteworthy differences were predicted. One of the striking results from the initial assembly of the ELA and extended regions from the equine genome project was that the size of the region was approximately 6.0 Mb in size, almost 1.3 Mb larger than the human HLA and extended regions. Furthermore, annotation of the ELA sequence identified 40 class I loci, many more than expected from serology, analysis of BAC clones, and gene expression studies (Ellis et al., 1995; Tallmadge et al., 2005, 2010).

Closer examination of the assembled sequence of ELA revealed that the increased size of ELA was largely due to two features apparently unique to the MHCs of horses when compared to other sequenced mammalian taxa. One feature is a gene desert of approximately 550 Kb at the boundary of the class II and class III regions from coordinates 20: 31,896,104..32,442,400 (Figure 5.1). This region contains a single annotated gene, C6orf10, also called testis-specific binding protein (TSBP), and two pseudogenes, one related to a tetraspanin 17-like sequence and the other to a Spi-c-like transcription factor. The chromosomal position and sequence homology of ELA C6orf10 to human C6orf10 indicates that these are orthologous loci. Alignment of the balance of the ELA desert sequence with whole genomic sequences of other species identified no comparable sequence at any location in any other species.

The second distinctive genomic feature in ELA is a large and strikingly conserved segmental duplication of at least 11 units, each about 45Kb in size, at the boundary of the ELA class I and III regions (Figure 5.2) (Brinkmeyer-Langford et al., 2010). Each unit in the segmental duplication contains one sequence related to a truncated form of the B-associated transcript 1 (Bat1) that aligns to the c-terminal domain of the helicase domain, and a second sequence with strong homology to class I sequences. The Bat-1 and class I sequences are regularly interspersed and arrayed largely in head to tail arrangement throughout the segmental duplication. The Bat-1 sequences are extraordinarily conserved within *Equus caballus* and among other *Perissodactyl* species (Brinkmeyer-Langford et al., 2010), arguing for some functional role(s) for these sequences. Twenty-four genes are predicted

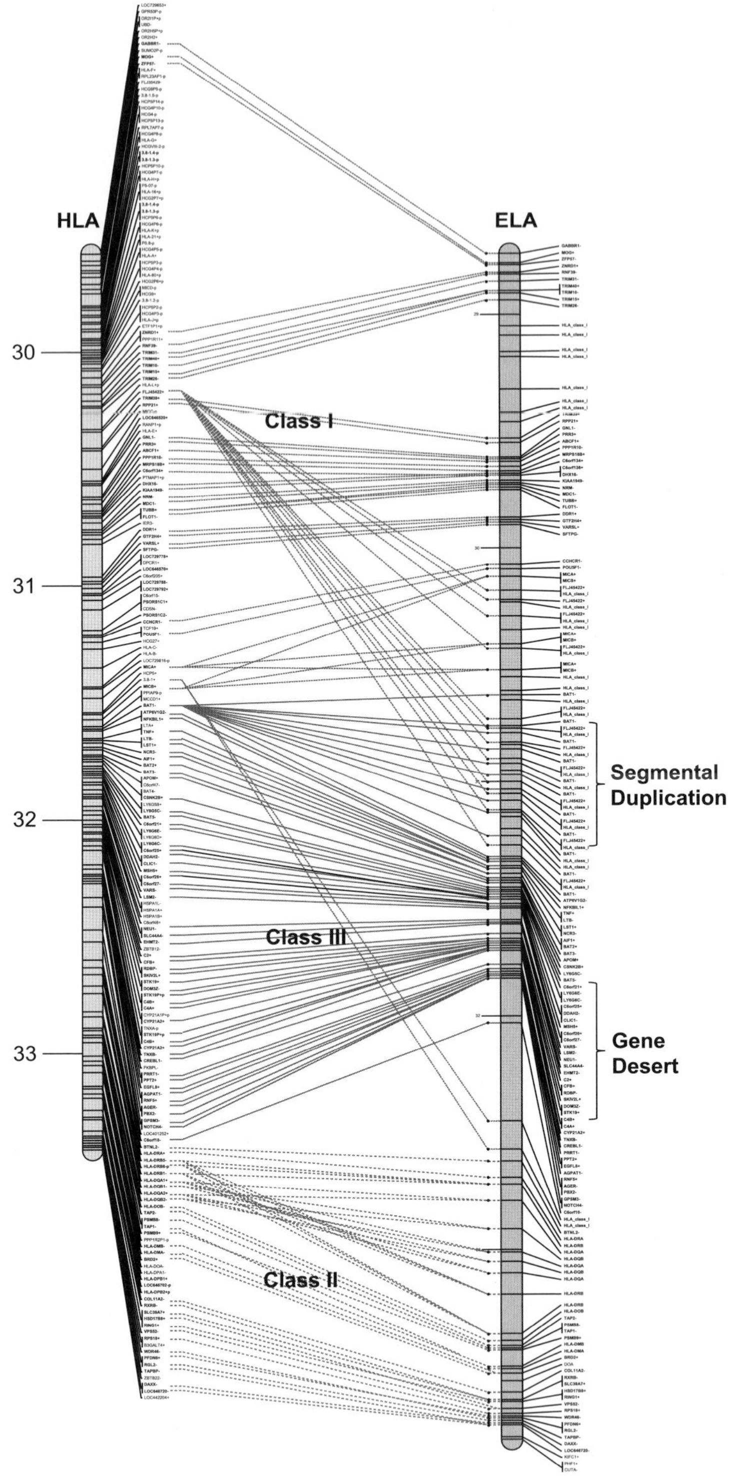

Figure 5.1 Alignment of genes between HLA and ELA based on best BLAST alignment of the horse sequence to the human sequence showing the locations of the Segmental Duplication and the Gene Desert. Class I and class II orthologues are difficult to identify across species, as evidenced by the several intersecting lines between genes in the class I and the class II regions. (*For color details, see color plate section.*)

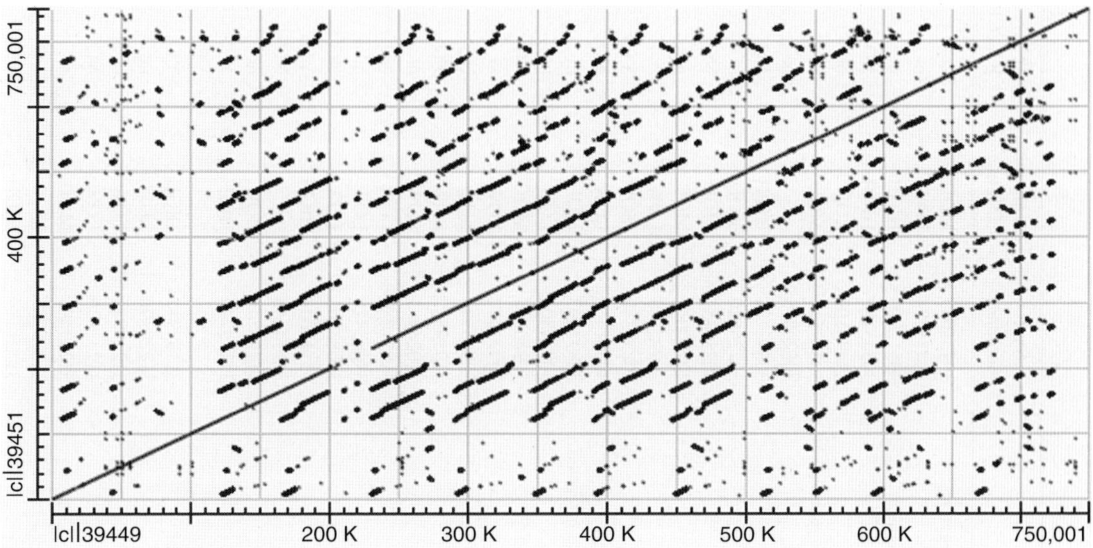

Figure 5.2 Dotplot of repeat-masked sequence in the ELA segmental duplication region on ECA20 (NW_001867389.1; positions 30600000-31350000bp) aligned against itself using the bl2seq option of the NCBI BLAST website. At least 11 regions contain related sequences generally aligned head to tail. Three additional closely related sequences are found in the ChrUn database of the horse, indicating that 14 repeat units may reside within the segmental duplication.

from the version 2.0 assembly to be contained within the segmental duplication. Three closely related sequences are chromosomally unassigned in the 2.0 assembly, indicating that the segmental duplication may include 14 repeating units and 30 annotatable genes.

The remarkable conservation of sequences within the segmental duplication feature among different *Perissodactyl* MHCs suggests that this region would contain functional genes. Evidence for gene expression has been sought using reverse transcription PCR and chromatin immunoprecipition sequencing (ChIP seq).

RT-PCR

We used locus-specific RT-PCR to seek transcripts from nine of the class I genes and four of the Bat-1 like genes in the segmental duplication (Table 5.1; Brinkmeyer-Langford et al., 2010). Sequence similarities precluded design of locus-specific primers for the remaining predicted genes. Transcripts for five of the nine class I sequences were successfully amplified from peripheral WBCs of at least some of several horses tested, while no transcripts were detected for any of the four tested BAT1-like sequences. Transcripts from the full-length BAT-1 sequence were identified as expected. These results indicate that at least some of the class I genes within the segmental duplication are transcriptionally active, including two genes previously identified (Ellis et al., 1995; Tallmadge et al., 2005). Consequently, it seems that most of the predicted genes within the segmental duplication are not expressed as mRNAs, although it is possible that some of these genes may demonstrate tissue-specific expression profiles not assessed by these studies.

Table 5.1 Summary of predicted genes and transcripts within the segmental duplication interrogated for gene expression by H3K4me3 ChIPseq and reverse transcription PCR.

NCBI ID	Ensembl gene ID	Ensembl transcript ID	Similar to	Start coordinate	Neutrophil H3K4me3	WBC RT-PCR
LOC100054397	ENSECAG00000018221	ENSECAT00000019472	class I classical fragment	30431045	MOD	nt
LOC100054303	—	—	pseudogene	30475773	LOW	nt
LOC100054254	ENSECAG00000022961	ENSECAT00000024683	class I classical fragment	30512900	HIGH	nt
LOC100054114	ENSECAG00000008977	ENSECAT00000011275	class I classical	30545182	HIGH	NO
—	ENSECAG00000007227	ENSECAT00000007205	transposase 22	30595102	LOW	nt
LOC100050550	ENSECAG00000019474	ENSECAT00000021134	class I classical fragment	30609517	MOD	nt
—	—	—	BAT1.1	30624815		nt
LOC100050473	ENSECAG00000017324	ENSECAT00000018358	class I classical	30642724	HIGH	NO
LOC100053918	ENSECAG00000020384	ENSECAT00000022326	class I classical	30725620	MOD	nt
LOC100053813	—	—	BAT1.2	30759456	MOD	NO
LOC100053764	ENSECAG00000019095	ENSECAT00000020228	class I classical	30770041	MOD	nt
LOC100053668	—	—	BAT1.3	30794138	LOW	?
MHCB3	ENSECAG00000021785	ENSECAT00000023847	class I classical	30832340	MOD	YES
LOC100057453	—	—	BAT1.4 pseudogene	30861460	LOW	NO
LOC100057490	ENSECAG00000020991	ENSECAT00000023286	class I non-classical truncated	30893697	MOD	nt
—	—	—	BAT1.5	30921801	LOW	nt
LOC100057647	ENSECAG00000000732	ENSECAT00000004388	class I classical	30950844	MOD	NO
—	—	—	BAT1.6	30971942	LOW	nt
LOC100057688	—	—	Pseudogene	30973146	LOW	NO
LOC100146808	—	—	Pseudogene	31019542	LOW	NO
—	—	—	BAT1.7	31024481	LOW	nt
LOC100057776	ENSECAG00000012217	ENSECAT00000013970	hypothetical	31051786	LOW	NO
—	—	—	BAT1.8	31082428	LOW	nt
LOC100057869	—	—	hypothetical	31083562	LOW	NO
LOC100057920	ENSECAG00000024771	ENSECAT00000026764	class I classical	31114658	HIGH	YES
		(ENSECAT0000026772)	class I fragment	31114546	?	NO
—	—	—	BAT1.9	31123741	LOW	nt
EQMHCC1	ENSECAG00000019928	ENSECAT00000021999	class I classical	31149749	HIGH	YES
—	—	(ENSECAT00000022135)	class I classical	"	?	YES
LOC100058098	ENSECAG00000021750	ENSECAT00000024180	class I non-classical	31193941	HIGH	YES
—	ENSECAG00000013160	**ENSECAT00000013660**	**MCCD1**	31229396	LOW	nt
—	—	—	BAT1.10	31229974	LOW	nt
—	ENSECAG00000013184	ENSECAT00000013697	BAT1 like	31231136	LOW	NO
LOC100049899	ENSECAG00000000826	ENSECAT00000000655	C13orf37	31249495	MOD	nt
LOC1000147575	ENSECAG00000013783	ENSECAT00000014949	class I classical	31270334	MOD	NO
LOC100146384	ENSECAG00000007750	ENSECAT00000008265	BAT1 full length	31321114	HIGH	YES

Chromatin Modifications Associated with Transcription

Another approach to identifying regions of transcribed DNA is by sequencing the DNA bound to nucleosomes immunoprecipitated with antibodies specific for histone modifications that are predictive of open chromatin (ChIP). Using unpublished data graciously provided by S. Dindot and N. Cohen of Texas A&M University, we examined the gene desert and segmental duplication regions of ELA in whole genome ChIP seq data from anti-H3K4me3-immunoprecipitated chromatin of neutrophils obtained from a newborn foal. Histone 3 trimethylated at lysine 4 (H3K4me3) is a histone modification predictably associated with the 5' transcribed regions of actively expressed genes in higher eukaryotes (Santos-Rosa et al., 2002; Schneider et al., 2003) and is useful as a surrogate for identifying expressed genes.

Representation of DNA sequences aligned to the gene desert coordinates ECA 20: 31,896,104..32,442,400 revealed low levels of H3K4me3-captured sequences over the entire span of the gene desert, consistent with the predictions of few or no expressed genes in this region of ELA. In contrast, the region of the segmental duplication (ECA 20:30,600,000..31,336,000) was highly enriched for H3K4me3 bound sequences, suggesting an abundance of actively transcribed DNA. Twenty-four predicted genes are located in the ELA segmental duplication and 17 of these sequences were located in regions that were moderately to highly represented in anti-H3K4me3 immunoprecipitates. These peaks corresponded to the locations of 13 predicted genes (Ensembl). Of the seven regions highly enriched in H3K4me3, four (57%) were also positive for RT-PCR transcripts. Of the six predicted genes in regions of low H3K4me3, none were detected as transcripts by RT-PCR. A summary of the evidence for gene expression in the segmental duplication is presented in Table 5.1.

Summary

As more detailed analyses of the vertebrate MHC become available, the picture emerging is of an organizationally constrained but structurally dynamic region evolving primarily by recombination and gene conversion. These processes act to sort and reshuffle combinations of genes that have been strongly selected for over evolutionary time frames. The early paradigm that mammalian MHCs were predictably arranged by gene content into three regions consisting of class I and class II genes flanking a class III region is proving to be overly simplistic. The disruption of the class II region of the ruminant MHC (Childers et al., 2006) and less dramatic rearrangements in the MHCs of cat (FLA) and dog (DLA) (Yuhki et al., 2007) promise to provide insights into the evolutionary processes at work in this important region of the genome. The unusual features now known to characterize the *Perissodactyl* MHCs add significantly to the list of diversifying structural changes present in the vertebrate MHC. Access to research material from a common domestic animal such as the horse, with deep pedigrees and well-developed formal biology, provides great potential to help unravel the puzzling structural and functional properties of the MHC.

References

Antczak, D. F., & Allen, W. R. (1989). Maternal immunological recognition of pregnancy in equids. *Journal of Reproduction & Fertility*, *37*(Suppl.), 69–78.

Bailey, E. (2010). Foreword: Horse genomics and the Dorothy Russell Havemeyer Foundation. *Animal Genetics 41*(Suppl. 2), 1.

Brinkmeyer-Langford C. L., Murphy, W. J., Childers, C. P., & Skow, L. C. (2010). A conserved segmental duplication within ELA. *Animal Genetics*, *41*(Suppl. 2), 186–195.

Childers, C. P., Newkirk, H. L., Honeycutt, D. A., Ramlachan, N., Muzney, D. M. et al. (2006). Comparative analysis of the bovine MHC class IIb sequence identifies inversion breakpoints and three unexpected genes. *Animal Genetics*, *37*(2), 121–129.

Chowdhary B. P., Raudsepp, T., Honeycutt, D., Owens, E. K., Piumi, F. et al. (2002). Construction of a 5000(rad) whole-genome radiation hybrid panel in the horse and generation of a comprehensive and comparative map for ECA11. *Mammalian Genome*, *13*(2), 89–94.

Chowdhary, B. P., Raudsepp, T., Kata, S. R., Goh, G., Millon, L. V. et al. (2003). The first-generation whole-genome radiation hybrid map in the horse identifies conserved segments in human and mouse genomes. *Genome Research*, *13*(4), 742–751.

Crump, A., Donaldson, W. L., Miller, J., Kydd, J. H., Allen, W. R., & Antczak, D. F. (1987). Expression of major histocompatibility complex (MHC) antigens on horse trophoblast. *Journal of Reproduction & Fertility*, *35*(Suppl.), 379–388.

de Bakker, P. I., McVean, G., Sabeti, P. C., Miretti, M. M., Green, T. et al. (2006). A high-resolution HLA and SNP haplotype map for disease association studies in the extended human MHC. *National Genetics*, *38*(10), 1166–1172.

Donaldson, W. L., Zhang, C. H., Oriol, J. G., & Antczak, D. F. (1990). Invasive equine trophoblast expresses conventional class I major histocompatibility complex antigens. *Development*, *110*(1), 63–71.

Doan, R., Cohen, N. D., Sawyer. J., Ghaffari, N., Johnson, C. D., & Dindot, S. V. (2012). Whole-genome sequencing and genetic variant analysis of a quarter horse mare. *BMC Genomics* 13:78.

Guérin, G., Bailey, E., Bernoco, D., Anderson, I., Antczak, D. F. et al. (1999). Report of the International Equine Gene Mapping Workshop: Male linkage map. *Animal Genetics*, *30*(5), 341–354.

Guérin, G., Bailey, E., Bernoco, D., Anderson, I., Antczak, D. F. et al. (2003). The second generation of the International Equine Gene Mapping Workshop half-sibling linkage map. *Animal Genetics*, *34*(3), 161–168.

Gustafson, A. L., Tallmadge, R. L., Ramlachan, N., Miller, D., Bird, H. et al. (2003). An ordered BAC contig map of the equine major histocompatibility complex. *Cytogenetic Genome Research*, *102*(1–4), 189–195.

Harris, E. E., & Meyer, D. (2006). The molecular signature of selection underlying human adaptations. *American Journal of Physiology & Anthropology*, *43*(Suppl.), 89–130.

Hedrick, P. W. (1998–1999). Balancing selection and MHC. *Genetica*, *104*(3), 207–214.

Klein, J. (1987). Origin of major histocompatibility complex polymorphism: The trans-species hypothesis. *Human Immunology*, *19*, 155–162.

Lazary, S., Antczak, D. F., Bailey, E., Bell, T. K., Bernoco, D. et al. (1988). Joint Report of the Fifth International Workshop on Lymphocyte Alloantigens of the Horse, October 31–November 1987, Baton Rouge, Louisiana. *Animal Genetics*, *19*(4), 447–456.

Lie, B. A., & Thorsby, E. (2005). Several genes in the extended human MHC contribute to predisposition to autoimmune diseases. *Current Opinions in Immunology*, *17*(5), 526–531.

McCue, M. E., Bannasch, D. L., Petersen, J. L., Gurr, J., Bailey, E. et al. (2012). A high density SNP array for the domestic horse and extant perissodactyla: Utility for association mapping, genetic diversity, and phylogeny studies. *PLoS Genetics*, *8*(1), e1002451.

McGuire, K. L., Duncan, W. R., & Tucker, P. W. (1985). Syrian hamster DNA shows limited polymorphism at class I-like loci. *Immunogenetics*, *22*(3), 257–268.

Miyata, T., Yasunaga, T., & Nishida, T. (1980). Nucleotide sequence divergence and functional constraint in mRNA evolution. *Proceedings of the National Academy of Sciences USA*, *77*(12), 7328–7332.

Nizetić, D., Stevanović, M., Soldatović, B., Savić, I., & Crkvenjakov, R. (1988). Limited polymorphism of both classes of MHC genes in four different species of the Balkan mole rat. *Immunogenetics*, *28*(2), 91–98.

Ohta, Y., Okamura, K., McKinney, E. C., Bartl, S., Hashimoto, K., & Flajnik, M. F. (2000). Primitive synteny of vertebrate major histocompatibility complex class I and class II genes. *Proceedings of the National Academy of Sciences USA*, *97*(9), 4712–4717.

Perrocheau, M., Boutreux, V., Chadi, S., Mata, X., Decaunes, P. et al. (2006). Construction of a medium-density horse gene map. *Animal Genetics*, *37*(2), 145–155.

Raudsepp, T., Gustafson-Seabury, A., Durkin, K., Wagner, M. L., Goh, G. et al. (2008). A 4,103 marker integrated physical and comparative map of the horse genome. *Cytogenetic Genome Research*, *122*(1), 28–36.

Santos-Rosa, H., Schneider, R., Bannister, A. J., Sherriff, J., Berstein, B. E. et al. (2002). Active genes are trimethylated at K4 of histone H3. *Nature*, *419*, 407–411.

Sato, A., Figueroa, F., Murray, B. W., Málaga-Trillo, E., Zaleska-Rutczynska, Z. et al. (2000). Nonlinkage of the major histocompatibility complex of class I and class II loci in bony fishes. *Immunogenetics*, *51*, 108.

Schneider, R., Bannister, A. J., Myers, F. A., Thorne, A. W., Crane-Robinson, C., & Kouzarides, T. (2004). Histone H3 lysine 4 methylation patterns in higher eukaryotic genes. *Nature Cell Biology*, *6*(1), 73–77.

Shiue, Y. L., Bickel, L. A., Caetano, A. R., Millon, L. V., Clark, R. S. et al. (1999). A synteny map of the horse genome comprised of 240 microsatellite and RAPD markers. *Animal Genetics*, *30*(1), 1–9.

Swinburne, J., Gerstenberg, C., Breen, M., Aldridge, V., Lockhart, L. et al. (2000). First comprehensive low-density horse linkage map based on two 3-generation, full-sibling, cross-bred horse reference families. *Genomics*, *66*(2), 123–134.

Tait, B. D. (2011). The ever-expanding list of HLA alleles: Changing HLA nomenclature and its relevance to clinical transplantation. *Transplant Review (Orlando)*, *25*(1), 1–8.

Tallmadge, R. L., Campbell, J. A., Miller, D. C., & Antczak, D. F. (2010). Analysis of MHC class I genes across horse MHC haplotypes. *Immunogenetics*, *62*(3), 159–172.

Tallmadge, R. L., Lear, T. L., & Antczak, D. F. (2005). Genomic characterization of MHC class I genes of the horse. *Immunogenetics*, *57*(10), 763–774.

Trowsdale, J. (2011). The MHC, disease and selection. *Immunology Letters*, *137*(1–2):1–8.

Tseng, C. T., Miller, D., Cassano, J., Bailey, E., & Antczak, D. F. (2010). Identification of equine major histocompatibility complex haplotypes using polymorphic microsatellites. *Animal Genetics*, *41*(Suppl 2), 150–153.

Wade, C. M., Giulotto, E., Sigurdsson, S., Zoli, M., Gnerre, S. et al. (2009). Genome sequence, comparative analysis, and population genetics of the domestic horse. *Science*, *326*(5954), 865–867.

Walsh, E. C., Mather, K. A., Schaffner, S. F., Farwell, L., Daly, M. J. et al. (2003). An integrated haplotype map of the human major histocompatibility complex. *American Journal of Human Genetics*, *73*, 580–590.

Yuhki, N., Beck, T., Stephens, R., Neelam, B., & O'Brien, S. J. (2007). Comparative genomic structure of human, dog, and cat MHC: HLA, DLA, and FLA. *Journal of Heredity*, *98*(5), 390–399.

6 Assembly and analysis of the equine genome sequence

Claire M. Wade

Introduction

In November 2009, the first genome sequence of the domestic horse (*Equus caballus*) was published in the international journal *Science* (Wade et al., 2009). This substantive achievement was the result of a concerted effort by a worldwide equine research community, which first lobbied the government and funding institutions to consider providing this significant resource and later gathered the large numbers of samples of horse DNA from many horse breeds around the globe, which were needed to undertake the sequencing and analysis.

The reasons that the horse was chosen as a priority species for genome sequencing were numerous. Foremost, the horse was sequenced to represent its mammalian order (*Perissodactyla*) with the objective of learning more about those parts of mammalian DNA that are conserved across all mammalian species and that enable us to function as living, breathing mammals.

As one of the earliest domesticated species, the horse has played an important role in human transportation, making it irreplaceable in exploration of novel territories and in warfare. The horse, *Equus caballus*, belongs to the order *Perissodactyla*, consisting of odd-toed ungulates (animals with hooves). The families within *Perissodactyla* are *equidae* (horses) including two horse species, three hemione species, two donkey species, and at least three zebra species, as well as five *rhinocerotidae* (rhinoceroses) and four *tapiridae* (tapirs). While the families of *Perissodactyla* separated circa 56 million years ago (Trifonov et al., 2008), the radiation of 8–9 species within the genus *Equus* is thought to have occurred closer to 3 million years ago (Carbone et al., 2006; Oakenfull & Clegg, 1998; Oakenfull et al., 2000). Members within the family *equidae* exhibit diverged karyotypes with chromosomal numbers ranging from 2N=32–66 (Trifonov et al., 2008), sometimes exhibiting common within-species variations (Houck et al., 1998) and inter-species centromeric repositioning (Carbone et al., 2006).

Among *Perissodactyla* there are two domesticated species: the horse (*Equus caballus*) and the ass (represented by *Equus africanus asinus* and *Equus africanus somalicus*). While the time of domestication of the horse is the subject of debate, it is thought to have occurred between 4,000 and 6,000 years ago (Levine, 1999). Based on mitochondrial evidence (Vila et al., 2001), domestic horses are derived from a wide genetic base, which is likely related to the domestication of large herds of female horses.

Horses have more than 90 hereditary conditions that may serve as models for human disorders (Chowdhary et al., 2008; Online Mendelian Inheritance in Animals, 2009) – for example, fertility,

Equine Genomics, First Edition. Edited by Bhanu P. Chowdhary.

inflammatory diseases, and muscle disorders – and have much to offer as a model species. Genetic studies in horses thus far have been aided by extensive record keeping of pedigrees within horse breeds and comprehensive genetic mapping resources (Lindgren et al., 1998; Swinburne et al., 2000; Penedo et al., 2005; Leeb et al., 2006; Perrocheau et al., 2006; Swinburne et al., 2006; Raudsepp et al., 2008). Here we discuss the recently assembled genome sequence and comparative genomic analysis of the horse, which strengthens the role of the horse as a model organism and provides insight into mammalian evolution.

Sequencing a Genome

DNA from a single Thoroughbred mare was used to construct the genome sequence. The mare, Twilight, was provided to the project from the herd managed by Cornell University in Ithaca, New York (Figure 6.1). In the genome-sequencing world it is common practice to prefer to sequence a female rather than a male because this enables adequate sequence coverage and assembly of the X chromosome. If a male had been chosen, then there would be only one X chromosome (but also, of course, a Y chromosome), and the amount of sequence covering each sex chromosome would be considerably less than is desirable for the purposes of assembling the genome after sequencing.

Another common practice among those who sequence and assemble genomes is to make use of animals that have high homozygosity. The reason that this is desirable is to minimize the considerable differences between different versions of the same chromosome that exist even within a species. The presence of intra-species variation presents extra challenges to the computational genome assembly process. We can find animals with high homozygosity by selecting animals with high levels of inbreeding. Twilight was chosen because she demonstrated the highest level of homozygosity among the horses tested for this purpose using a set of DNA markers.

The equine genome was sequenced using the whole-genome shotgun method. This process involves shearing the DNA into fragments and inserting these into clone libraries for replication

Figure 6.1 The Thoroughbred mare "Twilight" – the DNA donor for the sequencing project.

and sequencing. Typically only the end-most 500–800 base pairs of sequence are read from these "insert" fragments and these are called paired-end reads. Different-sized fragments are used to create bridges of different lengths that can join assembled "contigs" together to span gaps in the assembly. Contigs that are joined together unambiguously across gaps of known size into larger sequences are termed scaffolds. In the horse genome, insert fragments of 5,000 bases, 10,000 bases (both termed cosmid clones), and 40,000 bases (fosmid clones) were used for the sequencing. Later Bacterial Artificial Chromosome (BAC) sequences of around 150,000 bases from a related male horse were included into the assembly at the final stage.

Features of the Equine Genome Assembly

A high-quality draft assembly was constructed, with added contiguity generated by the inclusion of BAC end sequences from a related male Thoroughbred horse, from which a BAC map had been produced by researchers in Germany (Leeb et al., 2006). The resulting assembly had sequence coverage of 6.8 fold. The generated assembly (designated EquCab2.0) is of high quality and contiguity. More than 50% of the genome is contained in contiguous sequences of longer than 112 kilobases (kb) and in scaffolds of longer than 46 megabases (Mb). More than 95% of the euchromatic sequence was able to be anchored to the 64 (2N) equine chromosomes. Many features of the equine genome were similar to those of other mammals, but there were a number of notable differences. The genomes of human, cow, dog, and mouse were used for comparison based on their interesting population structures (Lander et al., 2001; Waterston et al., 2002; Lindblad-Toh et al., 2005; Liu et al., 2009).

The estimated euchromatic genome size of *Equus caballus* based on the total lengths of scaffolds lies between those of the dog (2.5 Gb) and human genomes (2.9 Gb) (Lander et al., 2001; Lindblad-Toh et al., 2005). Segmental duplications in the assembly, determined using standard methods (Mikkelsen et al., 2007), comprise less than 1% of the equine genome, and the majority of these (with ~80% mapped to chromosomes) are intra-chromosomal duplications such as those seen in the dog and mouse genomes. So, while the size of the genome based on the scaffold structure appears to be around 2.5 Gb, the assembly has many unplaced sequence reads suggesting that the true genome size may be up to 2.7 Gb. The unassembled sequences are highly repetitive in nature.

Comparison with Genetic Maps

The task of the genome assembly team is to construct contiguous sequences and scaffolds from sequencing reads. The general equine genetic researcher is interested in not only the amalgamation of sequence reads into contigs and contigs into scaffolds, but also to know the chromosomal location of these scaffolds on the equine karyotype. The way that this is accomplished is by using the existing genetic maps for the horse (Guérin et al., 2003; Penedo et al., 2005; Swinburne et al., 2006; Raudsepp et al., 2008) to place markers with known chromosomal locations onto the equine assembly. This serves two purposes: (1) to ensure that the assembly is accurate and (2) to establish the order and orientation of sequences onto the chromosomes. To correctly order and orient the contigs and scaffolds onto chromosomes, it is necessary that two markers with known physical locations should be available for each scaffold. Where mapped markers are unavailable and the scaffold is of sufficient size (typically > 2 Mb), a different method of ordering and orienting the scaffold is carried out. This uses fluorescence in situ hybridization (FISH) mapping. In this process, fluorescently labeled genomic sequences are hybridized to equine karyotypes and the chromosomal assignment is made by the observation of the fluorescence by an expert cytogeneticist (Lear et al., 1998). Ordering and orienting of the scaffolds is the ultimate process in the creation of the genome assembly. Once the genome is assembled, it becomes available to researchers for analysis.

Repetitive Elements

Repetitive elements occur at high frequency in all mammalian genomes. The typical mammalian genome has more than 40% of its DNA sequence derived from small mobile DNA elements known as transposons. We commonly term these elements as repeats. Such elements have been active in genomes throughout evolutionary history. As a result of this activity, some transposons are shared among animal species, whereas others have evolved uniquely subsequent to speciation. Using standard mammalian repeat libraries that can identify common elements across mammals, 39% of the equine genome assembly was annotated as comprising repetitive transposon-derived sequences. By applying customized libraries that were designed to include horse-specific repeats, 46% of the assembled sequence was identified as repetitive, a quantity comparable with that seen in the human genome. The predominant repeat classes present in the equine genome included long interspersed nuclear elements (LINEs) that were dominated by L1 and L2 types (19% of bases), and short interspersed nuclear elements (SINEs) such as the recent Equine Repeat Elements 1/2 (ERE1/2) and the ancestral Mammalian Interspersed Repeats (MIRs) (7% of bases). Both of these element classes are common in many mammalian genomes. Examination with the horse-specific libraries found that novel equine repeats accounted for a large fraction of the observed consensus transposon element sequences in the horse, but that only 48 of these consensus elements were present in significant numbers (more than 100 copies).

Chimeric repeats appear to stem from the random placement of new repeat sequences within existing repeats. Chimeric repeats were also identified as an important source of repetitive elements in the horse but they were difficult to classify unambiguously. The different repeat classes seemed to occur within the chimeras in proportion to their relative overall frequency in the genome.

Synteny with Humans

Comparative genomics is a term used to describe the use of information gained from the genome of one species to inform our understanding of another. Conserved linkage describes a preservation of gene order across species and it suggests an evolutionary commonality of sequence origin. If we compare horse and human chromosomes we observe surprisingly strong conserved linkage between these species given that we and horses seem phenotypically (in appearance) to be quite dissimilar. In fact, it might be surprising to learn that horses are a little closer to humans on an evolutionary scale than they are to cows.

Syntenic blocks are regions where a portion of any chromosome from the first species is present in the second species without another chromosome sequence intervening. Syntenic segments describe similar regions but do not allow for directional rearrangements between sections of the sequence. More than 2.76 Gb (out of 2.9 Gb) of the human genome sequence is covered by horse syntenic blocks that are at least 100 kb long. This implies that there has been relatively little rearrangement between the species and their common ancestor. Compare this with the weak synteny observed between human and mouse (Waterston et al., 2002). At the resolution of 100 kb-sized sequence windows, we find only 86 syntenic blocks and 425 syntenic segments in the alignment of the horse

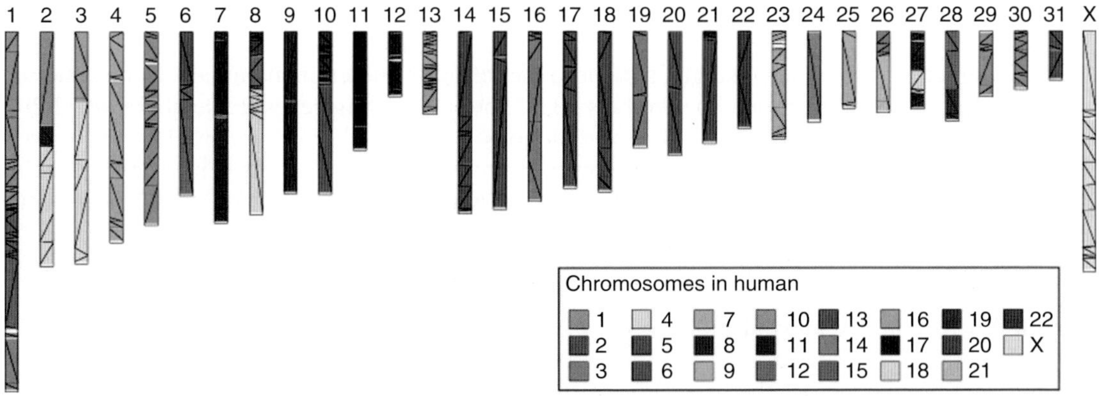

Figure 6.2 Conserved synteny between horse and human. Horse chromosomes are colored by corresponding human chromosomes. (*For color details, see color plate section.*)

and human genomes. While the human syntenic segments defined relative to horse and relative to dog have similar average sizes, 50% of the human genome sequence resides in blocks 26 Mb or larger on the horse genome, while on the dog genome the equivalent figure is a smaller 20 Mb. This larger block size between human and horse is strongly influenced by 12 large human-horse segments that exceed the maximum size observed in the dog-human comparison. In fact, 17 chromosomes in horse comprise material from a single human corresponding chromosome (53% of horse chromosomes and 29% of dog chromosomes correspond to a single counterpart human chromosome) (Figure 6.2).

Special Centromeres

Perhaps one of the most interesting and unexpected findings of the horse genome landscape was the identification of an evolutionary new centromere (ENC), captured in an immature state. A previous study (Carbone et al., 2006) showed that during the last 3 million years, several ENCs were generated in the genus *Equus* by centromere repositioning (shift of centromeric position without chromosome rearrangement). In particular, it was shown that the centromere of Chromosome 11 is an ENC. Mammalian centromeres are typically complex structures characterized by the presence of satellite tandem repeats. It is commonly thought that ENCs, following seeding of satellite tandem repeat DNA into a single-copy DNA region, progressively acquire the characteristic extended arrays of satellite tandem repeats and that presumably these stabilize the centromeric function (Ventura et al., 2007). However, centromeres lacking satellite sequence have been described only in occasional human clinical cases, where they stabilize an acentric fragment (Amor & Choo 2002; Warburton, 2004) or occasionally in normal appearing chromosomes of a normal individual, in which the centromere was repositioned (Amor et al., 2004; Ventura et al., 2004; Capozzi et al., 2009) as is the case in evolution. The centromere of Chromosome 11 was the only horse centromere lacking any hybridization signal in FISH experiments performed using the two major horse satellite sequences as probes. The absence of satellite signals in the Chromosome 11 centromere suggests that this ENC may not have yet "matured" to the point of being endowed with satellite DNA. It probably represents the first sequenced example of an evolutionary "immature" centromere, supporting the theory that horse centromeres are in a rapidly evolving state and providing a good model for future studies of centromeric function.

Genes

The equine gene set is, not surprisingly, similar to that of other eutherian mammals. Gene structure annotation by the ENSEMBL pipeline predicts 20,322 protein-coding genes (Ensembl build 52.2b). The number of genes orthologous to these predictions is comparable in human (16,617), mouse (17,106), and dog (16,159). The remainder comprises projected protein coding genes, novel protein-coding genes, and pseudogenes. One-to-one orthologs with human account for 15,027 horse gene predictions.

A Single Nucleotide Polymorphism Map

An important component of the assembly of the equine genome was the provision of genomic resources to enable better equine genomic research. As part of the genome project, partial genomic sequences were obtained from seven additional horses to provide a database of genetic markers. A single nucleotide polymorphism (SNP) map of more than 1 million markers was generated, comprising more than 700,000 SNPs discovered by comparing the two chromosomes within the "Twilight" genome assembly and ~400,000 SNPs that were discovered from ~100,000 whole genome shotgun reads from each of seven horse breeds (Akhal-Teke, Andalusian, Arabian, Icelandic Horse, Quarterhorse, Standardbred, and a second Thoroughbred). The resulting SNP map has an average density of one SNP in every 2,000 bases of DNA.

The average similarity between horses of different breeds is much like that observed among humans and dogs with around 1 SNP per 1,000 bp. Within Twilight herself, 46% of the DNA in the genome was found to be homozygous. This amount of homozygosity is expected to be higher than for the average horse, because Twilight was chosen for her high level of homozygosity.

Genomic Attributes of Equine Breeds

To discover the effects of population history on the genome structure, we characterized the landscape of genetic variation within and across breeds. To do this we genotyped 1,007 newly discovered SNPs from ten random regions of the genome in 12 horse populations, including 11 breed sets (each with 24 representatives), and 1 across-breed set of individual representatives from 24 other breeds and equids. The breeds used in the analysis aimed to reflect global breed diversity. Ninety-eight percent of the SNPs from the above SNP map were found to be polymorphic within this breed panel and on average 69% of the markers were polymorphic within any third breed. Among all horses, the mean minor allele frequency for polymorphic SNPs was 0.22(+/-0.15) in a flat distribution from 0 to 0.5.

The level of connectivity between nearby markers is described by linkage-disequilibrium (LD). When LD values are high, fewer markers are required in genotyping panels to accurately reflect the underlying genetic variation that may be responsible for a disease or other trait of interest. In horses, the within-breed LD is moderate, dropping to twofold the background (unlinked) levels at approximately 100–150kb. This is approximately five times shorter than the within-breed LD for dog and five times longer than in a human population of European ancestry (Centre d'Etude du Polymorphisme Humain). The LD is of similar extent to that observed in the bovine (Gibbs et al., 2009). While the majority of horse breeds show similar patterns of LD, the Thoroughbred appears to have LD akin to that of dog breeds. This was expected because the Thoroughbred breeding population has maintained a closed breeding structure for more than a century and was derived from relatively few founders (Ridgeway, 1905). Interestingly, the across-breed LD in the horse is only slightly shorter (50–70kb) than the within-breed values, likely reflecting the absence of strong bottlenecks during breed formation and the requirement of many mares to maintain population size due to the limited number of offspring per mare.

The horse's particular history is further illustrated by the frequent sharing of major haplotypes among diverse horse populations, emphasizing their common history. On average, we observe 5 haplotypes for each breed within a 100kb segment and 19 haplotypes across breeds. The 2 most common haplotypes across breeds, for a given segment, are typically present in the majority of breeds.

The marker density required for effective genome-wide association mapping can be estimated from the length of LD in the horse, the number of haplotypes within haplotype blocks, and the polymorphism rate. We estimate that effective tagging of haplotypes for genetic mapping purposes in the average breed will require a map density of greater than 100,000 SNP (for 5 haplotypes in 18,000 blocks of 150kb when 70% of SNP are polymorphic within a given breed), and more markers will be required for effective mapping in ancient breeds as well as in those with a large effective population size (such as the Quarter horse).

The outcome of this preliminary population genetic analysis is to suggest that mapping traits in horses can be expected to be a little less easy than in some other species, such as dog. The feature of the horse population, however, is that mapping across equine breeds is much more feasible than for other species described thus far.

Summary

The sequencing and assembly of the first equine genome in 2009 has provided an amazing resource to assist genetic research in horses. The genome project provided not only the DNA sequence for a Thoroughbred mare, but also a dense mapping resource in the form of a genotyping array and a comprehensive annotation of the genome that is available through various public genome browsing sites such as those hosted by the University of California Santa Cruz (UCSC), a joint project by the European Bioinformatics Institute and the Wellcome Trust Sanger Institute (Ensembl), and the National Center for Biotechnology Information (NCBI). This comprehensive resource sets the stage for teaching us more about mammalian evolution, population genetics, and disease gene mapping, and has already begun to do so for traits relating to medical conditions and performance in the horse.

Acknowledgments

We thank everyone who contributed to the sequencing and analysis of the equine genome, including all the researchers and horse owners who contributed samples. The sequencing and assembly were carried out entirely by The Broad Institute of Harvard and Massachusetts Institute of Technology and the Broad Institute Whole Genome Sequencing and Assembly Platforms, led in this project by Kerstin Lindblad-Toh as the director of vertebrate genomes. The Broad Institute produced a sequence of exceptionally high contiguity and quality. In particular, Sante Gnerre took charge of the genome assembly and Manuel Garber carried out the syntenic analysis. Work for the centromeric analysis of equine chromosome 11 was conducted by Elena Giulotto from the University of Pavia, and the analysis of repetitive elements was led by David Adelson from The University of Adelaide (formerly of Texas A & M University) with assistance from Joy Raison. Jan Vogels from Ensembl conducted the annotation of protein-coding sequences. The project could not have commenced without the white paper that was initiated by Ernest Bailey and Kerstin Lindblad-Toh with the assistance of many of the researchers who are chapter authors in this volume.

References

Amor, D. J., Bentley, K. et al. (2004). Human centromere repositioning "in progress". *Proceedings of the National Academy of Sciences USA*, *101*(17), 6542–6547.

Amor, D. J., & Choo, K. H. (2002). Neocentromeres: Role in human disease, evolution, and centromere study. *American Journal of Human Genetics*, *71*(4), 695–714.

Capozzi, O., Purgato, S. et al. (2009). Evolutionary descent of a human chromosome 6 neocentromere: A jump back to 17 million years ago. *Genome Research*, *19*(5), 778–784.

Carbone, L., Nergadze, S. G. et al. (2006). Evolutionary movement of centromeres in horse, donkey, and zebra. *Genomics*, *87*(6), 777–782.

Chowdhary, B. P., Paria, N. et al. (2008). Potential applications of equine genomics in dissecting diseases and fertility. *Animal Reproductive Science*, *107*(3–4), 208–218.

Gibbs, R. A., Taylor, J. F. et al. (2009). Genome-wide survey of SNP variation uncovers the genetic structure of cattle breeds. *Science*, *324*(5926), 528–532.

Guérin, G., Bailey, E. et al. (2003). The second generation of the International Equine Gene Mapping Workshop half-sibling linkage map. *Animal Genetics*, *34*(3), 161–168.

Houck, M. L., Kumamoto, A. T. et al. (1998). Chromosomal rearrangements in a Somali wild ass pedigree, Equus africanus somaliensis (Perissodactyla, Equidae). *Cytogenetic Cell Genetics*, *80*(1–4), 117–122.

Lander, E. S., Linton, L. M. et al. (2001). Initial sequencing and analysis of the human genome. *Nature*, *409*(6822), 860–921.

Lear, T. L., Adams, M. H. et al. (1998). Assignment of the horse progesterone receptor (PGR) and estrogen receptor (ESR1) genes to horse chromosomes 7 and 31, respectively, by in situ hybridization. *Cytogenetic Cell Genetics*, *82*(1–2), 110–111.

Leeb, T., Vogl, C. et al. (2006). A human-horse comparative map based on equine BAC end sequences. *Genomics 87*(6), 772–776.

Levine, M. A. (1999). Investigating the origins of horse domestication. *Equine Veterinary Journal*, *28*(Suppl.), 6–14.

Lindblad-Toh, K., Wade, C. M. et al. (2005). Genome sequence, comparative analysis and haplotype structure of the domestic dog. *Nature*, *438*(7069), 803–819.

Lindgren, G., Sandberg, K. et al. (1998). A primary male autosomal linkage map of the horse genome. *Genome Research*, *8*(9), 951–966.

Liu, Y., Qin, X. et al. (2009). Bos taurus genome assembly. *BMC Genomics*, *10*, 180.

Mikkelsen, T. S., Wakefield, M. J. et al. (2007). Genome of the marsupial Monodelphis domestica reveals innovation in non-coding sequences. *Nature*, *447*(7141), 167–177.

Oakenfull, E. A., & Clegg, J. B. (1998). Phylogenetic relationships within the genus Equus and the evolution of alpha and theta globin genes. *Journal of Molecular Evolution*, *47*(6), 772–783.

Oakenfull, E. A., Lim, H. N. et al. (2000). A survey of equid mitochondrial DNA: Implications for the evolution, genetic diversity and conservation of *Equus*. *Conservation Genetics*, *1*, 341–355.

Online Mendelian Inheritance in Animals (2009). Reprogen, Faculty of Veterinary Science, University of Sydney and Australian National Genomic Information Service (ANGIS) Online Mendelian Inheritance in Animals, OMIA., University of Sydney.

Penedo, M. C., Millon, L. V. et al. (2005). International Equine Gene Mapping Workshop Report: A comprehensive linkage map constructed with data from new markers and by merging four mapping resources. *Cytogenetic Genome Research*, *111*(1), 5–15.

Perrocheau, M., Boutreux, V. et al. (2006). Construction of a medium-density horse gene map. *Animal Genetics*, *37*(2), 145–155.

Raudsepp, T., Gustafson-Seabury, A. et al. (2008). A 4,103 marker integrated physical and comparative map of the horse genome. *Cytogenetic Genome Research*, *122*(1), 28–36.

Ridgeway, W. (1905). *The origin and influence of the Thoroughbred horse*. London: Cambridge University Press.

Swinburne, J., Gerstenberg, C. et al. (2000). First comprehensive low-density horse linkage map based on two 3-generation, full-sibling, cross-bred horse reference families. *Genomics*, *66*(2), 123–134.

Swinburne, J. E., Boursnell, M. et al. (2006). Single linkage group per chromosome genetic linkage map for the horse, based on two three-generation, full-sibling, crossbred horse reference families. *Genomics*, *87*(1), 1–29.

Trifonov, V. A., Stanyon, R. et al. (2008). Multidirectional cross-species painting illuminates the history of karyotypic evolution in Perissodactyla. *Chromosome Research*, *16*(1), 89–107.

Ventura, M., Antonacci, F. et al. (2007). Evolutionary formation of new centromeres in macaque. *Science*, *316*(5822), 243–246.

Ventura, M., Weigl, S. et al. (2004). Recurrent sites for new centromere seeding. *Genome Research*, *14*(9), 1696–1703.

Vila, C., Leonard, J. A. et al. (2001). Widespread origins of domestic horse lineages. *Science*, *291*(5503), 474–477.

Wade, C. M., Giulotto, E. et al. (2009). Genome sequence, comparative analysis, and population genetics of the domestic horse. *Science*, *326*(5954), 865–867.

Warburton, P. E. (2004). Chromosomal dynamics of human neocentromere formation. *Chromosome Research*, *12*(6), 617–626.

Waterston, R. H., Lindblad-Toh, K. et al. (2002). Initial sequencing and comparative analysis of the mouse genome. *Nature*, *420*(6915), 520–562.

7 Genomic tools and resources: Development and applications of an equine SNP genotyping array

Molly McCue and Jim Mickelson

Introduction

A major focus of the equine genomics research community has been the development of research tools required for applied equine genetics research. Specifically, work has centered on the completion of a comprehensive map of the equine genome necessary to complete whole genome linkage and association analyses and to identify positional candidate genes responsible for disease and performance traits of interest. Prior to sequencing of the equine genome, two medium-density equine genetic linkage maps (Swinburne et al., 2000; Swinburne et al., 2006) and two whole genome radiation hybrid maps (Chowdhary et al., 2002; Chowdhary et al., 2003; Raudsepp et al., 2008) were constructed. In both types of maps linkage groups on each chromosome were anchored by multiple FISH-mapped markers (see Chapters 2 and 3 in this volume for details). Subsequently, the coverage of several chromosomes and chromosome segments was increased to an average of one marker every Mb or less. A high-resolution comparative RH map containing data from ~4,200 loci, which integrated both genetic linkage maps, was generated (Raudsepp et al., 2008). Lastly, a whole genome physical map has been developed by BLAST analysis of paired BAC end sequences from the 150,000 clones that represent the CHORI-241 equine BAC library onto the human genome (Leeb et al., 2006). Details regarding these resources are available elsewhere in this volume. The focus of this chapter is on the newly developed moderate-density SNP genotyping array that, for the first time, has allowed high-throughput SNP genotyping in the horse.

Development of an Equine SNP Genotyping Array

Sequencing of the horse genome by the National Human Genome Research Institute (NHGRI) and the subsequent SNP discovery effort led to the identification of ~750,000 SNPs from the sequenced genome (Twilight) and ~400,000 SNPs from seven horses, each of a different breed, providing sufficient markers to construct a whole genome SNP panel for use in the domestic horse and related species (Wade et al., 2009). The first-generation whole genome SNP array termed the EquineSNP50 Beadchip has been available to researchers since July 2008, and several recent papers have described its application (Brooks et al., 2010; Hill et al., 2010; Orr et al., 2010). Here we describe the properties of this SNP array and its use in the horse as well as 18 related species (McCue et al., 2012). These

Equine Genomics, First Edition. Edited by Bhanu P. Chowdhary.

data come from two sources: (1) a collaborative study termed the Gentrain Project, consisting of researchers at the University of Minnesota, The Broad Institute and the University of California at Davis, along with several members of the equine genomics community (ref McCue-Gentrain-pending publication); and (2) recently published applications of the tool.

Equine SNP50 Beadchip Design and Validation in the Domestic Horse

Sixty thousand SNPs that gave suitable design scores for the Illumina Infinium II assay were selected for inclusion in the first-generation SNP array. This initial assay performance was assessed in a group on 354 horses representing 14 breeds (Andalusian, Arabian, Belgian, Franches-Montanges, French Trotter, Hanoverian, Icelandic, Mongolian, Norwegian Fjord, Quarter Horse, Saddlebred, Standardbred, Swiss Warmblood, Thoroughbred), including 6 of the 7 breeds represented in the SNP discovery effort. Of the 60,000 SNPs included in assay design, 54,602 reliably produced genotypes, resulting in a SNP conversion rate of 91.0%, which was slightly lower than the conversion rate reported for a similar assay in the bovine (Matukumalli et al., 2009). In contrast, 53,524 of the ~54 k genotyped SNPs were polymorphic (having at least one heterozygote within the sample set), resulting in a validation rate of 98.0%, which is higher than the validation rate of 95.1% reported for the bovine (Matukumalli et al., 2009). Approximately half of the converted SNPs (26,473) were polymorphic across all breeds, while the number of polymorphic SNPs within a breed ranged from 43,287 to 52,085. The average spacing between functional SNPs across the 31 autosomes was 43.1 kb, with only 12 gaps of greater than 500 kb, indicating that the assay provides good coverage of the genome. Coverage on ECAX (average inter-SNP spacing of 48.88 kb) was lower than the rest of the genome, reflecting fewer SNPs to choose from for assay design. This is likely due to the repetitive nature of the X chromosome in most mammals and the use of a SNP discovery algorithm that rejects sequences that align equally to multiple locations.

Along with polymorphism rates, both informativeness (MAF $\geq$ 5%) and mean minor allele frequency (MAF) across the SNP set varied between breeds. The mean MAF across all breeds was 0.236, with mean minor allele frequencies within a given breed ranging from 0.180 to 0.232. The numbers of informative SNPs within breeds ranged from 37,053 in the Norwegian Fjord to 47,669 in the Quarter Horse. Overall, breeds with recent or ongoing admixture, such as the Quarter Horse, Hanoverian, and Swiss Warmblood, had the highest mean MAF and the largest numbers of informative SNPs, while the lowest mean MAF were in the Norwegian Fjord, Belgian, and Icelandic horse. On average the number of SNPs informative in any given domestic horse breed was higher than the informativeness of similar assays within given cattle and dog breeds (Karlsson et al., 2007; Matukumalli et al., 2009). Only 17,428 SNPs were informative in every breed, which may limit some across-breed mapping applications; however, 49,603 SNPs were informative within the entire sample set (across all breeds), indicating that the tool will be useful for many population-based applications.

The Broad Institute's SNP discovery effort for the domestic horse was based on light re-sequencing of seven individuals, each from a different breed (Akhal Teke, Andalusian, Arabian, Icelandic, Quarter Horse, Standardbred, Thoroughbred), selected to represent a diverse set of populations (Wade et al., 2009). To reduce Thoroughbred bias on the SNP assay, preference was given to SNPs that were discovered in alternate breeds versus SNPs discovered during the assembly of Twilight's genome. Greater than two-thirds of the SNPs selected were identified in one of the seven SNP discovery breeds in reference to Twilight's sequence. The other SNPs were present in at least one breed as well as in Twilight (McCue et al., 2012). As anticipated, SNP validation rate was highest for SNPs that were discovered in two or more breeds versus SNPs discovered within a single breed. Overall, the SNPs demonstrated good polymorphism rates, even in the eight breeds not included in the SNP discovery effort, suggesting that the use of a diverse, although limited, set of individuals for SNP discovery was a valid approach. Interestingly, in the face of low genetic diversity and high levels of inbreeding, mean MAF in the Thoroughbred was higher than any other non-admixed breed. Furthermore, the fraction of informative SNPs in the Thoroughbred exceeded that of any other breed included in the SNP discovery effort, with the exception of the Quarter Horse. This suggests that despite efforts to limit SNP bias, the use of the Thoroughbred in SNP discovery and as the reference genome sequence is reflected in the high level of SNP informativeness in this breed.

Recent studies in humans have demonstrated that the source population used for SNP discovery can have a dramatic impact on the fraction of informative SNPs in other populations (No author, 2010). Therefore, the validation rate for SNPs discovered within a single breed was assessed to determine if some breeds may be more useful for SNP discovery than others. Overall, the proportion of informative SNPs was similar regardless of discovery breed, with the proportion of SNPs identified in a single discovery breed that were polymorphic across all breeds ranging from 0.85 in the Icelandic to 0.88 in the Arabian. This data is in contrast to a previous report that suggested that Standardbred SNPs were the most polymorphic and Akhal-Teke SNPs the least polymorphic in other breeds (Wade et al., 2009).

Utility of the Equine SNP50 Beadchip for Genome-Wide Mapping Strategies

A major application of this technology will be in genome-wide association mapping of economically important and disease traits (Binns, Boehler, & Lambert, 2010; Brooks et al., 2010; Cook, Gallagher, & Bailey, 2010; Lykkjen et al., 2010). The success of such studies will depend on LD within the mapping population, properties of the loci themselves, population structure, and the mode of inheritance. Initial estimates of marker numbers and density necessary for genome-wide association mapping in the domestic horse suggested that 100,000 SNP markers would allow association at a mean maximum $r^2 > 0.5$ within and across horse breeds, but would only allow mapping at a mean maximum of $r^2 > 0.8$ in breeds with low effective population size/ high LD (Wade et al., 2009). This suggests that even with 100,000 SNPs there would be insufficient power for mapping in ancient breeds or breeds with larger effective population sizes. Thus, the number of SNP markers available on this Beadchip represents about one-half of the estimated marker density required for highly powered association studies.

To better estimate the power of association mapping within breeds, the extent of average LD across the genome was characterized in the Gentrain 14 breeds (Figure 7.1). LD was estimated by computing the r^2 value for all pairs of SNPs less than 4 Mb apart both within and across breeds. Similar to findings in dogs and cattle, LD was higher within a breed than across all breeds (Karlsson et al., 2007; The Bovine HapMap Consortium, 2009). Initial LD declined rapidly across all horses, with mean r^2 dropping below 0.2 by 50 kb. In general, within breed, r^2 values dropped below 0.2 within 100 to 150 kb in most breeds. Notable exceptions were the Thoroughbred, where LD was initially the highest; r^2 did not drop below 0.2 until 400 kb and remained higher than other breeds until approximately 1,200 kb, and the Standardbred and French Trotter where long-range LD was the highest between approximately 2,000 and 4,000 kb. The Mongolian and Quarter Horse were notable in their relatively short LD in comparison to other breeds, which likely reflects a long breed history and large effective population size in the Mongolian, whereas in the Quarter Horse this likely reflects recent rapid population expansion and large effective population size (Pritchard &

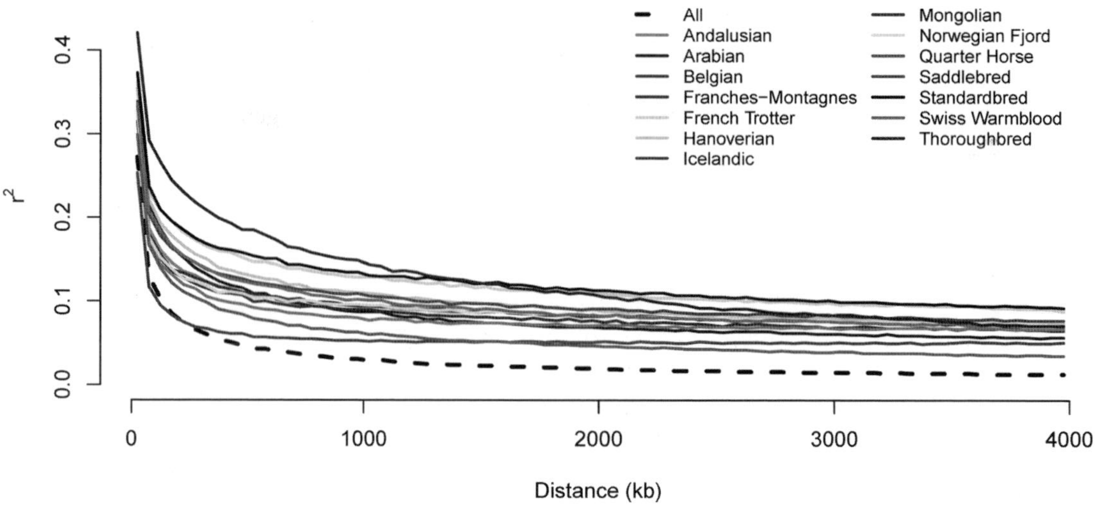

Figure 7.1 Decline in genome-wide linkage disequilibrium across and within breeds. Genome-wide linkage disequilibrium (LD) was estimated both within a given breed and across all breeds by calculating r^2 values between all pairs of SNPs with inter-SNP distances of less than 4 Mb. (*For color details, see color plate section.*)

Przeworski, 2001; Reich et al., 2001). Thus, LD in the 14 breeds evaluated in this study tended to fall into three categories. In the first category are breeds with high LD, such as the Thoroughbred and Standardbred, which would likely result in good power for association mapping. In the second category are those breeds with moderate LD that may have good power for association mapping depending on marker density and locus characteristics. And in the third category are those breeds with low LD that will likely result in poor power for association mapping with the 54,000 marker set. Despite these limitations, initial work has demonstrated the utility of this chip for several mapping applications.

Utility of the Beadchip for Mapping Simple Traits within Breeds

As a proof of principle, mapping of three known coat color loci – the recessive chestnut coat color (*MC1R*) (Marklund, Moller, Sandberg, & Andersson, 1996), the recessive black coat color locus (*ASIP*, agouti) (Rieder, Taourit, Mariat, Langlois, & Guerin, 2001), and the dominant gray locus caused by a 4.6 kb duplication within the *STX17* gene (Rosengren et al., 2008) – were attempted as a part of the Gentrain Project. For the 354 horses included in this study, coat color phenotype was not necessarily recorded; therefore, the phenotype of each individual was predicted in one of two ways: (1) based on that individual's genotype of all nine published coat color loci or (2) based only on the genotype at the coat color locus of interest. These two alternate phenotyping schemes were chosen by the authors to acknowledge the complexity of the epistatic interactions of the known coat color loci. The first phenotyping scenario most closely represents the actual phenotype of the horse: for example, a horse that is homozygous for the recessive *MC1R* allele, and also heterozygous for the dominant *MATP* cream dilution allele, would be visually phenotyped as palomino, not chestnut. The second phenotyping scenario was chosen to represent a simple recessive or dominant trait, thus ignoring known epistatic interactions (although in the case of the gray, which is dominant to all other known loci, the resulting phenotype is the same).

Within-breed mapping of coat color loci was not attempted unless there was a minimum of 6 cases and 6 controls. When coat color phenotype was inferred using all 9 coat color genotypes, the chestnut locus on ECA3 was successfully mapped within 2 of the 6 breeds in which it was attempted (Quarter Horses [22 cases and 24 controls] and Thoroughbreds [11 cases and 26 controls]). In one of the other four breeds (Arabian [7 cases and 17 controls]), no association was found; in the Saddlebred (13 cases and 18 controls), the lowest p-value, while not significant at the genome wide level, did assign the association to the correct region on ECA3. In the remaining 2 breeds (Hanoverian [7 cases and 12 controls] and Swiss Warmblood [8 cases and 11 controls]), however, the suggestive association for the chestnut locus was assigned to the wrong chromosome. The use of the *MC1R* genotype alone to infer coat color did allow for identification of the chestnut locus in one additional breed (Arabian [14 cases and 10 controls]) but did not correct the misassignment in the Hanoverian or Swiss Warmblood. The lack of association and the misassignment of the chestnut locus largely reflect small sample sizes in the study; researchers should be cautioned about drawing conclusions from data with low sample sizes, as the associations of the chestnut locus to ECA8 and ECA9 in the Hanoverian and Swiss Warmblood breeds clearly stood out above background levels. Neither black nor gray was mapped within any one breed when phenotype was inferred from all coat color loci. Black was mapped successfully in the Andalusian (6 cases and 10 controls) when the *ASIP* genotype was considered alone.

The ease by which the chestnut locus was mapped even with small samples sizes reflects the extended homozygosity surrounding the locus due to its centromeric location, which limits recombination, selection for the chestnut trait in many breeds, as well as sufficient SNP density on the chip in this region. In contrast, neither the *ASIP* nor *STX17* loci had high SNP density or large conserved haplotypes, making the mapping of these loci with small sample sizes impractical. Nevertheless, these results demonstrate the utility of whole genome mapping within breeds when studies are sufficiently powered and that power varies among breeds, often in relation to LD. It is also important to note that the rate of false positives likely increases with small sample sizes.

Several reports have demonstrated the use of the SNP50 Beadchip for association mapping. In a study of the genetic basis of Lavender Foal syndrome, an autosomal recessive disorder characterized by foals born with a dilute coat color and a spectrum of neurologic abnormalities. A sample set of 6 affected individuals as well as 30 first- and second-degree relatives was successful in identifying the chromosomal location harboring the disease causing mutation in the gene *MYO5A* (Brooks et al., 2010). The success of this mapping with a small sample cohort likely reflects the long-range LD in the population of Egyptian Arabians utilized, which is similar to that of the Thoroughbred, as well as the simple recessive mode of inheritance of this trait (Brooks et al., 2010). The SNP50 Beadchip has also been used to map another recessive condition in the horse – extreme lordosis in the American Saddlebred. In this study, 20 affected and 20 unaffected individuals were used in a GWAS to identify a chromosomal segment on ECA20 highly associated with the disease phenotype. This association was replicated in an independent sample of 13 affected and 166 unaffected individuals, providing strong evidence for association at this locus, although the causal variant has yet to be identified (Cook et al., 2010).

Use of the Equine SNP50 Beadchip for Mapping of Complex Traits in the Horse

Two independent studies have demonstrated the utility of this assay in mapping performance related traits in the Thoroughbred racehorse [Binns et al., 2010; Hill et al., 2010). In both studies, the goal was to identify chromosomal locations harboring variants that were associated with optimal racing distance in Thoroughbreds, given that previous work has indicated that it is highly heritable (Williamson & Beilharz, 1998). In the first study, 118 elite race horses from Great Britain, Ireland, and New Zealand were used to map chromosomal location using either case control (categorized as short versus long distance) or best racing distance as a quantitative trait. Regardless of whether racing distance was considered as a binary or quantitative trait, a clear strong association was detected on ECA18 (Hill et al., 2010). The authors in this study subsequently identified a SINE insertion in the myostatin gene, which is a strong candidate for differences in racing distance due to its role in muscle development (Hennebry et al., 2009). The second independent study utilized 189 elite Thoroughbreds from North America where racing distance was considered as a categorical trait. This study also demonstrated the highest association on ECA18 in the region of the myostatin locus (Binns et al., 2010), thus corroborating the first study's findings in an independent sample of horses.

Mapping across Breeds

Initial analyses of haplotype sharing in horses demonstrated that domestic horses share a much larger proportion of haplotypes across breeds than other species, including the domestic dog (Wade et al., 2009). Therefore, it is reasonable to hypothesize that for certain traits that are highly conserved across breeds, an across-breeds mapping approach may be reasonable. To demonstrate the utility of this approach in the horse, the Gentrain dataset was used to map the three coat color loci mapped within breeds, mentioned earlier in the chapter. The chestnut locus was easily mapped across breeds to ECA3 using a basic chi-square case-control allelic association analysis, regardless of how phenotype was inferred in a sample population of >100 cases and >200 controls (Figure 7.2). The black coat color locus was identified most clearly using the *ASIP* genotype in a population of >50 cases and >250 controls. Attempts to map the gray coat color locus across breeds by allelic association resulted in a large number of false-positive associations and failure to identify the true locus on ECA25. In all across-breed simple association analyses there was a high false discovery rate and significant inflation of the p-values (based on quantile-quantile plots), likely due to the across-breed population structure. In these analyses, false discovery rates and inflation of p-values dramatically improved when the Cochran-Mantel- Haenszel (CMH) association test was used. The CMH test allowed both the chestnut and black coat color loci to be unambiguously mapped across breeds with no false positives. However, the gray locus (28 cases, 310 controls) was not mapped across all breeds using CMH, or structured association mapping using principal components or mixed-model analyses to control for underlying population structure. The failure to map this locus likely resulted from confounding by population substructure, sparse marker density in the region,

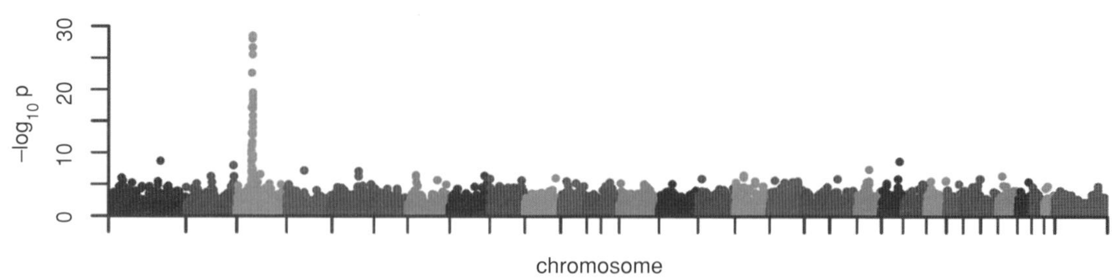

Figure 7.2 Allele association analysis for chestnut coat color loci. Mapping of the chestnut locus across breeds based on coat color phenotype inferred from 9 genotyped loci. (*For color details, see color plate section.*)

and poor power to detect a dominant locus due to low sample sizes both within and across breeds, and was not necessarily a limitation of the across-breed mapping approach in itself. Thus in certain scenarios, in particular in disease association studies where well-phenotyped individuals are difficult to acquire, across-breed mapping may be a viable option for researchers.

Initial successes in association mapping with the SNP chip are encouraging, but the reported success thus far is limited to mapping simple recessive traits, and/or mapping in breeds with high LD. The authors are aware of several ongoing mapping projects for complex traits or mapping in low LD breeds that have yet to be successful using this tool. Thus, ideally, increased genome coverage with more highly informative SNPs would be more effective for mapping studies, particularly in admixed and/or low LD breeds.

Utility of the Equine SNP50 Beadchip for Population Genetic Analysis in the Horse

The availability of a large set of genomic markers in the horse will also allow for population genetic analyses to be completed with a wealth of information about the autosomal genome that has not been previously possible. Initial use of the Equine SNP50 Beadchip for population genetic analyses examined the relationship between breeds with MDS and pair-wise genetic distances, as well as estimates of inbreeding, heterozygositiy (genetic diversity), and LD. Inbreeding and genetic diversity were estimated across the 15 breeds included in the Gentrain data set. In general, inbreeding coefficients and genetic diversity reflected population history and LD patterns, with inbreeding highest and genetic diversity the lowest in the Thoroughbred and Standardbred, two breeds with closed studbooks under intense selection for phenotype. Inbreeding was lowest and genetic diversity highest in the Hanoverian, Quarter Horse, and Mongolian breeds characterized by admixture (Quarter Horse and Hanoverian), rapid population expansion (Quarter Horse), or long breeding history (Mongolian).

The relationships between domestic horse breeds were evaluated by calculating the pair-wise genetic distances between individuals both within and across breeds (McCue et al. 2012). The mean genetic distance (D) between pairs of individuals from different breeds was 0.270, compared to the mean distance of 0.240 between pairs of individuals from the same breed. The mean distance between individuals within a given breed is higher than similar calculations that have been performed in cattle, but lower than those reported in sheep (Kijas et al., 2009). However when the pair-wise distance matrix is partitioned by breed, three distinct peaks are seen (Figure 7.1); with admixed breeds had greater than average pair-wise distances (Quarter Horse and Swiss Warmblood 0.26), while breeds with a history of population bottlenecks, such as the Norwegian Fjord and Icelandic horse, had smaller pair-wise distances (0.21 in both breeds). The clustering of breeds and separation between breeds can be visualized in an MDS plots demonstrated that individuals within most breeds were tightly clustered in relation to other breed groups (Figure 7.2). This was true even for the Thoroughbred population where two geographically distinct sample origins were represented (United Kingdom and United States). The exceptions to this were the three breeds with recent and/or ongoing admixture: the Quarter Horse, Hanoverian, and Swiss Warmblood. In addition, the Hanoverian and Quarter Horse, and to a lesser extent the Swiss Warmblood, had larger variation along dimension 1 than other breeds, suggesting that the admixture may be resulting in significant population substructure. ANOVA of this data showed a significant proportion of the variation (14.3%) is accounted for among the breeds.

Pair-wise genetic distances were also calculated between all domestic horse breeds and the Przewalski's Horse. The Przewalski's Horse, also known as the Asiatic wild horse *Equus przewalskii*, is thought to be a sister species to the Tarpan, the European wild horse *Equus ferus*, which gave rise to the domestic horse (Olsen, 2006). The average distance (D) between Przewalski's Horses and domestic horses was greater than the average D between pairs of individuals drawn from any two different domestic horse breeds; however, there is significant overlap in the distribution of D values in the Przewalski's-domestic pair-wise distances and the pair-wise distances between two distinct domestic horse breeds (Figure 7.1) (McCue et al., 2012). Pair-wise calculations between Przewalski's Horse and individual domestic horse breeds show genetic distances between certain breeds such as the Mongolians, Norwegian Fjords, Belgians, and Icelandics that are less than the average distances (0.27) between domestic horse breeds (McCue et al., 2012). This finding is likely the result of known introgressions from the domestic horse in an effort to prevent extinction (Bowling et al., 2003; Mohr, 1973), and likely interbreeding between *Equus przewalskii* and *Equus caballus* in the wild, due to overlapping range in China, Russia, and Mongolia (Geyer et al., 1989).

Although few publications outside the Gentrain project are available to date, this tool is actively being used by the equine research community to answer population genetics questions, for example the study of LD and estimation of effective population size in a large cohort of Thoroughbred horses (Corbin et al., 2010), and an evaluation of genetic diversity in the Maremmano horse (Felicetti et al., 2010). Furthermore, as an extension of the initial Gentrain analyses, the Equine Genetic Diversity Consortium has been formed to perform a large-scale study of breed diversity in more than 35 breeds using the equine SNP50 Beadchip. At the time of this writing these analyses are underway.

Utility of the Equine SNP50 Beadchip in Extant *Perissodactyla*

The utility of the Equine SNP50 Beadchip was also evaluated in 16 species evolutionarily related to the domestic horse, including domestic and wild asses, zebras, tapirs, and rhinoceroses. The *Perissodactyla* species comprise the odd-toed hoofed mammals separated into three families. The domestic horse, as well as wild horses, asses, and zebras, represents the *Equidae*, the rhino species comprise the *Rhinocerotidae*, and tapirs comprise the *Tapiridae*. The *Perissodactyla* are divided into two suborders, the *Hippomorpha* (horses, asses, and zebras) and the *Ceratomorpha* (rhinos and tapirs) (Price & Bininda-Emonds, 2009). As a component of the Gentrain project, 53 individuals from these 16 different species were genotyped on the SNP50 Beadchip. Although sample numbers in several species were low, and two of the individuals were removed from further analysis due to poor genotyping quality, at least one individual was genotyped in each of the 16 species, allowing for some conclusions to be drawn about the utility of this tool in these species. As expected with SNPs ascertained in the modern horse, the number of SNPs that produced a genotype declines in the species more evolutionarily distant to *Equus caballus*. The number of genotypes was high in the Przewalski's Horse (*Equus przewalskii*), with 54,410 SNPs producing genotypes and a SNP conversion rate of 0.997, whereas the conversion rate and number of genotypes was low in the South African Black Rhino (*Diceros bicornis minor*, 10,661, 0.195).

While assay conversion rate was fairly high in the more closely related species, the number of polymorphic loci was often low and ranged from 346 (0.7%) in the Domestic Ass (*Equus asinus*) to 27,675 (50.9%) in the Przewalski's Horse. Polymorphism rates in this case may reflect species divergence, but were also likely impacted by the very limited number of individuals genotyped in most species ($n = 2$ to $n = 9$), and in two cases only a single representative of the species was genotyped. In all cases, genotyping a larger cohort within each species would be necessary to determine the true polymorphism rates of converted SNPs. However, there was a trend toward SNP validation rate to be lowest in the equids other than the Przewalski's Horse (mean MAF 0.13).

Because this trend likely reflected low numbers of samples in the equids rather than evolutionary distances, the amount of allele sharing between the other species and the domestic horse was determined. For all SNPs that were converted but not validated in each species, the proportion of instances in which the genotyped allele was also the major allele of the domestic horse was calculated. These results show the high proportion of allele sharing in the *Hippomorpha* with allele sharing decreasing in the *Ceratomorpha*, with the proportions ranging from 0.10 to 0.82 in the Great Indian Rhino and Przewalski's Horse, respectively.

Despite variable conversion and polymorphism rates of the 54,602 SNPs validated in *Equus caballus*, the conversion rate across species allowed for visualization of the relationships between *Perissodactyla* using multidimensional scaling (MDS) plots. MDS plots clearly separated the species into four main clusters: (1) the domestic and Przewalski's Horse, (2) the zebras and asses, (3) the rhinos, and (4) the tapirs (Figure 7.3). Parsimony analysis using over 50,000 SNPs in the *Equus* spp. mirrored the relationships seen with MDS – distinguishing *Equus caballus* from *Equus przewalskii* as well as distinguishing those species from the asses and zebras. The older domestic horse breeds, including the Mongolian, Norwegian Fjord, and Icelandic, fell out with strong bootstrap support along with the Belgian and Franches Montagnes, which was seen in MDS analyses. Interestingly,

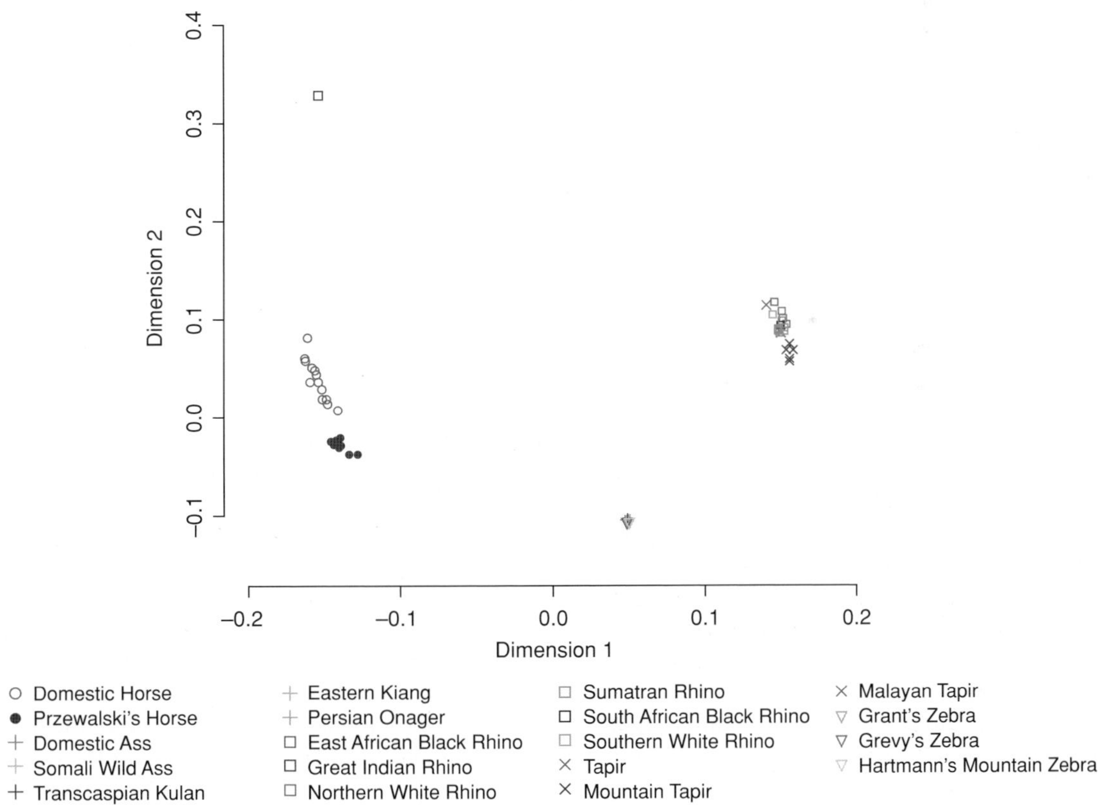

Figure 7.3 Multidimensional scaling with 14 domestic horses and all other extant *Perissodactyla*. Metric multidimensional scaling analysis of pair-wise genetic distance (up to 6 dimensions) was used as described in Materials and Methods to identify relationships between one individual drawn at random from each domestic horse breed and the individuals from each of the 17 other *Perissodactyl* species. (*For color details, see color plate section.*)

high bootstrap values also suggested population substructure within the Przewalski's Horse as well as within the zebras and asses.

The close clustering of the domestic and Przewalski's Horse is not surprising as the horse is thought to have been domesticated from the Tarpan (*Equus ferus ferus*) (Olsen, 2006), a sister species to the Przewalski's Horse (*Equus ferus przewalskii).* It is also important to note that the high proportion of allele sharing between the domestic horse and the Przewalski's Horse may reflect relatively recent gene flow between these lineages (Bowling et al., 2003), as the two species can interbreed despite a difference in chromosome number (2n = 64 and 2n = 66, respectively). This gene flow may have occurred in the wild, as the range of the Przewalski's Horse and the domestic horse overlapped in China, Russia, and Mongolia (Geyer, Thompson, & Ryder (1989), and is also the result of conservation efforts where known introgressions took place during the propagation program that prevented the extinction of the Przewalski's Horse (Mohr, 1973). The known and suspected gene flow between these two species likely resulted in the inability to separate the Przewalski's Horse from the domestic horse in previous phylogenic analyses (Wade et al., 2009). However, using more than 50,000 SNPs, *Equus caballus* and *Equus przewalskii* are clearly distinguished with high bootstrap support (McCue et al., pending publication).

Approximately 4,500 SNPs produced genotypes of at least a 90% genotyping rate across all species, allowing for phylogenetic analysis in the *Perissodactyla*. This application of the tool to other *Perissodactyla* is particularly exciting because, while many prior phylogenetic analyses have examined some of the *Perissodactyl* species (Oakenfull & Clegg, 1998; Oakenfull, Lim, & Ryder, 2000; Weinstock et al., 2005), none have benefited from whole-genome, nuclear SNP data. Interestingly, when phylogenetic analyses were applied to all 16 species included in this data set, the data did not support monophyly of *Equus* spp., with the domestic horse and Przewalski's Horse falling into a separate clade from the remaining *Equus* species. This paraphyly was attributed to bias in SNP ascertainment toward more modern variation. However, within the other species, the relationships among taxa reflect previously reported analyses of mitochondrial and limited nuclear data sets, with the exception of the Asian asses. The classification of the Asian asses has been unclear in prior analyses (Oakenful, Lim, & Ryder, 2000), and the Gentrain data set supported a more recent common ancestry between *Equus hemionus onager* and *Equus kiang holdeneri* than between either and *Equus hemionus kulan*.

Design of a Second-Generation Equine SNP Genotyping Array

After 30 months of availability, the Equine SNP50 Beadchip was removed from the commercial market and has been replaced in early 2011 with a new Illumina Infinium array containing ~74,500 SNP markers. The markers on the second-generation Beadchip have an average of 1.5 SNPs per 50 kb bin and therefore represent a substantial increase in the number of markers across the genome. The SNP markers included on this new genotyping array include ~53,500 markers that were validated on the Equine SNP50 Beadchip, which had a minimum minor allele frequency of ≥0.005 across the 354 samples analyzed in the Gentrain dataset. The ~21,000 additional markers included in the new assay design were chosen to address as many gaps in coverage from the Equine SNP50 Beadchip as possible, and to globally improve coverage across the genome. As there were insufficient SNPs from the seven discovery breeds alone to achieve these goals, ~3,900 SNPs from Twilight and ~2,800 SNPs from RNAseq data (Coleman et al., 2010) were used. In instances where few SNPs were available resulting in gaps in coverage, increased numbers of SNPs were included in bins that flanked the gaps. In addition, the new SNP panel includes enhanced coverage of the MHC region on ECA20 and the X chromosome, as well as several SNPs from coat color loci to use for sample validation purposes.

References

Albrechtsen, A., Nielsen, F. C., & Nielsen, R. (2010). Ascertainment biases in SNP chips affect measures of population divergence. *Molecular Biology & Evolution*, *27*, 2534–2547.

Binns, M. M., Boehler, D. A., & Lambert, D. H. (2010). Identification of the myostatin locus (MSTN) as having a major effect on optimum racing distance in the Thoroughbred horse in the USA. *Animal Genetics*, *41*, 154–158.

Bowling, A. T., Zimmermann, W., Ryder, O., Penado, C., Peto, S. et al. (2003). Genetic variation in Przewalski's horses, with special focus on the last wild caught mare, 231 Orlitza III. *Cytogenetic and Genome Research 102*, 226–234.

Brooks, S. A., Gabreski, N., Miller, D., Brisbin, A., Brown, H. E. et al. (2010). Whole-genome SNP association in the horse: Identification of a deletion in myosin Va responsible for lavender foal syndrome. *PLoS Genetics*, *6*, e1000909.

Chowdhary, B. P., Raudsepp, T., Honeycutt, D., Owens, E. K., Piumi, F. et al. (2002). Construction of a 5000(rad) whole-genome radiation hybrid panel in the horse and generation of a comprehensive and comparative map for ECA11. *Mammalian Genome 13*, 89–94.

Chowdhary, B. P., Raudsepp, T., Kata, S. R., Goh, G., Millon, L. V. et al. (2003). The first-generation whole-genome radiation hybrid map in the horse identifies conserved segments in human and mouse genomes. *Genome Research*, *13*, 742–751.

Coleman, S. J., Zeng, Z., Wang, K., Luo, S., Khrebtukova, I. et al. (2010) Structural annotation of equine protein-coding genes determined by mRNA sequencing. *Animal Genetics*, *41*, 121–130.

Cook, D., Gallagher, P. C., & Bailey, E. (2010). Genetics of swayback in American Saddlebred horses. *Animal Genetics*, *41*, 64–71.

Corbin, L. J., Blott, S. C., Swinburne, J. E., Vaudin, M., Bishop, S. C., & Woolliams, J. A. (2010). Linkage disequilibrium and historical effective population size in the Thoroughbred horse. *Animal Genetics*, *41*(Suppl. 2), 8–15.

Felicetti, M., Lopes, M. S., Verini-Supplizi, A., Da C+ómara Machado, A., Silvestrelli, M. et al. (2010). Genetic diversity in the Maremmano horse and its relationship with other European horse breeds. *Animal Genetics*, *41*, 53–55.

Geyer, C. J., Thompson, E. A., & Ryder, O. A. (1989). Gene survival in the Asian wild horse (Equus przewalskii): II. Gene survival in the whole population, in subgroups, and through history. *Zoology & Biology*, *8*, 313–329.

Guerin, G., Bailey, E., Bernoco, D., Anderson, I., Antczak, D. F. et al. (1999). Report of the International Equine Gene Mapping Workshop: Male linkage map. *Animal Genetics*, *30*, 341–354.

Guerin, G., Bailey, E., Bernoco, D., Anderson, I., Antczak, D. F. et al. (2003). The second generation of the International Equine Gene Mapping Workshop half-sibling linkage map. *Animal Genetics*, *34*, 161–168.

Hennebry, A., Berry, C., Siriett, V., O'Callaghan, P., Chau, L. et al. (2009). Myostatin regulates fiber-type composition of skeletal muscle by regulating MEF2 and MyoD gene expression. *American Journal of Physiology & Cell Physiology*, *296*, C525–C534.

Hill, E. W., McGivney, B. A., Gu, J., Whiston, R., & Machugh, D. E. (2010). A genome-wide SNP-association study confirms a sequence variant (g.66493737C>T) in the equine myostatin (MSTN) gene as the most powerful predictor of optimum racing distance for Thoroughbred racehorses. *BMC Genomics*, *11*, 552.

Karlsson, E. K., Baranowska, I., Wade, C. M., Salmon Hillbertz, N. H., Zody, M. C. et al. (2007). Efficient mapping of mendelian traits in dogs through genome-wide association. *National Genetics*, *39*, 1321–1328.

Leeb, T., Vogl, C., Zhu, B., de Jong, P. J., Binns, M. M. et al. (2006). (2006). A human-horse comparative map based on equine BAC end sequences. *Genomics 87*, 772–776.

Lykkjen, S., Dolvik, N. I., McCue, M. E., Rendahl, A. K., Mickelson, J. R., & Roed, K. H. (2010). Genome-wide association analysis of osteochondrosis of the tibiotarsal joint in Norwegian Standardbred trotters. *Animal Genetics*, *41*, 111–120.

Marklund, L., Moller, M. J., Sandberg, K., & Andersson, L. (1996). A missense mutation in the gene for melanocyte-stimulating hormone receptor (MC1R) is associated with the chestnut coat color in horses. *Mammalian Genome*, *7*, 895–899.

Matukumalli, L. K., Lawley, C. T., Schnabel, R. D., Taylor, J. F., Allan, M. F. et al. (2009) Development and characterization of a high density snp genotyping assay for cattle. *PLOS One*, *4*, e5350.

McCue, M. E., Bannasch, D. L., Petersen, P. L., Gurr, J., Bailey, E. et al. (2012). A high density SNP array for the domestic horse and extant Perissodactyla: utility for association mapping, genetic diversity, and phylogeny studies. *PLoS Genetics*, *8*, e1002451.

Mohr, E. (1973). *The Asiatic wild horse*. New York: J. A. Allen.

No author (2010). Integrating common and rare genetic variation in diverse human populations. *Nature*, *467*, 52–58.

Oakenfull, E. A., & Clegg, J. B. (1998). Phylogenetic relationships within the genus *equus* and the evolution of + l and + + globin genes. *Journal of Molecular Evolution*, *47*, 772–783.

Oakenfull, E. A., Lim, H., & Ryder, O. (2000). A survey of equid mitochondrial DNA: Implications for the evolution, genetic diversity and conservation of Equus. *Conservation Genetics*, *1*, 341–355.

Olsen, S. O. (2006). Early horse domestication on the eurasian steppe. In M. Zeder, D. Bradley, E. Emshwiller & B. Smith (eds.), *Documenting domestication: New genetic and archaeological paradigms* (pp. 245–269). Berkeley: University of California Press.

Orlando, L., Metcalf, J. L., Alberdi, M. T., Telles-Antunes, M., Bonjean, D. et al. (2009). Revising the recent evolutionary history of equids using ancient DNA. *Proceedings of the Nationa Academy of Science USA*, *106*, 21754–21759.

Orr, N., Back, W., Gu, J., Leegwater, P., Govindarajan, P. et al. (2010) Genome-wide SNP association-based localization of a dwarfism gene in Friesian dwarf horses. *Animal Genetics*, *41*, 2–7.

Penedo, M. C., Millon, L. V., Bernoco, D., Bailey, E., Binns, M. et al. (2005). International Equine Gene Mapping Workshop Report: A comprehensive linkage map constructed with data from new markers and by merging four mapping resources. *Cytogenetic Genome Research*, *111*, 5–15.

Price S. A., & Bininda-Emonds, O. R. P. (2009). A comprehensive phylogeny of extant horses, rhinos and tapirs (Perissodactyla) through data combination. *Zool Reihe*, *85*, 277–292.

Pritchard, J. K., & Przeworski, M. (2001). Linkage disequilibrium in humans: Models and data. *American Journal of Human Genetics*, *69*, 1–14.

Raudsepp, T., Gustafson-Seabury, A. et al. (2008). A 4103 marker integrated map of the horse genome. *Cytogenetic Genome Research*, *122*, 28–36.

Reich, D. E., Cargill, M., Bolk, S., Ireland, J., Sabeti, P. C. et al. (2001). Linkage disequilibrium in the human genome. *Nature*, *411*, 199–204.

Rieder, S., Taourit, S., Mariat, D., Langlois, B., & Guerin, G. (2001). Mutations in the agouti (ASIP), the extension (MC1R), and the brown (TYRP1) loci and their association to coat color phenotypes in horses (Equus caballus). *Mammalian Genome*, *12*, 450–455.

Rosengren, P. G., Golovko, A., Sundstrom, E., Curik, I., Lennartsson, J. et al. (2008). A cis-acting regulatory mutation causes premature hair graying and susceptibility to melanoma in the horse. *National Genetics*, *40*, 1004–1009.

Swinburne, J., Gerstenberg, C., Breen, M., Aldridge, V., Lockhart, L. et al. (2000). First comprehensive low-density horse linkage map based on two 3-generation, full-sibling, cross-bred horse reference families. *Genomics*, *66*, 123–134.

Swinburne, J. E., Boursnell, M., Hill, G., Pettitt, L., Allen, T. et al. (2006). Single linkage group per chromosome genetic linkage map for the horse, based on two three-generation, full-sibling, crossbred horse reference families. *Genomics*, *87*, 1–29.

The Bovine HapMap Consortium (2009). Genome-wide survey of SNP variation uncovers the genetic structure of cattle breeds. *Science*, *324*, 528–532.

Wade, C. M., Giulotto, E., Sigurdsson, S., Zoli, M., Gnerre, S. et al. (2009). Genome sequence, comparative analysis, and population genetics of the domestic horse. *Science*, *326*, 865–867.

Weinstock, J., Willerslev, E., Sher, A., Tong, W., Ho, S. Y. W. (2005). Evolution, systematics, and phylogeography of pleistocene horses in the new world: A molecular perspective. *PLoS Biology*, *3*, e241.

Williamson, S. A., & Beilharz, R. G. (1998). The inheritance of speed, stamina and other racing performance characters in the Australian Thoroughbred. *Journal of Animal Breeding and Genetics*, *115*, 1–16.

8 Functional genomics

Stephen J. Coleman, Michael J. Mienaltowski, and James N. MacLeod

Introduction: From Genotype to Phenotype

Functional genomics is the field of science focusing on the biology of an organism's full complement of genetic information, including the specific details of how this information is related to physiological and pathological processes. As the amount of available data on the equine genome has increased through technological advances in high-throughput molecular biology, the focus of functional genomics has expanded to consider not only the roles of individual genes, but also the cooperative and dynamic nature of their regulation and interactions. Fundamental questions in the horse are similar to those in other species and primarily concern the relationship between genotype and phenotype. How does the single genotype defined in an individual at conception generate such a diverse array of cell types, tissues, and organ systems during prenatal development and postnatal maturation? How does the genome regulate cellular and tissue physiology to maintain normal homeostasis? How do sequence variants and other aberrations of the genome contribute to the etiology and pathogenesis of different diseases?

The central dogma of molecular biology provides a useful framework for functional genomics. Each component – the genome, transcriptome, and proteome – yields insight into gene function and ultimately the generation of distinct phenotypes. The genome defines the structural organization of nucleotide sequences and provides a valuable reference for functional genomics. Nearly every aspect of biological structure and function in an organism can trace much of its phenotype directly or indirectly to information encoded in the genome. The transcriptome and proteome are expressions of the informational content of the genome. They enable adaptability and variability in gene expression that result in the observed diversity and complexity of biological phenotypes (Figure 8.1). This chapter concentrates on concepts of functional genomics and discusses (1) the functional contributions and components of the equine genome, transcriptome, and proteome, (2) network analysis and data integration to capture biological understanding from functional genomics, and (3) the modeling of complex biological systems. These sections are used to demonstrate how functional genomics can be used to describe the relationship between genotype and phenotype and improve our understanding of equine biology on a molecular, cellular, organ system, and organismal level.

Equine Genomics, First Edition. Edited by Bhanu P. Chowdhary.

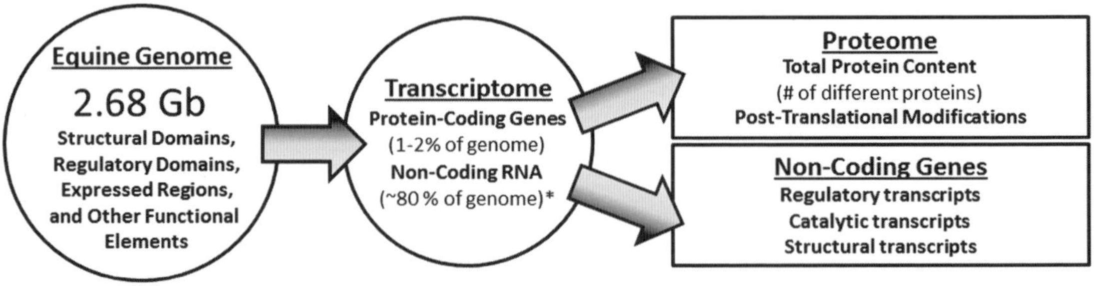

Figure 8.1 Overview of functional genomics (as covered in the current chapter). Functional genomics in the horse denotes how the information encoded within the 2.68 billion bases of genomic DNA is expressed and regulated in equine biology. Biological functions are the product of combined contributions from the genome, transcriptome, and proteome together with non-genetic (epigenomic/environmental) variables. Functional genomics research works toward an understanding of the relationship between genotype and phenotype. *ENCODE Project Consortium (2007).

Functional Impact of the Genome

Genomic DNA contains the "informational blueprint" of an organism, which has now been elucidated in the horse by sequencing of the equine genome (Wade et al., 2009). Understanding the relationship between the primary DNA sequence and biological function, however, requires the identification and characterization of functional sequence elements together with their regulation and interactions.

Functional elements of the encoded genome

Genomic DNA contains many different functional elements. Three broad functional classes are: (1) protein-coding genes, (2) non-coding RNA genes, and (3) regulatory sequences associated with transcription (The ENCODE Project Consortium, 2004). Annotation of these elements is critical to interpreting the information contained within the genome sequence and applying it to the study of equine biology.

Protein-coding genes are often the best-characterized functional elements in a genome. The first level of structural annotation is a gene's location or genomic interval. Important features within this interval include the transcriptional start and stop sites, exon/intron boundaries, translational start and stop sites, conserved domain sequences, polyadenylation signals, and the 5' and 3' untranslated regions. Any change to these elements or their specific combination has the potential to alter the structure and function of the encoded gene product or expression characteristics. The gold standard for the annotation of protein-coding gene structure is the generation and alignment of full-length cDNA sequence (Brent, 2008). Alternatively, sequence homology and *in silico* sequence analyses can be used to accurately predict and annotate gene structures. After sequencing and assembly of the equine genome (Wade et al., 2009), these strategies were used with the Ensembl Automatic Gene Annotation System (Curwen et al., 2004) and NCBI Gnomon eukaryotic gene prediction tool (http://www.ncbi.nlm.nih.gov/genome/guide/gnomon.shtml) to predict equine gene structure. Both approaches are computational pipelines that rely on heuristic sequence alignment algorithms (BLAST) and *ab initio* gene modeling. The resulting predicted gene sets represented an important development in horse genomics, describing for the first time the "whole" set of equine protein-coding genes. At the time of these analyses, however, there were very limited equine expressed sequence data available in public databases to facilitate gene structure predictions (only 35,702 ESTs and 1,236 mRNA transcripts, UCSC Genome Browser). As a result, a substantial majority of the predictions (20,436 genes from Ensembl; 17,610 genes from NCBI) were based on the projection of gene structure from other mammalian species. Most recently, messenger RNA sequencing (RNA-seq) experiments have expanded available equine transcriptome data several hundred fold and enabled many refinements to the *in silico* predictions (Coleman et al., 2010b). Revised consensus models of equine protein-coding gene structure have been generated from these efforts (http://macleod.uky.edu/equinebrowser/) and will continue to improve.

Protein-coding genes account for only a small portion of the entire equine genome sequence (1–2%). Recent findings suggest that a majority of the remaining genome may also be transcribed into a diverse population of different RNA species. These RNAs are collectively referred to as non-coding (i.e., do not encode a protein), though they clearly have a number of important functions. Several of the non-coding RNA types are well known (rRNA, tRNA, snoRNA, and miRNA), with functions that include involvement in the translation of messenger RNA into protein. However, the population of identified non-coding RNAs has grown substantially in recent years to comprise more than 30 different classes of transcripts. Most are believed to have regulatory functions, though their specific roles remain an area of very active investigation. Annotation of these elements will be required for a full understanding of transcriptional and translational dynamics and how information stored in the genome is accessed, regulated, and functionally important. Commonly used approaches for locating and annotating non-coding transcripts include the alignment of experimentally derived sequence, either from a tiling microarray or direct sequencing method, and computational sequence analysis to identify conserved sequence motifs associated with a specific class of non-coding RNA (Adelson & Raison, 2010). In the horse, Zhou et al. (2009) used a predictive and comparative approach to identify more than 400 novel equine microRNAs.

In addition to the transcribed regions, the genome contains a number of sequence elements that participate in the regulation of gene expression. These regulatory elements are categorized primarily as "cis-acting" elements (regulatory sequences in close proximity to the gene or genes they affect) and long-range elements (regulatory elements that exert influence at a distance). Transcriptional regulatory elements in the genome are recognized by, and in most cases interact with, "trans-acting" factors (transcription factors, accessory proteins, RNAs, enzymes, and metabolites) to direct (regulate, influence, modify) transcriptional parameters. Names commonly applied to specific regulatory sequence elements include promoters, enhancers, repressors, insulators, locus control regions, and positive or negative molecular response elements. Although not expressed directly, annotation of these regions is an important consideration in functional genomics.

Genetic interactions

Primary annotation of a segment of DNA within the genome often represents only the first level of functional assessment as it relates to phenotype and the biology of an organism. Higher-level understanding includes how individual DNA segments (or their encoded products) functionally interact. These interactions can be divided into two broad types. First, there are the interactions of different functional elements (regulatory sequences, protein-coding genes, non-coding RNAs) to generate a specific set of transcripts. Second, there are functional interactions involving two or more genes or their encoded products (epistasis). Systematic analysis of these genetic interactions helps provide a more complete understanding of pathways and networks involved in different biological processes.

Epigenetics/epigenomics

The genome also carries information that is not directly encoded by the nucleotide sequence, but which can clearly influence gene expression and other functional parameters. This information is generally referred to as epigenetic or epigenomic because it is carried outside of the primary genome sequence. Examples of two main types of epigenetic modifications that can influence gene expression include DNA methylation and chromatin structure. DNA methylation is the addition of a methyl group by a DNA methyl transferase to convert cytosine nucleotides to 5-methylcytosine. This methylation most commonly occurs at GC repeats (called CpG islands), with high levels of methylation correlated to low levels of transcriptional activity. A subset of epigenomic parameters, including some specific patterns of methylation, can be transferred from parent to offspring. This process is referred to as genomic imprinting. Factors such as uterine environment can influence imprinting, as suggested by studies in mule and hinny pregnancies that result in different levels of equine chorionic gonadotrophin (Allen et al., 1993; Antczak et al., 2011). Epigenetic effects involving chromatin structure are often the result of histone modifications. The close association of histone structure and the DNA strand influence physical access of the transcriptional machinery to the nucleotides. DNA sequence stabilized in close association with histones can be sequestered from transcription while DNA in loose association is physically more accessible, leading to differential patterns of expression. Patterns of histone modification can be inherited, though specific mechanisms are not fully known. Indeed, many aspects of epigenetic modifications and their impact on functional genomics and phenotype remain a very active area of research (Beisel & Paro, 2011).

The Transcriptome

The transcriptome is the total sum of all RNAs transcribed from the genome. Until recently, the transcriptome was considered mainly as a "bridge," serving as a mechanism to transfer information between genomic DNA and proteins (Costa et al., 2010). High-throughput sequencing and other technological advancements have dramatically expanded our understanding of transcriptome complexity (Lindberg & Lundeberg 2010), which includes not just the messenger, ribosomal, and transfer RNAs, but also an expanding list of additional transcript classes (Table 8.1).

Profiling the equine mRNA transcriptome

The mRNA transcriptome refers to all the RNAs transcribed from the genome that code for proteins. Quantitative and qualitative assessment of gene expression on a transcriptome level enables broad analyses of gene expression and their collective impact on cellular activity in normal and pathological conditions (Hoheisel, 2006). The original "one gene → one transcript → one protein" model has been replaced by knowledge that a single protein-coding gene can in fact produce multiple distinct transcripts, explaining the apparent discrepancy in the number of different proteins compared to the number of protein-coding gene loci. Results from the ENCODE projects have shown that, on average, there are 5.4 transcripts generated from every protein-coding gene locus (Lindberg & Lundeberg 2010). A number of genome-wide techniques have been developed to assess the mRNA transcriptome. These include the generation of expressed sequence tags (ESTs; Nagaraj et al., 2007), serial analysis of gene expression (SAGE; Velculescu et al., 1995), microarrays (Schena et al., 1995), and the analysis of mRNA by next-generation sequencing (RNA-seq; Wang et al., 2009).

Table 8.1 Summary of identified transcript types

Protein Synthesis	
Messenger RNA	mRNA
Ribosomal RNA	rRNA
Signal Recognition Particle RNA	SRP-RNA
Transfer RNA	tRNA
Transfer Messenger RNA	tmRNA
Promoter Associated Short RNA	PASR
Transcription Start Site RNA	TSSa-RNA
Transcription Initiation RNA	tiRNA
Termini Associated Short RNA	TASR
Transcriptional Modification and DNA Synthesis	
Small Nuclear RNA	snRNA
Small Nucleolar RNA	snoRNA
SmY RNA	SmY
Small Cajal Body-Specific RNA	scaRNA
Guide RNA	gRNA
Y RNA	–
Telomerase RNA	–
Regulatory	
Antisense RNA	aRNA
Natural Antisense Transcripts	NAT
Natural Antisense Transcripts Small Interfering RNA	natsiRNA
Long Non-coding RNA	lncRNA
Micro RNA	miRNA
snoRNA-derived RNA	sdRNA
Piwi-interacting RNA	piRNA
Small Interfering RNA	siRNA
Transacting Small Interfering RNA	tasiRNA
Repeat Associated Small Interfering RNA	rasiRNA
Long-Interspersed Non-coding RNA	lincRNA
Promoter Upstream Transcripts	PROMPT
Cryptic Unstable Transcripts	CUT
Catalytic (Ribozymes)	
Ribonuclease P	RNase P
Group I Self-Catalytic Intron	–
Group II Self-Catalytic Intron	–
Mammalian CPEB3 Ribozyme	CPEB3 Ribozyme
Glucosamine-6-phosphate Activated Ribozyme	glmS Ribozyme
Beta-globin Co-transcriptional Cleavage Ribozyme	CotC Ribozyme

To date, microarrays have served as a primary experimental method for analyzing gene expression in the horse on a transcriptome level. Objectives of these studies have been the identification of transcripts with differential patterns of expression associated with specific equine physiological or pathological traits (reviewed by Chowdhary & Raudsepp 2008; Ramery et al., 2009). The earliest reported application of microarrays for equine gene expression analysis was conducted using a non-equine array, relying on sequence conservation across species to detect gene-specific hybridization. Mousel et al. (2002) profiled equine PBMC gene expression using a human cDNA microarray. Human microarrays were subsequently used to profile steady state mRNA levels in equine testicular tissue (Ing et al., 2004), equine superficial digital flexor tendon (Nomura et al., 2007), chronic equine respiratory disease (Ramery et al., 2008), and in equine brain, liver, and articular chondrocytes (Graham et al., 2009). Mouse-specific microarrays have been used to profile gene expression in equine blood and muscle cells (Barrey et al., 2005; Barrey et al., 2006; Mucher et al., 2006). Budak et al. (2009) used a Bovine GeneChip to analyze laminar tissues in the hoof during the developmental phase of carbohydrate-overload-induced laminitis.

Concurrently, efforts were under way in multiple laboratories to develop equine-specific microarray platforms constructed with probes representing a subset of the equine mRNA transcriptome. The first published report was by Gu and Bertone (2004) and included probes for 3,098 expressed equine sequences on an Affymetrix platform. Performance was evaluated using lipopolysaccharide stimulated synoviocytes, and the array was subsequently used to investigate the pathology of equine musculoskeletal diseases and treatment strategies, and stem cell differentiation (Smith et al., 2006; Zachos et al., 2006; Santangelo et al., 2007; Yuan et al., 2008; Murray et al., 2010). The second equine microarray, a spotted cDNA array, had probes representing 1,000 expressed equine sequences and was used to examine the gene expression profiles of pro-inflammatory conditions (Vandenplas et al., 2005a) with particular emphasis on the in vitro effects of bacterial cell wall toxins on leukocyte gene expression (Vandenplas et al., 2005b). This array was subsequently expanded to represent 3,076 expressed equine sequences and used to study temporal aspects of equine laminitis (Noschka et al., 2009). The third equine-specific array had 9,322 spotted cDNA sequences representing 5,307 different genes (Figure 8.2; Macleod, 2005). The cDNA sequences were isolated from an articular cartilage library with sequence identity determined using BLAST to define gene homology and DAVID to discern gene ontology. This array has a fairly broad representation of expressed equine genes (MacLeod et al., 2003; Coleman et al., 2007; Mienaltowski et al., 2008b) and has been used for a number of transcriptional profiling experiments, including articular cartilage maturation and repair (Mienaltowski & MacLeod, 2006; Mienaltowski et al., 2008; Mienaltowski et al., 2009; Mienaltowski et al., 2010), optimal culture conditions for articular chondrocytes (Miura & MacLeod, 2006), identification of both stable and highly differentiated patterns of gene expression (Zhu et al., 2007; Tremblay et al., 2009, Vanderman et al., 2011), and muscle exercise physiology (McGivney et al., 2007; McGivney et al., 2009). Data produced by this array has also been used to develop statistical methods for the analysis of microarray data (Huang et al., 2008a, 2008b). Another equine-specific microarray developed on the Affymetrix platform with 12,320 probe sets has been used to investigate gene expression in articular cartilage from young horses (Nixon et al., 2008). This array was also used to compare articular cartilage to other equine tissues (Glaser et al., 2009). Finally, Barrey et al. (2009) constructed an array that included 334 probe sets for nuclear transcripts and 50 probe sets representing mitochondrial genome features. This array was used to study gene expression in skeletal muscle biopsies from horses with polysaccharide storage myopathy.

With completion of the equine reference genome sequence (Wade et al., 2009) and subsequent *in silico* gene structure predictions (Ensembl and NCBI), several groups initiated efforts to construct improved equine-specific microarrays, which for the first time approached an assessment of all protein-coding genes in the equine genome. Three arrays have been generated using the Agilent Custom Array platform. Miller et al. (2009) used available equine EST and UniGene sequences combined with structural predictions from NCBI to produce an array with 14,357 probe sequences. Array performance was assessed by comparing expression patterns between invasive and noninvasive trophoblast cells in the equine chorionic girdle and membrane. Klein et al. (2010) used the Ensembl gene structure predictions and expressed sequences generated by 454 pyrosequencing to construct an array with 43,803 probe sequences. The array was used for transcriptional profiling of equine endometrium during maternal recognition of pregnancy. Independently, Agilent developed a commercially available array using the Ensembl gene structure predictions. This array included

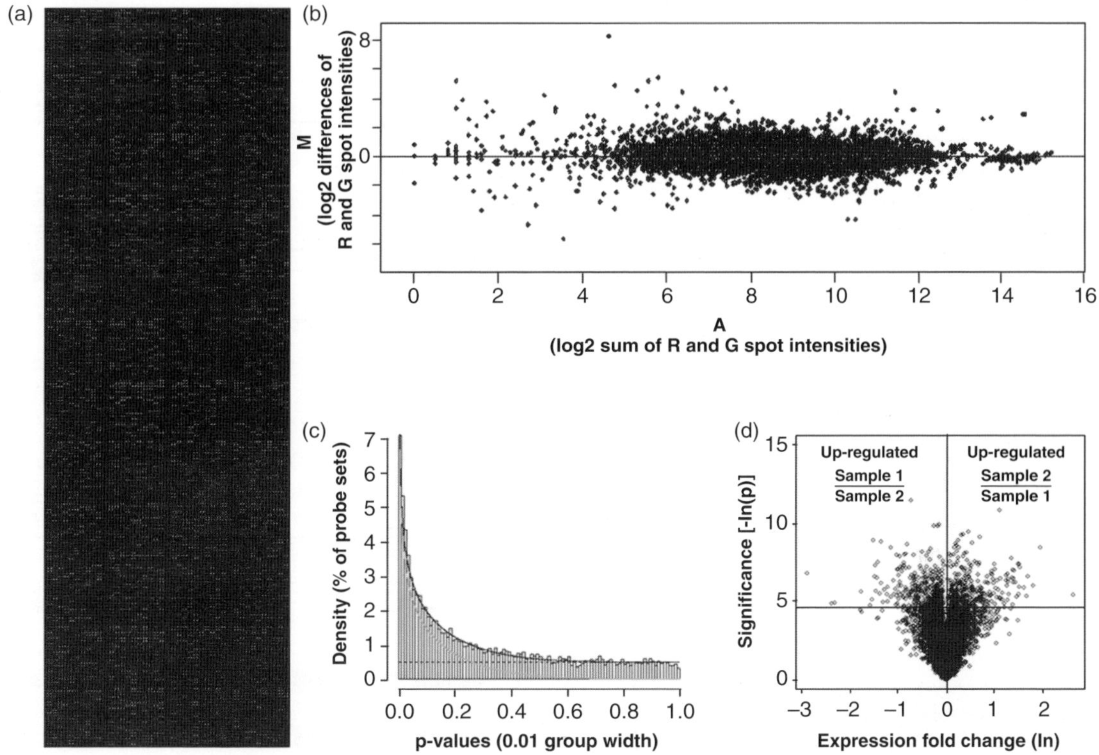

Figure 8.2 Presentation of two-color microarray data. Two-color array data are visualized by detection of the fluorescence emission of labeled cDNAs (either green or red wavelengths, depending on label assignment), which are excited by lasers (a). In this example, labeled equine cDNAs in the experimental sample hybridized to known target cDNAs (probes) within spots on the microarray slide. Resulting fluorescence data for array spots are then reported on an *MA plot* (b), where $M = \log_2 R - \log_2 G$ and $A = \frac{1}{2} (\log_2 R + \log_2 G)$ [R = red fluorescence units and G = green fluorescence units]. MA plots provide information on the degree of fluorescence difference between samples, M, and the level of fluorescence intensity on the slide, A. Data are then statistically analyzed, and a *p-value histogram* (c) is generated to visualize the number of spots with significant expression differences. Typically, spots are divided into 100 bins along the x-axis with 0.01 significance units per bin; thus, bins to the far left represent small p-values reflecting transcripts with significant differential expression between the experimental groups. Expression data can also be depicted in a *volcano plot* (d) where expression differences are plotted along the x-axis and significance of the differences along the y-axis. Images are based on data from Mienaltowski and MacLeod (2006)

43,553 probe sequences and was also used to compare transcriptional profiles of equine endometrium during early pregnancy (Merkl et al., 2010). A fourth whole-genome array was developed using 70mer oligonucleotide probes designed from the equine reference genome sequence and mapped to gene ontology entries (Bright et al., 2009). This array has been used to generate expression profiling data in studies of laminitis (Wang et at., 2009; Wang et al., 2010) and recurrent airway obstruction (Kachroo et al., 2010).

Not all transcriptional profiling studies in the horse have been microarray-based. Suppression subtractive hybridization has been used to profile differentially expressed genes in wound repair (Lefebvre-Lavoie et al., 2005) and equine neonatal growth cartilage (Johannessen et al., 2007). Cappelli et al. (2005) used cDNA-amplified fragment length polymorphism techniques to study transcript profiles in different equine tissues. Illumina digital gene expression has been used to profile gene expression of leukocytes from horses with osteochondrosis (Serteyn et al., 2010).

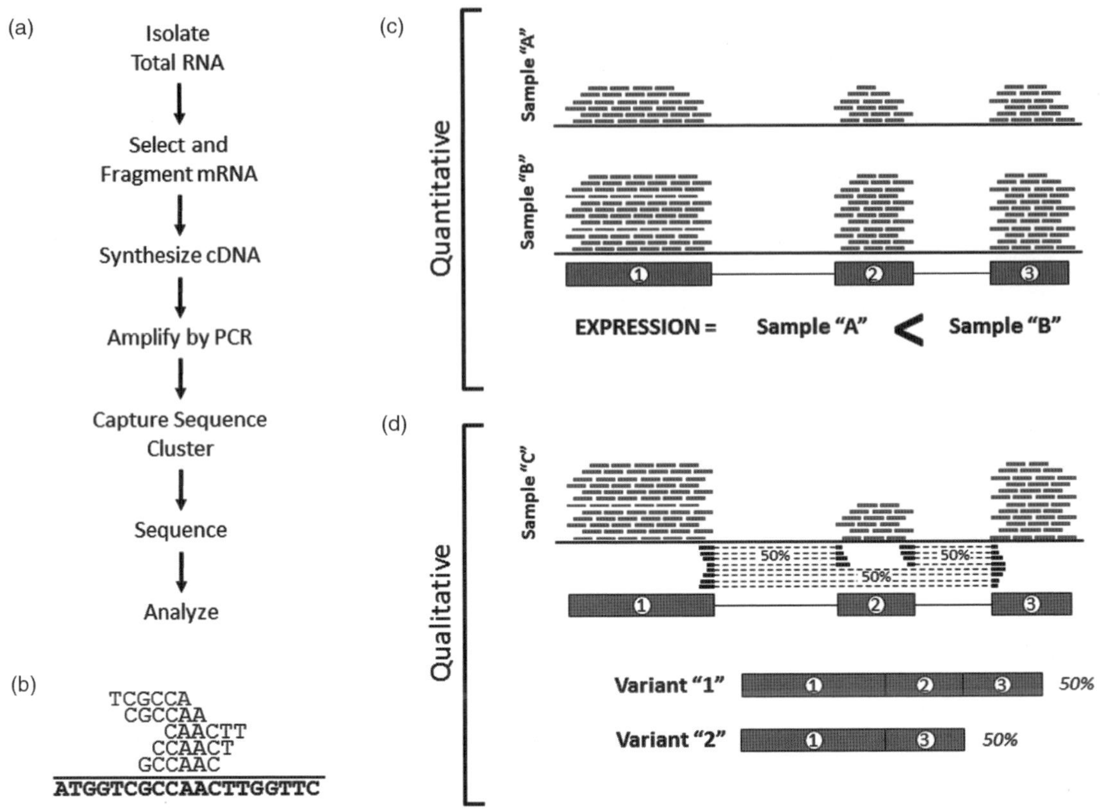

Figure 8.3 Presentation of RNA-sequencing data. RNA-sequencing data can be generated using several next-generation sequencing platforms. Second-generation sequencing platforms (454, Illumina, and ABI SOLID) use different sequencing reactions (either sequencing-by-synthesis or sequencing-by-ligation), but share common protocol features (a). The so-called third-generation or single-molecule sequencing platforms generally bypass the PCR amplification step. Read lengths vary by platform and sequencing assay, but are typically 20 to several hundred nucleotides long. The first step in the analysis of RNA-seq data is to map the sequence reads by aligning them to the appropriate reference genome (b). Relative gene expression can be determined by counting the number of tags that align to the annotated coding region of a gene relative to gene size and comparisons can be made between different tissues or samples (c). Tags that map to the reference genome with a gapped alignment represent splice junctions in the mature mRNA transcripts and can be used to track splicing and identify patterns of alternative splicing within or between samples (d).

Most recently, RNA-sequencing methods have been used to refine equine gene structure models and evaluate expression patterns across eight equine tissue samples (Coleman et al., 2010a; Coleman et al., 2010b, Vanderman et al., 2011). RNA sequencing or transcriptome shotgun sequencing (RNA-seq; Wang et al., 2009b) has been used to investigate the transcriptomes of several species including human, mouse, yeast, and Arabidopsis (Cloonan et al., 2008; Lister et al., 2008; Morin et al., 2008; Mortazavi et al., 2008, Nagalakshmi et al., 2008; Pan et al., 2008; Rosenkranz et al., 2008; Sultan et al., 2008; Wang et al., 2008; Wilhelm et al., 2008). It has been a transformative technology, generating quantitative and qualitative data concurrently on a transcriptome scale (Figure 8.3). A key benefit of RNA-seq is the unprecedented view it provides of alternative transcripts from the same gene, particularly in the ability to map splicing variants (Lindberg & Lundeberg, 2010). An intensive study of splicing patterns demonstrated that more than 90% of multi-exon human genes are subject to alternative splicing and that more than half of all splicing events occur in a tissue-restricted pattern (Wang et al., 2008). Significant efforts have been made to develop methods capable of identifying RNA splicing events. Computational tools now exist that use RNA-seq tags to detect splice junctions, notably MapSplice (Wang et al., 2010b), TopHat (Trapnell et al., 2009), MMES (Wang et al., 2010c), SplitSeek (Ameur et al., 2010), and SpliceMap (Au et al., 2010).

The non-coding RNA transcriptome

The non-coding RNA (ncRNA) transcriptome is made up of all the RNA transcribed from the genome that does not code for a protein. Originally, the non-coding transcriptome was thought to be limited to only ribosomal and transfer RNA transcripts. The functional roles of these transcripts in protein synthesis are fairly well described. This narrow view suggested a limited complexity of the non-coding transcriptome and a belief that a majority of the genome sequence was "junk" DNA having no major biological impact. The identification of micro RNA (miRNA), small nuclear RNA (snRNA), and small nucleolar RNA (snoRNA) transcripts began to challenge this belief by revealing additional ncRNA species and suggesting functional roles for the non-coding transcriptome in post-transcriptional and translational regulation. The application of tiling arrays and next-generation sequencing to transcriptome analyses has revealed an even broader level of transcriptional activity and has led to a substantial expansion in the number of named classes of RNA transcripts (Table 8.1). These newly identified transcript classes (numbering close to 30) can be divided into long and short non-coding RNA (lRNA and sRNA) with many having defined roles in protein synthesis through the regulation of transcription, post-transcriptional modifications, and translation (Stefani & Slack, 2008; Mercer et al., 2009). Additional transcript types are known to exist, but their functions are not currently defined. This newly identified level of complexity is reshaping the perception of the non-coding transcriptome, suggesting that it plays a major role in regulating and fine-tuning cellular processes (Barrandon et al., 2008). While these transcript types are observed across species with some level of sequence conservation, important details may be species-specific. Studies of the non-coding transcriptome are are being initiated in the horse (Zhou et al., 2009; Adelson & Raison, 2010).

The Proteome

Complexity of the proteome is comparable to that of the transcriptome and is generated by a combination of translational and post-translational mechanisms. Although the primary amino acid sequence is defined by the nucleotide sequence of codons in mRNA transcripts, post-translational modifications are variable and include enzyme-catalyzed additions of chemical groups (e.g., glycosylation, methylation, phosphorylation, and metal ion binding) and the proteolytic cleavage of precursor proteins at specific sites (Walsh et al., 2005; Table 8.2). These post-translational modifications are important determinants of protein function and stability. Published proteomic studies in the horse include analyses of milk components (Miranda et al., 2004; Zava et al., 2009), recurrent uveitis (Deeg, 2009), blood doping (Barton et al., 2009), rhabdomyolysis (Bouwman et al., 2010), and airway disease (Bright et al., 2011). Proteomic approaches are being further enabled on equine samples by mRNA nucleotide data providing accurate predictions of primary amino acid sequence, thereby facilitating annotation of peptide fragments.

Proteomic data will be an essential part of equine functional genomics analyses. Transcriptional and translational parameters are not always closely coupled, such that differential levels of steady

Table 8.2 Post-translational protein modifications

Covalent Modifications
(enzyme-catalyzed addition of a chemical group)
- Phosphorylation
- Acylation
- Alkylation
- Glycosylation
- Oxidation

Cleavage of Peptide Backbone
(enzyme-catalyzed or autocatalytic cleavage of a protein)
- Extracellular proteolysis
- Cosubstrate-catalyzed cleavage (H_20, alcohol, or other adduct)
- Intramolecular nucleophile autocatalytic cleavage

Typical Secretory Protein Processing
- Signal peptide cleavage in ER
- Pro-protein convertase cleavage in Golgi
- Extracellular domain removal at plasma membrane
- Intramembrane proteolysis

Post-translational modifications include the covalent addition of various biochemical functional groups and peptide cleavage events.

state mRNA do not always parallel changes in the encoded protein. Further, while some post-translational modifications can be predicted from the nucleotide sequence of the mRNA, it is far less certain than the primary amino acid sequence. Finally, proteomic data often can contribute directly to diagnostic or therapeutic objectives. An example is biomarker discovery. Originally, protein biomarkers were either studied individually to determine if a specific candidate protein was associated with the molecular physiology or pathology of a system, or they were determined by subtractive studies of bands on a gel or blot that compared samples from treated or affected individuals versus untreated or unaffected samples (Cox & Mann, 2007; Ruiz-Romero & Blanco, 2010). In contrast, biomarker screens using proteomic approaches can be highly automated and enable high-throughput analyses (Rabilloud et al., 2010; Ruiz-Romero & Blanco, 2010; Tunon et al., 2010). As with transcriptional profiling, proteomic analyses can be both qualitative (identity, form, tissue localization) and quantitative (abundance of forms). Results are then interpreted in context with other functional genomics parameters to understand more fully the phenotype of interest.

Networks

Genes and their encoded functional products, RNA or protein, act as components within pathways and networks where coordinated functions lead to biological processes. In linear pathways, genetic relationships relevant to a specific cellular process or function are considered in series. This pathway concept reduces the complexity of cellular functions, but is limited because it overlooks the fact that many genes have roles in multiple processes. As such, it is often biologically more accurate to represent functional relationships between genes as a network. Networks attempt to model interactions as a set of "nodes" and "edges." Nodes represent the specific genes, transcripts, or proteins. Edges depict the functional relationship(s) between them. Some nodes are further characterized as "hubs," indicating a biologically active component that has multiple functional relationships (edges)

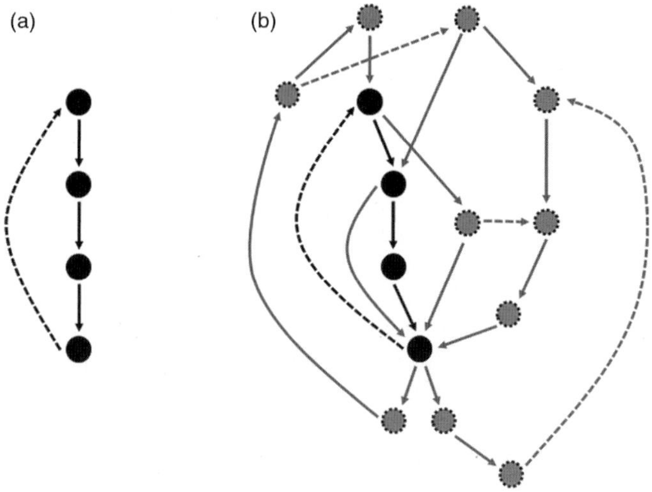

Figure 8.4 Pathways and networks. Most biological processes are based on a set of individual components that interact in a coordinated and regulated manner. Pathways model interactions in a simplified linear framework (a). Networks maintain the "canonical" pathway, but also consider interactions and integration with other components of the biological system (b).

to other nodes. As such, hubs are potential decision points or pathway divergence points that can be regulated and alter functional outcomes (Figure 8.4).

Networks can be developed as highly integrated models that attempt to illustrate relationships between multiple functional layers of a biological system (i.e., genes, mRNA, ncRNA, proteins, etc.), or they can focus on specific subcomponents of the whole system. Some of the more common types of biological networks constructed for analyses in functional genomics include: gene regulatory networks, co-expression networks, functional interaction networks, and social networks. Gene regulatory networks describe parameters that regulate expression, like the transcription of a gene or translation of a protein. Co-expression networks are assembled by grouping genes together that have a similar expression pattern. Functional interaction networks model genetic relationships based on established functional annotation and canonical pathways. Social network analyses attempt to describe gene function in terms of both protein-protein and protein-metabolite interactions. In all cases, however, network analyses as applied to functional genomics attempt to integrate individual components and processes in a biological context in order to better understand the relationship of genotype to phenotype.

Capturing Biological Understanding from Functional Genomics

This chapter has overviewed analyses of the equine genome, transcriptome, and proteome in a functional context and introduced conceptual models of integration through pathways and networks. The goal of functional genomics is to define the relationship between genotype and phenotype. For example, evaluation of the genome sequence may describe causative sequence variants (mutations, indels, SNPs, microsatellites, chromosome morphology, etc.) that help establish and refine patterns of inheritance. Analysis of the transcriptome and proteome can help identify features of gene expression associated with a particular disease. A key point is that while the analysis of each component provides valuable insight into deciphering biological processes, each of these individual parts needs to be integrated with the others to achieve a comprehensive understanding. The number of different approaches available for evaluating function and making biological interpretation are not restricted to the genome, transcriptome, and proteome. Some additional approaches are: metabolomics (analysis of all cellular metabolites and their processing), the regulome (the entire repertoire of a cell's regulatory elements), and the interactome (all possible molecular interactions within a cell). These alternative "omics" approaches can offer unique perspectives to specific aspects of the central dogma of molecular biology and in some cases genetic-epigenetic-environmental interactions. What they all share is a system or global level of data collection with a focus on integrative analysis for developing biological knowledge and understanding.

Potential benefits of integrative analyses can be illustrated with the orthopaedic disease osteochondrosis (OC), a major biomedical problem in the horse. It has a relatively high rate of occurrence (Grondahl & Dolvik 1993; McIlwraith, 2002) and is defined as a "focal disturbance of endochondral ossification" (Ytrehus et al., 2007, p. 429), resulting from abnormal development at the articular cartilage/subchondral bone interface. Severity can be categorized based on the level of progression, ranging from lesions restricted to the epiphyseal cartilage to lesions that include extensive structural defects of the articular surface. Osteochondrosis is a multifactorial disease and highlights the complex research challenges that limit progress on many important equine health problems. At the same time, however, the application of broad data-driven approaches developed for functional genomics provides important new opportunities for scientific investigation. Figure 8.5 illustrates a systems-level approach to the investigation of osteochondrosis using the full spectrum of available "omics" tools. As the studies are performed, it will be important to consider these datasets together

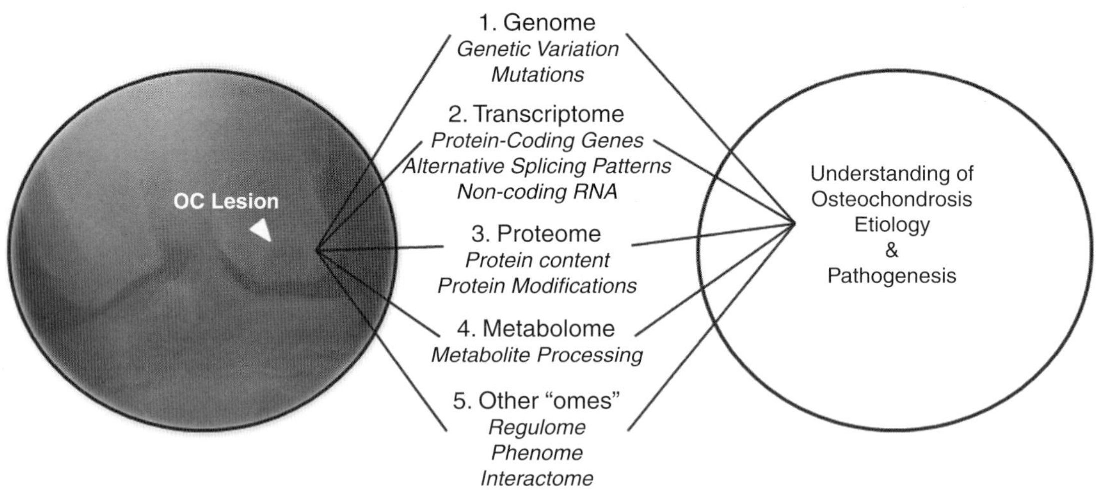

Figure 8.5 Using functional genomics to investigate and model complex disease phenotypes. Major biomedical problems are frequently complex and multifactorial. Osteochondrosis is an example of an important equine disease being studied using broad and data-driven "-omic" approaches. Analysis of the genome may reveal a mutation or sequence variant associated with the disease. Transcriptome and proteome analyses may identify molecular mechanisms as manifested through altered patterns of gene expression. Changes in the metabolome (or other "omic" category) might then reflect the downstream pathophysiological consequences on a cellular or tissue level. Taken together, the goal is to generate a "network of understanding" about the etiology, pathogenesis, and biomedical consequences of a disease, providing a basis for developing new diagnostic, therapeutic, and management strategies. (Osteochondrosis image courtesy of Dr. Katie Garrett, used with permission).

in order to facilitate an understanding of how changes in one component impact the others. This level of integration will elucidate not only potential diagnostic and therapeutic targets, but also breeding and/or management strategies that may limit or even prevent the disease.

Conclusion

The application of functional genomics to equine biology represents a transformative new research opportunity. The initial discovery phase of science is shifting to an open and unbiased profile of all elements (genes, regulatory regions, transcripts, proteins, metabolites, etc.), with consideration of how they interact on a systems level. This enables a data-driven approach for the selection of specific targets for more detailed study. Many questions in equine biology remain to be addressed and many more need to be reexamined. The concepts of functional genomics described in this chapter together with emergent and powerful analytical methods are providing exciting opportunities to greatly advance our understanding of the horse.

References

Adelson, D. L., & Raison, J. (2010). High-throughput sequencing analysis for equine miRNA discovery. *Equine Genome Workshop, Plant and Animal Genome XVIII*, San Diego, CA (W208).

Allen, W. R., Skidmore, J. A., Stewart, F., & Antczak, D. F. (1993). Effects of fetal genotype and uterine environment on placental development in Equids. *Journal of Reproduction and Fertility*, *98*(1), 55–60.

Ameur, A., Wetterbom, A., Feuk, L., & Gyllenstein, U. (2010). Global and unbiased detection of splice junctions from RNA-seq data. *Genome Biology*, *11*(3), R34.

Antczak, D. F., Wang, X., Miller, D., & Clark, A. G. (2011). Imprinted genes in placenta and fetus of horse x donkey hybrids. *Animal Epigenetics Workshop, Plant and Animal Genome XIX*, San Diego, CA (W018).

Au, K. F., Jiang, H., Lin, L., Xing, Y., & Wong, W. H. (2010). Detection of splice junctions from pair-end RNA-seq data by SpliceMap. *Nucleic Acid Research*, *38*(14), 4570–4578.

Barrandon, C., Spiluttini, B., & Bensaude, O. (2008). Non-coding RNAs regulating the transcriptional machinery. *Biology of the Cell*, *100*(2), 83–95.

Barrey, E., Mucher, E., Jeansoule, N., Larcher, T., Guigand, L. et al. (2009). Gene expression profiling in equine polysaccharide storage myopathy revealed inflammation, glycogenesis inhibition, hypoxia, and mitochondrial dysfunctions. *BMC Veterinary Research*, *5*, 29–51.

Barrey, E., Mucher, E., Maltere, P., Amiot, F., & Gidrol, X. (2005). Expression profiling in blood cells of exercising horses using mouse and human cDNA microarrays. *6th Dorothy Russell Havemeyer International Equine Genome Mapping Workshop*, Ireland.

Barrey, E., Mucher, E., Robert, C., Amiot, F., & Gidrol, X. (2006). Gene expression profiling in blood cells of endurance horses completing competition or disqualified due to metabolic disorder. *Equine Exercise Physiology*, *7*(Suppl. 36), 43–49.

Barton, C., Beck, P., Kay, R., Teale, P., & Roberts, J. (2009). Multiplexed LC-MS/MS analysis of horse plasma proteins to studying doping in sport. *Proteomics*, *9*(11), 3058–3065.

Beisel, C., & Paro, R. (2011). Silencing chromatin: Comparing modes and mechanisms. *Nature Reviews Genetics*, *12*(2), 123–135.

Bouwman, F. G., van Ginneken, M. E. M., van der Kolk, J. H., van Breda, E., & Mariman, E. C. M. (2010). Novel markers for tying-up in horses by proteomics analysis of equine muscle biopsies. *Comparative Biochemistry and Physiology Part D*, *5*(2), 178–183.

Brent, M. R. (2008). Steady progress and recent breakthroughs in the accuracy of automated genome annotation. *Nature Reviews Genetics*, *9*(1), 62–73.

Bright, L. A., Burgess, S. C., Chowdhary, B. P., Swiderski, C. E., & McCarthy, F. M. (2009). Structural and functional-annotation of an equine whole genome oligoarray. *BMC Bioinformatics*, *10*(Suppl. 11), S8.

Bright, L. A., Mujahid, N., Nanduri, B., McCarthy, F. M., Costa, L. R. R. et al. (2011). Functional modeling of an equine bronchoalveolar lavage fluid proteome provides experimental confirmation and functional annotation of equine genome sequences. *Animal Genetics*, *42*(4), 395–405.

Budak, M. T., Orsini, J. A., Pollitt, C. C., & Rubinstein, N. A. (2009). Gene expression in the lamellar dermis-epidermis during the developmental phase of carbohydrate overload-induced laminitis in the horse. *Veterinary Immunology and Immunopathology*, *131*(1–2), 86–96.

Cappelli, K., Porceddu, A., Verini-Supplizi, A., Capomaccio, S., De Marchis, F. et al. (2005). cDNA AFLP-based techniques for studying transcript profiles in horses. *Research in Veterinary Science*, *79*(2), 105–112.

Chowdhary, B. P., & Raudsepp, T. (2008). The horse genome derby: Racing from map to whole genome sequence. *Chromosome Research*, *16*(1), 109–127.

Cloonan, N., Forrest, A. R., Kolle, G., Gardiner, B. B., Faulkner, G. J. et al. (2008). Stem cell transcriptome profiling via massive mRNA sequencing. *Nature Methods*, *5*(7), 613–619.

Coleman, S. J., Clinton, R., & MacLeod, J. N. (2007). Construction of a master gene list for a 9,322 feature equine cDNA microarray. *Equine Poster Section, Plant and Animal Genome XV*, San Diego, CA (P587).

Coleman, S. J., Zeng, Z., Liu, J., & MacLeod, J. N. (2010a). Analysis of equine protein-coding gene structure and expression by RNA-sequencing. *BMC Bioinformatics*, *11*(Suppl. 4), O8.

Coleman, S. J., Zeng, Z., Wang, K., Luo, S., Khrebtukova, I. et al. (2010b). Structural annotation of equine protein-coding genes determined by mRNA sequencing. *Animal Genetics*, *41*(Suppl. 2), 121–130.

Costa, V., Angelini, C., De Feis, I., & Ciccodicola, A. (2010). Uncovering the complexity of transcriptomes with RNA-seq. *Journal of Biomedicine and Biotechnology*, *2010*, 853–916.

Cox, J., & Mann, M. (2007). Is proteomics the new genomics? *Cell*, *130*(3), 395–398.

Curwen, V., Eyras, E., Andrews, T. D., Clarke, L., Mongin, E. et al. (2004). The Ensembl automatic gene annotation system. *Genome Research*, *14*(5), 942–950.

Deeg, C. A. (2009). A proteomic approach for studying the pathogenesis of spontaneous equine recurrent uveitis (ERU). *Veterinary Immunology and Immunopathology*, *128*(1–3), 132–136.

Diatchenko, L., Lau, Y. F., Campbell, A. P., Chenchik, A., Moqadam, F. et al. (1996). Suppression subtractive hybridization: A method for generating differentially regulated or tissue-specific cDNA probes and libraries. *PNAS*, *93*(12), 6025–6030.

Elliott, M. H., Smith, D. S., Parker, C. E., & Borchers, C. (2009). Current trends in quantitative proteomics. *Journal of Mass Spectrometry*, *44*(12), 1637–1660.

The ENCODE Project Consortium (2004). The Encode (ENCyclopedia of DNA Elements) Project. *Science*, *306*(5696), 636–640.

Glaser, K. E., Sun, Q., Wells, M. T., & Nixon, A. J. (2009). Development of a novel equine whole transcript oligonucleotide GeneChip microarray and its use in gene expression profiling of normal articular-epiphyseal cartilage. *Equine Veterinary Journal*, *41*(7), 663–670.

Graham, N. S., Clutterbuck, A. L., James, N., Lea, R. G., Mobasheri, A. et al. (2009). Equine transcriptome quantification using human GeneChip arrays can be improved using genomic DNA hybridization and probe selection. *The Veterinary Journal*, *186*(3), 323–327.

Griffith, M., Griffith, O. L., Mwenifumbo, J., Goya, R., Morrissy, A. S. et al. (2010). Alternative expression analysis by RNA sequencing. *Nature Methods*, *7*(10), 843–847.

Grondahl, A. M., & Dolvick, N. I. (1993). Heritability estimations of osteochondrosis in the tibiotarsal joint and of bony fragments in the palmar/plantar portion of the metacarpo- and metatarsophalangeal joint of horses. *Journal of the American Veterinary Medical Association*, *203*, 101–104.

Gu, W., & Bertone, A. L. (2004). Generation and performance of an equine-specific large-scale gene expression microarray. *American Journal of Veterinary Research*, *65*(12), 1664–1673.

Hoheisel, J. D. (2006). Microarray technology: Beyond transcript profiling and genotype analysis. *Nature Reviews Genetics*, *7*(3), 200–210.

Huang, L., Zhu, W., Saunders, C. P., MacLeod, J. N., Zhou, M. et al. (2008a). A novel application of quantile regression for identification of biomarkers exemplified by equine cartilage microarray data. *BMC Bioinformatics*, *9*, 300–308.

Huang, L., Miura, N., Mienaltowski, M., MacLeod, J., Stromberg, A. et al. (2008b). Rediscovering the power of well-planned comparisons: Normalization and analysis of cDNA microarray using linear combinations. *Proceedings of the Joint Statistics Meeting*, Denver, CO.

Ing, N. H., Laughlin, A. M., Varner, D. D., Welsh Jr., T. H., Forrest, D. W. et al. (2004). Gene expression in the spermatogenically inactive "dark" and maturing "light" testicular tissues of the prepubertal colt. *Journal of Andrology*, *25*(4), 535–544.

Johannessen, M. K., Skretting, G., Ytrehus, B., & Roed, K. H. (2007). Neonatal growth cartilage: Equine tissue specific gene expression. *Biochemical and Biophysical Research Communications* *354*(4), 975–980.

Kachroo, P., Gustafson-Seabury, A., Lefebvre-Lavoie, J., Lavoie, J. P., Ivanov, I., & Chowdhary, B. P. (2010). Gene expression profiling in equine recurrent airway obstruction (RAO): Detection of differentially expressed genes and their role in pathogenesis of the disease. *Equine Genome Workshop, Plant and Animal Genome XVIII*, San Diego, CA (W211).

Kapranov, P., Cawley, S. E., Drenkow, J., Bekiranov, S., Strausberg, R. L. et al. (2002). Large-scale transcriptional activity in chromosomes 21 and 22. *Science*, *296*(5569), 916–919.

Keren, H., Lev-Maor, G., & Ast, G. (2010). Alternative splicing and evolution: Diversification, exon definition, and function. *Nature Reviews Genetics*, *11*(5), 345–355.

Klein, C., Scoggin, K. E., Ealy, A. D., & Troedsson, M. H. T. (2010). Transcriptional profiling of equine endometrium at the time of maternal recognition of pregnancy. *Biology of Reproduction*, *83*(1), 102–113.

Lefebvre-Lavoie, J., Lussier, J. G., & Theoret, C. L. (2005). Profiling of differentially expressed genes in wound margin biopsies of horses using suppression subtractive hybridization. *Physiological Genomics*, *22*(2), 157–170.

Lindberg, J., & Lundeberg, J. (2010). The plasticity of the mammalian genome. *Genomics*, *95*(1), 1–6.

Lister, R., O'Malley, R. C., Tonti-Filippini, K., Gregory, B. D., Berry, C. C. et al. (2008). Highly integrated single-base resolution maps of the epigenome in arabidopsis. *Cell*, *133*(3), 523–536.

MacLeod, J. N. (2005). Equine articular cartilage microarray. *Equine Genome Workshop, Plant and Animal Genome XIII*, San Diego, CA (W092).

MacLeod, J. N., Hernandez, A. G., Sun, Q., Schweitzer, P. A., & Band, M. R. (2003). Generation of a normalized equine articular cartilage cDNA library and characterization of 13,962 ESTs. *5th Dorothy Russell Havemeyer International Equine Genome Mapping Workshop*, Kruger National Park, South Africa.

McGivney, B. A., Eivers, S. S., MacHugh, D. E., MacLeod, J. N., O'Gorman, G. M. et al. (2009). Transcriptional adaptations following exercise in thoroughbred horse skeletal muscle highlights molecular mechanisms that lead to muscle hypertrophy. *BMC Genomics*, *10*, 638–656.

McGivney, B., Gu, J., Eivers, S., Katz, L., & Hill, E. (2007). Population and functional genomics investigations of performance associations in thoroughbred horses. *7th Dorothy Russell Havemeyer International Equine Genome Mapping Workshop*, Tahoe City.

McIlwraith, C. W. (2002). Diseases of joints, tendons, ligaments, and related structures. In T. S. Stashak (ed.), *Adams' lameness in horses, 5th Edition* (pp. 459–644). Baltimore: Lippincott Williams & Wilkins.

Mendes Soares, L. M., & Valcarcel, J. (2006). The expanding transcriptome: The Genome as the 'Book of Sand'. *The EMBO Journal*, *25*(5), 923–931.

Mercer, T. R., Dinger, M. E., & Mattick, J. S. (2009). Long non-coding RNAs: Insights into functions. *Nature Reviews Genetics*, *10*(3), 155–159.

Merkl, M., Ulbrich, S. E., Otzdorff, C., Herbach, N., Wanke, R. et al. (2010). Microarray analysis of equine endometrium at days 8 and 12 of pregnancy. *Biology of Reproduction*, *83*(5), 874–886.

Mienaltowski, M. J., Huang, L., Bathke, A. C., Stromberg, A. J., & MacLeod, J. N. (2010). Transcriptional comparisons between equine articular repair tissue, neonatal cartilage, culture chondrocytes, and mesenchymal stromal cells. *Briefings in Functional Genomics*, *9*(3), 238–250.

Mienaltowski, M. J., Huang, L., Frisbie, D. D., McIlwraith, C. W., Stromberg, A. J. et al. (2009). Transcriptional profiling differences for articular cartilage and repair tissue in equine joint surface lesions. *BMC Medical Genomics*, *2*, 60–74.

Mienaltowski, M. J., Huang, L., Stromberg, A. J., & MacLeod, J. N. (2008). Differential gene expression associated with postnatal equine articular cartilage maturation. *BMC Musculoskeletal Disorders*, *9*, 149–163.

Mienaltowski, M. J., & MacLeod, J. N. (2006). Transcriptional profiling of equine chondrocyte de-differentiation in culture. *Microarray Poster Section, Plant and Animal Genome XIV*, San Diego, CA (P772).

Miller, D., Wang, W., & Antczak, D. F. (2009). Estimation of gene expression in equine placental tissues using a custom long oligo array. *Equine Genome Workshop, Plant and Animal Genome XVII*, San Diego, CA (W145).

Miranda, G., Mahe, M. F., Leroux, C., & Martin, P. (2004). Proteomic tools to characterize the protein fraction of Equidae milk. *Proteomics*, *4*(8), 2496–2509.

Miura, N., & MacLeod, J. N. (2006). Transcriptional profiling of equine chondrocytes under hypoxic culture conditions. *Equine Poster Section, Plant and Animal Genome XIV*, San Diego, CA (P605).

Morin, R., Bainbridge, M., Fejes, A., Hirst, M., Kryzwinski, M. et al. (2008). Profiling the HeLa S3 transcriptome using randomly primed cDNA and massively parallel short read sequencing. *Biotechniques*, *45*(1), 81–94.

Mortazavi, A., Williams, B. A., McCue, K., Schaeffer, L., & Wold, B. (2008). Mapping and quantifying mammalian transcriptomes by RNA-seq. *Nature Methods*, *5*(7), 621–628.

Mousel, M. R., Moody, D., & Bailey, E. (2002). Gene expression of the horse detected by human microarrays. *Equine Poster Section, Plant and Animal Genome X*, San Diego, CA.

Mucher, E., Jayr, L., Rossignol, F., Amiot, F., Gidrol, X., & Barrey, E. (2006). Gene expression profiling in equine muscle tissues using mouse cDNA microarrays. *Equine Exercise Physiology*, *7*(Suppl. 36), 359–364.

Murray, S. J., Santangelo, K. S., & Bertone, A. L. (2010). Evaluation of early cellular influences of bone morphogenic proteins 12 and 2 on equine superficial digital flexor tenocytes and bone marrow-derived mesenchymal stem cells *in vitro*. *American Journal of Veterinary Research*, *71*(1), 103–114.

Nagalakshmi, U., Wang, Z., Waern, K., Shou, C., Raha, D. et al. (2008). The transcriptional landscape of yeast genome defined by RNA sequencing. *Science*, *320*(5881), 1344–1349.

Nagaraj, S. H., Gasser, R. B., & Ranganathan, S. (2007). A hitchhiker's guide to expressed sequence tag (EST) analysis. *Briefings in Bioinformatics*, *8*(1), 6–21.

Nixon, A. J., Glaser, K. E., Sun, Q., & Well, M. E. (2008). Development and characterization of a novel large-scale whole transcript oligonucleotide exon array for the horse. *Equine Genome Workshop, Plant and Animal Genome XVI*, San Diego, CA (W152).

Nomura, M., Hosaka, Y., Kasashima, Y., Ueda, H., Takehana, K. et al. (2007). Active expression of matrix metalloproteinase-13 mRNA in the granulation tissue of equine superficial digital flexor tendinitis. *Journal of Veterinary Medical Science*, *69*(6), 637–639.

Noschka, E., Vandenplas, M. L., Hurley, D. J., & Moore, J. N. (2009). Temporal aspects of laminar gene expression during the developmental stages of equine laminitis. *Veterinary Immunology and Immunopathology*, *129*(3–4), 242–253.

Pan, Q., Shai, O., Lee, L. J., Frey, B. J., & Blencowe, B. J. (2008). Deep surveying of alternative splicing complexity in the human transcriptome by high-throughput sequencing. *Nature Genetics*, *40*(12), 1413–1415.

Rabilloud, T., Hochstrasser, D., & Simpson, R. J. (2010). Is a gene-centric human proteome project the best way for proteomics to serve biology? *Proteomics*, *10*(17), 3067–3072.

Ramery, E., Closset, R., Art, T., Bureau, F., & Lekeux, P. (2009). Expression microarrays in equine sciences. *Veterinary Immunology and Immunopathology*, *127*(3–4), 197–202.

Ramery, E., Closset, R., Bureau, F., Art, T., & Lekeux, P. (2008). Relevance of using a human microarray to study gene expression in heaves-affected horses. *The Veterinary Journal*, *177*(2), 216–221.

Rosenkranz, R., Borodina, T., Lehrach, H., & Himmelbauer, H. (2008). Characterizing the mouse ES cell transcriptome with Illumina sequencing. *Genomics*, *92*(4), 187–194.

Ruiz-Romero, C., & Blanco, F. J. (2010). Proteomics role in the search for improved diagnosis, prognosis, and treatment of osteoarthritis. *Osteoarthritis and Cartilage*, *18*(4), 500–509.

Santangelo, K. S., Johnson, A. L., Ruppert, A. S., & Bertone, A. L. (2007). Effects of hyaluronan treatment on lipopolysaccharide-challenged fibroblast-like synovial cells. *Arthritis Research & Therapy*, *9*, R1.

Schena, M., Shalon, D., Davis, R. W., & Brown, P. O. (1995). Quantitative monitoring of gene expression patterns with a complementary DNA microarray. *Science*, *270*(5235), 467–470.

Serteyn, D., Piquemal, D., Vanderheyden, L., Lejeune, J. P., Verwilghen, D., & Sandersen, C. (2010). Gene expression profiling from leukocytes of horses affected by osteochondrosis. *Journal of Orthopeadic Research* *28*(7), 965–970.

Smith, K. J., Bertone, A. L., Weisbrode, S. E., & Radmacher, M. (2006). Gross, histologic, and gene expression characteristics of osteoarthritic articular cartilage of the metacarpal condyle of horses. *American Journal of Veterinary Research*, *67*(8), 1299–1306.

Stefani, G., & Slack, F. J. (2008). Small non-coding RNAs in animal development. *Nature Reviews Molecular Cell Biology*, *9*(3), 219–230.

Sultan, M., Schulz, M. H., Richard, H., Magen, A., Klingenhoff, A. et al. (2008). A global view of gene activity and alternative splicing by deep sequencing of the human transcriptome. *Science*, *321*(5891), 956–960.

Talavera, D., Orozco, M., & de la Cruz, X. (2009). Alternative splicing of transcription factors' genes: Beyond the increase of proteome diversity. *Comparative and Functional Genomics*, *2009*, e905894.

Trapnell, C., Pachter, L., & Salzberg, S. L. (2009). TopHat: Discovering splice junctions with RNA-seq. *Bioinformatics*, *25*(9), 1105–1111.

Tremblay, M., Huang, L., Zhu, W., Mienaltowski, M. J., Bathke, A., & MacLeod, J. N. (2009). Ontology of genes with stable and constitutive expression across eleven equine tissues. *Equine Poster Section, Plant and Animal Genome XVII*, San Diego, CA (P539).

Tunon, J., Martin-Ventura, J. L., Blanco-Colio, L. M., Lorenzo, O., Lopez, J. A., & Egido, J. (2010). Proteomics strategies in the search of new biomarkers in atherothrombosis. *Journal of the American College of Cardiology*, *55*(19), 2009–2016.

Vandenplas, M. L., Moore, J. N., Hurley, D. J., Last, L., Grafton, M. et al. (2005a). Equine cDNA microarrays for examining gene expression profiles in proinflammatory conditions and other diseases. *Equine Genome Workshop, Plant and Animal Genome XIII*, San Diego, CA (W091).

Vandenplas, M. L., Moore, J. N., Hurley, D. J., Last, L., Grafton, M. et al. (2005b). Use of an equine-specific cDNA microarray to examine the *in vitro* effects of bacterial cell wall toxins on leukocyte gene expression profiles. *Equine Genome Workshop, Plant and Animal Genome XIII*, San Diego, CA (P583).

Vanderman, K. S., Tremblay, M., Zhu, W., Shimojo, M., Mienaltowski, M. J. et al. (2011). Brother of CDO (BOC) expression in equine articular cartilage. *Osteoarthritis and Cartilage*, *19*(4), 435–438.

Velculescu, V. E., Zhang, L., Vogelstein, B., & Kinzler, K. W. (1995). Serial analysis of gene expression. *Science*, *270*(5235), 484–487.

Wade, C. M., Guilotto, E., Sigurdsson, S., Imsland, F., Lear, T. L. et al. (2009). Genome sequence, comparative analysis, and population genetics of the domestic horse (*Equus caballus*). *Science*, *326*(5954), 865–867.

Walsh, C. T., Garneau-Tsodikova, S., & Gatto, G. J. (2005). Protein posttranslational modifications: The chemistry of proteome diversifications. *Angewandte Chemie (International Edition in English)*, *44*(45), 7342–7372.

Wang, E. T., Sandberg, R., Luo, S., Khrebtukova, I., Zhang, L. et al. (2008). Alternative isoform regulation in human tissue transcriptomes. *Nature*, *456*(7221), 470–476.

Wang, J., Gustafson-Seabury, A., Raudsepp, T., & Chowdhary, B. P. (2009a). Gene expression profiling in Equine laminitis. *Equine Poster Section, Plant and Animal Genome XVII*, San Diego, CA (P542).

Wang, J., Gustafson-Seabury, A., Zhou, H., Galantino-Homer, H. L., Carter, R. et al. (2010a). Genome-wide transcriptome analysis of laminar tissue during early stages of carbohydrate overload and insulin induced equine laminitis. *Equine Genome Workshop, Plant and Animal Genome XVIII*, San Diego, CA (W209).

Wang, K., Singh, D., Zeng, Z., Coleman, S. J., Huang, Y. et al. (2010b). MapSplice: Accurate mapping of RNA-seq reads for splice junction discovery. *Nucleic Acids Research*, *38*(18), e178.

Wang, L., Xi, Y., Yu, J., Dong, L., Yen, L., & Li W. (2010c). A statistical method for the detection of alternative splicing using RNA-seq. *PLoS ONE*, *5*, e8529.

Wang, Z., Gerstein, M., & Snyder, M. (2009b). RNA-seq: a revolutionary tool for transcriptomics. *Nature Reviews Genetics*, *10*(1), 57–63.

Ytrehus, B., Carlson, C. S., & Ekman, S. (2007). Etiology and pathogenesis of osteochondrosis. *Veterinary Pathology*, *44*(4), 429–448.

Yuan, Z. Q., Nicolson, L., Marchetti, B., Gault, E. A., Campo, M. S., & Nasir, L. (2008). Transcriptional changes induced by bovine papillomavirus type 1 in equine fibroblasts. *Journal of Virology*, *82*(13), 6481–6491.

Zachos, T. A., Shields, K. M., & Bertone, A. L. (2006). Gene-mediated osteogenic differentiation of stem cells by bone morphogenetic proteins-2 or -6. *Journal of Orthopeadic Research*, *24*(6), 1279–1291.

Zava, S., Barello, C., Pessione, A., Garoffo, L. P., Fattori, P. et al. (2009). Mare's colostrums globules stimulate fibroblast growth in vitro: a biomechanical study. *Journal of Medicinal Food*, *12*(4), 836–845.

Zhou, M., Wang, Q., Sun, J., Li, X., Xu, L. et al. (2009). *In silico* Detection and Characteristics of Novel microRNA genes in the Equus caballus genome using an integrated *ab initio* and comparative genomics approach. *Genomics*, *94*(2), 125–131.

Zhu, W., Huang, L., Saunders, C. P., Bathke, A., Stromberg, A. J., & MacLeod, J. N. (2007). Identification of genes with a cartilage-restricted pattern of expression. *Equine Poster Section, Plant and Animal Genome XV*, San Diego, CA (P605).

9 Coat color genomics

Samantha A. Brooks and Rebecca R. Bellone

Introduction

The study of color has long been an important tool for the field of genetics. Retrospectively, as early as Gregor Mendel and his studies on peas, the use of color traits has led to many valuable discoveries. Additionally, during domestication of the horse approximately 5,000–6,000 years ago, it seems likely that color was one of the first phenotypic changes to result from positive selection by humans (Ludwig et al., 2009; NCBI Map Viewer, n.d.). The horse has continued to play an iconic role in the history of humanity. As such, their color has captured our interest and, through selection, has expanded into nearly infinite variations. It is therefore a natural progression driven by both science and the horse industry that coat color has dominated the discovery of alleles in the horse. However, recent discoveries show that this variation in color has not always come without a price. Pleiotropic effects resulting in conditions that are either detrimental to health or simply undesirable to the horse owner are not uncommon among coat color variants in the horse. Knowledge of these effects provides valuable clues to the role of coat color genes in biological pathways important for health and disease, as well as valuable tools to the horse breeder.

Base Colors (Black, Chestnut, Bay, and Seal Brown)

The base color of a horse is most frequently described as black, chestnut, bay, or seal brown. Black horses have black bodies and black points (mane, tail, ear tips, and lower legs), whereas chestnut horses have red bodies and nonblack points. Bay horses have a red body with black points, while seal brown horses also have black points but their body is black with red or tan areas in parts of the coat (flank, upper legs, muzzle, and around the eyes). The base coat color is determined by two interacting loci (*extension* and *agouti*) that affect melanocyte function (Barsh, 1996; Rees, 2003). These loci determine if eumelanin (black and brown pigment) or phaeomelanin (red and yellow pigment) will be produced in their corresponding melanosome (eumelansome or phaeomelansome), the organelle where pigment productions occurs. The *extension* locus (*E*) encodes the melanochortin-1 receptor (MC1R) and the *agouti* locus (*A*) codes for the MC1R antagonist, agouti signaling protein (ASIP). Signaling through MC1R with the agonist (α-MSH) results in the production of eumelanin, while antagonist binding (ASIP) results in phaeomelanin production.

Equine Genomics, First Edition. Edited by Bhanu P. Chowdhary.

Bay, the most frequent base color in most breeds, is the product of the dominant wild type alleles at both loci (*A,E*).

Chestnut coloration results from a recessive mutation at the *extension locus* (*e*); it was the first coat color to be characterized at the molecular level in the horse (Marklund et al., 1996). This phenotype results from a loss of function mutation in MCIR (p.Ser83Phe). Another recessive allele (e^a) has since been described in the Black Forest Breed (p.Asp84Asn) (Wagner & Reissmann, 2000).

Black results from a recessive loss of function mutation in *ASIP* (*a*). An 11 bp deletion (g.2174-2184del) in exon 2 is thought to cause a frame shift and thus a modified protein (Rieder et al., 2001). The *seal brown* phenotype is thought to be caused by a different mutation at the *agouti* locus (A^t) (Sponenberg, 2009). A DNA test is commercially available for the seal brown mutation; however, it has not been described in the scientific literature. Chestnut is epistatic to all other colors, thus non-chestnut horses must have one wild-type allele (*E*) at the *extension* locus.

Dilutions (Cream, Pearl, Champagne, Silver, Dun, and Lavender Foal)

Four genes that dilute base coat color have been described in the horse (*SLC45A2, SLC36A1, PMEL17, and MYO5A*); one additional phenotype (dun) has been mapped but the causative mutation remains unknown (Table 9.1). All of the known dilution genes (with the exception of pearl) have a dominant or incomplete dominant mode of inheritance. Dilution genes reduce the amount of pigment being produced directly by inhibiting pigment production or indirectly by inhibiting melanosome transfer. Dilution genes can affect either eumelanin, phaeomelanin, or both.

Cream (*C*) is inherited as an incompletely dominant gene. In the heterozygous condition cream dilutes phaeomelanin from red to yellow, which results in palominos, buckskins, and smoky seal browns. Homozygotes are lighter in color (cremello and perlino). In addition, in the homozygous condition eumelanin is also diluted, which results in smoky cream (dilute black) horses (Sponenberg, 2009). The *cream* locus was mapped to ECA21q and a single base substitution in exon 2(G457A) of the *solute carrier 45 family A2* (SLC45A2, also known as membrane-associated transporter protein or MATP) is thought to be responsible for the cream dilution phenotypes (Locket et al., 2001; Mariat, Taourit, & Guerin, 2003). The precise function of *SLC45A2* is unknown, but in mice, mutations in this gene disrupt the processing and trafficking of tryosinase (a key enzyme in pigment production) to the melanosome (Costin et al., 2003).

Pearl color dilution in breeds of Iberian origin, or "Barlink factor" as it is known in Paints and Quarter Horses, is also caused by a mutation in SLC45A2 (Penedo et al., 2011). However unlike *cream,* the coat (both eumelanin and phaeomelanin) is diluted only in the homozygous condition, thus *pearl* is described as a recessive mutation (Sponenberg, 2009). However, in the heterozygous condition, while the hair remains unchanged, the skin is lighter and small pale spots may be present. The causative mutation for *pearl* was discovered only recently (Penedo et al., 2011), and a DNA test is available.

Champagne (*CH*) dilutes phaeomelanin and eumelanin in both the heterozygous and homozygous condition and is therefore caused by a dominant allele. Often homozygotes and heterozygous are phenotypically indistinguishable from each other and from that of cream and pearl dilutes. In addition to the diluted coat color, champagne horses often have amber-colored eyes and pumpkin-colored skin, which can sometimes help delineate these animals from other dilute phenotypes. *CH* was mapped to ECA14 and a SNP in exon 2 (resulting in an amino acid substitution T63R) of another solute carrier, *SLC36A1* (*Solute Carrier 36 family A1*), is thought to be responsible (Cook et al.,

Table 9.1 Summary of the currently known coat color genes, alleles, pleiotropic effects, and available tests in horses

Coat color phenotype	Genes	Allele	Pleiotropic conditions	DNA test available	References
Chestnut	*MC1R*	*e*		caustive	Marklund, 1996
Bay	*ASIP*	*A*		caustive	Rieder, 2001
Black	*ASIP*	*a*		caustive	Rieder, 2001
Seal brown	*ASIP*	*At*		associated*	–
Cream	*SLC45A2*	*CR*		caustive	Mariat, 2003
Pearl	*SLC45A2*	*P*		associated*	–
Silver	*SILV*	*Z*	multiple congenital ocular anomalies	caustive	Brunburg, 2006
Champange	*SLC36A1*	*CH*		caustive	Cook, 2008
Coat color dilution lethal	*MYO5A*	*L*	lethal lavendar foal syndrome	caustive	Brooks, 2010
Dun	*?*	*D*		zygosity in some breeds*	Bricker, 2003
Frame overo	*EDNRB*	*O*	lethal white foal syndrome, deafness?	caustive	Metallinos, Santschi, & Yang, 1998
Tobiano	*KIT*	*TO*		caustive	Brooks, 2007
Sabino-1	*KIT*	*SB*		caustive	Brooks, 2005
Dominant white	*KIT*	*W1-17*	potential homozygous lethal	caustive	Haase, 2007, 2009, 2011; Holl, 2010
Grey	*STX17*	*G*	melanoma	caustive	Pielberg, 2008
Leopard complex	*TRPM1*†	*LP*	congenital stationary night blindness	associated	Bellone, 2008
Roan	*KIT*†	*RN*	potential homozygous lethal	zygosity in some breeds*	Marklund, 1999
White face and leg markings	*KIT*, *MC1R*, and *MITF*	*?*		none	Rieder, 2008; Woolf, 1992; Haase, 2009b

*Denotes that mutation tested has not been published and therefore cannot be independently verified.
†Association with this gene, but causative polymorphism unknown.

2008). While the precise function of *SLC36A1* remains to be determined, the authors postulated that this gene may help in regulating pH and thus in the maturation of the melanosome.

The *silver* dilution (*Z*), also known as silver dapple, dilutes eumelanin only (phaeomelanin is unaffected) to a "silver" or "chocolate" color, usually leaving the main and tail much lighter than the body hairs. In addition, horses with silver dilution frequently have "dapples" or darker pigment outlining lighter areas. *Silver* is inherited as a dominant allele and homozygotes may be more dilute (Brunberg et al., 2006). *Silver* was mapped to ECA6 and the positional and functional candidate gene *PMEL17* (pre-melanosomal protein 17) or *SILV* (silver homolog) and was subsequently sequenced by two separate groups (Brunberg et al., 2006; Reissmann, Bierwolf, & Brockmann, 2007). A missense mutation in exon 11 (DQ665301:g.1457C>T) resulting in an amino acid substitution (Arg618Cys) is believed to cause this dilution (Brunberg et al., 2006). *PMEL17* is thought to be involved in the biogenesis of the pre-melanosome (Yasumoto et al., 2004; Hoashi et al., 2005). Equine multiple congenital ocular anomalies (MCOA) have been associated with the silver phenotype. MCOA is characterized by a diverse set of ocular phenotypes, with large cysts being the predominant phenotype (Ewart et al., 2000; Grahn et al., 2008). Incomplete penetrance of this disorder has made studying the molecular mechanism difficult, and it remains unknown whether *PMEL17* and/or the mutation Arg617Cys also causes MCOA (Andersson et al., 2008; Grahn et al., 2008).

Dun (*D*) is a dominantly inherited dilution of all base coat colors (Adelsteinsson, 1978). Similarly to the *CH*, both eumelanin and phaeomelanin pigments are lightened and heterozygotes are indistinguishable from homozygotes. In contrast to any of the other dilute phenotypes, Dun horses also have dorsal stripes and other primitive markings. While the exact genetic mechanism remains unknown, the genetic locus for *D* was mapped to ECA8 (Bricker, P. C. M., Millon, & Murray, 2003). A genetic test based on the zygosity of linked markers is commercially available and is useful for identifying homozygotes.

Coat Color Dilution Lethal (CCDL), also known as Lavender Foal syndrome (LFS), is a recessive condition that reduces pigmentation in homozygous individuals, creating a softened hue described as pale gray, pewter, and light chestnut, as well as lavender (Bowling, 1996). Unfortunately, it is also responsible for a neurological disorder that is lethal soon after birth (Bowling, 1996). Affected foals have various neurological signs including tetanic-like seizures, opisthotonus, stiff or paddling leg movements, and nystagmus (Fanelli, 2005). Mild leucopenia is also occasionally present (Fanelli, 2005; Page et al., 2006). These neurological impairments prevent the foal from standing and nursing normally and, if not lethal on their own, are often the causes for euthanasia. LFS is most frequently reported in the Egyptian subgroup of the Arabian horse (Bowling, 1996).

Mapping and mutation detection for LFS was accomplished in 2010 as the first published successful use of the Equine SNP50 SNP set in the horse (Brooks et al., 2010). The discovered deletion in exon 30 of myosin VA (*MYO5A*) leads to a frame shift and premature termination of transcription (ECA1 g.138235715del). Loss of the C-terminus of the protein, which encodes a portion of the secretory vesicle-specific binding domain of the globular tail, would likely impair binding of the myosin 5a motor to specific cargo organelles bearing the appropriate receptors (Pashkova et al., 2005). The resulting loss of vesicle traffic is likely to interfere with the function of dendritic cells like melanocytes and neurons, thus creating the associated phenotypes.

White Spotting and Depigmentation Patterns (Frame, Tobiano, Sabino, Dominant White, Leopard Complex, Gray, Roan, and White face and leg markings)

The genetics of white spotting in horses was among the first traits to be studied. White spotting can occur on any base color and can occur in combination with dilution and other white spotting patterns. White spotting often results from the absence of melanocytes due to defects in melanocyte differentiation and/or proliferation during embryonic development. It can also occur by the loss of pigment and/or depletion of pigment cells throughout life (depigmentation). Several white spotting and/or depigmentation genes have similar phenotypes and often horses are erroneously classified; thus DNA testing is a useful tool, and new tests will undoubtedly aid in correct classification. Most white spotting patterns vary in the amount of white present on the horse and much of this variation is thought to be caused by as yet uncharacterized modifier genes. Similarly to other species, several white spotting patterns are associated with pleiotropic effects (reviewed in Bellone, 2010; Rieder, 2009).

Frame overo patterning, inherited by a dominant allele (*O*), defines white spotting that typically occurs in the middle of sides of the flank, neck, and ventrally but not dorsally, so that pigmentation "frames" the horse. Homozygosity results in a lethal condition known as lethal white foal syndrome (LWFS or OLWS). This syndrome is characterized by foals that are born with a nearly complete white coat and are affected with intestinal aganglionosis. This lack of enervation causes intestinal obstruction and death soon after birth (Hultgren, 1982; McCabe et al., 1990). In Quarter Horses the frame pattern has also been associated with deafness (Magdesian, 2009). Using a candidate gene approach, three research groups independently identified the dinucleotide missense mutation (NM_001081837.1:c.353_354delinsAG) in the first exon of *endothelin receptor type B* (*EDNRB*) (Metallinos, Bowling, & Rine, 1998; Santschi et al., 1998; Yang et al., 1998). The resulting protein substitution (Ile118Lys) is thought to disrupt functioning of melanocytes and enteric ganglia cells, but the exact mechanism is not completely understood.

The *Tobiano* pattern (*TO*), a dominant trait, is characterized by depigmented patches of skin and the associated hair, usually crossing the dorsal midline and covering the legs (Figure 9.1a). The phenotype can vary from minimal body white with some leg markings to depigmentation of all but perhaps 10% of the body surface. Homozygous *TO/TO* horses often have dime- to quarter-sized spots of pigment interspersed among the larger white patches (frequently termed "ink spots" or "paw prints"). However, these markings are not a reliable indicator of zygosity (Sponenberg, 2009). *Tobiano* is present in many diverse breeds of the horse. Linkage of *TO* to the blood protein locus *Albumin (ALB)* was first reported in a pony family (Trommershausen-Smith, 1978). Further work extended this association to a conserved haplotype on ECA3: *Albumin* (*ALB*)*-B* and *Vitamin D binding factor* (*GC*)*-S* (Bowling, 1987). To explain the unusually strong linkage disequilibrium between the *TO, ALB* and *GC* loci Bowling et al. proposed that an inversion on ECA3 could be preventing recombination in this region (Bowling, 1987).

Early work by Raudsepp et al. (1999) did not detect such an inversion. However, in 2007, subsequent *FISH* studies with additional markers discovered a large pericentric inversion on ECA3q

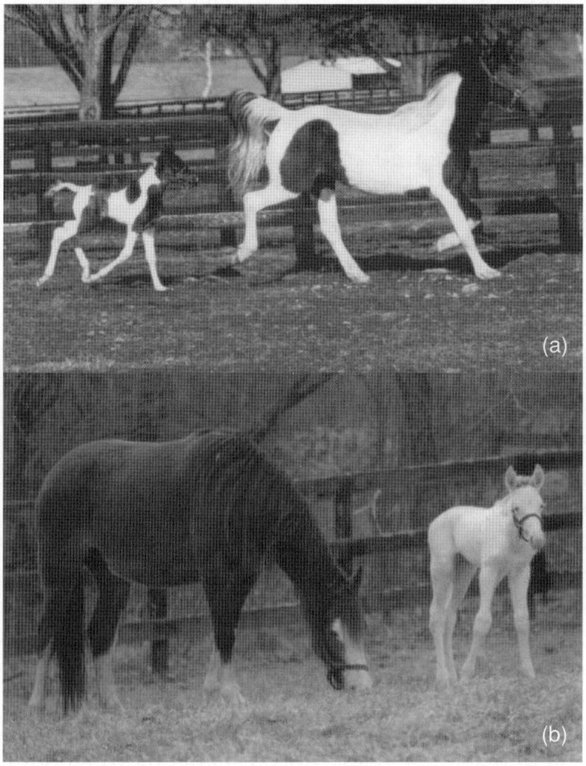

Figure 9.1 The *Tobiano* pattern (a) and *Sabino* pattern (b) are both attributed to the *KIT* locus in the horse.

in *TO* horses (Brooks et al., 2007). This inversion spans more than 47 Mb beginning approximately 100 kb from the distal end of the *KIT* gene, a gene involved in white spotting phenotypes in several species, and does not appear to disrupt any coding sequence. However, similar to inversions in the mouse producing the *Rump-white* and *Sash* spotting patterns, disruption of sequences near *KIT* are thought to cause aberrant special or temporal expression of the receptor and thus result in improper migration of melanocytes (Stephenson et al., 1994; Nagle et al., 1995). The inversion is completely associated with *TO* in several American and European breeds of the horse (Haase et al., 2008).

Another white spotting pattern, termed *Sabino*, is characterized by irregularly bordered white patches (hair and skin) that begin at the extremities and face (Figure 9.1a). These white patches often extend to include the belly and midsection, either as distinct areas of white hair or as a diffuse scattering of white hairs resembling roan (Kunst et al., 1992). While phenotypically similar, there appears to be several different genetic mechanisms for *Sabino*, and the genetics of one type of *Sabino* patterning, *Sabino 1 (SB1)*, is incompletely dominant. Heterozygotes have the typical *Sabino* pattern while homozygotes are white or nearly white with some pigmentation persisting around the topline. In the United States, *SB1* is found among many light horse breeds, notably in the Tennessee Walking Horse and Missouri Foxtrotter (Sponenberg, 2009). Genomic DNA sequencing of the *KIT* gene revealed a base substitution in intron 16 (KI16+1037A) responsible for *SB1* (Brooks & Bailey, 2005). When the KI16+1037A SNP is present, *KIT* is transcribed lacking exon 17, making this trait a notable example of polymorphic exon-skipping. While exon-skipping is not complete, as some normal gene products are produced, it has been proposed that this SNP reduces the overall strength of the 3′ splicing consensus sequence and that loss of exon 17 results in impaired protein function (Brooks & Bailey, 2005). Interestingly, unlike *KIT* mutant mice strains, *SB1* horses appear to have no health defects (Geissler, McFarland, & Russell, 1981).

It has been observed that the *Sabino*-type pattern has heterogeneous genetic origins as *SB1* does not explain all *Sabino* phenotypes (Brooks & Bailey, 2005). Presumably, variation at other genetic sites within *KIT* or another gene is responsible for other *Sabino*-type patterns. For example, Clydesdale and Shire draft horses are well known for their *Sabino* phenotype, but the *SB1* mutation was not found among a sample group of those breeds (Brooks & Bailey, 2005).

"White" horses uniquely possess a predominantly white coat and dark eyes. *Dominant White* (*W*), so named for its mode of inheritance and the orthologous murine locus, was first identified as a homozygous lethal among a horse herd in Nebraska (Pulos & Hutt, 1969). Mau et al. (2004) studied dominant white in Swiss Franches-Montagnes Horses and identified linkage with the *KIT* gene. In sharp contrast to the other known coat color loci in the horse, many novel mutations were subsequently identified in one gene (*KIT*) and attributed to the same phenotype (Haase et al., 2007; Haase et al., 2009; Holl, Brooks, & Bailey, 2010). The twelve total *W* mutations (discovered in several breeds) encompass phenotypes ranging from a splotched pattern possessing much of the underlying base color to a completely white coat. In each case the mutation occurred fairly recently; either a founder was identified based on breed records or completely *de novo*, with neither confirmed parent carrying the variant allele. The *W* alleles are therefore breed or, in some cases, family specific. For this reason, although DNA testing is available, it is not broadly applicable. To date no horse has been identified as homozygous for any *W* allele (or compound heterozygous for two *W* alleles), suggesting that the hypothesis of homozygous lethality put forward by Pulos and Hutt (1969) is true. No hematopoietic abnormalities have been identified in the one *W* allele that has been scrutinized for pleiotropic effects (Haase et al., 2010).

Leopard complex spotting (*LP*; also referred to as appaloosa spotting) defines a group of white spotting patterns that occur in several breeds of horses including Appaloosa, Noriker, Knabstrupper, American Miniature Horse, British Spotted Pony, and Pony of the Americas (Bellone et al. 2010;

Haase et al., 2010). Notably, *LP* is among the few ancient coat color patterns that have been detected in ~25,000-year-old ancient DNA samples originating from wild, pre-domestic horses of Western and Eastern Europe (Pruvost et al., 2011). Leopard spotting is inherited by a single incompletely dominant gene. Homozygotes tend to have few spots of pigment (termed leopard spots) in their white patterned areas (Figure 9.2a), while the opposite is true of heterozygotes (Figure 9.2b). In either zygotic condition, the amount of white on the coat can range from very minimal (white flecks on rump, Figure 9.2c) to a coat that is almost entirely white (few spot, Figure 9.2a), and this variation is thought to be determined by several modifier genes. One of such modifiers, which is associated with the "Leopard"-specific pattern (denoted as *PATN1*, Figure 9.2b), was recently mapped to ECA3 (Archer et al., 2007). In addition to the patterning in the coat, *LP* is thought to be responsible for the associated traits of mottled skin, striped hooves, white sclera, and *LP*-specific progressive roaning (also termed varnish roan). This roaning is separate and distinct from classical roaning (described below), thus making *LP* both a white spotting and depigmentation pattern.

Homozygosity for *LP* has been associated with congenital stationary night blindness (CSNB) in the Appaloosa breed (Sandmeyer et al., 2007). These horses have visual problems under low-light condition, which result from a defect in the ON bipolar cell signaling pathway as evidenced by a "negative ERG."

LP was mapped to a 6 cM region on ECA1 and *Transient Receptor Potential Cation Channel, Subfamily M, Member 1* (*TRPM1*) was chosen as the positional and functional candidate for further investigation (Terry et al., 2004; Bellone et al., 2006). Quantitative real-time PCR analysis and fine-mapping strongly supported *TRPM1* as the candidate gene for both *LP* and CSNB (Bellone et al., 2008; Bellone et al., 2010a). While a mutation in the known coding region was not detected, three SNPs within or near *TRPM1* (ECA1 g.108281765T>C, ECA1 g.108288853C>T, and ECA1 g.108337089T>G) have been completely associated with *LP* and CSNB and could be used as a DNA test until the causative mutation is identified (Bellone et al., 2010b). To our knowledge, at this time no labs are commercially running these markers as a DNA test.

Traditionally grouped with white spotting but more correctly classified as depigmentation, gray horses will progressively acquire white hairs throughout the coat as they age. This rate and location of depigmentation (in the hair, or in the hair and skin) varies from horse to horse and results from depletion of stem cells (Rosengren Pielberg et al., 2008; Sponenberg, 2009). Gray horses often develop melanomas as a result of hyperproliferation of melanocytes in the skin (Valentine, 1995; Rosengren Pielberg et al., 2008). The gray gene was independently mapped by three groups to ECA25 (Henner et al., 2002; Locke et al., 2002; Swinburne, Hopkins, & Binns, 2002). The map position was further refined and a 4.6 kb duplication in intron 6 of *syntaxin 17* (*STX17*) (ECA25 g.6575277_6579862dup) was identified as the cause of gray (Pielberg et al., 2005; Rosengren Pielberg et al., 2008). This duplication is thought to act as a cis regulator mutation upregulating both *STX17* and *nuclear receptor subfamily 4, group A, member 3* (*NR4A3*) that results in the hyperproliferation and depletion of melanocyte stem cells. Homozygosity for the duplication is associated with a faster rate of graying, less pigment retention in the coat, more skin depigmentation, and a higher incidence of melanoma (Rosengren Pielberg et al., 2008).

Roan (*RN*) in the horse describes an intermingling of white hairs throughout the coat, while the head, main, tail, and lower legs do not have white hairs unless another white spotting pattern is present (Castle, 1954). The amount of white hair does not appreciably change with age, in contrast to *Gray* and *LP* progressive roaning. Thus this phenotype is also known as classic roan to delineate it from others with similar appearance. Roan was first mapped to ECA3, and the position later refined to near the *KIT* gene (Andersson & Sandberg, 1982; Marklund et al., 1999). Roan is thought to be lethal in the homozygous state similar to the alleles of the *W* series (Castle, 1954). The precise

Figure 9.2 Leopard complex spotting: (a) *LP/LP* homozygote (b) *LP/lp* heterozygote (c) minimally marked *LP* horse whose genotype is difficult to distinguish by phenotype alone.

mutation responsible for Roan is unknown, but research is ongoing and currently only a zygosity test is available.

The most frequent appearance of white in the coat of the horse is that found on the extremities. White markings are described in lay terms most frequently as "socks" on the limbs and "blaze" on the head. Although multiple genes and a complex mode of inheritance are thought to play a role (Woolf, 1989), association for a locus of major effect near the *KIT* gene has been established in the Franches-Montagnes breed (Rieder et al., 2008).

To date, in the horse, more alleles have been genetically identified for coat color traits than all other known alleles combined. This exemplifies their utility in genetic study, as well as the large number of variants available and the keen interest of horse owners to maintain these phenotypes by providing samples for research. Many of these traits are also under negative selection, either to suit breed standards or to reduce detrimental pleiotropic effects. The discovery of causal mutations has provided powerful tools to horse breeders. They can now use these to make informed breeding decisions and alter the frequency of these alleles in the population as they see fit (Table 9.1). For now it remains the responsibility of the individual or the breed organization to determine how the results of these tests will be applied.

Sequencing of the equine genome and the identification of new markers will soon make mapping of complex traits a reality in the horse. Undoubtedly, these will include additional coat color phenotypes. As the molecular mechanisms of additional phenotypes are unraveled, connections between these genes and biological pathways unrelated to pigmentation will add to our understanding of the gene-function relationship.

References

Adalsteinsson, S. (1978) Inheritance of yellow dun and blue dun in the Icelandic toelter horse. *Journal of Heredity*, *69*(3), 146–148.

Andersson, L., & Sandberg, K. (1982). A linkage group composed of three coat color genes and three serum protein loci in horses. *Journal of Heredity*, *73*(2), 91–94.

Andersson, L. S. et al. (2008). Equine multiple congenital ocular anomalies maps to a 4.9 megabase interval on horse chromosome 6. *BMC Genetics*, *9*, 88.

Archer, S. et al. (2007). Investigation of ECA3 as the Location of the Gene for White Patterning (PATN1) in Leopard Spotted Appaloosas. In International Plant & Animal Genomes Conference XV, San Diego, CA.

Barsh, G. S. (1996). The genetics of pigmentation: From fancy genes to complex traits. *Trends in Genetics*, *12*(8), 299–305.

Bellone, R. (2010). Pleiotropic effects of pigmentation genes in horses. *Animal Genetics*, *41*(Suppl. 2), 100–110.

Bellone, R. et al. (2006). Comparative mapping of oculocutaneous albinism type II (OCA2), transient receptor potential cation channel, subfamily M member 1 (TRPM1) and two equine microsatellites, ASB08 and 1CA43, among four equid species by fluorescence in situ hybridization. *Cytogenetic Genome Research*, *114*(1), 93A.

Bellone, R. R. et al. (2008). Differential gene expression of TRPM1, the potential cause of congenital stationary night blindness and coat spotting patterns (LP) in the Appaloosa horse (Equus caballus). *Genetics*, *179*(4), 1861–1870.

Bellone, R. R. et al. (2010a). Fine-mapping and mutation analysis of TRPM1: A candidate gene for leopard complex (LP) spotting and congenital stationary night blindness in horses. *Briefing on Functioning Genomics*, *9*(3), 193–207.

Bellone, R. et al. (2010b). Association analysis of candidate SNPs in TRPM1 with leopard complex spotting (LP) and congenital stationary night blindness (CSNB) in horses. *Animal Genetics*, *41*(Suppl. 2), 207.

Bowling, A. T. (1987). Equine linkage group II: Phase conservation of To with AlB and GcS. *Journal of Heredity*, *78*(4), 248–250.

Bowling, A. T., & CAB International (1996). *Horse genetics*. Wallingford, Oxon: CAB International.

Bricker, S. J., P. C. M., Millon, L. V., & Murray, J. D. (2003). Linkage of the dun coat color locus to microsatellites on horse chromosome 8. In *International Plant & Animal Genomes Conference XI*, San Diego, CA.

Brooks, S. A., & Bailey, E. (2005). Exon skipping in the KIT gene causes a Sabino spotting pattern in horses. *Mammalian Genome*, *16*(11), 893–902.

Brooks, S. A. et al. (2007). A chromosome inversion near the KIT gene and the Tobiano spotting pattern in horses. *Cytogenetic Genome Research*, *119*(3–4), 225–230.

Brooks, S. A. et al. (2010). Whole-genome SNP association in the horse: Identification of a deletion in myosin Va responsible for Lavender Foal Syndrome. *PLoS Genetics*, *6*(4), e1000909.

Brunberg, E. et al. (2006). A missense mutation in PMEL17 is associated with the silver coat color in the horse. *BMC Genetics*, *7*, 46.

Castle, W. E. (1954). Coat color inheritance in horses and in other mammals. *Genetics*, *39*(1), 35–44.

Cook, D. et al. (2008). Missense mutation in exon 2 of SLC36A1 responsible for champagne dilution in horses. *PLoS Genetics*, *4*(9), e1000195.

Costin, G. E. et al. (2003). Tyrosinase processing and intracellular trafficking is disrupted in mouse primary melanocytes carrying the underwhite (uw) mutation. A model for oculocutaneous albinism (OCA) type 4. *Journal of Cellular Science*, *116*(15), 3203–3212.

Ewart, S. L. et al. (2000). The horse homolog of congenital aniridia conforms to codominant inheritance. *Journal of Heredity*, *91*(2), 93–98.

Fanelli, H. H. (2005). Coat colour dilution lethal ('lavender foal syndrome'): a tetany syndrome of Arabian foals. *Equine Veterinary Education*, *17*(5), 260–263.

Geissler, E. N., McFarland, E. C., & Russell, E. S. (1981). Analysis of pleiotropism at the dominant white-spotting (W) locus of the house mouse: A description of ten new W alleles. *Genetics*, *97*(2), 337–361.

Grahn, B. H. et al. (2008). Congenital ocular anomalies in purebred and crossbred Rocky and Kentucky Mountain horses in Canada. *Canadian Veterinary Journal*, *49*(7), 675–681.

Haase, B. et al. (2007). Allelic heterogeneity at the equine KIT locus in dominant white (W) horses. *PLoS Genetics*, *3*(11), e195.

Haase, B. et al. (2008). An equine chromosome 3 inversion is associated with the tobiano spotting pattern in German horse breeds. *Animal Genetics*, *39*(3), 306–309.

Haase, B. et al. (2009). Seven novel KIT mutations in horses with white coat colour phenotypes. *Animal Genetics*, *40*(5), 623–629.

Haase, B. et al. (2010). Haematological parameters are normal in dominant white Franches-Montagnes horses carrying a KIT mutation. *Veterinary Journal*, *184*(3), 315–317.

Henner, J. et al. (2002). Genetic mapping of the (G)-locus, responsible for the coat color phenotype "progressive greying with age" in horses (Equus caballus). *Mammalian Genome*, *13*(9), 535–537.

Hoashi, T. et al. (2005). MART-1 is required for the function of the melanosomal matrix protein PMEL17/GP100 and the maturation of melanosomes. *Journal of Biology & Chemistry*, *280*(14), 14006–14016.

Holl, H., Brooks, S., & Bailey, E. (2010). De novo mutation of KIT discovered as a result of a nonhereditary white coat color pattern. *Animal Genetics*, *41*(Suppl. 2), 196–198.

Hultgren, B. D. (1982). Ileocolonic aganglionosis in white progeny of overo spotted horses. *Journal of American Veterinary Medicine Association*, *180*(3), 289–292.

Kunst, A. et al. (1992). *International variation in socio-economic inequalities in self-reported health: A comparison of the Netherlands with other industrialised countries*. The Hague: SDU Publishers.

Locke, M. M. et al. (2001). The cream dilution gene, responsible for the palomino and buckskin coat colours, maps to horse chromosome 21. *Animal Genetics*, *32*(6), 340–343.

Locke, M. M. et al. (2002). Linkage of the grey coat colour locus to microsatellites on horse chromosome 25. *Animal Genetics*, *33*(5), 329–337.

Ludwig, A. et al. (2009). Coat color variation at the beginning of horse domestication. *Science*, *324*, 485.

Magdesian, K. G. et al. (2009). Evaluation of deafness in American Paint Horses by phenotype, brainstem auditory-evoked responses, and endothelin receptor B genotype. *Journal of American Veterinary Medicine Association*, *235*(10), 1204–1211.

Mariat, D., Taourit, S., & Guerin, G. (2003). A mutation in the MATP gene causes the cream coat colour in the horse. *Genetics & Selective Evolution*, *35*(1), 119–133.

Marklund, L. et al. (1996). A missense mutation in the gene for melanocyte-stimulating hormone receptor (MC1R) is associated with the chestnut coat color in horses. *Mammalian Genome*, *7*(12), 895–899.

Marklund, S. et al. (1999). Close association between sequence polymorphism in the KIT gene and the roan coat color in horses. *Mammalian Genome*, *10*(3), 283–288.

McCabe, L. et al. (1990). Overo lethal white foal syndrome: Equine model of aganglionic megacolon (Hirschsprung disease). *American Journal of Medicine & Genetics*, *36*(3), 336–340.

Metallinos, D. L., Bowling, A. T., & Rine, J. (1998). A missense mutation in the endothelin-B receptor gene is associated with Lethal White Foal Syndrome: An equine version of Hirschsprung disease. *Mammalian Genome*, *9*(6), 426–431.

Nagle, D. L. et al. (1995). Physical mapping of the Tec and Gabrb1 loci reveals that the Wsh mutation on mouse chromosome 5 is associated with an inversion. *Human Molecular Genetics*, *4*(11), 2073–2079.

NCBI Map Viewer. Online at: http://www.ncbi.nlm.nih.gov/mapview/

Page, P. et al. (2006). Clinical, clinicopathologic, postmortem examination findings and familial history of 3 Arabians with lavender foal syndrome. *Journal of Veterinary Internal Medicine*, *20*(6), 1491–1494.

Pashkova, N. et al. (2005). Myosin V attachment to cargo requires the tight association of two functional subdomains. *Journal of Cellular Biology*, *168*(3), 359–364.

Penedo, C. et al. (2011). A mutation in SLC45A2 is associated with pearl coat color dilution in the horse. *The 9th Dorothy Russell Havemeyer Foundation International Equine Genome Mapping Workshop*, July 27–29, Minneapolis, MN.

Pielberg, G. et al. (2005). Comparative linkage mapping of the Grey coat colour gene in horses. *Animal Genetics*, *36*(5), 390–395.

Pruvost, M. et al. (2011). Genotypes of predomestic horses match phenotypes painted in Paleolithic works of cave art. *Proceedings of the National Academy of Sciences of the United States of America*, *108*(46), 18626–18630.

Pulos, W. L., & Hutt, F. B. (1969). Lethal dominant white in horses. *Journal of Heredity*, *60*(2), 59–63.

Raudsepp, T. et al. (1999). Comparison of horse chromosome 3 with donkey and human chromosomes by cross-species painting and heterologous FISH mapping. *Mammalian Genome*, *10*(3), 277–282.

Rees, J. L. (2003). Genetics of hair and skin color. *Annual Review of Genetics*, *37*, 67–90.

Reissmann, M., Bierwolf, J., & G.A. Brockmann, G. A. (2007). Two SNPs in the SILV gene are associated with silver coat colour in ponies. *Animal Genetics*, *38*(1), 1–6.

Rieder, S. et al. (2008). Genetic analysis of white facial and leg markings in the Swiss Franches-Montagnes horse breed. *Journal of Heredity*, *99*(2), 130–136.

Rieder, S. (2009). Molecular tests for coat colours in horses. *Journal of Animal Breeding & Genetics*, *126*(6), 415–424.

Rieder, S. et al. (2001). Mutations in the agouti (ASIP), the extension (MC1R), and the brown (TYRP1) loci and their association to coat color phenotypes in horses (Equus caballus). *Mammalian Genome*, *12*(6), 450–455.

Rosengren Pielberg, G. et al. (2008). A cis-acting regulatory mutation causes premature hair graying and susceptibility to melanoma in the horse. *Natural Genetics*, *40*(8), 1004–1009.

Sandmeyer, L. S. et al. (2007). Clinical and electroretinographic characteristics of congenital stationary night blindness in the Appaloosa and the association with the leopard complex. *Veterinary Ophthalmology*, *10*(6), 368–375.

Santschi, E. M. et al. (1998). Endothelin receptor B polymorphism associated with lethal white foal syndrome in horses. *Mammalian Genome*, *9*(4), 306–309.

Sponenberg, D. P. (2009). *Equine color genetics* (3rd ed). Ames, IA: Wiley-Blackwell.

Stephenson, D. A. et al. (1994). Mouse rump-white mutation associated with an inversion of chromosome 5. *Mammalian Genome*, *5*(6), 342–348.

Swinburne, J. E., Hopkins, A., & Binns, M. M. (2002). Assignment of the horse grey coat colour gene to ECA25 using whole genome scanning. *Animal Genetics*, *33*(5), 338–342.

Terry, R. B. et al. (2004). Assignment of the appaloosa coat colour gene (LP) to equine chromosome 1. *Animal Genetics*, *35*(2), 134–137.

Trommershausen-Smith, A. (1978). Linkage of tobiano coat spotting and albumin markers in a pony family. *Journal of Heredity*, *69*(4), 214–216.

Valentine, B. A. (1995). Equine melanocytic tumors: A retrospective study of 53 horses (1988 to 1991). *Journal of Veterinary Internal Medicine*, *9*(5), 291–297.

Wagner, H. J., & Reissmann, M. (2000). New polymorphism detected in the horse MC1R gene. *Animal Genetics*, *31*(4), 289–290.

Woolf, C. M. (1989). Multifactorial inheritance of white facial markings in the Arabian horse. *Journal of Heredity*, *80*(3), 173–178.

Yang, G. C. et al. (1998). A dinucleotide mutation in the endothelin-B receptor gene is associated with lethal white foal syndrome (LWFS): A horse variant of Hirschsprung disease. *Human Molecular Genetics*, *7*(6), 1047–1052.

Yasumoto, K. et al. (2004). Epitope mapping of the melanosomal matrix protein gp100 (PMEL17): Rapid processing in the endoplasmic reticulum and glycosylation in the early Golgi compartment. *Journal of Biology & Chemistry*, *279*(27), 28330–28338.

10 Genomics of skin disorders

Amy E. Young, Stephen D. White, and Danika L. Bannasvch

Introduction

The skin is the largest organ of the body. Flexible, elastic, and tough, it provides a barrier between the body and the environment. It protects the body from physical, chemical, and microbiologic injury and prevents the loss of water, electrolytes, and macromolecules.

The skin is important for temperature regulation and its sensory components perceive sensations such as heat, cold, pain, itch, touch, and pressure. It produces glands and keratinized structures such as hair, hoof, and the horny layer (*stratum corneum*) of the epidermis. Melanin formation takes place in the skin and helps determine the color of the skin and coat. The skin is synergistic with internal organ systems and is often a reflection of pathologic processes occurring elsewhere in the body (Scott & Miller, 2003).

A subset of skin diseases in the horse is recognized as hereditary and a heritable basis has been suggested in another handful of dermatologic phenotypes. These conditions range from cosmetic annoyances that do not cause any discomfort to harmful phenotypes that have no treatment and are ultimately lethal.

A genetic component is often suspected for a disease because it is recognized in distinct breeds. By definition, a breed implies a more restricted gene pool and line breeding is common in certain breeds. Recognizing a pattern of inheritance in half-sibling families, which are the norm in horses, is challenging, but within a breed, increased frequency of disease likely indicates a heritable component.

Some of these disorders have been extensively studied in part because the phenotypes can be very severe, treatment options are virtually nonexistent, and the majority of cases have to be euthanized. The availability of equine genetic resources has led to the discovery of the underlying genetic mutations and the availability of DNA tests for some of these conditions.

Other dermatological diseases are presumed to have an inherited component but at present, a molecular cause or causes remain elusive. Candidate gene approaches based on similar phenotypes in other species may be effective at identifying causative mutations in horses. Advancements in equine genomics will hopefully facilitate investigations into the molecular bases for these less well-studied equine skin disorders.

This chapter includes descriptions of cutaneous and skin-related phenotypes in horses and relays the known genetic information with relation to each condition.

Equine Genomics, First Edition. Edited by Bhanu P. Chowdhary.

Skin Diseases with Identified Mutations

The availability of DNA samples from half-sibling families as well as accurate phenotypic information have allowed for the genetic characterization of two inherited skin diseases in the horse. Genetic tests are now available to breeders for both diseases.

Hereditary equine regional dermal asthenia

Hereditary equine regional dermal asthenia (HERDA) is a skin disease that affects young horses primarily of the American Quarter Horse breed (White et al., 2004). The phenotype is described as fragile, thin skin that exhibits slow-healing wounds and scarring. Lesions, either single or multiple, are commonly observed on the dorsal aspect of the horse (Lerner et al., 1978). The skin of affected horses is uniformly weak and abnormal as compared to unaffected horses and has been shown to have a two- to threefold reduction in tensile strength except on the abdomen, which may account for the less common presentation of clinical signs at this site (Grady et al., 2009).

Swellings, typically in the form of seromas and/or hematomas, open wounds or sloughing skin, easily damaged skin upon trauma, and poor healing of wounds are reported by owners and veterinarians (Figure 10.1). Lesions are common on the dorsum but have also been reported on other parts of the body (White et al., 2004). Once lesions appear, their frequency and severity typically increase with time (Tryon et al., 2007). Environmental factors such as saddling, increased exposure to UV light, and normal pasture-related activities such as rolling and scratching may impact the presence of wounds (Rashmir-Raven, 2004; White et al., 2004; Grady et al., 2009).

Noticeable lesions indicative of the disease generally appear at approximately 1.5 years of age (White et al., 2004; White et al., 2007). This means that HERDA-affected horses rarely show signs at birth and may not exhibit severe signs of clinical disease for the first year and a half (White et al., 2007). However, it has been noted that skin trauma can cause the clinical signs to appear as early as six months (Tryon et al., 2005). Once lesions develop, affected horses can deteriorate quickly as there is no proven treatment (Tryon et al., 2007); the majority of affected horses are euthanized (Tryon et al., 2005).

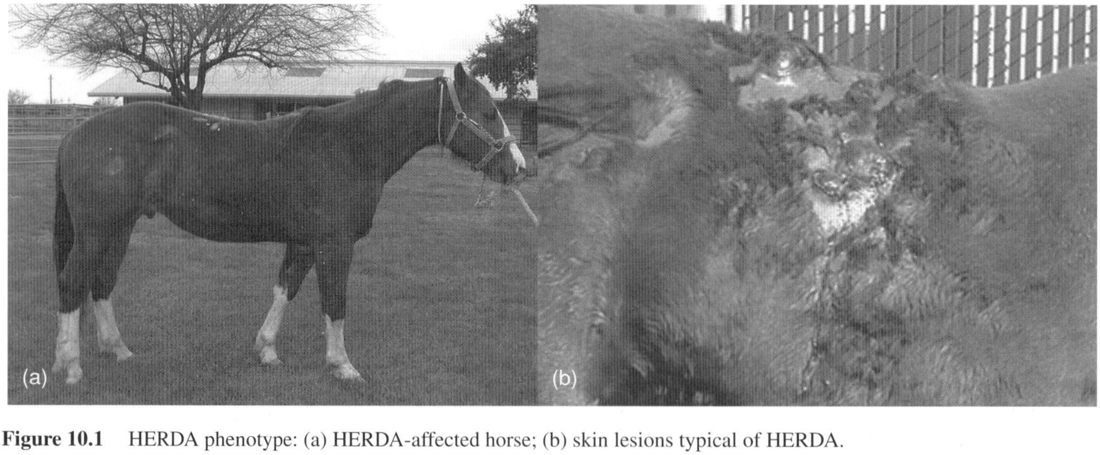

Figure 10.1 HERDA phenotype: (a) HERDA-affected horse; (b) skin lesions typical of HERDA.

HERDA is most commonly described in Quarter Horses, but cases have been reported in Appaloosas and American Paint Horses with Quarter Horse bloodlines (White et al., 2004) and in a single Arabian crossbred horse. Males and females are equally affected (Tryon et al., 2005), and a common ancestor appears in both the sires' and dams' pedigrees of the vast majority of affected horses (White et al., 2004).

The heritability for HERDA has been calculated at 0.38 +/– 0.13, and it is inherited in an autosomal recessive manner. It was been proposed that HERDA most closely resembles Ehlers-Danlos syndrome (EDS) in people. Although the clinical signs are not exactly the same between HERDA and EDS, genes known to be associated with EDS were proposed as potential candidates for HERDA (Tryon et al., 2005).

Identifying the HERDA locus by traditional linkage mapping methods was challenging. Horse families usually consist of half-siblings and horses have long generation times, meaning that each individual has relatively few offspring in a lifetime as compared to other species. Additionally, by the time the HERDA phenotype manifests itself, affected horses are frequently no longer owned by the breeder and full siblings of affected horses are not produced (Tryon et al., 2007). Based on the available pedigree information in which affected horses had a common ancestor on both sides of their pedigrees, a homozygosity mapping approach was successfully undertaken to map the HERDA locus to equine chromosome 1 (ECA1) (Tryon et al., 2007).

The genomic interval localized by homozygosity mapping contained 20 genes. A non-synonymous mutation in exon 1 of the equine PPIB gene was identified that segregates with the HERDA phenotype. The SNP (c.115G>A), which causes a glycine-to-arginine change in the putative N-terminal domain of the protein, maps to a syntenic block that has been strictly maintained throughout vertebrate evolution. This was the first whole-genome scan used to identify a novel disease gene in the horse (Tryon et al., 2007).

Genotyping of the identified mutation led to an estimated carrier frequency of 3.5%. Pedigree analysis suggests that the HERDA mutation is concentrated within particular lines of cutting horses. It is likely that the disproportionate breeding of highly successful stallions has led to a number of obligate carriers of the HERDA mutation, contributing significantly to the breeding population of Quarter Horses within the past two decades (Tryon et al., 2007).

A study that followed on the heels of the identification of the HERDA locus examined the mutation in a variety of Quarter Horse subgroups delineated by discipline. The HERDA allele was unsurprisingly high in cutting horses, with an allele frequency of 0.142. It was also identified in the reining (0.046), working cow horse (0.057), and Western pleasure (0.064) subgroups. The overall allele frequency was found to be 0.021 in Quarter Horses and 0.008 in Paint horses (Tryon et al., 2009).

Junctional epidermolysis bullosa

Junctional epidermolysis bullosa (JEB) is a skin disease that has been reported in two distinct breeds, the Belgian draft horse and American Saddlebred. Although phenotypically similar, it is caused by discrete genetic mutations in each breed.

First identified in both breeds as epitheliogenesis imperfecta (EI), the disease is characterized by blistering of the skin immediately after birth, primarily at pressure points such as the limbs (Spirito et al., 2002). Hoof loss and oral cavity involvement have been reported (Lieto et al. 2002; Spirito et al., 2002) (Figure 10.2). The progression of the disease is fatal and affected foals are usually euthanized shortly after birth (Lieto et al., 2003). The disease was reclassified as JEB

Figure 10.2 JEB phenotype: (a) complete loss of skin from the distal extremities of an 18-hour-old Belgian draft horse foal with JEB. (b) 24-hour-old Belgian foal with JEB showing premature eruption of incisor teeth, enamel hypoplasia (pitting of enamel, irregular serrated edges); (c) 8-day-old Belgian draft horse foal with JEB with skin lesions over pressure points and exungulation (loss of hoof capsule) of right front hoof. Courtesy Dr. J. D. Baird.

based on similarities to ultrastructural features of human JEB (Johnson et al., 1988; Goureau et al., 1989).

JEB has been described in Belgians from the United States (Kohn et al., 1989) and Canada (Shapiro & McEwen, 1995) and exhibits an autosomal recessive mode of inheritance (Spirito et al., 2002). The known crossing of closely related draft breeds means that sporadic cases of JEB have also been reported in other draft breeds such as the Comtois and Breton (Gourreau et al., 1989).

Research into the molecular cause of JEB in Belgians uncovered a change in laminin 5 expression in tissues from affected foals as indicated by immunofluorescence experiments (Spirito et al., 2002). Further analysis discovered the absence of immunoreactivity to laminin $\gamma 2$ chain antibodies in the basement membrane of the dermal-epidermal junction in affected foals. In humans and mice, absence of laminin 5 compromises the adhesion of keratinocytes and causes significant areas of skin detachment (Aberdam et al., 1994).

These observations indicated that *LAMC2*, a gene that resides on equine chromosome 5 (ECA5) and encodes the laminin $\gamma 2$ chain, was a promising candidate gene for equine JEB. Upon sequencing of the equine $\gamma 2$ cDNA, a homozygous basepair insertion (1368insC) was discovered in affected horses. This mutation results in a premature stop codon (TGA) in the N-terminal portion of domain III. The location and nature of this mutation results in the absence of the C-terminal domain I/II that is important for the assembly of laminin 5.

Further testing of the identified mutation in more than 150 horses confirmed that JEB in Belgians is autosomal recessive and determined that 48% of the horses tested were carriers (Spirito et al., 2002). Affected individuals from the Trait Breton and Trait Comtois draft breeds were also found to be homozygous for the same mutation (Milenkovic et al., 2003).

Prior to the identification of the equine *LAMC2* mutation in Belgians, an identical phenotype had been observed in Saddlebred foals (Lieto et al., 2002), which was thought to have an autosomal recessive mode of inheritance. Approximately 4% of the American Saddlebred breeding population was estimated to carry the disease allele, and the construction of a partial Saddlebred pedigree suggested that the spread of the disease was likely due to a single founder (Lieto, 2002).

Genetic mapping experiments localized Saddlebred JEB to equine chromosome 8 (ECA8), suggesting *LAMA3* as a candidate gene. Sequencing of the *LAMA3* cDNA identified a deletion that eliminates exons 24, 25, 26, and 27, a total deletion of 6,589 bases and a predicted 169 amino acids, in affected horses. This deletion is predicted to result in the absence of functional laminin 5 molecules in affected horses. Further testing of this mutation in 175 random Saddlebreds identified nine heterozygotes, showing the carrier frequency at 0.051. Belgian foals were tested and none had the *LAMA3* mutation (Graves et al., 2008).

The identification of these two laminin 5-associated genes and their association with equine JEB is not surprising considering that the Herlitz variant of JEB in humans is usually caused by a premature stop codon in one of the genes coding for the laminin 5 heterotrimer which results in a lack of expression of the protein chain (Aberdam et al., 1994; Vidal et al., 1995). The identification of these genes in horses supports the idea that equine JEB most closely resembles the Herlitz form of JEB in humans (Lieto et al., 2002). Equine JEB is the first reported hereditary disease in domestic animals with more than one identified causative mutation (Spirito et al., 2002).

Skin Diseases with Suspected Heritable Basis

For some skin diseases in horses a heritable basis is likely but currently unproven. These diseases are often specific to a certain breed, or a couple of closely related breeds, and may have been reported in small families. In most cases, additional work is needed to confidently determine the mode of inheritance and no definite causative genetic mutations have as yet been described.

Insect bite dermal hypersensitivity

Insect bite (dermal) hypersensitivity, or IBH, has been well documented in horses throughout the world (Braverman et al., 1983, Andersson et al., 1988, Littlewood, 1998). This seasonal, chronic dermatitis is also known as summer dermatitis/eczema/sores, sweet itch, and culicoides hypersensitivity (Riek, 1953), and is caused by IgE-mediated reactions to the bites of midges of the genus *Culicoides* (Quinn et al., 1983, Larsen et al., 1988). Affected horses scratch excessively, which can lead to secondary skin lesions. The most common signs are scratching, thickening of the skin, alopecia, excoriation, scaling, and wounds (Bjornsdottir et al., 2006). No consistently effective treatment is currently available (Marti et al., 2008) aside from environmental changes to aid horses in avoiding biting midges (Bjornsdottir et al., 2006) and corticosteroids to ameliorate the pruritus. Hyposensitization is controversial but has been effective in some cases (Anderson et al., 1996). A heritable basis has been proposed for IBH in certain horse breeds (Marti et al., 1992).

A study that examined the occurrence of IBH in half-sibling Swiss Warmblood families strongly suggested a hereditary component to susceptibility to IBH and proposed a recessive mode of inheritance. The same study indicated that a gene or genes of the equine major histocompatibility complex (MHC), along with other genes and environmental factors, play a role in susceptibility to this condition in some families (Marti et al., 1992).

A "slight but significant" association of the equine MHC with the occurrence of hypersensitivity dermatitis had previously been reported in Icelandic horses (Lazary et al., 1985). IBH generally does not occur in horses in Iceland as *Culicoide*s midges are not found in that region. However, it has been reported to have a particularly high prevalence in Icelandic horses that are exported to areas where *Culicoides* are present. One study reported the diagnosis of IBH in 34.5% of 330 horses exported from Iceland to Denmark, Sweden, or Germany. The number of affected horses subsequently increased to 49.5% at two years post-export and 47–54% in horses living in areas heavily infested with *Culicoides* (Bjornsdottir et al., 2006). Consequently, IBH has been recognized as a serious problem in Icelandic horses that are exported, but it is also of concern in Icelandic horses that are born abroad.

It has been suggested that the fact that only 50% of exported Icelandic horses become IBH-affected means that genetic factors are involved. An overall prevalence of IBH in Icelandic horses born in Germany was calculated to be 4.6%, but the prevalence increased to 12.2% when both parents were IBH-affected. However, one study did not find evidence of a sire effect and noted that the estimated heritability of IBH in their sample set was not significantly different from zero (Bjornsdottir et al., 2006). Currently, it is assumed that IBH in Icelandic horses is a complex trait with a large environmental component and it is likely that multiple genes with major effects are involved. Whole genome association analysis has been proposed to identify underlying genetic causes of IBH (Marti et al., 2008).

Genetic studies of IBH have already been undertaken in other breeds. A genome scan for IBH using 50 microsatellites was performed in individuals from the Old Kladruber breed. Significant associations were identified to the microsatellite AHT04 and to SNPs in the *FcεR1* alpha and the *IFNγ* genes. The study also found that total IgE levels were significantly different between individuals heterozygous for an *IFNγ* intron 1 SNP and individuals homozygous for the same SNP (Marti et al., 2008).

IBH has some features similar to atopic dermatitis in humans, which is recognized to have a heritable basis and for which associated genetic variants have been identified (Barnes et al., 2010). In humans, the *SPINK5* gene has been linked to Netherton syndrome, a disease that has atopic clinical signs (Chavanas et al., 2000, Nishio et al., 2003). Atopic dermatitis has also been reported in horses (Lorch et al., 2001a, 2001b, 2001c) and may have a heritable basis (Reese, 2001). However, analysis of SNPs in equine *SPINK5* in Icelandic horses revealed no significant association with IBH (Andersson et al., 2009). Additional studies are required to identify molecular changes associated with this complex disorder.

Chronic progressive lymphedema/Chronic pastern dermatitis

Chronic progressive lymphedema (CPL), also known as chronic pastern dermatitis (CPD), is an inflammatory skin disease common in heavily feathered draft horses. It is widely recognized in

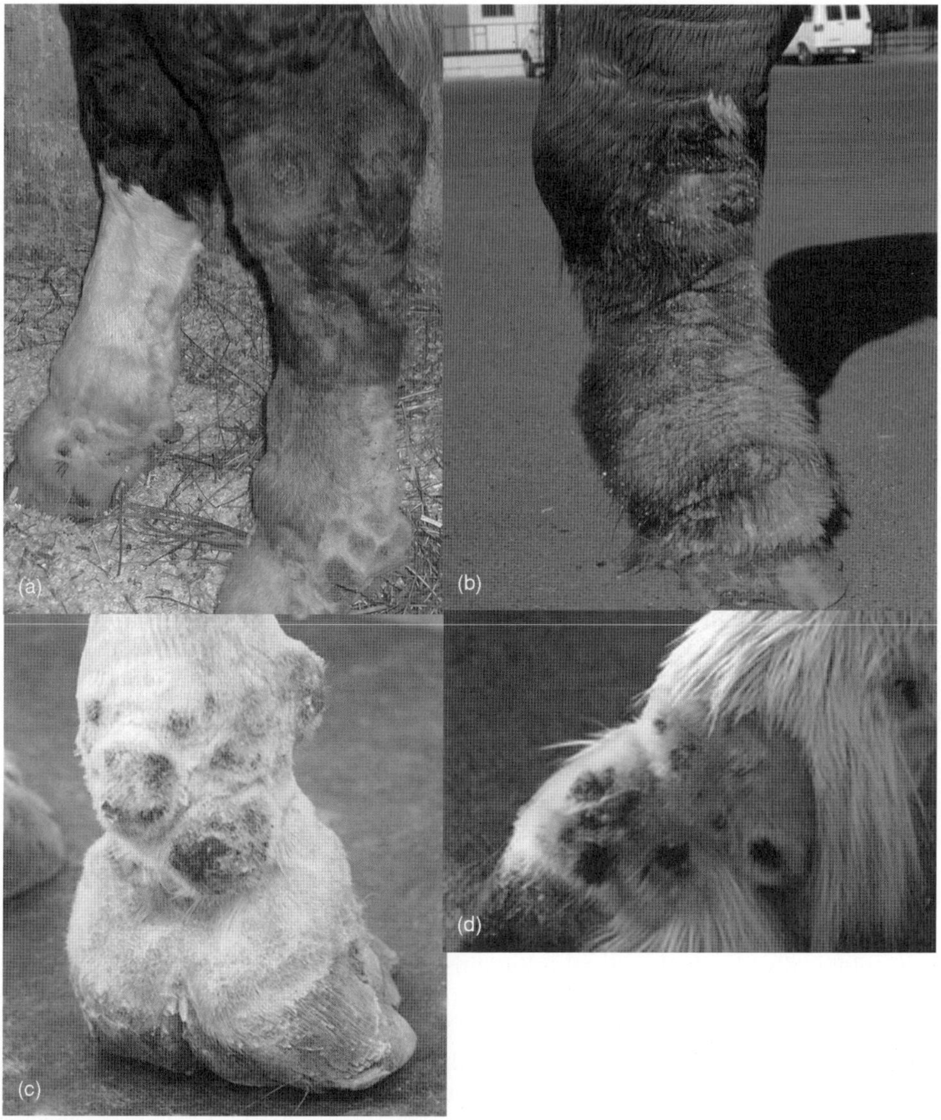

Figure 10.3 CPL phenotype: (a) hind limb of a CPL-affected horse; (b) lower limb swelling; (c) nodule formation; (d) lesions characteristic of CPL. Courtesy of the UC Davis Center for Equine Health.

horses from the Clydesdale, Shire, and Belgian draft horse breeds, as well as in other draft breeds around the world.

Diagnosis of CPL is mainly based on clinical signs including progressive and painful swelling, hyperkeratosis, nodule formation, thick skin folds, and fibrosis of the distal limbs (Figure 10.3), although lymphoscintigraphy has also been used to a limited extent (De Cock et al., 2006). Symptoms progress throughout the horse's life and there is currently no known effective treatment. The clinical signs in horses can lead to severe disfigurement and interfere with movement, causing significant discomfort (De Cock et al., 2003). Horses with CPL are susceptible to secondary infestations by

Chorioptes spp mites and bacterial infections that are very difficult to treat (Rufenascht et al., 2010), both of which can exacerbate lower-leg swelling. The most severe symptoms of the disease are generally observed later in life and severely affected horses are euthanized (De Cock et al., 2003).

Due to the prevalence of this disease in closely related draft breeds, it is assumed that genetic factors play a role (Schaper, 1950; Wallraf et al., 2004). Heritability across different German draft breeds has been calculated at 0.21, with heritability estimates within breeds ranging from 0.14 in South German draft horses to 0.98 in the Rhenish German (Wallraf et al., 2004). Environmental factors are presumed to influence clinical signs and the age at which these signs are observed (Mittmann et al., 2010).

CPL in horses closely resembles chronic lymphedema or elephantiasis nostras verrucosa in humans. Candidate gene approaches to identifying the molecular cause for CPL have been unsuccessful to date (Momke & Distl, 2007a, 2007b; Young et al., 2007), and the mode of inheritance of CPL is currently undetermined.

A study using a whole-genome scan with microsatellites in the aforementioned German draft breeds identified quantitative trait loci (QTL) for CPL (Mittmann et al., 2010). Seven chromosomes were implicated in various breeds and combinations of breeds. Potential candidate genes are present in these regions, but further studies are needed to identify causative mutations.

Determining the underlying genetic cause of CPL in horses is challenging. The prevalence of CPL is known to be high in many breeds and it has a variable age of onset. This makes it difficult to identify appropriate controls for genome-wide association studies. Secondly, a low number of microsatellite polymorphisms have been observed in draft breeds as compared to Thoroughbreds and warm-blooded horses, and it is assumed that the level of polymorphic SNPs will be even lower (Mittmann et al., 2010). Consequently, a large number of SNPs may be required to identify a causal genetic mutation, or mutations, for CPL.

Linear keratosis

Linear keratosis is characterized by firm, elevated, circumscribed, and linear areas of excessive keratin production (Rook et al., 1979; Muller et al., 1983). One of the authors (SDW) has seen a case that histologically also had a mural folliculitis, as seen in linear alopecia (see below). A rare dermatosis in the horse, it has been primarily reported in closely related horses of the Quarter Horse breed, although cases have occurred in a Morgan, Standardbred, and Percheron (Scott, 1985), and linear epidermal nevi has been reported in a family of Belgian horses (Paradis et al., 1993). This breed predilection suggests a hereditary basis (Scott, 1985).

Single or multiple linear, vertically oriented bands of alopecia and hyperkeratosis over the neck and lateral part of the thorax characterize the disease. It can begin as early as 6 months of age and usually persists throughout the life of the horse (Stannard et al., 1976; Ihrke et al., 1983). Topical therapies are reportedly effective to minimize the surface hyperkeratosis (Scott, 1985).

Equine linear keratosis is believed to closely resemble a linear epidermal nevus in humans, which may have a hereditary basis (Scott et al., 1984; Scott 1985). Currently, no genetic loci have been associated with equine linear keratosis.

Skin/Hair Color Phenotypes

Melanocytes, the cells that contain melanosomes, which in turn produce melanin pigments in the skin and hair, are found in the skin as well as in hair follicles (Thomsett, 1991). Because the skin is often a good indicator of processes taking place in other parts of the body, skin and coat color and/or condition can be characteristic of early or outward signs of diseases that primarily impact other parts or organ systems. In-depth discussions of most of these diseases are presented elsewhere in this book (see Chapter 11 in this volume), so they are only briefly mentioned here.

Gray/melanoma

Gray horses are born colored and gradually lose hair pigment as they age but maintain dark skin pigment. Gray coat color in horses is dominant. It is well known that gray horses have a high incidence of dermal melanomas that generally take the form of firm, black nodules under the tail base and in anal, perianal, and genital regions, as well as the perineum, lips, and eyelids (Swinburne et al., 2002). Although some melanomas are benign, others, depending on the age of the horse and the histopathology of the tumor (Valentine, 1998), can metastasize to internal organs. Melanomas are reported in 70–80% of gray horses older than 15 years of age (Sutton et al., 1997, Fleury et al., 2000).

Other characteristics that often accompany the gray phenotype include skin depigmentation that appears as patches of red pigmentation (Fleury et al., 2000), speckling of the coat, and the speed with which the coat turns gray. These features are widely variable among gray horses. Reduced longevity has also been reported in gray horses (Comfort et al., 1958).

The gray coat color locus in horses is associated with a 4.6 kilobase (kb) insertion in the *STX17* gene (Rosengren Pielberg et al., 2008). The duplication was detected in all gray horses tested but was not observed in non-gray horses. A high level of expression of *STX17* was seen in melanomas from gray horses. Horses that are homozygous for the *STX17* duplication are more homogenously white as compared to heterozygotes and have a higher incidence of melanoma, patches of depigmentation, and exhibit almost no speckling.

In addition to *STX17*, there is a highly significant association between a horse's genotype for the *ASIP* gene, an *MC1R* antagonist, and the incidence of melanoma (Rosengren Pielberg et al., 2008). However, *ASIP* genotype has no significant effects on graying, patchy depigmentation, or speckling. These results show that the *ASIP* gene influences dermal melanocytes and implies that melanoma in gray horses is promoted by increased MC1R signaling.

Researchers also noted in melanomas from gray horses a markedly higher expression of *NR4A3*, a gene located in the same genomic interval as *STX17*. Expression of *NR4A3* in horses heterozygous for gray occurs only from the gray haplotype, suggesting that a *cis*-acting regulatory mutation, possibly the *STX17* duplication, underlies the upregulation of expression.

Albinism

The lack of pigment in the skin, hair, and iris is known as albinism. Albinos are recognized by white hair, pink skin, and pink or light blue eyes. Albinism is an autosomal dominant trait: all albinos are heterozygous, while homozygotes are not viable. Aside from their appearance, albinos are normal but have a tendency to sunburn, are prone to squamous cell carcinoma, and might suffer from photophobia and light-related retinal damage (Knottenbelt, 2009).

In humans, oculocutaneous albinism (OCA) has a genetic basis; more than 36 mutations have been identified in the tyrosinase gene for type I OCA (Giebel et al., 1990, Oetting & King, 1993). As in horses, the affected individuals in these studies were heterozygotes. Tyrosinase mutations have also been implicated in murine albinism (Kwon et al., 1988, Jackson & Bennett, 1990).

The genetics of equine depigmentation phenotypes, including albinism, is heterogeneous, and is discussed in detail in Chapter 11 in the "White spotting and depigmentation patterns" section.

Lethal white foal syndrome (LWFS)

Foals affected with this disease are all white or nearly white and may be deaf and/or have blue eyes. The white coat is caused by the absence of melanocytes in the skin rather than a biochemical absence of pigment, as is the case in albinism. In addition to effects on melanoctyes, other cells derived from the neural crest, such as the enteric ganglia, are also absent, causing severe intestinal blockage/colic and resulting in the death of affected foals within 12 hours of birth. There is no treatment for LWFS.

The condition is seen primarily in American Paint horses, especially those with the frame overo white spotting pattern (Vonderfecht et al., 1983; McCabe et al., 1990), and is similar to Hirschsprung disease in humans. A mutation in the *EDNRB* gene is known to cause Hirschsprung disease and is also associated with lethal white spotting in mice and rats. Likewise, a missense mutation in the equine *EDNRB* gene is responsible for LWFS (Metallinos et al., 1998; Santschi et al., 1998; Yang et al., 1998).

Lavender foal syndrome (LFS)

Like LWFS, Lavender foal syndrome (LFS) has no effective treatment and the disease is lethal. Outwardly, affected individuals exhibit a characteristic dilute coat color that is described as lavender, pale gray, pewter, or light chestnut. Foals with LFS have multiple neurologic abnormalities that prevent them from standing and nursing normally. The condition is most commonly reported in Egyptian Arabians where it has an autosomal recessive mode of inheritance. A single base pair deletion in the *MYO5A* gene, which is associated with Griscelli syndrome in humans, causes a frameshift and a premature stop codon. The identification of the LFS mutation represents the first successful use of whole genome SNP association study in the horse (Brooks et al., 2010).

Arabian fading syndrome (AFS)/pinky Arab syndrome

Arabian fading syndrome (AFS) is most often reported in horses from the Arabian breed, but it has also been observed in Welsh mountain ponies and Clydesdales. It is characterized by a loss of pigmentation and hair. Skin biopsies from affected horses reveal a loss of melanin from the epidermal basal cells. It can occur spontaneously at any age and is more frequently observed in gray horses (McMullan, 1982; Mullowny, 1985; Yager & Scott, 1985). Depigmentation of the muzzle, lips, periorbital regions, perineum, sheath, and hooves has been noted, and patches of depigmentation on the body are usually permanent. These regions of depigmentation generally do not show inflammatory or traumatic changes and do not appear to be excessively pruritic. The condition is not harmful to the horse, although the pink skin is more susceptible to sunburn and affected horses have an increased susceptibility to squamous cell carcinoma.

As the disease is most often encountered in a certain breed, a heritable basis has been proposed. Similar cases of hypopigmentation have been reported in certain family lines in Belgian Tervuren dogs (Mahaffey, Yardrough, & Munnell, 1978), and an autosomal recessive form has been reported in black Angus cattle (Foreman et al., 1994). Susceptibility to patchy depigmentation of the skin and hair in humans is genetically complex and has been associated with more than one locus (Jin et al., 2010). The molecular cause of AFS is currently unknown.

Foal immunodeficiency syndrome (FIS)

Foal immunodeficiency syndrome (FIS), also known as Fell pony syndrome or Fell pony immunocompromise disorder, affects young purebred Fell (Scholes et al., 1988; Butler et al., 2006) and Dales ponies (Anonymous, 2009). Primary cutaneous signs, including a rough, russet-colored coat, are characteristic of the disease and can provide an early warning of a serious condition. Coat hairs can be abnormally long, especially around the head and neck, which gives the foal a "halo-like appearance" (Richards et al., 2000). Opportunistic infections typically develop on the skin and mucosal surfaces but can occur in different organ systems. There is no treatment for FIS and it is fatal at an early age (Knottenbelt, 2009).

The disease has an autosomal recessive mode of inheritance (Thomas et al., 2003) and the carrier rate in the United Kingdom and the Netherlands has been estimated at 40–60% (Butler et al., 2006). Recent genome-wide SNP analysis associated FIS with a 2.6 Mb region on ECA26, and proposed a mutation in *SLC5A3* responsible for the compromise of the immune system (Fox-Clipsham et al., 2011).

Hair Phenotypes

Hairs are produced in hair follicles, emerge from ostia in the surface of the skin, are important for thermal insulation and sensory perception, and protect the skin against injury. The length, thickness, and density of hair correlate with the ability of the hair coat to regulate body temperature (Scott & Miller, 2003).

Hypotrichosis/mane and tail dystrophy/follicular dysplasia

Hypotrichosis, also known as follicular dysplasia or mane and tail dystrophy, is a condition in which the amount and density of hair is significantly reduced, most commonly accompanied by thinning or lack of hair around the eyes and muzzle. The skin of affected horses is reported to be scaly and thin. Although severe forms are rare, it is presumed to be a hereditary condition in some lines of Arabians and has also been seen in other breeds to a lesser extent. It is generally not uncomfortable for the horse and there is no treatment.

Some forms of this condition primarily affect the hairs of the mane and tail. Affected tail hairs are characteristically sparse and brittle. Biopsies can be used to confirm the presence of a reduced number of hair follicles. The condition is well recognized, and even accepted as normal, in some lines of Appaloosas, suggesting a heritable component (Knottenbelt, 2009).

Comparatively, autosomal recessive forms of hypotrichosis in humans are associated with the *LIPH* (Ali et al., 2007) and *DSG4* (Schaffer et al., 2006; Shimomura et al., 2006) genes. These genes may represent potential candidates for the analogous condition in horses.

Curly coat syndrome

Foals with curly coat syndrome are born with an abnormally long, curly hair coat and a curly mane and tail. The affected horses often have hypotrichosis that causes "string tail" or "scanty tail" and alopecia caused by follicular dysplasia that is presumed to be genetically determined. The curliness of the coat varies from tight ringlets to waves in the winter to a smoother, slightly wavy summer coat. It has mainly been described in Paint horses in which a dominant mode of inheritance has been suggested (Thomas, 1989). Curly coat syndrome has also been described in other breeds including Quarter Horse, Percheron, Arabian, Appaloosa, Missouri Fox Trotter, Tennessee Walking Horse, Morgan, and Paso Fino. It is thought to be recessive in these breeds (Knottenbelt, 2009).

There is currently no available treatment for curly coat syndrome and the phenotype is primarily considered cosmetic, with no direct negative effects on the affected horse. To date, registries that represent the curly horse include the American Bashkir Curly Registry, the American Curly Horse Association, and the International Curly Horse Organization (Scott, 2004).

Molecular causes for the curly coat syndrome are not well understood. A genomic scan for the regions known to carry candidate genes for hair texture was undertaken to identify genetic markers associated with dominant curly coat (Cothran et al., 2009). Evidence of linkage or association was not observed. It has been proposed that Curly coat syndrome in horses is similar to Wilson's disease and non-photosensitive trichothiodystrophy in humans, which are associated with mutations in the *TTDN1* gene (Nakabayashi et al., 2005). Additional research is needed to uncover loci associated with this trait in either species.

Linear alopecia

Linear alopecia is characterized by circular areas of baldness that are oriented vertically, most commonly in areas of the neck, shoulder, and lateral thorax. Histologically it is typified by a mural folliculitis. The condition is rare in horses but has been seen in different breeds, most predominantly the Quarter Horse (Fadok, 1995). The lesions do not appear to be painful or pruritic (Scott & Miller, 2003).

Alopecia areata

Alopecia areata is one of the most common human autoimmune diseases (Gilhar & Kalish, 2006). It is known to be genetically determined and several loci have been implicated in susceptibility. Similarly, multiple loci are involved in the disease in the mouse model (Sundberg et al., 2004). In the horse, circular areas of alopecia are seen. Histologically, a lymphocytic infiltrate surrounding the base of the hair follicle is seen; however, in long-standing cases this infiltrate may not be present. Presence of antibodies targeting the hair follicles have been documented (Tobin, 1998). Lesions may be focal or progress to whole body alopecia (often sparing the mane and tail). Most horses regrow the hair but this may take up to two to three years. Corticosteroids *may* hasten the hair regrowth.

Conclusion

As described here, evidence exists for a hereditary basis for a number of equine skin conditions. Advances in genetic tools have facilitated the discovery of the causative mutations for some phenotypes. It is anticipated that further improvement of genomics tools will enable researchers to gain a better understanding of the more complex conditions and develop genetic tests to help breeders produce foals that are free from serious dermatological issues.

References

Aberdam, D., Galliano, M. F., Vailly, J., Pulkkinen, L., Bonifas, J. et al. (1994). Herlitz's junctional epidermolysis bullosa is linked to mutations in the gene (LAMC2) for the gamma 2 subunit of nicein/kalinin (LAMININ-5). *Nature Genetics*, *6*, 299–304.

Ali, G., Chishti, M. S., Raza, S. I., John, P., & Ahmad, W. (2007). A mutation in the lipase H (LIPH) gene underlie autosomal recessive hypotrichosis. *Human Genetics*, *121*(3–4), 319–325.

Andersson, G. S., Belton, P., & Kleider, N. (1988). The hypersensitivity of horses to Culicoides bites in Brittish Columbia. *Canadian Veterinary Journal*, *29*, 718–723.

Anderson, G. S. et al. (1996). Immunotherapy trial for horses in British Columbia with *Culicoides* hypersensitivity. *Journal of Medical Entomology*, *33*, 458–466.

Andersson, L. S., Hogstrom, C., Mikko, S., Eriksson, S., Grandinson, K. et al. (2009). Polymorphisms in SPINK5 do not associate with insect bite hypersensitivity in Icelandic horses born in Sweden. *Animal Genetics*, *40*, 788–794.

Anonymous (2009). Fell pony syndrome: Mutation identified. *Veterinary Records*, *165*, 732.

Barnes, K. C. (2010). An update on the genetics of atopic dermatitis: Scratching the surface in 2009. *Journal of Allergy and Clinical Immunology*, *125*(1), 16–29.

Bjornsdottir, S., Sigvaldadottir, J., Brostrom, H., Langvad, B., & Sigurdsson, A. (2006). Summer eczema in exported Icelandic horses: influence of environmental and genetic factors. *Acta Veterinaria Scandinavia*, *48*, 3.

Braverman, Y., Ungar-Waran, H., Frith, K., Adler, H., Danielli, Y. et al. (1983). Epidemiological and immunological studies of sweet itch in horses in Israel. *Veterinary Records*, *112*, 521–524.

Brooks, S. A., Gabreski, N., Miller, D., Brisbin, A., Brown, H. E. et al. (2010). Whole-genome SNP association in the horse: Identification of a deletion in myosin Va responsible for Lavender Foal Syndrome. *PLoS Genetics*, *6*(4), e1000909.

Butler, C. M., Westermann, C. M., Koeman, J. P., & van Oldriutenbortgh-Oosterbaan, S. (2006). The Fell Pony Immunodeficiency Syndrome also occurs in the Netherlands: A review and six cases. *Tijdschr Diergeneeskd*, *131*, 114–118.

Chavanas, S., Bodemer, C., Rochat, A., Hamel-Teillac, D., Ali, M. et al. (2000). Mutations in SPINK5, encoding a serine protease inhibitor, cause Netherton syndrome. *Nature Genetics*, *25*, 141–142.

Comfort, A. (1958). Coat colour and longevity in thoroughbred mares. *Nature*, *182*, 1531–1532.

Cothran, E. G., Juras, R., & Bruce, A. (2009). Mapping of the Dominant Gene for the Curly Hair Trait in the Horse. *Plant & Animal Genomes XVII Conference*, January 10–14, San Diego, CA.

DeCock, H. E. V., Affolter, V. K., Wisner, E. R., Ferraro, G. L., & Machlachlan, N. J. (2003). Progressive swelling, hyperkeratosis, and fibrosis of distal limbs in Clydesdales, Shires, and Belgian draft horses, suggestive of primary lymphedema. *Lymphatic Research and Biology*, *1*(3), 191–199.

DeCock, H. E. V., Affolter, V. K., Wisner, E. R., Larson, R. F., & Ferraro, G. L. (2006). Lymphoscintigraphy of draught horses with chronic progressive lymphoedema. *Equine Veterinary Journal*, *38*(2), 148–151.

Fadok, V. A. (1995). Update on four unusual equine dermatoses. *Veterinary Clinician in North America*, *11*, 105–110.

Fleury, C., Bérard, F., Leblond, A., Faure, C., Ganem, N., & Thomas, L. (2000). The study of cutaneous melanomas in Camargue-type gray-skinned horses (2): epidemiological survey. *Pigment Cell Research*, *13*, 47–51.

Foreman, M. E., Lamoreux, M. L., Kwon, B., & Womack, J. E. (1994). Mapping the bovine albino locus. *Journal of Heredity*, *85*(4), 318–320.

Fox-Clipsham, L. Y., Carter, S. D., Goodhead, I., Hall, N., Knottenbelt, D. C. et al. (2011). Identification of a mutation associated with fatal foal immunodeficiency syndrome in the Fell and Dales Pony. *PLoS Genetics*, *7*(7), e1002133.

Giebel, L. B., Strunk, K. M., King, R. A., Hanifin, J. M., & Spritz, R. A. (1990). A frequent tyrosinase gene mutation in classic, tyrosinase-negative (type IA) oculocutaneous albinism. *Proceedings of the National Academy of Sciences of the United States of America*, *87*, 3255–3258.

Gilhar, A., & Kalish, R. S. (2006). Alopecia areata: A tissue specific autoimmune disease of hair follicle. *Autoimmunity Review*, *5*, 64–69.

Goureau, J. M., Feillou, C., Morand, M., Courreau, J. F., Dupere, A. M., & Alliot A. (1989). Épidermolyse bulleuse jonctionelle léthale chez le cheval de trait en France. *Bulletin Academe Véterinaire Francaise*, *62*, 345–353.

Grady, J. G., Elder, S. H., Ryan, P. L., Swiderski, C. E., & Rashmir-Raven, A. M. (2009). Biomechanical and molecular characteristics of hereditary equine regional dermal asthenia in Quarter Horses. *Veterinary Dermatology*, *20*, 591–599.

Graves, K. T., Henney, P. J., & Ennis, R. B. (2008). Partial deletion of the LAMA3 gene is responsible for hereditary junctional epidermolysis bullosa in the American Saddlebred Horse. *Animal Genetics*, *40*, 35–41.

Ihrke, P. J. (1983). Diseases of abnormal keratinization (seborrhea). In *Current Therapy in Equine Medicine* (pp. 546–547). Philadelphia: WB Saunders Co.

Jackson, I. J., & Bennett, D. C. (1990). Identification of the albino mutation of mouse tyrosinase by analysis of an in vitro revertant. *Proceedings of the National Academy of Sciences of the United States of America*, *87*, 7010–7014.

Jin, Y., Birlea, S. A., Fain, P. R., Mailloux, C. M., Riccardi, S. L. et al. (2010). Common variants in FOXP1 are associated with generalized vitiligo. *Nature Genetics*, *42*(7), 576–578.

Johnson, G. C., Kohn, C. W., Johnson, C. W., Garry, F., Scott, D., & Martin, S. (1988). Ultrastructure of junctional epidermolysis bullosa in Belgian foals. *Journal of Comparative Pathology*, *98*, 329–336.

Kohn, C. W., Johnson, G. C., Garry, F., Johnson, C. W., Martin, S., & Scott D. W. (1989). Mechanobullous disease in two Belgian foals. *Equine Veterinary Journal*, *21*, 297–301.

Knottenbelt, D. C. (2009). *Pascoe's principles and practice of equine dermatology* (2nd ed). Edinburgh: Saunders Elsevier.

Kwon, B. S., Wakulchik, M., Haq, A. K., Halaban, R., & Kestler, D. (1988). Sequence analysis of mouse tyrosinase cDNA and the effect of melanotropin on its gene expression. *Biochemistry and Biophysics Research and Communication*, *153*, 1301–1309.

Larsen, H. J., Bakke, S. H., & Mehl, R. (1988). Intradermal challenge of Icelandic horses in Norway and Iceland with extracts of *Culicoides spp*. *Acta Veterinaria Scandinavia*, *29*, 311–314.

Lazary, S., Larsen, H., Glatt, P. A., Gerber, H., Kristensen, B., & Bakke, S. (1985). ELA in horses afflicted by sweet itch. *Animal Blood Groups and Biochemical Genetics* (Suppl. 1), 93.

Lerner, D. J., & McCracken, M. D. (1978). Hyperelastosis in 2 horses. *Journal of Equine Medicine and Surgery*, *2*, 350–352.

Lieto, L. D., & Cothran, E. G. (2003). The epitheliogenesis imperfecta locus maps to equine chromosome 8 in American Saddlebred horses. *Cytogenetic and Genome Research*, *102*, 207–210.

Lieto, L. D., Swerczek, T. W., & Cothran, E. G. (2002). Equine epitheliogenesis imperfecta in two American Saddlebred foals is a lamina lucida defect. *Veterinary Pathology*, *39*, 576–580.

Littlewood, J. D. (1998). Incidence of recurrent pruritus ("sweet itch") in British and German shire horses. *Veterinary Records*, *142*, 66–67.

Lorch, G., Hillier, A., Kwochka, K.W. et al. (2001a). Comparison of immediate intradermal test reactivity with serum IgE quantitation by use of a radioallergosorbent test and two ELISA in horses with and without atopy. *Journal of American Veterinary Medicine*, *218*, 1314–1322.

Lorch, G., Hillier, A., Kwochka, K. W. et al. (2001b). Results of intradermal tests in horses without atopy and horses with chronic obstructive pulmonary disease. *American Journal of Veterinary Research*, *62*, 389–397.

Lorch, G., Hillier, A., Kwochka K. W. et al. (2001c).Results of intradermal tests in horses without atopy and horses with atopic dermatitis or recurrent urticaria. *American Journal of Veterinary Research*, *62*, 1051–1059.

Mahaffey, M. B., Yarbrough, K. M., & Munnell, J. F. (1978). Focal loss of pigment in the Belgian Tervuren dog. *Journal of American Veterinary Medicine Association*, *173*(4), 390–396.

Marti, E., Gerber, H., & Lazary, S. (1992). On the genetic basis of equine allergy diseases: II. Insect bite dermal hypersensitivity. *Equine Veterinary Journal*, *24*(2), 113–117.

Marti, E., Gerber, V., Wilson, A. D., Lavoie, J. P., Horohov, D. et al. (2008). Report of the 3rd Havemeyer workshop on allergic diseases of the Horse. Holar, Iceland, June 2007. *Veterinary Immunology and Immunopathology*, *126*, 351–361.

McCabe, L., Griffin, L. D., Kinzer, A., Chandler, M., Beckwith, J. B., & McCabe, E. R. (1990). Overo lethal white foal syndrome: equine model of aganglionic megacolon (Hirschsprung disease). *American Journal of Medical Genetics*, *36*(3), 336–340.

McCaig, J. (1973). A survey to establish the incidence of sweet itch in ponies from the United Kingdom. *Veterinary Records*, *93*, 444–446.

McMullan, W. C. (1982). The skin. In R. Mansmann (ed.), *Equine Medicine and Surgery* (p. 789) Santa Barbara, CA: American Veterinary Publications Inc.

Metallinos, D. L., Bowling, A. T., & Rine, J. (1998). A missense mutation in the endothelin-B receptor gene is associated with Lethal White Foal Syndrome: an equine version of Hirschsprung disease. *Mammalian Genome*, *9*(6), 426–431.

Milenkovic, D., Chaffaux, S., Taorit, S., & Guerin, G. (2003). A mutation in the LAMC2 gene causes the Herlitz junctional epidermolysis bullosa (H-JEB) in two French draft horse breeds. *Genetics of Selective Evolution*, *35*, 249–256.

Mittmann, E. H., Mumke, S., & Distl, O. (2010). Whole-genome scan identifies quantitative trait loci for chronic pastern dermatitis in German draft horses. *Mammalian Genome*, *21*, 95–103.

Momke, S., & Distl, O. (2007a). Molecular characterization of the canine ATP2 A2 gene. *Cytogenetic and Genome Research*, *116*, 256–262.

Momke, S., & Distl, O. (2007b). Molecular analysis of the ATP2A2 gene as candidate for chronic pastern dermatitis in German draft horses. *Journal of Heredity*, *98*, 267–271.

Muller, G. H., Kirk, R. W., & Scott, D. W. (1983). *Small animal dermatology III*. Philadelphia: WB Saunders Co.

Mullowney, P. C. (1985). Dermatologic disease of horses. Part V. Alergic, immune mediated and miscellaneous skin diseases. *Compendium of Continuing Education Practices in Veterinary*, *7*, 217–228.

Nakabayashi, K., Amann, D., Ren, Y., Saarialho-Kere, U., Avidan, N. et al. (2005). Identification of C7orf11 (TTDN1) gene mutations and genetic heterogeneity in nonphotosensitive trichothiodystrophy. *American Journal of Human Genetics*, *76*, 510–516.

Nishio, Y., Noguchi, E., Shibasaki, M., Kamioka, M., Ichikawa, E. et al. (2003). Association between polymorphisms in the SPINK5 gene and atopic dermatitis in the Japanese. *Genes Immunology*, *4*, 515–517.

Oetting, W. S., & King, R. A. (1993). Molecular basis of type I (tyrosinase-related) oculocutaneous albinism: mutations and polymorphisms of the human tyrosinase gene. *Human Mutations*, *2*, 1–6.

Paradis, M., Rossier, Y., & Rosseel, G. (1993). Linear epidermal nevi in a family of Belgian horses. *Equine Practice*, *15*(7), 10–14.

Rosengren Pielberg, G. R., Golovko, A., Sundstrom, E., Curik, I., Lennartsson, J. et al. (2008). A cis-acting regulatory mutation causes premature hair graying and susceptibility to melanoma in the horse. *Nature Genetics*, *40*(8), 1004–1009.

Quinn, P. J., Baker, K. P., & Morrow, A. N. (1983). Sweet itch: Responses of clinically normal and affected horses to intradermal challenge with extracts of biting insects. *Equine Veterinary Journal*, *15*, 266–272.

Rashmir-Raven, A. M., Winand, N. J., & Read, R. W. (2004). Equine hyperelastosis cutis update. *Proceedings of the 50th Annual Convention of the Association of Equine Practitioners*, Houston, TX, 47–57.

Rees, C. A. (2001). Response to immunotherapy in six related horses with urticaria secondary to atopy. *Journal of American Veterinary Medicine Association*, *218*, 753–755.

Riek, R. F. (1953). Studies on allergic dermatitis (Queensland itch) of the horse. I. Description, distribution, symptoms and pathology. *Australian Veterinary Journal*, *29*, 177–184.

Rufenacht, S., Roosje, P. J., Sager, H., Doherr, M. G., Straub, R. et al. (2010). Combined moxidectin and environmental therapy do not eliminate Chorioptes bovis infestation in heavily feathered horses. *Veterinary Dermatology*, *22*(1), 17–23.

Santschi, E. M., Purdy, A. K., Valberg, S. J., Vrotsos, P. D., Kaese, H., & Mickelson, J. R. (1998). Endothelin receptor B polymorphism associated with lethal white foal syndrome in horses. *Mammalian Genome*, *9*(4), 306–309.

Schaffer, J. V., Bazzi, H., Vitebsky, A., Witkiewicz, A., Kovich, O. I. et al. (2006). Mutations in the desmoglein 4 gene underlie localized autosomal recessive hypotrichosis with monilethrix hairs and congenital scalp erosions. *Journal of Investigative Dermatology*, *126*(6), 1286–1291.

Schaper, W. (1950). Die Verbesserung der Konstitution unserer Haustiere. *Z Tierzuchtg Zuchtgsbiologie*, *58*, 1–57.

Scholes, S. F. E., Holliman, A., May, P. D. F., & Holmes, M. A. (1998). A syndrome of anemia, immunodeficiency and peripheral ganglionopathy in Fell pony foals. *Veterinary Record*, *142*, 128–134.

Scott, D. W. (1985). Equine linear keratosis. *Equine Practice*, *7*(7), 39–42.

Scott, D. W. (2004). Skin of the neck, mane and tail of the curly horse. *Equine Veterinary Education*, *16*(4), 201–206.

Scott, D. W., & Miller, Jr., W. H. (2003). *Equine dermatology*, St. Louis, MO: Saunders Elsevier.

Scott, D. W., & Yager-Johnson, J. A. (1984). Nevi in the dog. *Journal of the American Animal Hospital Association*, *20*, 505–512.

Shapiro, F., & McEwen, B. (1995). Mechanobullous disease in a Belgian foal in eastern Ontario. *Canadian Veterinary Journal*, *36*, 572.

Shimomura, Y., Sakamoto, F., Kariya, N., Matsunaga, K., & Ito, M. (2006). Mutations in the desmoglein 4 gene are associated with monilethrix-like congenital hypotrichosis. *Journal of Investigative Dermatology*, *126*(6), 1281–1285.

Spirito, F., Charlesworth, A., Linder, K., Ortonne, J. P., Baird, J., & Meneguzzi, G. (2002). Animal models for skin blistering conditions: Absence of laminin 5 causes hereditary junctional mechanobullous disease in the Belgian horse. *Journal of Investigative Dermatology*, *119*, 684–691.

Stannard, A. A. (1976). Equine dermatology. *Proceedings of the American Association of Equine Practitioners*, *22*, 273–292.

Sundberg, J. P., Silva, K. A., Li, R., Cox, G. A., & King, L. E. (2004). Adult-onset alopecia areata is a complex polygenic trait in the C3H/HeJ mouse model. *Journal of Investigative Dermatology*, *123*, 294–297.

Sutton, R. H., & Coleman, G. T. (1997). Melanoma and the graying horse. *RIRDC Research Paper Series*, 1–34 (Barton, Australia).

Swinburne, J. E., Hopkins, A., & Binns, M. M. (2002). Assignment of the horse gray coat colour gene to ECA25 using whole genome scanning. *Animal Genetics*, *33*, 338–342.

Thomas, G. W., Bell, S. C., Phythian, C., Taylor, P., Knottenbelt, D. C., & Carter, S. D. (2003). Aid to the antemortem diagnosis of Fell pony syndrome by the analysis of B lymphocytes. *Veterinary Record*, *152*(20), 618–621.

Thomas, S. (1989). *Myth and mystery: The Curly Horse in America*. Freestone: C. S. Fund.

Thomsett, L. R. (1991). Pigmentation and pigmentary disorders of the equine skin. *Equine Veterinary Education*, *3*, 131–134.

Tobin, D. J., Alhaidari, Z., & Olivry, T. (1998). Equine alopecia areata autoantibodies target multiple hair follicle antigens and may alter hair growth. A preliminary study. *Exploratory Dermatology*, *7*, 289–297.

Tryon, R. C., Penedo, M. C. T., McCue, M. E., Valberg, S. J., Mickelson, J. R. et al. (2009). Evaluation of allele frequencies of inherited disease genes in subgroups of American Quarter Horses. *Journal of the American Veterinary Medical Association*, *234*, 120–125.

Tryon, R. C., White, S. D., & Bannasch, D. L. (2007). Homozygosity mapping approach identified a missense mutation in equine cyclophilin B (PPIB) associated with HERDA in the American Quarter Horse. *Genomics*, *90*, 93–102.

Tryon, R. C., White, S. D., Famula, T. R., Schultheiss, P. C., Hamar, D. W., & Bannasch, D. L. (2005). Inheritance of hereditary equine regional dermal asthenia in Quarter Horses. *American Journal of Veterinary Research*, *66*, 437–442.

Valentine, B. A. (1998). Equine melanocytic tumors: a retrospective study of 53 horses (1988–1991). *Journal of Veterinary Internal Medicine*, *9*, 291–297.

Vidal, F., Baudoin, C., Miquel, C., Galliano, M. F., Christiano, A. M. et al. (1995). Cloning of the laminin alpha 3 chain gene (LAMA3) and identification of a homozygous deletion in a patient with Herlitz junctional epidermolysis bullosa. *Genomics*, *30*, 273–280.

Vonderfecht, S. L., Bowling, A. T., & Cohen, M. (1983). Congenital intestinal aganglionosis in white foals. *Veterinary Pathology*. *20*(1), 65–70.

Wallraf, A., Hamann, H., Deegen, E., Ohnesorge, B., & Distl, O. (2004). Analysis of the prevalence of pastern dermatitis in German Coldblood horse breeds. *Berliner und Munchner Tierarztliche Wochenschrift*, *117*, 148–152.

White, S. D., Affolter, V. K., & Bannasch, D. L. (2004). Hereditary equine regional dermal asthneia ("hyperelastosis cutis") in 50 horses: clinical, histological, immunohistological and ultrastructural findings. *Veterinary Dermatology*, *15*, 207–217.

White, S. D., Affolter, V. K., Schulteiss, P. C., Ball, B. A., Wessel, M. T. et al. (2007). Clinical and pathological findings in a HERDA-affected foal for 1.5 years of life. *Veterinary Dermatology*, *18*, 36–40.

Yager, J. A., & Scott, D. W. (1985). The skin and appendages. In K. V. F. Jubb, P. C. Kennedy & N. Pamer (eds.), *Pathology of domestic animals* (p. 432). Orlando, FL: Academic Press.

Yang, G. C., Croaker, D., Zhang, A. L., Manglick, P., Cartmill, T., & Cass, D. (1998) A dinucleotide mutation in the endothelin-B receptor gene is associated with lethal white foal syndrome (LWFS); a horse variant of Hirschsprung disease. *Human Molecular Genetics*, *7*(6), 1047–1052.

Young, A. E., Bower, L. P., Affolter, V. K., DeCock, H. E., Ferraro, G. L., & Bannasch, D. L. (2007). Evaluation of FOXC2 as a candidate gene for chronic progressive lymphedema in draft horses. *Veterinary Journal*, *174*, 397–399.

11 Genomics of muscle disorders

James R. Mickelson, Stephanie J. Valberg, Carrie J. Finno, and Molly E. McCue

Introduction

Many horses are bred for performance traits such as speed, endurance, and jumping ability, for temperament and ability to work with cattle, or for specific physical traits such as coat color. Optimal performance of skeletal muscle, which comprises up to 55% of their body mass, is a necessity for equine athletes. As in other domestic animal species, the selection of horses for beneficial physical and performance traits often has the unintended consequence of increasing the frequency of heritable diseases due to geographic isolation, artificial insemination, the use of popular sires, and resultant inbreeding. Heritable muscle diseases in horses disturb pathways involving energy metabolism, muscle contractility, and muscle structure resulting in a readily recognizable phenotype that disrupts athletic performance.

The most common muscle disorder in horses is exertional rhabdomyolysis (ER), which denotes the dissolution of muscle fibers with exercise, and affects many breeds (Hodgson & Parish, 1990). This syndrome is often called "tying-up" because of the stiff gait, muscle cramping, pain, and reluctance to move that may develop following mild to moderate exercise. Subsets of horses have specific heritable forms of tying-up. Other heritable muscle disorders are recognized at rest with signs of weakness, muscle fasciculations, or hyperthermia. A unifying goal of much equine veterinary research has been to define specific forms of muscle disease and ER through careful definition of inciting causes, patient characteristics, muscle histopathology, biochemistry, and genetic analysis.

The goal of this chapter is to review the known bases for heritable muscular disorders in the horse, with the understanding that this effort really only addresses known diseases of quite high frequency for which sufficient financial resources were garnered to perform in-depth investigations over a period of many years. The techniques thus far used to identify the genetic basis of muscle and metabolic disorders have evolved over time, from candidate genes selected by comparative medicine analogy to human conditions to whole genome approaches with microsatellite markers, and have been most successful in Mendelian genes with almost complete penetrance that are major problems in a limited number of breeds. Current and future studies to identify genes for muscle and metabolic disorders in horses will now most often use the equine SNP chips and study traits that likely have a more complex genetic basis, and may not necessarily reside solely within breeds deemed to be closely related.

Equine Genomics, First Edition. Edited by Bhanu P. Chowdhary.

Hyperkalemic Periodic Paralysis

Hyperkalemic periodic paralysis (HYPP) was the first disease in horses for which the molecular defect was determined (Rudolph et al., 1992). This groundbreaking achievement was a result of an in-depth clinical investigation that revealed large parallels in clinical signs between affected Quarter Horses and humans, coupled with the more advanced genetic tools and knowledge available to human geneticists at the time. The HYPP myopathy is seen clinically as abnormal skeletal muscle membrane excitability that leads to episodes of muscle twitching, weakness, and temporary paralysis. Genetically susceptible horses may be asymptomatic and have periodic muscle fasciculations or daily episodes of fasciculations, which may progress to recumbency or respiratory insufficiency due to paralysis of the upper respiratory muscles.

The genetic cause of HYPP is a missense mutation resulting in a phenylalanine/leucine substitution in the alpha-subunit of the voltage-dependent skeletal muscle sodium channel alpha subunit (*SCN4A*) (Rudolph et al., 1992). This channel normally functions to allow transient Na^+ entry into muscle cells during the initial phase of the action potential and surface membrane depolarization. The equine *SCN4A* mutation results in a failure of the channel inactivation mechanism that, in turn, causes bursts of persistent activity during which the open state probability is significantly increased (Cannon et al., 1995). Impairment appears to be more pronounced during times of elevated extracellular K^+, such as can occur with certain diets containing high levels of potassium. Regardless, failure of inactivation results in membrane depolarization, irritability, and potentially depolarization block and muscular paralysis (Spier & Hoffman, 2008). Clinical episodes can be aborted by treatment with intravenous calcium gluconate, glucose, or sodium bicarbonate. Prevention of episodes involves strict regulation of dietary potassium to less than 1% of the total diet.

HYPP affects Quarter Horses, American Paint Horses, Appaloosas, and Quarter Horse crossbred animals worldwide. The mutation has been linked to a popular Quarter Horse sire named Impressive. This prolific sire, born in 1969, has many hundreds of thousands of descendants, and these offspring dominate the halter horse industry. Heterozygosity for the *SCN4A* mutation is sufficient to cause susceptibility to HYPP. A co-dominant mode of inheritance is indicated as a considerable number of affected homozygotes exist in the Quarter Horse breed, and episodes are more severe and more frequent in these horses with the potential fatal complication of upper airway obstruction (Finno, Spier, & Valberg, 2009). Current estimates indicate that 4% of the Quarter Horse breed as a whole are affected (Tryon et al., 2009), with a prevalence as high as 58% in halter horses. Unfortunately, the gene frequency has not decreased in the past 14 years, since genetic testing has been available to breeders. This is the result of the fact that affected horses appear to have been preferentially selected as halter horse breeding stock due to pronounced muscle development, and there is evidence of selection of HYPP-affected horses as superior halter horses by show judges (Naylor, 1994). To our knowledge, the mechanism by which improper Na^+ channel inactivation causes the appearance of increased muscling at the cellular and whole muscle level is not known. In other words, it has not been determined whether the mutation causes increased muscle tone, myofiber numbers, or increased fiber diameters, or both, and if certain muscles are affected preferentially. The AQHA has officially recognized HYPP as a genetic defect or undesirable trait and ruled that foals born in 2007 and later testing homozygous for HYPP will not be eligible for registration.

Malignant Hyperthermia

Malignant hyperthermia (MH) in swine and humans was one of the first heritable muscle diseases affecting skeletal muscle contractility for which causative mutations were discovered. In 2005, a similar heritable form of MH was identified in horses (Aleman et al., 2005). MH episodes are initiated in susceptible individuals upon exposure to halogenated anesthetics or succinylcholine, or in some cases upon stress or excitement. Classic episodes of MH are diagnosed based on clinical signs of lactic acidosis and hyperthermia >40°C under halothane or isoflurane anesthesia, or following succinylcholine injection, accompanied by muscle contracture (rigidity). The most common known cause of MH susceptibility in any species are mutations in the *RYR1* gene encoding the calcium release channel of the skeletal muscle sarcoplasmic reticulum (Betzenhauser & Marks, 2010). The muscle contracture, cell damage, and elevated temperature indicative of MH are all explainable by defective regulation of the mutant calcium channel that is exacerbated by halothane to result in excessive calcium release. This uncontrollable rise in myoplasmic calcium results in a stimulation of contraction and the metabolism to sustain it; this in turn generates heat, lactic acidosis, increased O_2 consumption, and increased CO_2 production (Betzenhauser & Marks, 2010).

A number of reports of unusual reactions to halothane anesthesia in horses eventually led to identification of two proband Quarter Horses with classic MH episodes from which a search for *RYR1* mutations was successful (Aleman et al., 2005). These horses were heterozygous for a mutation in exon 46 of the *RYR1* gene, which suggests that the disorder is inherited in a dominant fashion (Aleman et al., 2004; Aleman et al., 2005). This region of the calcium release channel protein apparently plays a regulatory role in channel gating and is one of the hotspots for MH-causing mutations in other species. It is unknown whether this is the only *RYR1* mutation in horses or whether there may be other yet unidentified mutations that cause signs of hyperthermia and metabolic acidosis during anesthesia. In humans, there are currently more than 200 reported *RYR1* mutations associated with MH and central core disease (Betzenhauser & Marks, 2010). The prevalence of the known equine *RYR1* mutation in a random sample of Quarter Horses was recently shown to be approximately 1% (Nieto & Aleman, 2009). Thus, MH is an example of a disorder that was considered insignificant until a molecular basis became known, but in reality has a considerable impact on the overall health of the population. Horses with the *RYR1* mutation are also reported to have intermittent clinical signs of exertional rhabdomyolysis accompanied by excessive sweating and an elevation in rectal temperatures. Some of the episodes of ER in MH-affected horses have been fatal (Alleman, Nieto, & Magdesian, 2009).

Glycogen Branching Enzyme Deficiency

Glycogen branching enzyme deficiency (GBED) was originally discovered from clinical, histochemical, and biochemical investigations of an aborted foal and several Quarter Horse foals that had survived to term but appeared weak and hypothermic and had flexural limb deformities. The foals would sometimes progress to sudden death from hypoglycemic seizures, cardiac arrest or respiratory failure, or be euthanized by 8 weeks of age due to the severity of muscle weakness (Render et al., 1999; Valberg et al., 2001). Despite intensive care in neonatal facilities, no foals with GBED have survived. The histochemical demonstration of abnormal globular polysaccharide and crystalline material in heart, Purkinje cells, liver and skeletal muscle, with little or no normal appearing glycogen, suggested a diagnosis of glycogen branching enzyme (GBE) deficiency, also known as Anderson's disease or glycogen storage disease Type IV (Figure 11.1) (Render et al., 1999; Valberg et al., 2001). Subsequent measurements of little detectable GBE enzymatic activity or immune-detectable GBE protein in affected foal tissues, accompanied by highly unbranched polysaccharide, was consistent with this hypothesis (Valberg et al., 2001).

Eventual sequencing of the *GBE1* gene coding sequences resulted in the discovery of a mutation in exon 1 that results in a premature stop codon (Ward et al., 2004). GBED was then demonstrated to be

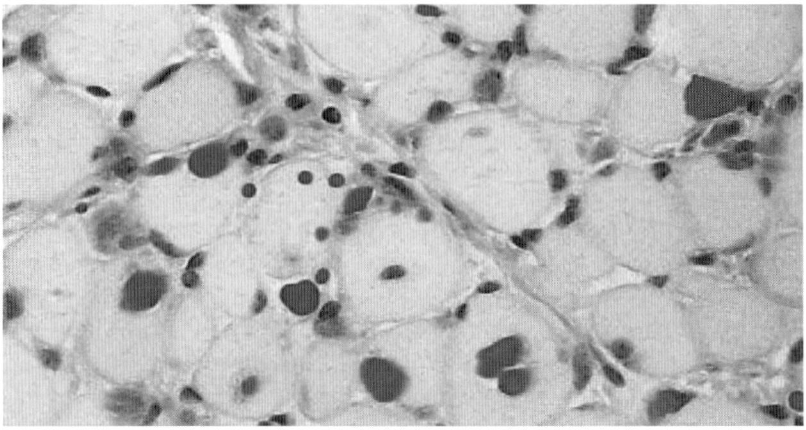

Figure 11.1 Abnormal polysaccharide observed in glycogen branching enzyme deficiency. Semimembranosus muscle biopsy from a foal affected with GBED stained with Periodic Acid Schiff's stain. Note the little normal background staining for glycogen and accumulation of large globular as well as smaller crystalline inclusions of polysaccharide. Adapted from Tryon et al. (2009).

an autosomal recessive disease currently known only to affect Quarter Horse and Paint Horse breeds. Glycogen is a required energy source in the rapidly growing fetus and neonate and is synthesized by glycogen synthase, which creates straight chains of glucose with alpha 1,4-glycosidic linkages, and by glycogen branching enzyme, which creates a branched structure through alpha 1,6-linkages. It is likely that the *GBE1* mutation and lack of GBE enzymatic activity results in cardiac and skeletal muscle, liver, and brain being unable to store and mobilize glycogen to maintain normal glucose homeostasis.

Carrier frequency estimates of 7.1% and 8.3% in the Paint and Quarter Horse breeds, respectively, have been reported (Wagner et al., 2006), which, along with the tracing of pedigrees of carriers back to founders of the breed in the early 1900s, would indicate that many GBED foals are, or should be, born every year. However, clinically, the number of GBED foals born alive appears less than expected. After the discovery of the *GBE1* mutation, genotyping of foals that were aborted or stillborn due to unknown causes demonstrated that up to 5% of such cases were the result of homozygosity for the *GBE1* mutation (Render et al., 1999; Valberg et al., 2001; Wagner et al., 2006). Thus, abortion may be the primary clinical presentation for GBED. Within Quarter Horse performance horse types, Western Pleasure horses have the highest prevalence of carriers of GBED, with 28% of horses being heterozygous (Tryon et al., 2009).

Polysaccharide Storage Myopathy Type 1

In 1932, Carlstrom recognized that episodes of rhabdomyolysis in Draft Horses were associated with increased glycogen storage; however, this condition was not yet recognized as a specific disease (Carlstrom, 1932). Polysaccharide storage myopathy (PSSM) in Quarter Horses, also characterized by increased muscle glycogen concentration, abnormal polysaccharide accumulation, and rhabdomyolysis, was first described in 1992 (Valberg et al., 1992). Subsequently, a similar combination of findings was also recognized in Belgian and Percheron horses (Valentine et al., 1997; Firshman et al., 2005).

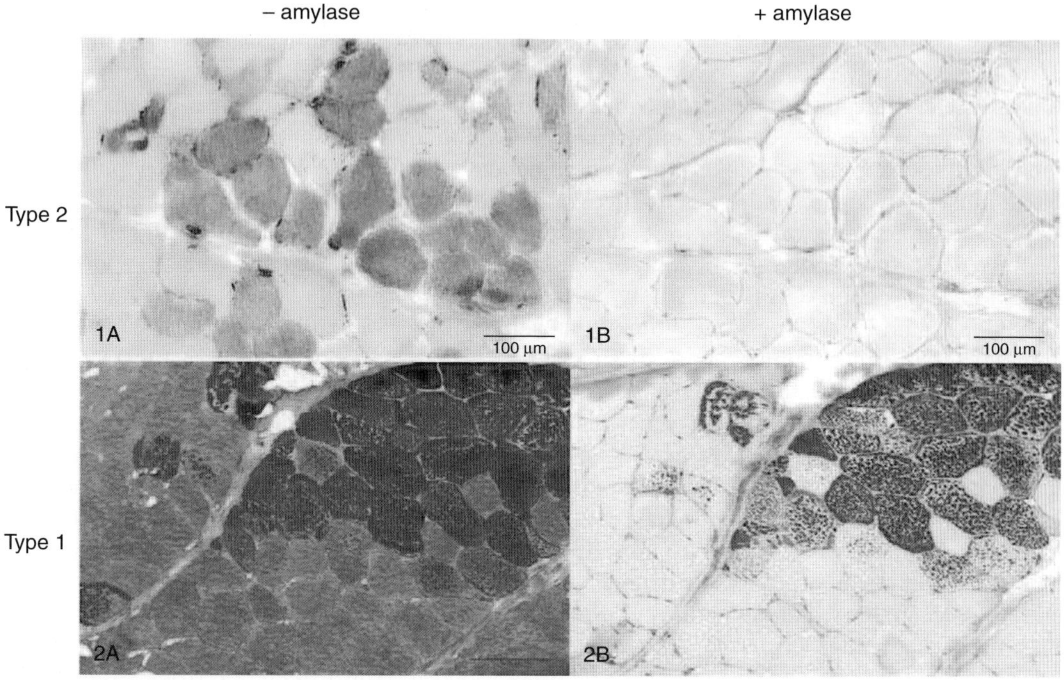

Figure 11.2 PSSM Types 1 and 2 distinguished by PAS staining of fresh-frozen muscle biopsy sections with and without amylase pre-incubation. Note abnormal dark-staining PAS-positive material. This material in Type 1 PSSM has a more aggregated and crystalline appearance throughout the sarcoplasm (2A), while in Type 2 PSSM this material is located more under the sarcolemma (1A). The PAS-positive material is typically removed by amylase digestion in Type 2 PSSM (1B), but not Type 1 PSSM (2B).

PSSM in any breed of horse has been diagnosed primarily by the histological demonstration of periodic acid Schiff's (PAS) positive inclusions of abnormal polysaccharide in as few as one or two fibers, to as many as 30% of the Type II skeletal muscle fibers in a fresh-frozen biopsy (McCue, Ribiero, & Valberg, 2006). Controversy existed as to whether diagnostic criteria for abnormal-appearing polysaccharide should be restricted to amylase-resistant polysaccharide or should also include increased amylase-sensitive glycogen (Figure 11.2). Abnormal polysaccharide has been described in the cardiac muscle of a severely affected Quarter Horse and a Belgian horse; however, it is not considered a typical feature of the disease (Valentine et al., 1997; Annandale et al., 2004). No histological abnormalities have been found in the liver or adipose tissue of affected horses.

Environmental factors clearly play a role in the clinical phenotype of horses with PSSM. Providing dietary fat as an alternative energy substrate to carbohydrate, limiting the daily consumption of starch and sugar, and gradually increasing daily exercise typically decrease muscle stiffness and eliminate episodes of ER in Quarter Horses with PSSM (Firshman et al., 2003; Ribiero et al., 2004). Phenotypic variability is also readily apparent between breeds, with PSSM Quarter Horses having a higher frequency of ER but less likely to have gait abnormalities than Draft Horses; however, Draft Horses are more likely to have muscle atrophy than Quarter Horses. Warmbloods with PSSM frequently present with a gait abnormality and lack of impulsion with infrequent episodes of rhabdomyolysis. Thus, Draft Horses and Quarter Horses have similar histopathological features of a glycogenosis but differences in clinical signs.

Considerable evidence indicated that a form of PSSM in Quarter Horses was highly heritable (Valberg et al., 1996). A genome-wide scanning approach was therefore considered possible, and at that time increasingly feasible. A study with 105 microsatellite markers spread across the genome enabled the identification of a region on ECA10 associated with PSSM in a population of 48 PSSM Quarter Horses descended from a founder stallion and 48 controls (McCue et al., 2008a). Following fine-mapping with local microsatellites and validation in an unrelated population, the skeletal muscle glycogen synthase gene (*GYS1*) was clearly the most plausible positional candidate gene. Sequence analysis of the entire coding region, as well as the 5′ and 3′ UTRs of the *GYS1* gene, revealed a single polymorphism in exon 6 that changes the normal arginine (R) codon (C<u>G</u>T) to a histidine (H) codon (C<u>A</u>T). This polymorphism is at amino acid residue 309 of the 737 amino acid equine GS protein. Almost the entire amino acid sequence of exon 6 is conserved in sequenced vertebrate *GYS1* genes, suggesting both an essential role in glycogen synthase function and the plausibility that this polymorphism is the causative PSSM mutation. Segregation of the H allele consistent with autosomal dominant inheritance was confirmed in a Quarter Horse herd developed as a controlled breeding trial (McCue, 2008a). The activity of the glycogen synthase enzyme was found to be increased both without and with the allosteric activator glucose 6 phosphate, indicating this is a gain of function mutation (McCue, 2008a). Subsequently, it has become clear that there is a reduced penetrance of the H allele in herds across the United States, consistent with the known environmental influences on disease expression. In any event, discovery of this disease involving a well-known gene, for which no naturally occurring gain of function mutations had previously been found, demonstrates the power of genetic studies in horses to define new mechanisms of muscle disease. It is interesting that two of the mutations found to date in Quarter Horses (GBED and PSSM1) involve the synthesis of glycogen.

Glycogen synthase catalyzes the rate-limiting step in muscle glycogen synthesis and is under stringent covalent and allosteric regulation (Pederson et al., 2000). Several possibilities exist as to how an Arg309His substitution would result in gain of muscle GS function and excessive glycogen accumulation. Perhaps the best possibility is that the mutant enzyme is resistant to negative regulation, or more sensitive to positive regulation, due either to altered phosphorylation/dephosphorylation by a multitude of protein kinases and phosphatases, altered affinity for its substrates (UDP-glucose and glycogen), or altered allosteric effects resulting from glucose 6-phosphate binding (Figure 11.3). At this time we know only that maximal GS activity assayed by the incorporation of ^{14}C UDP-Glucose into glycogen is significantly higher in PSSM than in control muscle homogenates in both the presence and the absence of a maximally activating concentration of glucose 6-phosphate (McCue et al., 2008a).

A number of studies have described physiological and biochemical effects of the mutation in Quarter Horses. For example, during maximal exercise tests, Quarter Horses with PSSM deplete 26% of the resting glycogen concentration and accumulate twice as much muscle lactate as control horses (Valberg et al., 1999). Thus, excessive glycogen accumulation in horses with PSSM is not due to an inability to utilize glycogen, and these horses actually have a higher rate of glycolysis than normal horses with maximal exertion. Nevertheless, horses with PSSM do have exercise intolerance and a cellular energy deficit with submaximal exercise (Valberg, Townsend, & Mickelson, 1998; Valberg et al., 1999). When ATP levels in muscle cannot be effectively restored by metabolic pathways, the myokinase reaction increasingly is used to produce ATP and AMP from ADP. AMP is then increasingly degraded to IMP by AMP deaminase. During submaximal exercise tests designed to mimic low workload in normal horses, lactate, ATP, ADP, and AMP concentrations are not significantly different in normal and PSSM horses; however, IMP concentrations are significantly elevated in horses with PSSM (Annandale, Valberg, & Essen Gustavsson, 2005). Premature degradation of

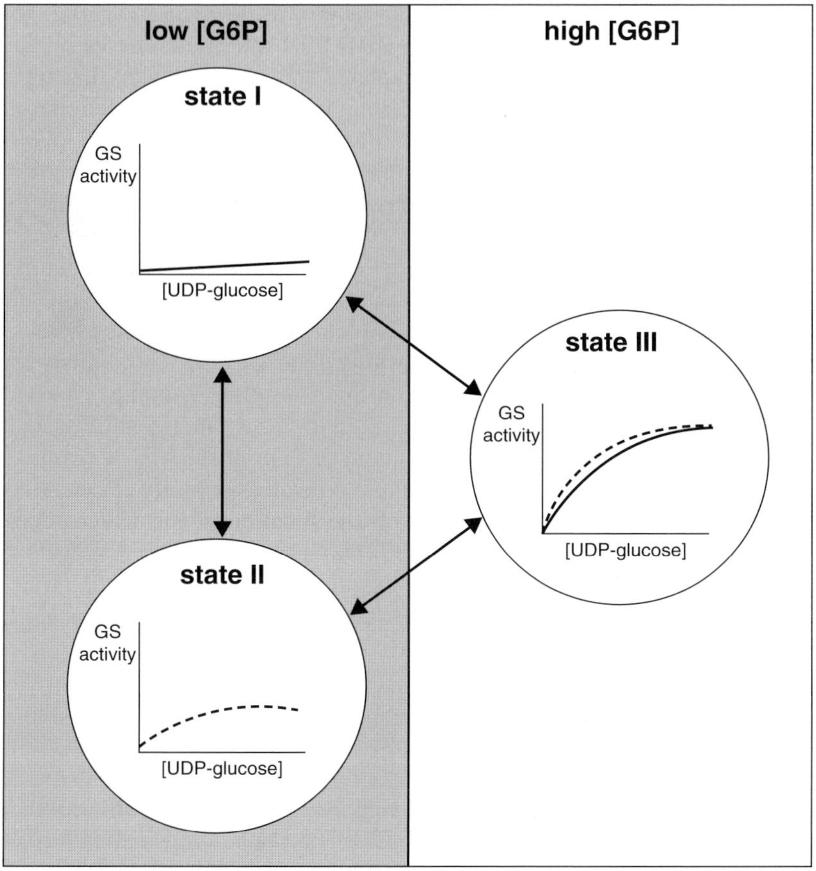

Figure 11.3 Regulation of glycogen synthase activity by glucose 6 phosphate and phosphorylation. Inserts represent the kinetics of synthase activity in relation to its UDP-glucose substrate in each state. The solid line is the phosphorylated enzyme, and the dashed line is the dephosphorylated enzyme. Adapted from Pederson et al. (2000).

adenine nucleotides to IMP in PSSM horses may be caused by abnormal regulation of the flux of substrates into aerobic metabolism.

Another approach to define the cellular responses involved in PSSM utilized a microarray containing ~750 oligonucleotide probes to measure expression of a subset of genes from the skeletal muscle of PSSM (*GYS1* mutation heterozygous and homozygotes) and control Normand Cob horses. The analysis revealed 129 genes significantly modulated, with 16 genes up-regulated over 1.5-fold and 37 genes down-regulated over 1.5-fold. Data mining showed that protein synthesis, apoptosis, cellular movement, growth, and proliferation were the main cellular functions significantly associated with the modulated genes. The authors concluded that PSSM mitochondrial dysfunction, glycogenesis inhibition, and chronic hypoxia may contribute to the PSSM disease process (Barrey et al., 2009).

A number of recent studies have identified the *GYS1* mutation in a large number of breeds in both North America and Europe (Herszberg et al., 2008; Stanley et al., 2009; Baird et al., 2010; McCue et al., 2010). Many Draft breeds originating in continental Europe are at particular risk, especially breeds that are related to the original Belgian breed. The *GYS1* genotype distribution in

Quarter Horses and Paint Horses is similar at 6–7% heterozygous or homozygous for the H allele. However, approximately 33% of Belgians and more than 50% of Percherons are either heterozygous or homozygous for the H allele. Assuming a dominant inheritance, this means that approximately 7% of all Quarter Horses and Paint horses, 33% of all Belgians, and 50% of all Percherons are actually genetically susceptible to PSSM. The prevalence of the *GYS1* mutation is very low in athletic light breeds such as Arabians, Thoroughbreds, and Standardbreds and relatively low in horses diagnosed with polysaccharide storage myopathy on the basis of amylase-sensitive accumulation of glycogen (McCue et al., 2008b).

We now know that there are likely multiple forms of PSSM that can be present to varying extents in different breeds, with potentially different penetrance and different clinical signs due to different genetic backgrounds. Phenotypic variability in the clinical presentation among horses with PSSM is also explained, at least in some instances, by a modifying gene or genes. Horses in several of the Type 1 PSSM Quarter Horse families used in the original microsatellite genome scan had a severe and occasionally fatal PSSM phenotype (McCue et al., 2008a). That same genetic association analysis was able to identify microsatellite markers on another segment of ECA10 as being associated with the severe PSSM phenotype in these families, but not in other, less severely affected families (McCue et al., 2009b). The ECA10 segment coincided with the location of the *RYR1* gene, in which the known equine *RYR1* mutation (Aleman et al., 2004) was also found to segregate in this family. Retrospective analysis of muscle biopsy submissions and a controlled exercise trial demonstrated that horses with both the *GYS1* and *RYR1* mutations had a more severe clinical phenotype than horses with the *GYS1* mutation alone (McCue et al., 2009b).

Polysaccharide Storage Myopathy Type 2

Approximately 20% of the Quarter Horses, 11% of Belgians, and 80% of Warmbloods with abnormal polysaccharide in muscle biopsies and clinical signs of PSSM examined thus far do not carry the *GYS1* H allele. Although it is possible that some horses receive a false-positive diagnosis of PSSM based on histopathology and clinical signs, it is far more likely that a population of horses with PSSM have separate, distinct glycogenosis(es) resulting from different molecular and cellular causes (McCue et al., 2008a; McCue et al., 2008b). This possibility has been examined in some detail in Quarter Horses. To distinguish the two forms of PSSM, we have proposed that the term "Type 1 PSSM" be used for horses that possess the *GYS1* mutation and "Type 2 PSSM" be used to distinguish the group of horses with abnormal polysaccharide in muscle biopsies that do not have the *GYS1* mutation (McCue et al., 2008b).

PSSM Types 1 and 2 can be histochemically and clinically distinguished by appropriate analyses (McCue et al., 2008b). The abnormal polysaccharide in Type 1 PSSM is often amylase-resistant and coarse granular in appearance and is typically located in the cytoplasm, whereas in Type 2 PSSM it is often amylase-sensitive, fine granular/ homogeneous in appearance, and located under the sarcolemma (Figure 11.2). Both forms of PSSM occur in young horses, but a high percentage of horses with Type 2 PSSM are younger than 1 year old. In contrast, the mean age of *GYS1*-positive horses with PSSM was greater than *GYS1*-negative PSSM, and less than 10% of Type 1 horses with PSSM were younger than 1 year old. Lastly, horses with Type 2 PSSM more commonly presented with an obscure or undiagnosed gait abnormality compared to horses with Type 1 PSSM.

The similarity in clinical presentation between the two forms of PSSM should not be unexpected as it is consistent with findings in other species. There are 11 known inherited glycogenoses affecting skeletal muscle in humans that can cause rhabdomyolysis, pain with exertion, muscle cramping, and myoglobinuria or fixed, progressive weakness (DiMauro & Lamperti, 2001; DiMauro, Hayes, & Tsujino, 2004). These clinical characteristics are shared by *GYS1*-positive and *GYS1*-negative horses with PSSM. A more thorough investigation of pedigrees, clinical history, clinical findings, and muscle biochemistry should help determine if there is a familial basis for *GYS1*-negative PSSM and what are the major distinguishing clinical and biochemical characteristics. Whole genome association studies with the EquineSNP50 BeadChip are also currently underway to identify significantly associated SNPs with Type 2 PSSM in Quarter Horses.

Recurrent Exertional Rhabdomyolysis

While ER is observed in many breeds of horse, the condition of recurrent exertional rhabdomyolysis (RER) is likely the most common form of tying-up and the most important muscular disorder of Thoroughbred horses (Valberg et al., 1992). About 5–10% of all Thoroughbreds in the United States and the United Kingdom develop ER at some point during a racing season, up to 75% of trainers have at least one horse with RER, and recurrence is so frequent in 17% of the affected horses that they do not race again that season (MacLeay et al., 1999a; McGowan, Fordham, & Christley, 2002). An increased prevalence of RER in young nervous fillies has been consistently identified. RER is clearly distinct from sporadic forms of ER that occur in otherwise healthy athletic horses due to overexertion, electrolyte depletion, or a dietary excess of carbohydrates. When horses with RER become riding horses later in life, they often have subtle signs of sore muscles, which, in our experience, results from RER.

Efforts have been made to determine if specific cellular or molecular defects can be associated with RER, and to explain the clinical findings of muscle cramping and rhabdomyolysis. Muscle histopathology in RER Thoroughbreds is distinguished only by the characteristic but nonspecific finding of increased numbers of central nuclei in horses with active clinical disease (Lentz et al., 1999). However, an in vitro alteration in the regulation of muscle contractility in RER horses has been demonstrated with intact electrically stimulated tendon-to-tendon muscle fiber bundles, dissected from surgically removed external intercostal muscles (Lentz et al., 1999). These preparations demonstrated a significantly increased sensitivity of RER muscles to the development of potassium-, caffeine- or halothane-induced contractures; all three types of treatment stimulate the release of calcium from the sarcoplasmic reticulum via the calcium release channel (Figure 11.4). In addition, a significantly increased rate of relaxation during a twitch was noted in the RER muscles (Figure 11.4). The fact that altered muscle contractility in RER could be recapitulated in vitro suggests an intrinsic muscle defect. This hypothesis was further supported by the observation that muscle cells grown in cell culture as myotubes displayed higher myoplasmic Ca^{2+} concentrations in response to caffeine (Lentz et al., 2002). It has also been demonstrated that the increased contracture sensitivity of RER muscles is not due to an enhanced Ca^{2+} sensitivity of the contractile apparatus (Mlekoday et al., 2001), and there was no support for altered activity of the sarcoplasmic reticulum Ca^{2+} release channel or Ca^{2+}-ATPase (Ward et al., 2000). An approach that uses a vector to force expression of MyoD and transform equine fibroblasts into myogenic cells that fuse to form myotubes in cell culture offers an exciting alternative approach to defining the possible RER muscle defects at a cellular level (Fernandez-Fuente et al., 2008).

There is strong evidence that RER susceptibility has an underlying genetic basis. An analysis of pedigrees containing the ancestors of RER-affected and clinically normal Thoroughbreds supported dominant inheritance with variable expression originating from a founder stallion (MacLeay et al., 1999b). Further, a multiyear controlled breeding trial demonstrated segregation of RER when

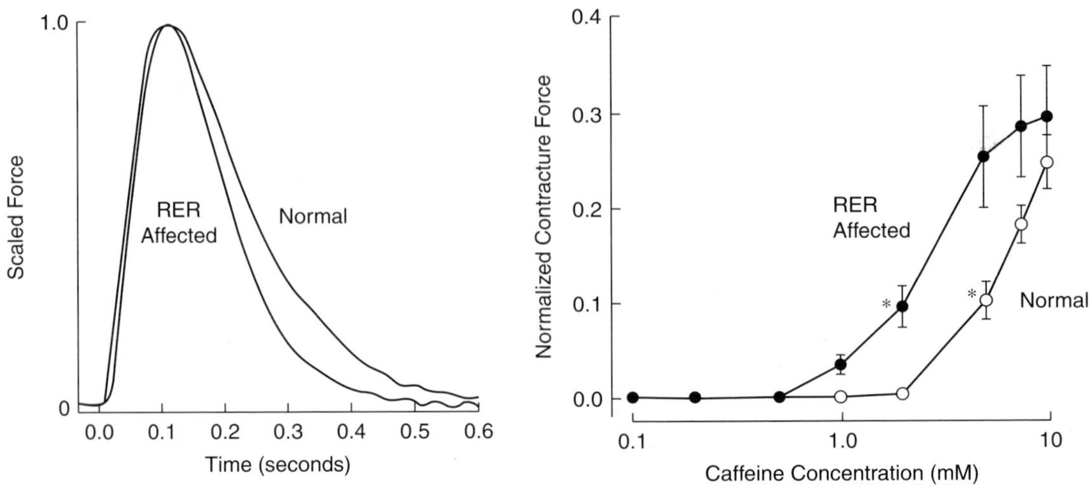

Figure 11.4 Abnormal sensitivity to in vitro contracture of muscle fiber bundles from RER-susceptible Thoroughbreds: (left) Representative twitch force records of muscle bundles from normal and RER susceptible Thoroughbreds. Note the faster return of the force to baseline in the RER muscle indicating a faster rate of relaxation. (right) Contracture force versus caffeine concentration for muscle bundles from normal and RER susceptible Thoroughbreds. Note the significantly greater contracture forces produced at lower caffeine concentrations in RER muscle, compared to normal, indicating an increased sensitivity to contracture. Adapted from Lentz et al. (1999).

individuals were phenotyped by the in vitro contracture test (Dranchak et al., 2005). Another study estimated the heritability of tying-up in Thoroughbreds to be approximately 0.42 (Oki et al., 2005). Clinical evaluation of RER horses from across the United States also indicate that gender, age, temperament, diet, and exercise routines all play roles in the sporadic expression of the clinical phenotype (MacLeay et al., 1999c; McGowan et al., 2002; MacLeay et al., 2000). Thus, RER is very likely to be a complex genetic disease resulting from one or more genes of major effect, as well as modifying genes, gender, and environmental factors. Some or all of these genes may also play a role in RER in other breeds, particularly in Standardbred horses, which display similar clinical signs and predisposing factors as seen in Thoroughbreds (Lindholm, Jonsson, & Kjaersgaard, 1974; Isgren et al., 2010). Interestingly, the most recent study in Standardbreds also observed an increased risk of RER in females and nervous horses, but also an increased performance of susceptible horses from standing starts and a higher winning percentage.

Sufficient evidence for a gene of major effect in RER Thoroughbreds was available to fuel attempts to identify the actual contributing gene loci. Genes encoding the sarcoplasmic reticulum Ca^{2+} release channel (*RYR1*) and Ca^{2+}-ATPase (*ATP2A1*), as well as the transverse tubule voltage sensor (*CACNA1S*), which are the three major regulatory proteins of excitation-contraction coupling, were genetically excluded from involvement with RER, although this genetic exclusion depended in large part on a simple Mendelian model (Dranchak et al., 2006). Our whole genome studies with microsatellite markers using a linkage approach in Thoroughbred pedigrees, or allelic association in populations, have not identified regions of the genome containing RER susceptibility loci (Dranchak, Valberg, & Mickelson, personal observations). However, a study in Japan with microsatellites that combined linkage and association approaches identified regions on ECA12 and ECA20 as candidate regions for containing RER loci (Tozaki et al., 2010). These 3–4 Mb regions need additional confirmation with more markers in additional populations. Whole genome association studies with the EquineSNP50 BeadChip are currently under way in at least two institutions to attempt to identify loci significantly associated with RER.

Conclusions and Future Directions

Focused research efforts over the past two decades have defined several specific forms of heritable equine muscle diseases through careful definition of inciting causes, patient characteristics, muscle histopathology, biochemistry, and genetic analysis. This research has led the way toward clinical diagnostic testing and specific treatments for highly prevalent muscle diseases such as HYPP and Type 1 PSSM. In contrast, the wide variety of remedies and treatments suggested for RER currently reflects the scarcity of much-needed information regarding its physiological and genetic basis. If our understanding of the genetic basis of other equine muscle disorders does not increase, they will continue to be unknowingly transmitted to future generations, and their management and prevention will continue to be empirically derived and largely ineffective.

It is appropriate here to use the known equine muscle disease genes to illustrate issues that have increasingly impacted our understanding of equine genetic diseases as a whole. This includes: the fact that horses can have multiple mutations; we often lack a complete understanding of a disease until a causative gene has been identified; phenocopies likely occur in horses that are predicted to have known mutations; variations occur in the frequencies of mutations within subdivisions of breeds and across diverse breeds; and inadvertently, some mutations are now and have been under selection.

The modification of the severity of the Type 1 PSSM phenotype in Quarter Horses by the presence of both the *GYS1* and *RYR1* mutations was mentioned earlier in the chaper (McCue et al., 2009b). The practical implication is that horses that carry both of these mutations have a more severe PSSM phenotype, which is less responsive to standard therapies such as consistent daily exercise and a low-starch fat-supplemented diet. Another recent study that examined three of the mutations reviewed here (*SCN4A*, *GYS1*, and *GBE1*) found that in a group of 299 Quarter Horses and Paint Horses that had one disease allele, 5 of them were heterozygous for two dominant disease-causing alleles (*SCN4A* and *GYS1*), and 1 was heterozygous for *GYS1* and *GBE1* (Tryon et al., 2009). Possible alteration of clinical phenotypes by interaction of these alleles is not known. Nevertheless, this more complex genetic picture might be more common than recognized and begs the question of how many horses have multiple mutations in disease genes that are yet to be discovered.

The coexistence of MH and Type 1 PSSM alleles within Quarter Horses also serves as an example of how certain pedigree lines – and beyond that, subpopulations within breeds – can have different frequencies of disease. The family of Quarter Horses with a high frequency of both mutations is not representative of the breed as a whole, but it exists nevertheless (McCue et al., 2009b). Although the *RYR1* and *GYS1* genes are both located on ECA10, they are far enough apart to be considered as unlinked. Quarter Horses are also categorized by performance groups with specific selection criteria for breeding, and this in turn has an impact on disease allele frequencies within these groups. For example, halter horses have higher frequencies of the *GYS1* (fourfold higher) and *SCN4A* (eightfold higher) alleles than the random average for the breed, whereas Western pleasure horses have a higher frequency of *GBE1* (Table 11.1) (Tryon et al., 2009). This in turn will impact diagnostic and genetic testing decisions by veterinarians and the public.

Our knowledge about certain muscle diseases continues to evolve. Here again, what was once essentially considered polysaccharide storage myopathy is now Type 1 and Type 2 (*GYS1* and *non-GYS1* mutation). After discovery of the *GYS1* mutation, it became possible to distinguish more

Table 11.1 Allele frequencies of muscle disease genes in Quarter Horse subpopulations.

Population	Number of horses	HYPP	GBED	Type 1 PSSM
Control	200	0.008	0.054	0.055
Halter	118	0.299	0.026	0.155
Western pleasure	39	0.013	0.132	0.043
Cutting	113	0	0.068	0.033
Working cow horse	96	0	0.047	0.028
Reining	97	0	0.016	0.022
Barrel racing	82	0.006	0.006	0.007
Racing	106	0	0	0.010

readily between these conditions histologically by the nature and cellular location of abnormal polysaccharide and by age of onset, and it would not be surprising if there were other forms yet to be genetically and phenotypically defined. Furthermore, the story of GBED, which started with infrequent cases of affected foals, led to the discovery that GBED mutation has an overall allele frequency of approximately 0.05, and is a significant cause of late-term abortion. Lastly, while simulation studies suggested that RER could be modeled as a Mendelian trait, it is now clear that is it more genetically complex.

Although a number of simple genetic mutations have now been described, it is highly likely that many genetic diseases in horses are complex and involve more than a single gene. Also of interest are reports of horses having clinical signs similar to HYPP but are negative for the known *SCN4A* mutation (e.g., Diakakis, Spanoudes, & Dessiris, 2008), and it is likely that there are many cases of MH that do not have the known mutation as well. The *SCN4A* and *RYR1* genes could easily harbor other mutations that adversely affect the protein function, albeit perhaps not exactly as do the commonly known mutations, but still result in a similar phenotype. When there are sometimes hundreds of disease-causing mutations known in a single human gene, it should not be surprising if there are multiple mutations within the equine homologs. Next-generation sequencing of transcriptomes, entire genes, and even genomes from carefully selected case and control populations could begin to address this genotypic and phenotypic heterogeneity.

The GYS1/PSSM story is also an example of a mutation that crosses a great many breeds. This is almost certainly attributable to its early origins, analogous to common coat color alleles that are present in multiple breeds due to founders many centuries ago. On the other hand, HYPP, and likely GBED, are examples of recent mutations that are known only to be in Quarter Horses and related breeds at this time. It is not known at this time whether MH is a Quarter Horse disease only, or present in other breeds that contributed to the founding of the Quarter Horse. If a disease has similar phenotypes across multiple breeds, it may be possible to use this to advantage in fine-mapping studies to locate positional candidate genes from whole genome-mapping data. Such approaches have been successful in canine studies and may help with ongoing studies to identify genetic association with RER.

Haplotype analysis of markers flanking the *GYS1* gene in diverse breeds with PSSM shows that all chromosomes with the *GYS1 A* allele have a single conserved haplotype (McCue et al., 2008a). The conserved haplotype was the shortest in Belgian draft horses (351 kb), consistent with the earlier origin of the Belgian breed and more decay of linkage disequilibrium (LD) around the mutation. Quarter Horses, Paint Horses, and Appaloosas shared a larger conserved segment (642 kb), likely because of the more recent origin of these breeds. We have estimated the mutation to have existed for approximately 150 generations (1,200–1,500 years old) and suggest that it was present in the domestic horse population before the establishment of the diverse breeds known today (McCue et al., 2008a). It has likely been passed from older established breeds to newer breeds by admixture at the time of new breed creation. It is curious that the *GYS1* gene can be at high frequency and in a large conserved haplotype, while also causing clinical disease. One possible explanation is that it is an example of a "thrifty gene" that was originally selected for its beneficial effects of enabling high muscle glycogen content despite constant exercise, but which is detrimental under current management conditions of denser energy content in the diet and infrequent exercise. It would be interesting to determine the region of shared haplotype around *SCN4A* (HYPP), which is a very recent mutation and apparently undergoing very rapid selection.

In conclusion, there have been some wonderful success stories on identification of equine muscle disease genes and mutations, and ongoing projects using whole genome SNP marker panels have the promise of identifying more. Although this provides important diagnostic tools for veterinarians, it will increasingly pose perplexing problems for breeders and breed associations when deciding how to use this information to improve the overall health and well-being of all horses. In any event, the study of muscle disease in such a highly athletic species will continue to provide exciting insights into the normal structure and function of muscle.

References

Aleman, M., Riehl, J., Aldridge, B. M., Lecouteur, R. A., Stott, J. L., & Pessah, I. N. (2004). Association of a mutation in the ryanodine receptor 1 gene with equine malignant hyperthermia. *Muscle Nerve*, *30*, 356–365.

Aleman, M., Brosnan, R. J., Williams, D. C., Lecouteur, R. A., Imai, A. et al. (2005). Malignant hyperthermia in a horse anesthetized with halothane. *Journal of Veterinary Internal Medicine*, *19*, 363–366.

Aleman, M., Nieto, J. E., & Magdesian, K. G. (2009). Malignant hyperthermia associated with ryanodine receptor 1 (C7360G) mutation in Quarter Horses. *Journal of Veterinary Internal Medicine*, *23*, 329–334.

Annandale, E. J., Valberg, S. J., Mickelson, J. R., & Seaquist, E. R. (2004). Insulin sensitivity and skeletal muscle glucose transport in horses with equine polysaccharide storage myopathy. *Neuromuscular Disorders*, *14*, 666–674.

Annandale, E. J., Valberg, S. J., & Essen Gustavsson, B. (2005). The effects of submaximal excercise onadenine nucleotide concentrations in skeletal muscle fibers of horses with polysaccharide storage myopathy. *American Journal of Veterinary Research*, *66*, 839–845.

Baird, J. D., Valberg, S. J., Anderson, S. M., McCue, M. E., & Mickelson, J. R. (2010). Presence of the glycogen synthase 1 (*GYS1*) mutation causing polysaccharide storage myopathy in Continental European Draught horse breeds. *Veterinary Record*, *167*, 781–784.

Barrey, E., Mucher, E., Jeansoule, N., Larcher, T., Guigand, L. et al. (2009). Gene expression profiling in equine polysaccharide storage myopathy revealed inflammation, glycogenesis inhibition, hypoxia and mitochondrial dysfunctions. *BMC Veterinary Research*, *7*(5), 29.

Betzenhauser, M. J., & Marks, A. R. (2010). Ryanodine receptor channelopathies. *Pflugers Arch*, *460*, 467–480.

Cannon, S. C., Hayward, L. J., Beech, J., & Brown, Jr., R. H. (1995). Sodium channel inactivation is impaired in equine hyperkalemic periodic paralysis. *Journal of Neurophysiology*, *73*, 1892–1899.

Carlstrom, B. (1932). Uber die atiologie und pathogenese der kreuzlahme des pferdes (Hemaglobinaemia paralytica). *Scandinavia Archiv*, *62*, 1–62.

Diakakis, N., Spanoudes, K., & Dessiris, A. (2008). Hyperkalaemic periodic paralysis-like syndrome in a Criollo Argentino horse. *Equine Veterinary Education*, *20*, 396–400.

DiMauro, S., & Lamperti, C. (2001). Muscle glycogenoses. *Muscle Nerve*, *24*, 984–999.

DiMauro, S., Hayes, A. P., & Tsujino, S. (2004). Nonlysosmal glycogenoses. In A. G. Engel & C. Franzini-Armstrong (eds.), *Myology* (3rd ed.) (pp. 1535–1558). New York: McGraw-Hill.

Dranchak, P. K., Valberg, S. J., Onan, G. W., Gallant, E. M., et al. (2005). Inheritance of recurrent exertional rhabdomyolysis in Thoroughbreds. *Journal of the American Veterinary Medical Association*, *227*, 762–767.

Dranchak, P. K., Valberg, S. J., Onan, G. W. et al. (2006). Exclusion of linkage of recurrent exertional rhabdomyolysis in Thoroughbred horses to the *RYR1*, *CACNA1S* and *ATP2A1* genes. *American Journal of Veterinary Research*, *67*, 1395–1400.

Fernandez-Fuente, M., Ames, E. G., Wagner, M. L., Zhou, H., Strom, M. et al. (2008). Assessment of the transformation of equine skin-derived fibroblasts to multinucleated skeletal myotubes following lentiviral-induced expression of equine myogenic differentiation. *American Journal of Veterinary Research*, *69*, 1637–1645.

Finno, C. J., Spier, S. J., & Valberg, S. J. (2009). Equine diseases caused by known genetic mutations. *The Veterinary Journal*, *179*, 336–347.

Firshman, A. M., Valberg, S. J., Bender, J. B., & Finno, C. J. (2003). Epidemiologic characteristics and management of polysaccharide storage myopathy in Quarter Horses. *American Journal of Veterinary Research*, *64*, 1319–1327.

Firshman, A. M., Baird, J. D., & Valberg, S. J. (2005). Prevalence and clinical signs of polysaccharide storage myopathy and shivers in Belgian draft horses. *Journal of the American Veterinary Medical Association*, *227*, 1958–1964.

Herszberg, B., McCue, M. E., Larcher, T., Mata, X., Vaiman, A. et al. (2008). A *GYS1* gene is highly associated with polysaccharide storage myopathy in Cob Normand draft horses. *Animal Genetics*, *40*, 94–96.

Hodgson, D. H., & Parish SM. (1990). Muscle diseases. In B. P. Smith (ed.), *Large animal internal medicine* (pp. 13–35). St. Louis, MO: CV Mosby Co.

Isgren, C. M., Upjohn, M. M., Fernandez-Fuente, M., Massey, C., Pollott, G. et al. (2010). Epidemiology of exertional rhabdomyolysis susceptibility in Standardbred horses reveals associated risk factors and underlying enhanced performance. *PLoS One*, *5*, e11594.

Lentz, L. R., Valberg, S. J., Balog, E. M., Mickelson, J. R., & Gallant, E. M. (1999). Abnormal regulation of muscle contraction in equine recurrent exertional rhabdomyolysis. *American Journal of Veterinary Research*, *60*, 992–999.

Lentz, L. R., Valberg, S. J., Herold, L. V., Onan, G. W., Mickelson, J. R., & Gallant, E. M. (2002). Myoplasmic calcium regulation in myotubes from horses with recurrent exertional rhabdomyolysis. *American Journal of Veterinary Research*, *63*, 1724–1731.

Lindholm, A., Jonsson, L., & Kjaersgaard, P. (1974). Acute rhabdomyolysis (tying-up) in Standardbred horses. A morphological and biochemical study. *Acta Veterinaria Scandinavia*, *15*, 325.

MacLeay, J. M., Sorum, S. A., Valberg, S. J. et al. (1999a). Epidemiological analysis of factors influencing exertional rhabdomyolysis in Thoroughbred racehorses. *American Journal of Veterinary Research*, *60*, 1562–1566.

MacLeay, J. M., Valberg, S. J., Pagan, J. D., de laCorte, F., Roberts, J. et al. (1999b). Effect of diet on thoroughbred horses with recurrent exertional rhabdomyolysis performing a standardised exercise test. *Equine Veterinary Journal*, *30*(Suppl.), 458–462.

MacLeay, J. M., Valberg, S. J., Sorum, S. A., Sorum, M. D., Kassube, T. et al. (1999c). Heritability of recurrent exertional rhabdomyolysis in Thoroughbred racehorses. *American Journal of Veterinary Research*, *60*, 250–256.

MacLeay, J. M., Valberg, S. J., Pagan, J. D., Xue, J. L., De La Corte, F. D., & Roberts, J. (2000). Effect of ration and exercise on plasma creatine kinase activity and lactate concentration in Thoroughbred horses with recurrent exertional rhabdomyolysis. *American Journal of Veterinary Research*, *61*, 1390–1395.

McCue, M. E., Ribiero, W. P., & Valberg, S. J. (2006). Prevalence of polysaccharide storage myopathy in horses with neuromuscular disorders. *Equine Veterinary Journal*, *36*, 340–344.

McCue, M. E., Valberg, S. J., Miller, M. B., Wade, C. M., DiMauro, S. et al. (2008a). Glycogen synthase (GYS1) mutation causes a novel skeletal muscle glycogenosis. *Genomics*, *91*, 458–466.

McCue, M. E., Valberg, S. J., Lucio, L., & Mickelson, J. R. (2008b). Glycogen synthase 1 (GYS1) mutation in diverse breeds with polysaccharide storage myopathy. *Journal of Veterinary Internal Medicine*, *22*, 1228–1233.

McCue, M. E., Armién, A. G., Lucio, M., Mickelson, J. R., & Valberg, S. J. (2009a). Comparative skeletal muscle histopathologic and ultrastructural features in two forms of polysaccharide storage myopathy in horses. *Veterinary Pathology*, *46*, 1281–1291.

McCue, M. E., Valberg, S. J., Jackson, M., Lucio, M., & Mickelson, J. R. (2009b). Polysaccharide Storage Myopathy phenotype in quarter horse-related breeds is modified by the presence of an RYR1 mutation. *Neuromuscular Disorders*, *19*, 37–43.

McCue, M. E., Anderson, S. M., Valberg, S. J., Piercy, R., Binns, M. M. et al. (2010). Estimated prevalence of the Type 1 polysaccharide storage myopathy mutation in selected North American and European breeds. *Animal Genetics*, *41*(Suppl. 2), 145–149.

McGowan, C. M., Fordham, T., & Christley, R. M. (2002). Incidence and risk factors for exertional rhabdomyolysis in Thoroughbred racehorses in the United Kingdom. *Veterinary Record*, *151*, 623–626.

Mlekoday, J., Mickelson, J. R., Horton, J. et al. (2001). Calcium sensitivity of muscle fibers and myofibrillar ATPase in equine recurrent exertional rhabdomyolysis. *American Journal of Veterinary Research*, *62*, 1647–1652.

Naylor, J. M. (1994). Selection of quarter horses affected with hyperkalemic periodic paralysis by show judges. *Journal of the American Veterinary Medical Association*, *204*, 926–928.

Nieto, J. E., & Aleman, M. (2009). A rapid detection method for the ryanodine receptor 1 (C7360G) mutation in Quarter Horses. *Journal of Veterinary Internal Medicine*, *23*, 619–622.

Oki, H., Miyake, T., Hasegawa, T., & Sasaki, Y. (2005). Estimation of heritability for tying-up syndrome in the Thoroughbred racehorse by Gibbs sampling. *Journal of Animal Breeding Genetics*, *122*, 289–293.

Pederson, B. A., Cheng, C., Wilson, W. A., & Roach, P. J. (2000). Regulation of glycogen synthase. Identification of residues involved in regulation by the allosteric ligand glucose-6-P and by phosphorylation. *Journal of Biology and Chemistry*, *275*(36), 27753–27761.

Render, J. A., Common, R. S., Kennedy, F. A., Jones, M. Z., & Fyfe, J. C. (1999). Amylopectinosis in fetal and neonatal Quarter Horses. *Veterinary Pathology*, *36*, 157–160.

Ribiero, W. P., Valberg, S. J., Pagan, J. D., & Gustavsson, B. E. (2004). The effect of varying dietary starch and fat content on serum creatine kinase activity and substrate availability in equine polysaccharide storage myopathy. *Journal of Veterinary Internal Medicine*, *18*, 887–894.

Rudolph, J. A., Spier, S. J., Byrns, G., Rojas, C. V., Bernoco, D., & Hoffman, E. P. (1992). Periodic paralysis in quarter horses: A sodium channel mutation disseminated by selective breeding. *Nature Genetics*, *2*, 144–147.

Spier S. J., & Hoffman, E. P. (2008). Hyperkaleemic periodic paralysis: Mother nature versus human nature. *Equine Veterinary Education*, *20*, 401–405.

Stanley, R. L., Valberg, S. J., McCue, M. E., Mickelson, J. R., McGowan, C. et al. (2009). A glycogen synthase 1 mutation associated with equine polysaccharide storage myopathy and exertional rhabdomyolysis occurs in a variety of UK breeds. *Equine Veterinary Journal*, *41*, 597–601.

Tozaki, T., Hirota, K., Sugita, S., Ishida, N., Miyake, T. et al. (2010). A genome-wide scan for tying-up syndrome in Japanese Thoroughbreds. *Animal Genetics*, *41*(Suppl. 2), 80–86.

Tryon, R. C., Penedo, M. C. T., McCue, M. E., Valberg, S. J., Mickelson, J. R. et al. (2009). Evaluation of allele frequencies of inherited disease genes in subgroups of American Quarter Horses. *Journal of American Veterinary Medical Association*, *234*, 120–125.

Valberg, S. J., Cardinet, III, G.H., Carlson, G. P., & DiMauro, S. (1992). Polysaccharide storage myopathy associated with recurrent exertional rhabdomyolysis in horses. *Neuromuscular Disorders*, *2*, 351–359.

Valentine, B. A., Credille, K. M., Lavoie, J. P., Fatone, S., Guard, C., Cummings, J. F. et al. (1997). Severe polysaccharide storage myopathy in Belgian and Percheron draught horses. *Equine Veterinary Journal*, *29*, 220–225.

Valberg, S. J., Geyer, C., Sorum, S. A., & Cardinet III, G. H. (1996). Familial basis of exertional rhabdomyolysis in quarter horse-related breeds. *American Journal of Veterinary Research*, *57*, 286–290.

Valberg, S. J., Townsend, D., & Mickelson, J. R. (1998). Skeletal muscle glycolytic capacity and phosphofructokinase regulation in horses with polysaccharide storage myopathy. *American Journal of Veterinary Research*, *59*, 782–785.

Valberg, S. J., MacLeay, J. M., Billstrom, J. A., Hower-Moritz, M. A., & Mickelson, J. R. (1999). Skeletal muscle metabolic response to exercise in horses with 'tying-up' due to polysaccharide storage myopathy. *Equine Veterinary Journal*, *31*, 43–47.

Valberg, S. J., Ward, T. L., Rush, B., Kinde, H., Hiraragi, H. et al. (2001). Glycogen branching enzyme deficiency in quarter horse foals. *Journal of Veterinary Internal Medicine*, *15*, 572–580.

Wagner, M. L., Valberg, S. J., Ames, E. G., Bauer, M. M., Wiseman, J. A. et al. (2006). Allele frequency and likely impact of the glycogen branching enzyme deficiency gene in Quarter Horse and Paint Horse populations. *Journal of Veterinary Internal Medicine*, *20*, 1207–1211.

Ward, T. L., Valberg, S. J., Gallant, E. M., & Mickelson, J. R. (2000). $Ca2^+$ regulation by skeletal muscle membranes of horses with recurrent exertional rhabdomyolysis. *American Journal of Veterinary Research*, *61*, 242–247.

Ward, T. L., Valberg, S. J., Adelson, D. L., Abbey, C. A., Binns, M. M., & Mickelson, J. R. (2004). Glycogen branching enzyme (GBE1) mutation causing equine glycogen storage disease IV. *Mammalian Genome*, *15*, 570–577.

12 Genomics of skeletal disorders

Ottmar Distl

Osteochondrosis

Diseases of the locomotor system are the most important causes of reduced performance and premature retirement of horses. Radiological examinations to diagnose skeletal diseases have become an integral part in veterinary orthopaedic practices and pre-purchase examinations. The outcome of radiological examinations has considerable impact on the market value of a horse and as a selection criterion for breeding stallions.

Most commonly observed orthopaedic problems in riding and race horses are osteochondrosis, degenerative joint diseases, and navicular disease. For all these different conditions, a potentially relevant genetic disposition could be shown. Genome-wide marker studies and molecular genetic analysis of candidate genes have been performed for osteochondrosis and navicular disease. Genome-wide association analyses are underway for osteochondrosis, and along with further re-sequencing of linked and associated genomic regions, mutations determining the pathogenesis of osteochondrosis should be unravelled.

Introduction

Osteochondrosis (OC) is caused by a disturbance of the endochondral ossification of growing cartilage of the articular/epiphyseal complex (Jeffcott, 1996). Focal areas of cartilage growth fail to undergo matrix calcification and therefore do not become converted to bone. The primary lesion is characterized by a focal area of necrosis. Cartilage cells appear to proliferate normally, but maturation and differentiation are abnormal in OC. The loss of normal differentiation of cartilage cells means that transitional calcification of the matrix does not occur, and that capillary sprouts cannot penetrate the distal region of the hypertrophic zone. The altered process of endochondral ossification leads to retention and thickening of cartilage, resulting in the development of a cartilage core. These primary lesions can progress to further damage within the joints, due to which subchondral fractures, subchondral cyst-like lesions, cartilage flaps, separation of osteochondral fragments (joint mice, chips, corpora libera, osteochondritis dissecans, OCD), osteochondritis, and synovitis, can occur. This developmental disease mainly affects articulation of the fetlock, hock, and stifling of joints. Radiographic signs consistent with OC are irregular texture of the bone with variable radiopacity (e.g., radioluceny of the subchondral bone) and changes of the regular bone contour, such as smoothly

Equine Genomics, First Edition. Edited by Bhanu P. Chowdhary.

flattened, irregularly flattened, and smaller or larger concavity or isolated radiodense areas at specific sites of the predisposed joints. The predilection sites in hock joints are the intermediate ridge of the distal tibia, the lateral/medial trochlear ridge of the talus, and the lateral/medial malleolus of the tibia. In the stifle joints, the predilection sites are the lateral/medial trochlear ridge of the femur, the trochlear groove of the femur, and the patella. The predilection site in the fetlock joint for OC is the dorsal aspect of the sagittal ridge of the third metacarpal/metatarsal bone. There is no common opinion whether or not the dorsoproximal rim of the proximal phalanx should be classified as predilection site for OC in horses. The intervertebral articular processes of the cervical spine are most commonly affected of the vertebral joints (Van Weeren & Barnefeld 1999; Ytrehus, Carlson, & Ekman, 2007).

Osteochondrosis belongs to those diseases of the locomotor system that are frequently detected radiographically in young horses. OC contributes three times more to economic losses compared to any other equine disease (Van der Harst et al., 2005). Reports on the prevalence indicate that OC is present in warmblood and trotter horse populations at frequencies between 10% and 64% across a range of different breeds (Wittwer et al., 2006; Van Grevenhof et al., 2009). The large differences may be attributed to the different definitions of the osteochondrotic findings, specific environmental conditions of the study populations, type of data sampling, and breed differences in susceptibility to OC. To date, with extant research resulting in controversial findings, there is no clear understanding of the gender effect on OC.

The prevalence of OC is related to body size. Taller warmblood horses and trotters have a predisposition for osseous fragments in fetlock and hock joints (Sandgren et al., 1993, Van Weeren, Sloet van Oldruitenborgh-Oosterbaan, & Barneveld, 1999; Stock, Hamann, & Distl, 2006). Taller warmblood foals with a big cannon bone circumference tend to exhibit more frequently signs of OC in hock joints (Wilke, 2003). Foals affected with hock OC have a higher body weight at birth and continue to be heavier, with significantly higher average daily weight gains up to an age of 12 months when compared to unaffected foals (Sandgren et al., 1993). The affected foals also have a larger frame, including a greater height at the croup and at the withers, and have a markedly larger circumference of the cannon bone and the carpus (Van Weeren et al., 1999). In contrast, foals with lower body weight were found to have a significantly higher prevalence of OC in fetlock joints (Schober, 2003).

The intensity of exercise in the first months of life of the foals influences the distribution of osteochondrotic lesions, with a tendency to more severe lesions in box-rested foals at an age of 5 month. It has been shown that OC in fetlock joints is clearly associated with a deficit in locomotion during the first months of life (Wilke, 2003). In stifle and hock joints, lesions were found mainly in the femoral condyles of box-rested foals and at the lateral trochlear ridge of the femur in trained foals (Barneveld & Van Weeren, 1999).

Heritability of OC, genetic correlations, and selection response

Genetic components play the most important role in the etiology of equine OC (Grøndahl & Dolvik, 1993; Philipsson et al., 1993; Winter et al., 1996; Willms, Röhe, & Kalm, 1999; Pieramati et al., 2003; Schober, 2003; Stock, Hamann, & Distl, 2005b; Wittwer et al., 2007a; Van Grevenhof et al., 2009). The relative importance of additive genetic variation is determined by the heritability in the narrow sense, which quantifies the degree to which phenotypic variance is explained by additive genetic variance. Heritability estimates are population-specific parameters and also depend on the analysis method used, family structure, and group of relatives used for estimation. The heritability estimates for OC in warmblood horses were between 0.10 and 0.34 when threshold models or scores for OC were applied, and the additive genetic effect of the animal was included in the model (Table 12.1). The heritability estimates in trotters vary between 0.17 and 0.52 (Schougaard, Ronne, & Philipsson, 1990; Grøndahl & Dolvik, 1993; Philipsson et al., 1993). The large difference may be because of the type of method applied for estimation (sire model instead of an animal model), the horse breed investigated, differing data structures, and, in some studies, the low numbers of animals included. Using a sire threshold model, a heritability of 0.26 was estimated, based on 325 yearlings representing the majority of the progeny of nine trotter stallions (Schougaard et al., 1990). Threshold model heritability estimates of 0.21 for hock joints were estimated for 644 trotters that were the progeny from 39 sires (Grøndahl & Dolvik, 1993). In Swedish trotters, a linear sire model applied to data from 793 horses showed a heritability of 0.27 for radiographic findings of OC in hock joints (Philipsson et al., 1993).

Genetic analyses among foals at an age between 4 and 9 months and in 2-year-old horses demonstrated genetic correlations between the signs for osteochondrotic lesions in foals and OCD (joint bodies at the predilection sites in horses aged 2 years) at $r_g = 0.84$ for fetlock joints and $r_g = 0.99$ for hock joints (Schober, 2003). The corresponding phenotypic correlations were $r_p = 0.66$ (fetlock joints) and $r_p = 0.68$ (hock joints), indicating a significant consistency of the radiological findings in the two different age classes of Hanoverian warmblood horses. Furthermore, genetic correlations among OC and OCD were $r_g = 0.80$ in Dutch warmblood (van Grevenhof et al., 2009). Osseous fragments in front and hind fetlock joints represent genetically uniform traits as high additive genetic correlations between these traits were found (Stock et al., 2005b). For male horses, a higher heritability of osseous fragments in front fetlock joints was more pronounced than in fetlock joints of the hind limbs (Stock et al., 2005b). The genetic correlations between osseous fragments in fetlock and hock joints were slightly-to-moderately negative in Hanoverian warmblood (Stock et al., 2005b), and slightly positive in Dutch warmblood (van Grevenhof et al., 2009) horses.

Osseous fragments in fetlock joints showed moderately positive genetic correlation with deforming arthropathies in hock joints, whereas osseous fragments and deforming arthropathies in hock joints were slightly negatively correlated (Stock & Distl, 2006a). Pathologic changes of navicular bones were not genetically correlated with osseous fragments in fetlock or hock joints (Stock & Distl, 2006a).

Positive genetic correlations were found between height at withers and osseous fragments in fetlock and hock joints (Stock et al., 2005b). Regarding hock and fetlock joints, negative genetic correlations were found between height at withers and the incidence of hock OC (Philipsson et al., 1993, Willms et al., 1999). Large carpal circumference was genetically associated with a high incidence of hock OC (Philipsson et al., 1993).

A positive genetic correlation was estimated between show jumping and the occurrence of OCD in sport horses (Winter et al., 1996). A study in 3,725 Hanoverian warmblood horses analyzed genetic correlation between performance in sports and osseous fragments in fetlock and hock joints (Stock & Distl, 2006b). Performance was measured as the number of annual tournament entries (TE) and placings (TP) in basic buildup, dressage and show-jumping competitions. High prevalence of osseous fragments in hock joints were genetically negatively correlated with TP and TE in dressage competitions. A similar tendency was found for osseous fragments in hock and fetlock joints with TP in show-jumping competitions (Stock & Distl, 2006b). Mare performance tests and inspections at riding horse auctions showed mostly close to zero genetic correlation between osseous fragments in fetlock and hock joints (Stock & Distl, 2007).

Heritability report indicates that selection for skeletal soundness may be feasible. When weighting radiographic traits with 30–60% as opposed to the respective performance traits, and selecting only

Table 12.1 Heritability estimates for equine osteochondrosis in different limb joints by horse breeds.

Population and number of investigated horses	Radiographic finding	Heritability estimate	Method of analysis	References
Danish trotters (n = 325)	OF[a] (hock)	0.26 ± 0.14	STM[f]	Schougaard et al. (1987)
Norwegian trotters (n = 644)	OC[b] (hock)	0.21	STM[f] (REML[g])	Grøndahl and Dolvik (1993)
Swedish trotters (n = 793)	OC[b] (hock)	0.27	LSM[h] (DL[i])	Philipsson et al. (1993)
Maremmano horses (n = 350)	DOF[d] (fetlock) OF[a] (hock)	0.13-0.14 ± 0.22-0.23	LAM[j] (REML[g], DL[i])	Pieramati et al. (2003)
Dutch warmblood (mares; n = 590)	OF[a] (hock)	0.14 ± 0.17	LAM[j] (REML[g], DL[i])	KWPN (1994)
Dutch warmblood (stallions; n = 1,965)	OC[b] (hock) OC[b] (stifle)	0.11 0.09	ATM[k] (REML[g], DL[i])	Der Kinderen (2005)
Dutch warmblood (foals; n = 811)	OC[b] (fetlock) OC[b] (hock) OC[b] (stifle) OCD[e] (fetlock) OCD[e] (hock) OCD[e] (stifle)	0.08 ± 0.10 0.15 ± 0.08 0.07 ± 0.06 0.06 ± 0.07 0.26 ± 0.09 0.02 ± 0.04	LAM[j] (REML[g])	Van Grevenhof et al. (2009)
Hanoverian warmblood (foals; n = 624) (yearlings; n = 396)	OC[b] (fetlock) OC[b] (hock) OC[b] (stifle) OCD[e] (fetlock) OCD[e] (hock) OCD[e] (stifle) OCD[e] (fetlock) OCD[e] (hock) OCD[e] (stifle)	0.07 ± 0.08 0.08 ± 0.05 0.07 ± 0.06 0.15 ± 0.07 0.10 ± 0.05 0.02 ± 0.04 0.06 ± 0.11 0.15 ± 0.07 0.00 ± 0.01	LAM[j] (REML[g])	Schober (2003)
Hanoverian warmblood (n = 3,725)	OF[a] (fetlock) OF[a] (hock)	0.19 ± 0.03 0.37 ± 0.06	LAM[j] (REML[g], DL[i])	Stock et al. (2005b)
German riding horses (n = 2,407)	OCD[e] (hock)	0.07 ± 0.03	LAM[j] (REML[g])	Winter et al. (1996)
Holsteiner horses (mares; n = 456)	OCD[e] (hock)	0.34 ± 0.06	ATM[k] (GS[l])	Willms et al. (1999)
Holsteiner horses (foals; n = 144)	OCD[e] (hock)	0.19 ± 0.02	ATM[k] (GS[l])	Willms et al. (1999)

[a]Osteochondral fragments.
[b]Osteochondrosis.
[c]Palmar/plantar osteochondral fragments.
[d]Dorsal osteochondral fragmentation.
[e]Osteochondrosis dissecans.
[f]Sire threshold model.
[g]Residual Maximum Likelihood.
[h]Linear sire model.
[i]Dempster Lerner transformation onto the liability model.
[j]Linear animal model.
[k]Animal threshold model.
[l]Gibbs sampling.

sires with above-average total indices, the prevalence of each radiographic finding can be lowered by up to 10% in the progeny. Considering only one radiographic finding at a time, the maximum attainable response to selection in a generation can be a decrease of prevalence of radiographic findings by 31–52% (Stock & Distl, 2005). Therefore, parameters of the health of the equine skeleton could be considered at this early stage of selection of breeding animals.

Quantitative trait loci (QTL) for OC

In recent years, there have been major advances in mapping the equine genome. Whole-genome radiation hybrid maps (Chowdhary et al., 2003; Raudsepp et al., 2008), comprehensive genetic linkage maps (Penedo et al., 2005; Swinburne et al., 2006), a medium-density horse gene map (Perrocheau et al., 2006), informative marker sets for QTL analyses (Mittman et al., 2010a), and a high-density microsatellite map (Mittman et al., 2010b) have been developed. The second release of the horse genome assembly became available in September 2007. EquCab2.0 is a high-quality draft sequence of the female Thoroughbred *Twilight*. A whole genome shotgun library was sequenced at the Broad Institute, Cambridge, Massachusetts, and BAC ends were sequenced at the University of Veterinary Medicine, Hannover and Helmholtz Centre for Infection Research, Braunschweig, Germany. A collection of more than 1 million single nucleotide polymorphisms (SNPs, one SNP per 2 kb) was generated mining the sequence from Twilight and additional sequences from seven horses of a variety of breeds, including Akhal-teke, Andalusian, Arabian, Icelandic, Quarter Horse, Standardbred, and Thoroughbred. Linkage disequilibrium is intermediate in length compared to dogs and humans, and dropping quickly down at 100–150 kb (Wade et al., 2009). With the horse genome assembly and SNP collection available, an equine beadchip comprising 54,602 SNPs has been developed and publicly released in 2009 (McCue et al., 2009).

Equine maps with a large number of highly polymorphic markers equidistantly distributed over the whole genome are prerequisites to perform linkage studies for identifying quantitative trait loci (QTL). QTL are genomic regions that are highly likely to harbor genes influencing the trait of interest. The larger the proportion of the phenotypic variation of the QTL, the more useful the QTL are for selection. QTL show linkage disequilibrium within families or in the whole population. An ideal marker for a QTL shows linkage disequilibrium in the whole population. After identification of QTL in population-wide linkage disequilibrium, marker-assisted selection provides the possibility to improve accuracy of selection in young horses and to shorten the generation interval in horse breeding.

Whole-genome scans for half-sib families with a high incidence of OC were employed to detect QTL significantly contributing to the phenotypic expression of OC. Data from 14 half-sib families of Hanoverian warmblood horses were analyzed using microsatellite markers to detect quantitative trait loci (QTL) with significant influence on the development of OC (Dierks et al., 2007). Genome-wide significant QTL for osteochondrosis in fetlock and hock joints were found on horse chromosomes (ECA) 2, 4, 5, and 16. Further, chromosome-wide significant QTL were located on ECA3, 15, 19, and 21. QTL for fetlock OC were on ECA2, 4, and 5. The locations of QTL for hock OC were on ECA2, 15, 16, and 21. To refine the identified QTL, an extended marker set containing published, as well as newly developed microsatellites and SNPs, was used. Fine-mapping confirmed and delimited QTL on ECA2 (15.65–31.91 Mb for fetlock OC and fetlock OCD; 26.89–33.05 for hock OC), ECA4 (7.42–13.10 Mb and 56.15–59.84 Mb for fetlock OC, 3.62–6.24 Mb for hock OC), ECA5 (78.03–90.23 Mb for fetlock OC and fetlock OCD), ECA16 (17.6–45.18 Mb for hock OCD), and ECA21 (5.45–17.14 Mb for hock OC and hock OCD) (Lampe, Dierks, & Distl, 2009a,

2009b). On ECA18, a new QTL for OC and hock OC at 74.94–82.25 Mb could be identified in Hanoverian warmblood horses (Lampe et al., 2009c). On ECA2, a SNP within the neurochondrin (*NCDN*) gene was significantly associated with OC traits. On ECA4, three significantly associated SNPs have been identified within the genome-wide significant OC-QTL region: two SNPs were located in intergenic regions and one SNP was in intron 2 of the HECT, C2, and WW domain containing E3 ubiquitin protein ligase 1 (*HECW1*) gene.

In 117 South German coldblood horses a whole genome scan for OC was carried out (Wittwer et al., 2007b). Chromosome-wide significant QTL were found on 10 horse chromosomes, including 7 QTL for fetlock OC and 1 QTL on ECA18 for hock OC and fetlock OC. On ECA4, a significant QTL for fetlock OCD was detected and within this QTL six candidate genes at 50–67 Mb were screened using 22 SNPs. A significantly associated SNP in the 3′-untranslated region of the acyloxyacyl hydrolase (*AOAH*) gene could be identified (Wittwer et al., 2008). All horses homozygous for this SNP were affected by fetlock OCD. The odds ratio of being affected for the homozygous genotype was 2.71–3.67 in comparison to the heterozygous and wildtype homozygous genotypes. On ECA18, the QTL for fetlock OCD and hock OC was refined to an interval of 35.5 to 47.1 Mb. Screening 18 candidate genes with 71 SNPs in this QTL revealed 2 significantly associated intronic SNPs in the xin actin-binding repeat containing 2 (*XIRP2*) gene (Wittwer, Hamann, & Distl, 2009). Homozygous and heterozygous horses were at a 1.3- to 2.4-fold higher risk for fetlock OC, fetlock OCD, and hock OC.

Gene expression analyses and candidate genes

Studies on the variation in gene expression of key chondrogenic genes and genes differentially expressed between normal and OC chondrocytes may also help identify candidate genes and their potential role in the pathogenesis of osteochondrosis and repair mechanisms in osteochondrotic lesions. Potential candidate genes may code for hormones, enzymes, metabolic factors, and/or their receptors involved in the process of cartilage maturation and differentiation during endochondral ossification. Differential expression studies between neonates and adults showed 7% of the 9,367 probe sets differentially expressed. Up-regulated genes primarily included ontologies related to extracellular matrix proteins, matrix modifying enzymes, skeletal development, and cell adhesion (Mienaltowski et al., 2008). Candidate genes could also be those involved in epiphyseal dysplasia such as collagen genes, cartilage oligomeric matrix protein genes, or diastrophic dysplasia sulfate transporter genes.

The endocrinological procedures of skeletal development and growth are controlled by hormones that are most likely to participate in enchondral ossification, such as insulin, thyroxine, growth hormone, parathyroid hormone, and calcitonin (Glade, 1986). Of the regulating proteins involved in enchondral ossification, the transforming growth factor ß (TGF-ß) plays an important role in growth cartilage metabolism, particulary in the control of chondrocyte differentiation and hypertrophy (Glade, 1986; Henson, Schofield, & Jeffcott, 1997; Jeffcott & Henson, 1998). Henson et al. (1997) described a reduced expression and immunoreactivity in focal lesions compared to normal cartilage, but strong expression of *TGFβ1* in the chondrocyte clusters immediately surrounding the lesion, and therefore a possible involvement in OC. Semevolos, Nixon, and Brower-Toland (2001) found a higher expression of *TGF-β1* in osteochondrosis cartilage and suggested that this might explain a healing response to the OC lesion. Hypertrophic differentiation and enchondral ossification of growth cartilage are regulated by a complex array of signaling peptides, including parathyroid hormone related protein (PTHrP), Indian hedgehog (Ihh), TGFß, runt-related transcription factor 2

(Runx2), and bone morphogenetic proteins (BMPs). PTHrP maintains chondrocytes in a proliferative state by limiting the number of cells capable of expressing Ihh, and thus participates in a negative feedback loop controlling the rate of hypertrophic differentiation (Chung, Lanske, & Lee, 1998). The main regulator of chondrocyte hypertrophic differentiation is the transcription factor Runx2. PTHrP inhibits *Runx2* expression, whereas *Ihh* expression is stimulated by Runx2. Further repressors of *Runx2* activity are Sox9 and the histon deacetylase-4 (HDAC4). Nkx3/Bpax1 mediates the actions of PTHrP and inhibits *Runx2* expression (Provot et al., 2006). A balance between the expression level of the transcription factor myocyte enhancer factor-2 (*MEF2C*) and repression by HDAC4 seems necessary for normal progression of chondrocyte hypertrophy (Arnold et al., 2007). Hedgehog signaling occurs through the transmembrane receptor, Patched (Ptc), which upon binding with Ihh releases its inhibitor, a transmembrane receptor, Smoothened (Smo). In turn, Smo activation results in stimulation of transcription factors, Gli1, Gli2, and Gli3. While a significant increase of *PTHrP* and *Ihh* expression in chondrocytes from OC-affected cartilage and a decrease of *Gli1* expression could be observed, no difference in expression patterns were identified for *BMP, Gli2, Gli3, Ptc*, and *Smo* (Semevolos et al., 2002; Semevolos, Nixon, & Strassheim, 2004; Semevolos et al., 2005). *RUNX2* showed the greatest differences of gene expression between lesions and normal cartilage (Mirams et al., 2008).

Insulin like growth factors (*IGFs*) play an important role in cartilage metabolism and growth, including the introduction of increasing cellular proliferation and the synthesis of cartilage aggrecan and collagen (Semevolos et al., 2001). An interdependency of OC in hock joints and plasma IGF-I levels has been observed (Sloet van Oldruitenborgh-Oosterbaan, Mol, & Barneveld, 1999). Foals with osteochondrotic findings showed significantly lower IGF-I levels than unaffected foals. Reduction in chondrocyte differentiation caused by lower plasma IGF-I concentrations may contribute to the development of osteochondrosis. Such an endocrine pathway to OC may be related with faster growth and earlier physical maturation in horses fed with high-energy diets and high genetic growth capacity. The significantly higher expression of *IGF-I* in cartilage obtained from osteochondrotic lesions in 8- to 12-month-old horses was correlated with the proliferation of cells repairing the cartilage damage, and therefore reflects a healing response to injured tissue rather than a primary alteration (Semevolos et al., 2001; Verwilghen et al., 2009).

The composition of the extracellular matrix has been targeted as another molecular mechanism involved in the development of OC. Various collagen types that are represented in the extracellular cartilage matrix are known to play a role in the development and maturation of cartilage. It is well known that the extracellular matrix of the articular cartilage goes through a phase of rapid remodelling in the neonatal animal (Van Weeren, 2005). Additional evidence for the crucial role of collagen was provided by the demonstration of differences in post-translational modifications of collagen type II in samples from early osteochondrotic lesions (Van de Lest et al., 2004). The expression of collagen type I, II and X in chondrocytes from OC cartilage was significantly higher than in normal cartilage (Garvican et al., 2008). These results could partly be confirmed by Mirams et al. (2008) who found a significantly higher expression of collagen type I and X in the osteochondrosis lesions, but no differences in the expression patterns of collagen type II. Also, Semevolos et al. (2001) could not find any significant differences in the expression of various collagen types I, II, and X between OC and normal joints. Cartilage-specific matrix genes (*COL2A1*, *COL3A1*, *COL11A1*, *COL1A2*) are abundantly expressed in normal articular cartilage (Gläser et al., 2009). Increased *COL2A1, SOX9*, and *AGGRECAN* expression was found in human articular chondrocytes during chondrogenesis (Miyaki et al., 2009). In osteoarthritic cartilage, *COL2A1* and *SOX9* expression was down-regulated, while cartilage-degrading enzymes (*ADAMTS5* and *MMP-13*) were up-regulated.

Peansukmanee et al. (2009) found a reduced glucose transporter 1 (*GLUT1*) expression in either OCD or osteoarthritis (OA) cartilage. The equine *GLUT1* gene is located on ECA2 at 15.56 to 15.57 Mb and consequently close to a detected QTL for OC in fetlock joints of Hanoverian warmblood horses (15.65 to 30.94 Mb). Peansukmanee et al. (2009) analyzed effects of hypoxia on glucose transport in equine chondrocytes and compared expression of the hypoxia responsive *GLUT1* gene in either OCD- or OA-affected and in normal cartilage. They suggested that reduced *GLUT1* might contribute to degenerative cartilage defects.

Metalloproteinases are considered to be a key feature in the loss of articular cartilage seen in many joint diseases. Different studies on the expression of matrix metalloproteinases *MMP-1, MMP-3*, and *MMP-13* came up with the same results showing there was no significant difference in the expression of either *MMP-1* or *MMP-3*, but a significant up-regulation of *MMP-13* in OC-chondrocytes (Garvican et al., 2008, Mirams et al., 2008). Brama et al. (2000) investigated the role of *MMP-3* activity in synovial fluid in common joint disorders in the horse and concluded that *MMP-3* activity in OC joints appears not to be different from normal joints, but was four times higher in osteoarthritis joints. The ADAM metallopeptidase with thrombospondin type 1 motif 4 (*ADAMTS4*) gene encodes an enzyme responsible for the degradation of aggrecan, a major proteoglycan of cartilage. Aggrecan degradation is an important factor in the erosion of articular cartilage in arthritic diseases, which is also reflected in a significantly higher expression of *ADAMTS4* in OC cartilage than in chondrocytes from normal cartilage (Garvican et al., 2008). However, aggrecan itself was not differently expressed among cartilage from OC lesions and normal cartilage (Semevolos et al., 2001; Garvican et al., 2008; Mirams et al., 2008). The proteins encoded by the tissue inhibitor of metalloproteinase (*TIMP*) gene family are natural inhibitors of the matrix metalloproteinases (MMPs). While *TIMP-1* showed a significant increase of expression in chondrocytes from OC cartilage in comparison to normal cartilage, the expression of *TIMP-2* and *TIMP-3* in OC chondrocytes was significantly less (Garvican et al., 2008).

Navicular Disease

Navicular disease is one of the main causes of chronic, often therapy-resistant forelimb lameness in middle-aged horses. Pathological alterations can primarily affect the navicular bone (*os sesamoideum distale*), the navicular bursa (*bursa podotrochlearis*), or the distal end of the deep digital flexor tendon. Radiographic findings have been associated with navicular bone pathology such as branched or lollipop-shaped *canales sesamoidales*, an irregular, sclerosed or radiolucent structure, and a contour with exostoses (Brunken, 1986; Hertsch & Steffen, 1986; Kaser-Hotz & Ueltschi, 1992). Several evaluation schemes have been developed for radiographs of the navicular bone. Dik and van den Broek (1995) classified navicular bones on the basis of the shape of the proximal border, whereas Lukas (1987) considered both shape of the proximal border and of the medial and lateral extremities of the navicular bones. The evaluation scheme of Brunken (1986) permits a classification of radiographic findings in the navicular bone according to size, shape, and distribution of *canales sesamoidales* and the structure and contour of the navicular bone. Riding horses show very often (up to 70–80%) slight or moderate changes in the navicular bone, whereas moderate to severe changes are at low frequencies of 2–5%. Heritabilities for radiological changes of different severity in the navicular bone were at $h^2 = 0.10$ to 0.34 (KWPN, 1994; Willms et al., 1999; Stock & Distl, 2006b). Chromosome-wide significant QTL were located on five different equine chromosomes (ECA2, 3, 4, 10, and 26). Genome-wide significant QTL on ECA2 and 10 (Diesterbeck, Hertsch, & Distl, 2007) were confirmed by fine-mapping revealing associated haplotypes and marker alleles. These studies represent the first step to get more insight into the molecular genetic determination of radiological alterations in the equine navicular bone.

Conclusions

Osteochondrosis and navicular disease are among the most common skeletal diseases in horses. The aetiology of these conditions is still not fully understood, but more research has focused on the molecular mechanisms involved. The substantial progress of comparative genomics and the horse genome sequencing project provide a very effective approach to unravelling the genetic basis of OC and navicular disease. The genomic regions containing QTL for OC, as identified by whole genome scans and genome-wide association studies, provide good chances for the detection of causal genes.

References

Arnold, M. A., Kim, Y., Czubryt, M. P., Phan, D., McAnally, J. et al. (2007). MEF2C transcription factor controls chondrocyte hypertrophy and bone development. *Developmental Cell, 12*, 377–389.

Barneveld, A., & Van Weeren, P. R. (1999). Conclusions regarding the influence of exercise on the development of the equine musculoskeletal system with special refernce to osteochondrosis. *Equine Veterinary Journal, 31*(Suppl.), 112–119.

Brama, P. A. J., TeKoppele, J. M., Van El, B., Barneveld, A., & Van Weeren, P. R. (2000). Influence of development and joint pathology on stromelysin enzyme activity in equine synovial fluid. *Annuals of the Rheumatic Diseases, 59*, 155–157.

Brunken, E. (1986). Röntgenologische Verlaufsuntersuchungen am Strahlbein des Pferdes. Master's thesis, University of Veterinary Medicine Hannover.

Chung, U. I., Lanske, B., & Lee, K. (1998). Parathyroid hormone/parathroid hormone-related peptide receptor coordinates endochondral bone development by directly controlling chondrocyte differentiation. *Proceedings of the National Academy of Sciences of the United States of America, 95*, 13030–13035.

Chowdhary, B. P., Raudsepp, T., Kata, S. R., Goh, G., Millon, L. V. et al. (2003). The first-generation whole-genome radiation hybrid map in the horse identifies conserved segments in human and mouse genomes. *Genome Research, 13*, 742–751.

Dierks, C., Löhring, K., Lampe, V., Wittwer, C., Drögemüller, C., & Distl, O. (2007). Genome-wide search for markers associated with osteochondrosis in Hanoverian warmblood horses. *Mammalian Genome, 18*, 739–747.

Diesterbeck, U., Hertsch, B., & Distl, O. (2007). Genome-wide search for microsatellite markers associated with radiological alterations in the navicular bone of Hanoverian warmblood horses. *Mammalian Genome, 18*, 373–381.

Dik, K. J., & Van den Broek, J. (1995). Role of navicular bone shape in the pathogenesis of navicular disease: A radiological study. *Equine Veterinary Journal, 27*, 390–393.

Garvican, E. R., Vaughan-Thomas, A., Redmond, C., & Clegg, P. D. (2008). Chondrocytes harvested from osteochondrosis dissecans cartilage are able to undergo limited in vitro chondrogenesis despite having pertubations of cell phenotype in vivo. *Journal of Orthopaedic Research, 26*, 1133–1140.

Glade, M. J. (1986). The control of cartilage growth in osteochondrosis: A review. *Journal of Equine Veterinary Science, 6*, 175–192.

Gläser, K. E., Sun, Q., Wells, M. T., & Nixon, A. J. (2009). Development of a novel equine whole transcript oligonucleotide genechip microarray and its use in gene expresson profiling of normal articular-epiphyseal cartilage. *Equine Veterinary Journal, 41*, 663–670.

Grøndahl, A. M., & Dolvik, N. I. (1993). Heritability estimations of osteochondrosis in the tibiotarsal joint and of bony fragments in the palmar/plantar portion of the metacarpo- and metatarsophalangeal joints of horses. *Journal of the American Veterinary Medical Association, 203*, 101–104.

Henson, F. M., Schofield, P. N., & Jeffcott, L. B. (1997). Expression of transforming growth factor-beta 1 in normal and dyschondroplastic articular growth cartilage of the young horse. *Equine Veterinary Journal, 6*, 434–439.

Hertsch, B., & Steffen, D. (1986). Röntgenologische und vergleichen patho-morphologische Untersuchungen an Strahlbeinen unter besonderer Berücksichtigung der Canales sesamoidales – Ein Beitrag zur Diagnose der Podotrochlose. *Deutsche Tierärztliche Wochenschrift, 93*, 353–359.

Jeffcott, L. B. (1996). Osteochondrosis – an international problem for the horse industry. *Journal of Equine Veterinary Science, 16*, 32–37.

Jeffcott, L. B., & Henson, F. M. D. (1998). Studies on growth cartilage in the horse and their application to aetiopathogenesis of dyschondroplasia (osteochondrosis). *The Veterinary Journal, 156*, 177–192.

Kaser-Hotz, B., & Ueltschi, G. (1992). Radiographic appearence of the navicular bone in sound horses. *Veterinary Radiology and Ultrasound, 33*, 9–17.

KWPN (1994). *The frequency and heredity of navicular disease, sesamoidosis, fetlock joint arthrosis, bone spavin, osteochondrosis of the hock. A radiographic progeny study*. KWPN (Koninklijke Vereniging Warmbloed Paardenstamboek Nederland), Zeist.

Lampe, V., Dierks, C., & Distl, O. (2009a). Refinement of a quantitative trait locus on equine chromosome 5 responsible for fetlock osteochondrosis in Hanoverian warmblood horses. *Animal Genetics, 40*, 553–555.

Lampe, V., Dierks, C., & Distl, O. (2009b). Refinement of a quantitative gene locus on equine chromosome 16 responsible for osteochondrosis in Hanoverian warmblood horses. *Animal, 3*, 1224–1231.

Lampe, V., Dierks, C., Komm, K., & Distl, O. (2009c). Identification of a new quantitative trait locus on equine chromosome 18 responsible for osteochondrosis in Hanoverian warmblood horses. *Journal of Animal Science, 87*, 3477–3481.

Lukas, B. 1987. *Radiographic demonstration of the development of the navicular bone (distal sesamoid bone) from birth to three years of age, with special reference to the sesamoid canals, in horses*. Hanover: Tierarztliche Hochschule.

McCue, M., Mickelson, J., Bannasch, D., Penedo, C., Bailey, E. et al. (2009). The horse gentrain project: Initial evaluation of the Equine SNP50 BeadCips. Paper presented at The 8th Dorothy Russell Havemeyer Foundation International Equine Genome Mapping Workshop, July 22–25, Suffolk, UK.

Mienaltowski, M. J., Huang, L., Stromberg, A. J., & MacLeod, J. N. (2008). Differential gene expression associated with postnatal equine articular maturation. *BMC Musculosceletal Disorders, 9*, 149.

Mirams, M., Tatarczuch, L., Ahmed, Y. A., Pagel, C. N., Jeffcott, L. B. et al. (2008). Altered gene expression in early osteochondrosis lesions. *Journal of Orthopaedic Research, 27*, 452–457.

Mittmann, E. H., Lampe, V., Momke, S., Zeitz, A., & Distl, O. (2010a). Characterization of a minimal microsatellite set for whole genome scans informative in warmblood and coldblood horse breeds. *Journal of Heredity, 101*: 246–250.

Mittmann, E. H., Momke, S., & Distl, O. (2010b). Whole-genome scan identifies quantitative trait loci for chronic pastern dermatitis in German draft horses. *Mammalian Genome, 21*: 95–103.

Miyaki, S., Nakasa, T., Otsuki, S., Grogan, S. P., Higashiyama, P. et al. (2009). MicroRNA-140 is expressed in differentiated human articular chondrocytes and modulates interleukin-1 responses. *Arthritis and Rheumatism, 60*, 2723–2730.

Peansukmanee, S., Vaughan-Thomas, A., Carter, S. D., Clegg, P. D., Taylor, S. et al. (2009). Effects of hypoxia on glucose transport in primary equine chondrocytes in vitro and evidence of reduced GLUT1 gene expression in pathologic cartilage in vivo. *Journal of Orthopaedic Research, 27*, 529–535.

Penedo, M. C. T., Millon, L. V., Bernoco, D., Bailey, E., Binns, M. et al. (2005). International equine gene mapping workshop report: a comprehensive linkage map constructed with data from new markers and by merging four mapping resources. *Cytogenetic and Genome Research, 111*, 5–15.

Perrocheau, M., Boutreux, V., Chadi, S., Mata, X., Decaunes, P. et al. (2006). Construction of a medium-density horse gene map. *Animal Genetics, 37*, 145–155.

Philipsson, J., Andreasson, E., Sandgren, B., Dalin, G., & Carlsten, J. (1993). Osteochondrosis in the tarsocrural joint and osteochondral fragments in the fetlock joints in Standardbred trotters. II. Heritybility. *Equine Veterinary Journal, 16*(Suppl.), 38–41.

Pieramati, C., Pepe, M., Silvestrelli, M., & Bolla, A. (2003). Heritability estimation of osteochondrosis dissecans in Maremmano horses. *Livestock Production Science, 79*, 249–255.

Provot, S., Kempf, H., Murtaugh, L. C., Chung, U. I., Chyung, J. et al. (2006). Nkx3.2/Bapx1 acts as a negative regulator of chondrocyte maturation. *Development, 133*, 651–662.

Raudsepp, T., Gustafson-Seabury, A., Durkin, K., Wagner, M. L., Goh, G. et al. (2008). A 4103 marker integrated physical and comparative map of the horse genome. *Cytogenetic Genome Research, 122*, 28–36.

Sandgren, B., Dalin, G., Carlsten, J., & Lundeheim, N. (1993). Development of osteochondrosis in the tarsocrural joint and osteochondral fragments in the fetlock joints of Standardbred trotters. II. Body measurements and clinical findings. *Equine Veterinary Journal, 16*(Suppl.), 48–53.

Schober, M. (2003). Schätzung von genetischen Effekten beim Auftreten von Osteochondrosis dissecans beim Warmblutpferd. Dissertation, Georg-August-Universität-Göttingen.

Schougaard, H., Ronne, J. F., & Philipsson, P. (1990). A radiographic survey of tibiotarsal osteochondrosis in a selected population of trotting horses in Denmark and its possible genetic significance. *Equine Veterinary Journal, 22*, 288–289.

Semevolos, S. A., Nixon, A. J., & Brower-Toland, B. D. (2001). Changes in molecular expression of aggrecan and collagen types I, II and X, insulin-like growth factor-I, and transforming growth factor-beta 1 in articular cartlage obtained from horses with naturally acquired Osteochondrosis. *American Journal of Veterinary Research, 7*, 1088–1094.

Semevolos, S. A., Brower-Toland, B. D., Bent, S. J., & Noxon, A. J. (2002). Parathyroid hormone-related peptide and Indian hedgehog expression patterns in naturally acquired equine osteochondrosis. *Journal of Orthopaedic Research*, *20*, 1290–1297.

Semevolos, S. A., Nixon, A. J., & Strassheim, M. L. (2004). Expression of bone morphogenetic protein-6 and -2 and bone morphogenetic antagonist in horses with naturally acquired osteochondrosis. *American Journal of Veterinary Research*, *1*, 110–115.

Semevolos, S. A., Strassheim, M. L. Haupt, J. L., & Nixon, A. J. (2005). Expression patterns of hedgehog signaling peptides in naturally acquired equine osteochondrosis. *Journal of Orthopaedic Research*, *23*, 1152–1159.

Sloet van Oldruitenborgh-Osterbaan, M. M., Mol, J. A., & Barneveld, A. (1999). Hormones, growth factors and other plasma variables in relation to osteochondrosis. *Equine Veterinary Journal*, *31*(Suppl.), 45–54.

Stock, K. F., & Distl, O. (2005). Prediction of breeding values for osseus fragments in fetlock and hock joints, deforming arthropathy in hock joints and pathologic changes in the nacivular bones of Hanoverian Warmblood horses. *Livestock Production Science*, *92*, 77–94.

Stock, K. F., & Distl, O. (2006a). Genetic correlations between osseus fragments in fetlock and hock joints, deforming arthropathy in hock joints and pathologic changes in the navicular bones of Warmblood riding horses. *Livestock Science*, *105*, 35–43.

Stock, K. F., & Distl, O. (2006b). Correlations between sport performance and different radiographic findings in the limbs of Hanoverian Warmblood horses. *Animal Science*, *82*, 83–93.

Stock, K. F., & Distl, O. (2007). Genetic correlations between performance traits and radiographic findings in the limbs of German Warmblood riding horses. *Journal of Animal Science*, *85*, 31–41.

Stock K. F., Hamann, H., & Distl, O. (2005a). Prevalence of osseous fragments in limb joints of Hanoverian Warmblood horses. *Journal of Veterinary Medicine*, *52*, 388–394.

Stock, K. F., Hamann, H., & Distl, O. (2005b). Estimation of genetic parameters for the prevalence of osseous fragments in limb joints of Hanoverian Warmblood horses. *Journal of Animal Breeding and Genetics*, *122*, 271–280.

Stock, K. F., Hamann, H., & Distl, O. (2006). Factors associated with the prevalence of osseous fragments in the limb joints of Hanoverian Warmblood horses. *The Veterinary Journal*, *171*, 147–156.

Swinburne, J. E., Boursnell, M., Hill, G., Pettitt, L., Allen, T. et al. (2006). Single linkage group per chromosome genetic linkage map for the horse, based on two three-generation, full-sibling, crossbred horse reference families. *Genomics*, *87*, 1–29.

Van de Lest, C. H., Brama, P. A., Van El, B., DeGroot, J., Van Weeren, P. R. (2004). Extracellular matrix changes in early osteochondrotic defects in foals: A key role for collagen? *Biochimica et biophysica Acta*, *1690*: 54–62.

Van der Harst, M. R., Van de Lest, C. H. A., DeGroot, J., Kiers, G. H., Brama, P. A. J., & Van Weeren, P. R. (2005). Study of cartilage and bone layers of the bearing surface of the equine metacarpophalangeal joint relative to different timescales of maturation. *Equine Veterinary Journal*, *37*, 200–206.

Van Grevenhof, E. M., Schurink, A., Ducro, B. J., van Weeren, P. R., van Tartwijk, J. M. F. M. et al. (2009). Genetic variables of various manifestations of osteochondrosis and their correlations between and within joints in Dutch warmblood horses. *Journal of Animal Science*, *87*, 1906–1912.

Van Weeren, P. R. (2005). Osteochondrosis: A challenging enigma. *Pferdeheilkunde*, *21*, 285–292.

Van Weeren, P. R., & Barneveld, A. (1999). The effect of exercise on the distribution and manifestation of osteochondrotic lesions in the Warmblood foal. *Equine Veterinary Journal*, *31*(Suppl.), 16–25.

Van Weeren, P. R., Sloet van Oldruitenborgh-Oosterbaan, M. M., & Barneveld, A. (1999). The influence of birth weight, rate of weight gain and final achieved height and sex on the development of osteochondrotic lesions in a population of genetically predisposed Warmblood foals. *Equine Veterinary Journal*, *31*(Suppl.), 26–30.

Verwilghen, D. R., Vanderheyden, L., Franck, T., Busoni, V., Enzerink, E. et al. (2009). Variations of plasmatic concentrations of Insulin-like growth factor-I in post-pubescent horses affected with developmental osteochondral lesions. *Veterinary Research Communications*, *33*, 701–709.

Wade, C., Giulotto, E., Sigurdsson, S., Zoli, M., Gnerre, S. et al. (2009). Genome sequence, comparative analysis, and population genetics of the domestic horse. *Science*, *326*, 865–867.

Wilke, A. (2003). Der Einfluss von Haltung und Aufzucht auf die Häufigkeit von Osteochondrose. Dissertation, Tierärztliche Hochschule Hannover.

Willms, F., Röhe, R., & Kalm, E. (1999). Genetische Analyse von Merkmalskomplexen in der Reitpferdezucht unter Berücksichtigung von Gliedmaßenveränderungen. *Züchtungskunde*, *71*, 330–345.

Winter, D., Bruns, E., Glodek, P., & Hertsch, B. (1996). Genetische Disposition von Gliedmaßenerkrankungen beim Reitpferd. *Züchtungskunde*, *68*, 92–108.

Wittwer, C., Hamann, H., Rosenberger, E., & Distl, O. (2006). Prevalence of osteochondrosis in the limb joints of South German Coldblood horses. *Journal of Veterinary Medicine*, *53*, 531–539.

Wittwer, C., Hamann, H., Rosenberger, E., & Distl, O. (2007a). Estimation of genetic parameters for the prevalence of osteochondrosis in the limb joints of South German Coldblood horses. *Journal of Animal Breeding and Genetics*, *124*, 302–307.

Wittwer, C., Löhring, K., Drögemüller, C., Hamann, H., Rosenberger, E., & Distl, O. (2007b). Mapping quantitative trait loci for osteochondrosis in fetlock and hock joints and palmar/plantar osseus fragments in fetlock joints of South German Coldblood horses. *Animal Genetics*, *38*, 350–357.

Wittwer, C., Dierks, C., Hamann, H., & Distl, O. (2008). Associations between candidate gene markers at a quantitative trait locus on equine chromosome 4 responsible for osteochondrosis dissecans in fetlock joints of South German Coldblood horses. *Journal of Heredity*, *99*, 125–129.

Wittwer, C., Hamann, H., & Distl, O. (2009). The candidate gene XIRP2 at a quantitative gene locus on equine chromosome 18 associated with osteochondrosis in fetlock and hock joints of South German Coldblood horses. *Journal of Heredity*, *100*, 481–486.

Ytrehus, B., Carlson, C. S., & Ekman, S. (2007). Etiology and pathogenesis of osteochondrosis. *Veterinary Pathology*, *44*, 429–448.

13 Genomics of reproduction and fertility

Terje Raudsepp, Pranab J. Das and Bhanu P. Chowdhary

Introduction

Fertility of the stallion and of the mare is the core components of the economy of the horse industry due to the economic impact they have on stud fees and foal crop. Several organized studies have been carried out during past decades to understand the role of environmental, behavioral, and physiological factors affecting horse fertility (Madill, 2002; Carnevale, 2008; Coutinho da Silva, 2008; Varner & Johnson, 2007; Varner, 2008). In contrast, knowledge about the role of genetic factors in equine reproduction remains limited (Leeb, Sieme, & Topfer-Petersen, 2005; Leeb, 2007; Giesecke, Sieme, & Distl, 2010b). This is because genetic regulation of male and female fertility and reproduction in mammals is complex. It involves a network of thousands of genes that undergo spatiotemporal differential regulation and collectively govern a cascade of events from sex determination to fertilization (Matzuk & Lamb, 2002, 2008; Carrell, 2007; Naz & Catalano, 2010). While a few hundred candidate genes for male and female fertility have been identified using transgenic, knockout, or mutant rodent models (Matzuk and Lamb, 2002, 2008; Carrell, 2008), only limited information is available for other species, including the horse (Leeb et al., 2005; Leeb, 2007; Durkin et al., 2010; Giesecke et al., 2010b). This also implies that very little is known about the underlying molecular causes of mare and stallion infertility. However, thanks to the unprecedented progress of equine genomics and the availability of new pan-genomic analysis tools for the horse, the situation is about to change.

In this chapter, we seek to provide a comprehensive overview of the current status of knowledge regarding the genetic component of equine reproduction. The first section summarizes the findings of classical and molecular cytogenetics and the role of chromosomal abnormalities in mare and stallion fertility. The second section introduces the reader to the latest achievements of molecular genetics of fertility, the analysis of candidate genes, and the discovery of genomic regions involved in reproductive disorders. The focus of the third section is on functional genomics. We review recent application of cutting-edge approaches, such as expression microarray analysis and RNA-Seq, to identify functional signatures of fertility in horses. It must, however, be emphasized that while cytogenetic studies equally address reproductive problems in mares and stallions, research in molecular and functional genomics has predominantly focused on stallions. Finally, future directions in this important area of research will be discussed.

Equine Genomics, First Edition. Edited by Bhanu P. Chowdhary.

Cytogenetics of Reproduction

Since determination of the normal diploid chromosome number ($2n$=64; Rothfels et al., 1959) in the domestic horse (*Equus caballus*, ECA) and the inception of basic cytogenetic analysis techniques in 1970s, clinical cytogenetics has been an integral component in assessing fertility in horses (for review, see Chowdhary & Raudsepp, 2000; Ducos et al., 2008; Lear & Bailey, 2008; Villagomez et al., 2009; Durkin et al., 2010). The introduction of molecular-hybridization-based methods in 1990s, such as fluorescence in situ hybridization (FISH), has further improved the depth and accuracy of clinical chromosome analysis (Lear & Bailey, 2008).

It is well established that structural and numerical chromosomal abnormalities interfere with meiosis leading to a reduced number of normal gametes. Abnormal gametes affect the viability of the zygote, or can cause a range of developmental defects if the fetus is carried to full term (King, 2008; Villagomez & Pinton, 2008; Villagomez et al., 2009). Furthermore, some defects, such as autosomal translocations, only reduce fertility, thus propagating the abnormal chromosome complement and subfertile phenotype to the next generations (Lear & Bailey, 2008; Durkin et al., 2010).

Equine clinical cytogenetics has been lately exhaustively reviewed (Chowdhary & Raudsepp, 2000; Ducos et al., 2008; King, 2008; Lear & Bailey, 2008; Villagomez & Pinton, 2008; Villagomez et al., 2009; Durkin et al., 2010). Thus, the following section aims to succinctly outline the most typical cytogenetic causes of equine reproductive disorders. The review does not include the Y chromosome, the role of which in stallion fertility is discussed in Chapter 5. Additional details about the methodology of clinical chromosome analysis can be found in Chapter 18.

Sex chromosomes

Numerical and structural abnormalities of the sex chromosomes cause abnormal sexual development and gonadal dysgenesis both in mares and stallions (Lear & Bailey, 2008; Villagomez et al., 2009). Majority of sex chromosome abnormalities have been detected in infertile mares (Bugno, Slota, & Koscielny, 2007; Lear & Bailey, 2008; Di Meo et al., 2009; Villagomez et al., 2009) and are estimated to affect about 2% of the general population (Bugno et al., 2007). The most common (about 34%) abnormality among sterile mares with gonadal dysgenesis is the X chromosome monosomy (Bowling, Millon, & Hughes, 1987). The condition is cytogenetically manifested either as X chromosome monosomy, deletion of the Xp (Bowling et al., 1987; our unpublished data), or isochromosome Xq (Makela, Gustavsson, & Hollmen, 1994). All three forms cause haploinsufficiency for the pseudoautosomal (PAR) genes and for the Xp genes that escape inactivation (Bondy & Cheng, 2009; Lopes et al., 2010).

Besides, there are a few reports about sterile mares with nonmosaic X chromosome trisomy (65,XXX) (Makinen et al., 1999; Bugno et al., 2003; de Lorenzi et al., 2010) and sterile stallions with either 65,XXY syndrome (Kubien, Pozor, & Tischner, 1993; Makinen et al., 2000; Kakoi et al., 2005; Durkin et al., 2010) or Y chromosome disomy (either XYY or isochromosome Yq) (Herzog, Höhn, & Hecht, 1989; Paget et al., 2001; Das et al., 2011a). The frequency of these conditions is very low. For example, a genotyping-based survey of a random population of 15,471 foals in Japan detected only 4 65,XXY males, which is 0.025% (Kakoi et al., 2005). In addition, there are numerous reports about various mosaic and chimaeric karyotypes (Chowdhary & Raudsepp, 2000; Lear & Bailey, 2008; Villagomez et al., 2009; Juras et al., 2010). The carriers expose a variety of phenotypes extending from normal fertile mares and stallions to sterile animals with severe abnormalities in sexual development.

Autosomes

Only a few reports describe viable autosomal abnormalities in the horse. Among these are trisomies of the smallest equine autosomes, namely ECA23, 26, 27, 28, 30, and 31, all of which are associated with sterility and variable degrees of other phenotypic abnormalities (Brito et al., 2008; Lear & Bailey, 2008; Durkin et al., 2010). Additionally, seven different autosomal or autosome/X chromosome translocations have been described in mares (Lear & Bailey, 2008; Lear et al., 2008) and one (ECA5/16 translocation) in a stallion (Durkin et al., 2010). Animals with autosomal translocations produce both balanced and unbalanced gametes. While the former can give rise to viable zygotes carrying either normal or translocated chromosomes, fertilization with unbalanced gametes results in repeated early embryonic loss (REEL) (Lear et al., 2008; Durkin et al., 2010).

Sex reversal, intersexuality, and hermaphroditism

Sex reversal is a situation where the genetic sex (the karyotype) disagrees with the gonadal and/or the phenotypic sex. The affected individuals are sterile, with various degrees of abnormalities in sexual development. The male-to-female sex reversal where phenotypic mares or mare-like individuals have 64,XY karyotypes (Raudsepp et al., 2010), described in detail in Chapter 5, is the only true sex reversal condition in horses. A reversed situation where phenotypic males have normal female karyotype, as it has been described in humans (Zenteno-Ruiz et al., 2001), probably does not exist in horses. So far, all cases reported as equine 64,XX sex reversal actually describe intersexual, hermaphroditic, or male pseudohermaphroditic animals (Meyers-Wallen et al., 1997; Buoen et al., 2000; Chowdhary & Raudsepp, 2000; Vaughan, Schofield, & Ennis, 2001; Bannasch et al., 2007; Villagomez et al., 2009). The chromosomes of these developmentally and reproductively abnormal horses are cytogenetically normal and the genetic causes of intersexuality and hermaphroditism remain as yet elusive.

Cytogenetics of gametes and embryos

Given the impact of chromosome abnormalities on meiotic synapsis and segregation and on the viability of gametes and embryos (Villagomez & Pinton, 2008), it comes as a surprise that the cytogenetics of equine gametes and embryos lags far behind that of species like pig and cattle (King, 2008; Pinton et al., 2008; Villagomez & Pinton, 2008; Pinton et al., 2009).

Meiotic recombination in mammalian oocytes occurs at the early stages of embryonic development (Hunt & Hassold, 2008) and is difficult to study. Therefore, meiotic prophase chromosomes have been analyzed only in the spermatocytes of stallions. Synaptonemal complexes have been characterized in the meiosis of normal stallions (Scott & Long, 1980; Safronova & Pimentova, 1988; von Nett, Jung, & Stranzinger, 1996), in interspecific hybrids between equids (Chandley et al., 1974; Short et al., 1974), and in one infertile stallion carrying trisomy ECA28 (Power et al., 1992). The latter demonstrated the presence of a trivalent, or a bivalent and a univalent, in primary spermatocytes. The abnormal pairing triggered meiotic arrest, causing the degeneration of spermatocytes and spermatids, and was considered as the main cause of observed azoospermia.

Sperm-FISH on decondensed sperm heads has become a state-of-art technique to analyze the chromosomal constitution of mature sperm in humans (Templado et al., 2010) and animals (Rubes, Vozdova, & Kubickova, 1999; Pinton, Ducos, & Yerle, 2004; Massip et al., 2009). The method was recently optimized for the stallion sperm (Bugno-Poniewierska, Jablonska, & Slota, 2009), and looks promising to be used soon for the evaluation of stallion fertility.

To date, only a few studies have examined the *karyotypes of equine oocytes and embryos*. Conventional cytogenetic methods have been used to distinguish between meiosis 1 (M1) and meiosis 2 (M2) cells (King et al., 1987) and to determine the overall frequency of chromosome abnormalities of in vitro matured oocytes (King et al., 1990; Lechniak, Wieczorek, & Sosnowski, 2002). While 5.5% of the oocytes had cytogenetic defects, no chromosome aberrations were detected in conceptuses (Blue, 1981; Haynes & Reisner, 1982; Romagnano et al., 1987). These findings were revised using FISH with ECA2 and ECA4 probes on in vivo produced equine morulae and blastocysts. It was revealed that approximately 18% of embryos contain one or more chromosomally abnormal, mainly tetraploid or triploid, cells (Rambags et al., 2005). This clearly indicates the need for a more extensive investigation of chromosome aberrations in equine embryos using advanced molecular cytogenetics methods.

Molecular Genetics of Reproduction

The findings of clinical cytogenetics indicate that only a fraction of genetic defects that cause reduced fertility or infertility in horses are confined to chromosomal abnormalities. Thus, the genetic causes of the majority of subfertile phenotypes remain undefined. This is largely because little is known about the molecular genetics of fertility in mammals. It is estimated that from fertilization until sexual maturation, more than 1,000 genes spatiotemporally control the development of mammalian testes and ovaries (Cederroth et al., 2007; Wilhelm, Palmer, & Koopman, 2007). Normal functioning of adult gonads is regulated by an additional 1,000 genes. Overall, it is estimated that about 10–20% of all genes in the mammalian genome are involved in male fertility (Carrell, 2007), while much less is known about the genetics of female reproduction (Barnett et al., 2006). For example, the roles of *RSPO1* and *WNT4* in female sex determination were discovered only recently (Matzuk & Lamb, 2008; Piprek, 2009), refuting the previous idea about the default determination of the female developmental pathway. Because of the limited knowledge, there are currently no good mammalian models for the study of the genetics of fertility and reproduction in the mare. This is one of the reasons why molecular studies in horses have primarily focused on stallion reproduction (Leeb et al., 2005; Leeb, 2007; Giesecke et al., 2010b) (Table 13.1). Also, the structure of horse breeding where one stallion typically covers many mares inclines the bias toward stallion research, because the performance of stallions affects the overall economy of the industry more than the fertility of mares.

Genetics of mare fertility

After REEL and gonadal dysgenesis that are caused by chromosomal abnormalities, the most common subfertile phenotype in mares is the failure to become pregnant (Coutinho da Silva, 2008). Genetic causes of pregnancy failure are unknown. Currently, there are no molecular tools for the diagnostics of mare fertility or for proper evaluation of oocytes and embryos prior their use for in vitro reproductive technologies.

Genetics of stallion fertility

Stallions are typically selected on the basis of their ancestry, athletic performance, and overall appearance, but not for their reproductive potential (Colenbrander, Gadella, & Stout, 2003). As a result, reduced fertility among breeding stallions has emerged as one of the major problems (Leeb

Table 13.1 Summary of currently known candidate genes and genomic regions corresponding to equine reproductive phenotypes; nd – not determined

Phenotype	Method	Candidate gene	Candidate genomic region, ECA, Mb	Affected breed	References
Male fertility – sperm-egg fusion	Sanger sequencing, SNP genotyping	*CRISP1 alias AEG1*	20:47.8	Multiple	(Giese et al., 2002)
Male fertility – paternal component for pregnancy rate per estrus	Sanger sequencing, SNP genotyping	*CRISP3*	20:47.7	Hanoverian	(Hamann et al., 2007)
Male fertility – reorganization of sperm chromatin during spermiogenesis	Sanger sequencing, Northern blotting, immunostaining	*STK31*	4:55.2	nd	(Sabeur et al., 2008)
Cryptorchidism, testicular tumors, intersexuality	Immunolabeling with AMH antibody	*AMH*	7:1.0	nd	(Ball et al., 2008)
Male fertility – paternal component for pregnancy rate per estrus	Sanger sequencing, SNP genotyping	*PRLR*	21:30.0	Hanoverian	(Giesecke et al., 2009a)
Male fertility – paternal component for pregnancy rate per estrus	Sanger sequencing, SNP genotyping	*SPATA1*	5:79.1	Hanoverian	(Giesecke et al., 2009b)
Abnormal sperm morphology	qRT-PCR	*TSPY*	Y	Arabian	(Paria, 2009)
Male fertility – paternal component for pregnancy rate per estrus	Sanger sequencing, SNP genotyping	*INHBA*	4:12.8	Hanoverian	(Giesecke et al., 2010a)
Stallion subfertility, REEL	Cytogenetics	5p/16 translocation	5p/16	Thoroughbred	(Durkin, Raudsepp, & Chowdhary, 2010)
XY SRY-negative sex reversal	Cytogenetics, STS analysis, sequencing	*SRY* and 21 kb flanking sequence	Y	Multiple	(Raudsepp et al., 2010)
Impaired acrosomal reaction	SNP50 genotyping, GWA, expression microarray, Sanger sequencing	*FKBP6*, *TRIM56*	13:11.0 13:9.0	TB	(Raudsepp et al., 2011)

et al., 2005; Leeb, 2007). For example, in two retrospective studies, 36–43% of prospective breeding stallions did not pass reproductive soundness test (Blanchard & Varner, 1997; Woods et al., 2000).

Genetic studies of stallion fertility started with microsatellite heterozygosity analysis (Aurich, Achman, & Aurich, 2003) and SNP search in candidate genes (Giese et al., 2002; Hamann et al., 2007; Giesecke et al., 2009a, 2010a; Giesecke et al., 2009b). The goal was to identify genetic markers associated with various reproductive phenotypes and semen parameters (Table 13.1). Not much success accompanied microsatellite analysis: genotyping of 12 markers in 110 Noriker stallions did not reveal any correlation between heterozygosity and semen parameters such as sperm count and motility (Aurich et al., 2003).

One of the first candidate gene studies dissected the molecular organization and function of *AEG1* locus, also known as cystein-rich secretory protein 1 or *CRISP1*, which is thought to be involved in sperm-egg fusion (Giese et al., 2002). It was determined that the gene comprises of eight exons and has three possible transcription start sites. Sequencing the exons and their flanking sequences in 16 horses of 9 different breeds identified 17 SNPs of which 3 were exonic and 1 changed the amino acid sequence. However, the role of *AEG1* in stallion fertility remains unclear because no attempts have been made to correlate *AEG1* polymorphism with various reproductive phenotypes.

Another group of studies analyzed association of sequence polymorphism of selected candidate genes with fertility data in a population of reproductively normal Hanoverian warmblood stallions (Giesecke et al., 2009a, 2010a; Giesecke et al., 2009b; Hamann et al., 2007). Reproductive performance of the stallions was evaluated based on pregnancy rate per estrus of mares. The genes *CRISP3* (Hamann et al., 2007), *PRLR* (Giesecke et al., 2009a), *INHBA* (Giesecke et al., 2010a), and *SPATA1* (Giesecke et al., 2009b) were examined and the results suggested association of sequence variants of all four genes with stallion fertility (Table 13.1).

Many more gene loci potentially involved in regulating male fertility have been proposed using extensive data mining from PubMed (http://www.ncbi.nlm.nih.gov /pubmed/) and Web of Science (http://apps.isiknowledge.com/), followed by *in silico* analysis of the most reliable candidate genes (Ogorevc, Dovc, & Kunej, 2010). Considering data from many mammalian species, 835 candidate genes for mammalian male fertility were identified. Of these, 33 genes were proposed as the most promising candidates because they were reported in at least two different species or using two or more approaches. According to this, one of the strongest candidate genes for male fertility in multiple species, including the horse, is prolactin receptor (*PRLR*). A similar data search was recently conducted to identify candidate genes specific for stallion fertility (Giesecke et al., 2010b). The authors proposed 37 genes with associated SNPs and microsatellites for linkage and GWA studies. Surprisingly, only 10 candidate male fertility genes were common between the two data sets, thus further underlining the complexity of the genetic component of male reproduction.

Due to the large number of genes involved, the molecular research of fertility requires genome-wide (GW) approaches. Thus, great expectations are placed on the use of horse whole genome sequence information (Wade et al., 2009), the SNP50/SNP74 beadchips (McCue et al., 2009; Mickelson, personal communication), and GW linkage and association analysis. However, subfertile phenotypes do not facilitate collection of large numbers of samples or pedigrees for statistically sound GWA and linkage studies. Difficulties are also encountered in precise clinical characterization of subfertile phenotypes because in most cases the causes of reduced fertility are unknown or idiopathic (Blanchard, Johnson, & Roser, 2000; Turner & Casas-Dolz, 2002). This is probably why GW molecular research has been initiated only for two conditions: cryptorchidism and impaired acrosomal reaction.

Cryptorchidism

Cryptorchidism, or failure of testes to descend, affects about 8% of live male births in horses, and is more frequent in ponies, draft horses, and breeds like Percheron, American Saddlebred, and Quarter Horse (Hayes, 1986). Cryptorchidism can occur as a single defect or as a part of complex congenital abnormalities accompanying intersex conditions or chromosomal aneuploidies (Constant et al., 1994; Chowdhary & Raudsepp, 2000). In all cases, failure of testes descent affects normal testicular functions and can result in reduced fertility or infertility. Etiology of cryptorchidism is complex, involves interaction of environmental, genetic, and epigenetic factors, and is, as yet, not well understood. Studies in mouse and human suggest complex genetic regulation of the condition involving genes for gubernaculum development and androgen signaling pathway (Ferlin et al., 2008; Foresta et al., 2008). Indeed, a missense mutation in relaxin/insulin-like family peptide receptor 2 (*RXFP2)* was recently associated with cryptorchidism in mouse (Harris et al., 2010). Genetic studies of the condition in horses are limited to a linkage analysis using a Thoroughbred pedigree of 23 cryptorchids and 24 unaffected horses (Diribarne et al., 2009). Gentoyping of 16 microsatellite markers for seven candidate genes, namely *AR, INSL3, RXFP2, CALCA, HOXA10, NR5A1*, and *ESR1*, did not reveal any association with the cryptorchid phenotype. In contrast, recent gene expression profiling showed significant ($P < 0.001$) up-regulation of *RXFP2, FGF10, HOXA10*, and down-regulation of *INSL3, HOXA11*, and *HSD3B1* in cryptorchid compared to normal testes (our unpublished data). It is likely that the observed differences in gene expression reflect the consequences rather than the primary genetic causes of abnormal testicular descent. Research is currently under way to explore the epigenetic landscape of cryptorchidism by comparing global DNA methylation levels of selected candidate genes between normal and cryptorchid testes, as well as to study the possible involvement of genomic copy number variation in the etiology of cryptorchidism (our ongoing research).

Impaired acrosomal reaction

Impaired acrosomal reaction (IAR) is a reproductive disorder characterized by the inability of sperm to perform acrosomal reaction, resulting in reduced fertility. The condition has been found in Thoroughbreds, and currently efforts are made to identify the genomic region(s) associated with the condition using the equine SNP50 (McCue et al., 2009) and the 21,000-element expression oligoarray (Bright et al., 2009). GW association analysis reveals statistically significant association of the subfertile phenotype to ECA13p around 8–9 Mb position (Raudsepp et al., 2011). Notably, some of the differentially expressed genes, as detected by microarray analysis, originate from the same region. Re-sequencing of two most significant candidate genes has identified a few statistically significant ($p<0.05$) amino acid changing SNPs in *FKBP6*, suggesting this as a susceptibility locus for IAR. With the exception of the Y chromosome (see Chapter 5), this is the first report about a direct association of a genomic region with a reproductive disorder in horses.

Functional Genomics of Reproduction

Functional genomics of reproduction studies gene expression profiles in male and female reproductive tissues and the zygote, to identify protein coding genes and other functional elements (i.e., small and long non-coding regulatory RNAs) that are critical for male and female fertility. The studies seek for understanding the functions and interactions of genes during different stages of sexual development, sexual maturation, and reproduction, both in normal and subfertile individuals. One of the main goals is to identify functional signatures that can lead to the development of molecular tests for fertility evaluation. Gene expression studies typically do not require the procurement of hundreds of samples, as is necessary for linkage and GWA analyses, and are therefore more feasible for reproduction-related research. Functional analysis can be carried out on individual genes of interest or across the whole genome. The former is typically based on reverse transcriptase (RT) and quantitative real-time (qRT) PCR (Das et al., 2010a; Das et al., 2010b) or differential display (Turner & Casas-Dolz, 2002). GW analysis, on the other hand, uses whole genome expression arrays (Ing et al., 2004; Das et al., 2010b) or direct mRNA sequencing, also known as RNA-Seq (Coleman et al., 2010; Das et al., 2011c).

Gene expression studies in the mare

Similarly to molecular genetics, functional genetics of mare reproduction lags far behind that of the stallion. A few studies have analyzed the expression of selected genes in ovaries and the uterus. It has been shown that high expression level of steroidogenic acute regulatory (*StAR*) gene in ovaries is required during the early stages of pregnancy (Watson, Thomson, & Howie, 2000; Watson et al., 2004). Ovarian expression of dopamine receptors 1 and 2 (*D1R* and *D2R*) suggests that dopamine can act directly on ovarian tissues through its interaction with dopamine receptors during fertilization (King et al., 2005). Additionally, there are preliminary studies on differential gene expression in blastocysts (Paris et al., 2010) and the endometrium during pregnancy recognition (Bruemmer et al., 2010), and the early stages of pregnancy (Merkl et al., 2010). Despite the important role of ovarian genes in follicular maturation and early embryogenesis in human, mouse, and chicken (Juengel et al., 2004; Elis et al., 2009), no gene expression studies have, as yet, been initiated in equine oocytes.

Functional genomics of the stallion

The majority of functional studies in the stallion have focused on a small number of selected candidate genes comparing their transcription and/or protein expression profiles at different time points in reproductive and somatic tissues. For example, it was determined that *AEG1* is highly expressed in epididymis but not in testis (Giese et al., 2002). Interesting findings were reported for the expression of anti-Müllerian hormone (*AMH*) which induces regression of Müllerian ducts during male fetal development (Ball et al., 2008). Equine *AMH* is highly expressed in Sertoli cells of fetal, neonatal, and prepubertal testes, but not in normal adult testes. Notably, *AMH* is also expressed in cryptorchid testes (in animals up to 3–4 years of age), in Sertoli cell tumors, and male intersex gonads, making it a useful biomarker for the detection of these conditions in the horse (Ball et al., 2008). A novel testis-specific equine serine/threonine kinase, *STK31* was recently discovered in stallion testes (Sabeur et al., 2008). Detailed characterization of the RNA transcript (3.1 kb), the functional domains of the protein (protein kinase and putative RNA-binding domains), and the protein localization in the equatorial segment of testicular spermatozoa suggest that *STK31* may have a role in reorganization of sperm chromatin during spermiogenesis. Systematic studies have also been carried out on horse Y chromosome genes (Paria, 2009) by determining their expression in testis and somatic tissues, and generating full-length cDNA sequences for selected genes of interest (see Chapter 5 for more details).

Promising solutions for functional genomics of reproduction are offered by global gene expression microarrays (Farber & Lusis, 2008) and RNA-Seq (Wang, Gerstein, & Snyder, 2009). The well-annotated 21,000-element oligoarray for the horse (Bright et al., 2009) is currently the most cost-effective tool for functional analysis of tens of thousands of genes from a small number of normal and affected individuals, providing a critical link between the genes and the phenotype. RNA-Seq technology, on the other hand, is the most powerful method for transcriptional profiling of any cell or tissue (Shendure, 2008; Wang et al., 2009; Montgomery et al., 2010). This includes the identification of novel genes, alternative transcripts, and small and long non-coding regulatory RNAs.

Testis transcriptome

Application of microarray technology for transcriptional profiling of stallion testis (Laughlin et al., 2002; Ing et al., 2004) can be considered the formal beginning of functional genomics in horses. In these studies, gene expression profiles in dark and light testicular tissues of three prepubertal colts were analyzed using a human WG cDNA array. Assessment of transcriptional profiles of 9,132 human genes showed that only 93 genes (1.3%) are differentially expressed between the light and dark testicular parenchyma.

Recently, testis transcriptome of reproductively normal stallions was studied using the 21,000-element equine whole genome expression oligoarray (Das et al., 2010b) and direct mRNA sequencing (Coleman et al., 2010). The two studies closely agree about the overall number of testis mRNA transcripts, which is between 11,000 and 15,000 according to microarray analyses and 12,013 according to RNA-Seq. Efforts are currently made to generate a comprehensive catalog of testis transcripts characteristic to a normal stallion. This includes annotations, expression profiles, and putative functions of all detected genes. The catalog will serve as an important foundation for the study of functional changes in the testis of subfertile animals, leading to the identification of genes that are critically involved in stallion fertility. Furthermore, recently optimized protocol for RNA isolation from stallion testis biopsies using a 14 gauge 22 mm deep needle and yielding about 12 micrograms of sequencing quality total RNA (Das et al., 2010a) offers an alternative to excision surgery for sample procurement.

Sperm transcriptome

Despite the achievements of testis transcriptome analysis, difficulties remain regarding the procurement of samples, because conducting excision surgery or biopsy on the testis of prospective breeding animals is potentially harmful. A promising solution is to use sperm instead of testis as the source of RNA. This builds on recent discoveries in humans showing that ejaculated sperm contain a rich repertoire of RNAs, including mRNAs, and small and long non-coding regulatory RNAs (Ostermeier et al., 2002; Krawetz, 2005; Ostermeier et al., 2005a; Ostermeier et al., 2005b; Boerke et al., 2007). Thus, the sperm's contribution to the oocyte is not only the paternal DNA but also a complex host of RNAs. Some sperm mRNAs might be needed for early embryonic development (Boerke et al., 2007) and for protein synthesis during capacitation using mitochondrial-type ribosomes (Gur & Breitbart, 2006, 2007, 2008). Sperm microRNAs might serve as transgenerational signals for heritable epigenetic modifications (Cuzin & Rassoulzadegan, 2010). While the exact functions of sperm RNAs are still a matter of debate, most researchers agree that the majority of sperm mRNAs are historical records of spermatogenesis, and that sperm RNA profiles can be used as a measure of male fertility (Krewetz, 2005; Boerke et al., 2007; Platts et al., 2007). Most importantly, the use of sperm instead of testis as the source of RNA will grant noninvasive and feasible procurement of samples.

Sperm transcriptome studies were recently initiated in stallions. As a first step, total RNA isolation methodology from sperm was optimized for the horse (Das et al., 2010a; Sudderth et al., 2010). This was necessary because sperm typically contain only 10–20 fg of RNA; the ejaculates contain somatic cells and immature sperm that must be removed before RNA extraction; sperm are depleted of ribosomal RNAs, thereby negating the use of rRNA integrity as a marker of quality; and differences between species in sperm attributes and packaging require adjustments in RNA isolation protocols. The optimized protocol allows isolation of about 2 micrograms of hybridization- and sequencing-quality total RNA from 100 million sperm (Das et al., 2010a). Comparison of the RNA quality and quantity with different sperm-processing and storage conditions showed that fresh semen remains the gold standard for sample submission. Processing methods utilizing an extender provide a useful alternative for shipping samples from a distance for evaluation, and flash-frozen semen is the least desirable submission type, but may be utilized if necessary (Sudderth et al., 2010).

Prior to the use of sperm RNA as a diagnostic tool for stallion fertility, the sperm RNA profiles that define the normal fertile male must be determined. Preliminary data about the number, identity, biological functions, and transcriptional profiles of mRNAs and regulatory non-coding RNAs in stallion sperm were recently obtained using microarray hybridization (Das et al., 2010b) and RNA-Seq (Das et al., 2011c).

Microarray analysis of stallion sperm identified more than 6,000 mRNA transcripts, which is roughly half of the number of mRNAs in stallion testis (Coleman et al., 2010; Das et al., 2010b). Notably, about 2.4% of sperm transcripts are sperm-specific and not detected in testis (Table 13.2), while about 200 sperm mRNAs show differential expression between the sperm and the testis (Table 13.3). Many sperm transcripts are associated with membranes and are involved in molecular functions related to G-protein coupled receptor (GPCR) activities. This might be important for fertilization because GPCRs are implicated in sperm chemotaxis, capacitation, and acrosome reaction.

Table 13.2 Twenty selected genes that are present in sperm (signal-to-noise ratio SNR $\geq$ 2) but not in testis mRNA (Das et al., 2010b; Das et al., 2011b).

Gene symbol	Gene name	Average sperm SNR	Likely function
TTLL7	tubulin tyrosine ligase-like family, member 7	49.6595	sperm morphology
C1R	complement component 1, r subcomponent	20.1041	plasma membrane of spermatozoa
C20orf4	chromosome 20 open reading frame 4	17.7532	spermatogenesis and male fertility, plays a role in the organization of the actin cytoskeleton
GAMT	guanidinoacetate N-methyltransferase	8.15417	basic proteins of sperm play a role in the sperm maturation
SFXN2	sideroflexin 2	7.72183	sperm mitochondria NADH- dependent dehydrogenase, present in germ cells where it is required for fertility
LAMC2	laminin, gamma 2	5.17825	embryonic development by interacting with other extracellular matrix
ATG12	ATG12 autophagy related 12 homolog (S. cerevisiae)	5.01901	sperm and egg fusion, required for autophagy
SRP68	signal recognition particle 68 kDa	4.98992	spermatogenesis related gene
SLC24A2	solute carrier family 24 (sodium/potassium/calcium exchanger), member 2	3.94958	solute carrier, ion channel
VAMP5	vesicle-associated membrane protein 5 (myobrevin)	3.86658	vesicle-associated membrane protein
TBCE	tubulin folding cofactor E	3.84092	sperm morphology
HYAL1	hyaluronoglucosaminidase 1	3.47808	including acrosomal reaction/ovum fertilization
TRAPPC6A	trafficking protein particle complex 6A	3.34733	spermatogenesis-associated protein, sperm-egg fusion
ACTR3B	ARP3 actin-related protein 3 homolog B (yeast)	3.21192	plays a role in the organization of the actin cytoskeleton
ARSF	arylsulfatase F	2.52342	outer dense fiber of sperm tails
SPAG11B	sperm associated antigen 11A; sperm associated antigen 11B	2.49075	fertilization, preimplantation, embryo implantation
OR6X1	olfactory receptor, family 6, subfamily X, member 1	2.38217	binding of sperm to zona pellucida
AKAP5	A kinase (PRKA) anchor protein 5	2.16017	prenatal development
REXO1	REX1, RNA exonuclease 1 homolog (S. cerevisiae)	2.05583	outer dense fiber of sperm tails
CASKIN2	CASK interacting protein 2	2.03525	required for spermatogenesis and male fertility

Table 13.3 Selected differentially expressed genes between stallion sperm and testis; FC – fold change (Das et al., 2010b; Das et al., 2011b).

SPERM UP-REGULATED GENES				
Gene symbol	Gene name	Log_2FC	*P*-value	Likely function
CLCA2	chloride channel accessory 2	1.228135	0.000834	accessory proteins necessary for putative ion channel
OR1A1	olfactory receptor, family 1, subfamily A, member 1	2.058675	0.001299	chemosensing and regulation of sperm motility
REEP6	receptor accessory protein 6	1.017873	0.001528	intrinsic sperm *protein* and embryonic development
PXMP3	peroxisomal membrane protein 3	1.615458	0.00318	*required for sperm* passageway through the epididymis
CTTN	cortactin	1.544705	0.006479	sperm morphology, sperm cytoskeleton
COL22A1	collagen, type XXII, alpha 1	1.742836	0.011753	basic protein of *sperm* head cytoskeleton
ARHGAP1	rho GTPase activating protein 1	1.192668	0.014931	sperm motility and adhesions
PADI6	peptidyl arginine deiminase, type VI	1.236651	0.022965	*sperm* capacitation
ELSPBP1	epididymal sperm binding protein 1	1.338774	0.047056	sperm-specific glycolytic enzyme required for sperm motility and male fertility

SPERM DOWN-REGULATED GENES				
Gene symbol	Gene name	Log_2FC	*P*-value	Likely function
UQCRB	similar to ubiquinol-cytochrome c reductase binding protein	−1.34178	3.03E-05	membrane protein expressed in *testis*
CALM3	calmodulin 3 (phosphorylase kinase, delta)	−1.05254	9.30E-05	*testis*-specific protein kinase link the signal transduction
PIAS2	protein inhibitor of activated STAT, 2	−1.37592	0.000162	*testis*-specific androgen receptor
SPA17	sperm autoantigenic protein 17	−2.19213	0.000257	novel cancer-*testis* antigen in multiple myeloma, expressed in somatic ciliated epithelia
DYNLT1	dynein, light chain, Tctex-type 1	−1.08358	0.000339	*t*-complex *testis*-specific protein
TPT1	similar to tumor protein, translationally-controlled 1	−2.11772	0.000728	mitotic growth integrator expressed in testes
GSTA1	glutathione S-transferase alpha 1	−1.02963	0.001349	Required for plasma concentration in *testis* microsomes
UBC	ubiquitin C	−2.32413	0.010426	reciprocal modulators of germ cell apoptosis in cryptorchid *testis*
DNAJA1	DnaJ (Hsp40) homolog, subfamily A, member 1	−1.02871	0.011753	*t*-complex-associated-*testis*-expressed

In another study, total RNA from the sperm of one reproductively normal stallion was sequenced using SOLiD (Applied Biosystems) platform (Das et al., 2011c). Altogether, 65 million 35-base-pair reads were generated and aligned with the horse draft assembly EcuCab2 (Wade et al., 2009). Sequence map locations were determined for 19,257 reads, of which 31% were identified as dbESTs, mRNAs and 33 known microRNAs, while 69% had no annotation in EcuCab2. Like in microarray analysis, many of the transcripts detected by RNA-Seq show direct relevance to a variety of known sperm functions (Table 13.4).

It is anticipated that transcriptional profiling of stallion sperm will lead to the identification of functional signatures for high- and low-fertility stallions, so that noninvasive molecular tests for stallion fertility can be developed. Global transcriptional profiling of stallion sperm will also

Table 13.4 Selected transcripts in stallion sperm as detected by RNA-Seq. (Average coverage corresponds to sequence tag count and shows the expression level) (Das et al., 2011b).

Sequence category	Gene symbol	Gene name	Average coverage	Likely function
mRNA	*MMP1*	matrix metallopeptidase 1 (interstitial collagenase)	11766.17	*Sperm motility*
	TNP2	Transition protein 2 (during histone to protamine replacement)	1730.25	*Sperm* chromatin structure
	PRM1	protamine 1	1730.25	*Sperm* nuclear proteins
	PKM2	pyruvate kinase, muscle	297.97	*Sperm* capacitation
	COL2A1	Collagen, type II, alpha 1	222.58	ion channel
dbEST	*LOC100058894*	similar to deoxynucleotidyltransferase, terminal, interacting protein 2	11420.97	*Morphology* of *sperm*
	LOC100050671	similar to LYR motif containing 4	5070.02	Major *sperm* native protein
	LOC100063232	similar to Glutamine-rich protein 1	4390.4	Fertilization transcriptional activation
	LOC100049794	similar to ubiquitin B	3577.89	*Sperm* acrosomal function, sperm capacitation
	LOC100060535	similar to protein disulfide-isomerase A4 precursor (Protein ERp-72) (ERp72)	3520.84	Embryonic fertilization
MicroRNA	*MIR34B*	microRNA mir-34b	336.58	Building up male *fertility*
	MIR191	microRNA mir-191	160.38	DNA methylation in abnormal *sperm*
	MIR449A	microRNA mir-449a	81.75	Negatively regulates E2F activity reduce sperm count
	MIR16	microRNA mir-16	70.84	Sperm egg fusion, early fertility
	MIR16-2	microRNA mir-16-2	32.96	Post-fertilization, early embryonic development

serve as a valuable non-rodent model for dissecting the genetics of male fertility in humans and other mammals.

Summary and Future Prospects

Genomics of equine reproduction is a challenging area of research that encompasses a wide range, from chromosomes to fine-tuning of gene expression. The stability and structural integrity of chromosomes is certainly a key component of normal sexual development and fertility in horses, and cytogenetic analysis will continue to be an essential part of clinical tests for subfertile animals. While the role of chromosomal abnormalities in equine fertility is well established, the knowledge about the fertility genes, their structure, function, and associated phenotypes is still limited. Furthermore, almost nothing is known about the role of structural genomic rearrangements, copy number variants (CNVs), gene expression regulation, and epigenetic mechanisms in male and female fertility. This is why great expectations are placed on the use of new genomics tools such as SNP beadchips (McCue et al., 2009; J. Mickelson, personal communication), the recently constructed WG tiling array (Qu et al., 2011), the expression microarrays (Bright et al., 2009), various next-generation sequencing platforms, and improved bioinformatics methods. It is anticipated that these tools will take equine genome analysis to levels that allow the detection of submicroscopic genomic rearrangements underlying the currently unknown causes of subfertility, will lead to the discovery of reproduction-related mutations in nuclear and mitochondrial genes, and will identify genetic and epigenetic regulatory mechanisms critical for stallion and mare reproduction. However, all the advanced tools will be of little value without the establishment of collections of phenotypically well-characterized samples. Thus, increased interaction between the geneticists and clinicians remains as one of the most critical factors for any success.

Acknowledgments

The research on the genomics of stallion fertility has been supported by the LINK Endowment and by the American Quarter Horse Research Foundation.

References

Aurich, C., Achmann, R., & Aurich, J. E. (2003). Semen parameters and level of microsatellite heterozygosity in Noriker draught horse stallions. *Theriogenology*, *60*, 371–378.

Ball, B. A., Conley, A. J., Grundy, S. A., Sabeur, K., & Liu, I. K. (2008). Expression of anti-Mullerian hormone (AMH) in the equine testis. *Theriogenology*, *69*, 624–631.

Bannasch, D., Rinaldo, C., Millon, L., Latson, K., Spangler, T. et al. (2007). SRY negative 64,XX intersex phenotype in an American saddlebred horse. *Veterinary Journal*, *173*, 437–439.

Barnett, K. R., Schilling, C., Greenfeld, C. R., Tomic, D., & Flaws, J. A. (2006). Ovarian follicle development and transgenic mouse models. *Human Reproduction Update*, *12*, 537–555.

Blanchard, T. L., Johnson, I., & Roser, A. J. (2000). Increased germ cell loss rates and poor semen quality in stallions with idiopathic testicular degeneration. *Journal of Equine Veterinary Science*, *20*, 263–265.

Blanchard, T. L., & Varner, D. D. (1997). Evaluation breeding soundness in stallions. 5. Predicting potential fertility. *Veterinary Medicine*, *97*, 815–818.

Blue, M. G. (1981). A cytogenetical study of prenatal loss in the mare. *Theriogenology*, *15*, 295–309.

Boerke, A., Dieleman, S. J., & Gadella, B. M. (2007). A possible role for sperm RNA in early embryo development. *Theriogenology*, *68*(Suppl. 1), S147–S155.

Bondy, C. A., & Cheng, C. (2009). Monosomy for the X chromosome. *Chromosome Research*, *17*, 649–658.

Bowling, A. T., Millon, L., & Hughes, J. P. (1987). An update of chromosomal abnormalities in mares. *Journal of Reproductive Fertility*, *35*(Suppl.) 149–155.

Bright, L. A., Burgess, S. C., Chowdhary, B., Swiderski, C. E., & McCarthy, F. M. (2009). Structural and functional-annotation of an equine whole genome oligoarray. *BMC Bioinformatics*, *10*(Suppl. 11), S8.

Brito, L. F. C., P. L., S., Durkin, K., Chowdhary, B. P., Turner, R. M. et al. (2008). Autosomic 27 trisomy in a standardbred colt. *Journal of Equine Veterinary Science*, *28*, 431–436.

Bruemmer, J. E., Bouma, G. J., Hess, A., Hansen, T. R., & Squires, E. L. (2010). Gene expresiion in the equine endometrium during maternal recognition of pregnancy. *Animal Reproduction Science*, *1215*, S285–S287.

Bugno-Poniewierska, M., Jablonska, Z., & Slota, E. (2009). Modification of equine sperm chromatin decondensation method to use fluorescence in situ hybridization (FISH). *Folia of Histochemical Cytobiology*, *47*, 663–666.

Bugno, M., Slota, E., & Koscielny, M. (2007). Karyotype evaluation among young horse populations in Poland. *Schweiz Archiv Tierheilkd*, *149*, 227–232.

Bugno, M., Slota, E., Wieczorek, M., Yang, F., Buczynski, J. et al. (2003). Nonmosaic X trisomy, detected by chromosome painting, in an infertile mare. *Equine Veterinary Journal*, *35*, 209–210.

Buoen, L. C., Zhang, T. Q., Weber, A. F., & Ruth, G. R. (2000). SRY-negative, XX intersex horses: The need for pedigree studies to examine the mode of inheritance of the condition. *Equine Veterinary Journal*, *32*, 78–81.

Carnevale, E. M. (2008). The mare model for follicular maturation and reproductive aging in the woman. *Theriogenology*, *69*, 23–-30.

Carrell, D. T. (2007). The genetics of male infertility in the era of genomics. In D. T. Carrell (ed.), *The genetics of male infertility* (pp. 3–27). Totowa, NJ: Humana Press.

Carrell, D. T. (2008). The clinical implementation of sperm chromosome aneuploidy testing: Pitfalls and promises. *Journal of Andrology*, *29*, 124–133.

Cederroth, C. R., Pitetti, J. L., Papaioannou, M. D., & Nef, S. (2007). Genetic programs that regulate testicular and ovarian development. *Molecular and Cellular Endocrinology*, *265–266*, 3–9.

Chandley, A. C., Jones, R. C., Dott, H. M., Allen, W. R., & Short, R. V. (1974). Meiosis in interspecific equine hybrids. I. The male mule (Equus asinus X E. caballus) and hinny (E. caballus X E. asinus). *Cytogenetics and Cell Genetics*, *13*, 330–341.

Chowdhary, B. P., & Raudsepp, T. (2000). Cytogenetics and physical gene maps. In A. T. Bowling and A. Ruvinsky (eds.), *The genetics of the horse* (pp. 171–242). Wallingford, Oxon: CABI.

Coleman, S. J., Zeng, Z., Wang, K., Luo, S., Khrebtukova, I. et al. (2010). Structural annotation of equine protein-coding genes determined by mRNA sequencing. *Animal Genetics*, *41*(Suppl. 2), 121–130.

Colenbrander, B., Gadella, B. M., & Stout, T. A. (2003). The predictive value of semen analysis in the evaluation of stallion fertility. *Reproduction of Domestic Animals*, *38*, 305–311.

Constant, S. B., Larsen, R. E., Asbury, A. C., Buoen, L. C., & Mayo, M. (1994). XX male syndrome in a cryptorchid stallion. *Journal of the American Veterinary Medical Association*, *205*, 83–85.

Coutinho da Silva, M. A. (2008). When should a mare go for assisted reproduction? *Theriogenology*, *70*, 441–444.

Cuzin, F., & Rassoulzadegan, M. (2010). Non-Mendelian epigenetic heredity: gametic RNAs as epigenetic regulators and transgenerational signals. *Essays in Biochemistry*, *48*, 101–106.

Das, P. J., Lyle, S. K., Beehan, D., Chowdhary, B. P., & Raudsepp, T. (2011a). Cytogenetic and molecular characterization of Y isochromosome in a 63XO/64Xi(Yq) mosaic karyotype of an intersex horse. *Sexual Development*, *6*(1–3), 117–127.

Das, P. J., McCarthy, F., Kachroo, P., Vishnoi, M., Gresham, C. et al. (2011b). Analysis of stallion sperm transcriptome by RNA sequencing. Paper presented at The 9th Dorothy Russell Havemeyer Foundation International Equine Genome Mapping Workshop, July 27–29, Minneapolis, Minnesota.

Das, P. J., McCarthy, F., Vishnoi, M., Li, G., Love, C. C. et al. (2011c). Characterizing stallion sperm transcriptome using RNA-Seq. Paper presented at Plant & Animal Genome XIX, January 15–19, San Diego, CA.

Das, P. J., Paria, N., Gustafson-Seabury, A., Vishnoi, M., Chaki, S. P. et al. (2010a). Total RNA isolation from stallion sperm and testis biopsies. *Theriogenology*, *74*, 1099–1106.

Das, P. J., Vishnoi, M., Kachroo, P., Wang, J., Love, C.C. et al. (2010b). Expression microarray profiling of sperm and testis mRNA of reproductively normal stallions. *Animal Reproduction Science*, *121S*, S175.

de Lorenzi, L., Molteni, L., Zannotti, M., Galli, C., & Parma, P. (2010). X trisomy in a sterile mare. *Equine Veterinary Journal*, *42*, 469–470.

Di Meo, G. P., Neglia, G., Perucatti, A., Genualdo, V., Iannuzzi, A. et al. (2009). Numerical sex chromosome aberrations and abnormal sex development in horse and sheep. *Sexual Development*, *3*, 329–332.

Diribarne, M., Vaiman, A., Péchayre, M., Pailhoux, E., Mata, X. et al. (2009). Polymorphism analysis of microsatellites associated with seven candidate genes for equine cryptorchidism. *Journal of Equine Veterinary Science*, *29*, 37–41.

Ducos, A., Revay, T., Kovacs, A., Hidas, A., Pinton, A. et al. (2008). Cytogenetic screening of livestock populations in Europe: An overview. *Cytogenetic & Genome Research*, *120*, 26–41.

Durkin, K., Raudsepp, T., & Chowdhary, B. P. (2011). Cytogenetic evaluation of the stallion. In A. O. McKinnon, W. E. Vaala & D. D. Varner (eds.), *Equine reproduction* (pp. 1462–1468). London: Wiley-Blackwell.

Elis, S., Blesbois, E., Couty, I., Balzergue, S., Martin-Magniette, M. L. et al. (2009). Identification of germinal disk region derived genes potentially involved in hen fertility. *Molecular Reproduction and Development*, *76*, 1043–1055.

Farber, C. R., & Lusis, A. J. (2008). Integrating global gene expression analysis and genetics. *Advances in Genetics*, *60*, 571–601.

Ferlin, A., Zuccarello, D., Zuccarello, B., Chirico, M. R., Zanon, G.F. et al. (2008). Genetic alterations associated with cryptorchidism. *Journal of the American Medical Association*, *300*, 2271–2276.

Foresta, C., Zuccarello, D., Garolla, A., & Ferlin, A. (2008). Role of hormones, genes, and environment in human cryptorchidism. *Endocrinology Review*, *29*, 560–580.

Giese, A., Jude, R., Kuiper, H., Piumi, F., Schambony, A. et al. (2002). Molecular characterization of the equine AEG1 locus. *Gene*, *292*, 65–72.

Giesecke, K., Hamann, H., Sieme, H., & Distl, O. (2010). Evaluation of prolactin receptor (PRLR) as candidate gene for male fertility in Hanoverian warmblood horses. *Reproduction of Domestic Animals*, *45*, 214–230.

Giesecke, K., Hamann, H., Sieme, H., & Distl, O. (2010a) INHBA-associated markers as candidates for stallion fertility. *Reproduction of Domestic Animals*, *45*, 342–347.

Giesecke, K., Hamann, H., Stock, K. F., Woehlke, A., Sieme, H. et al. (2009b) Evaluation of SPATA1-associated markers for stallion fertility. *Animal Genetics*, *40*, 359–365.

Giesecke, K., Sieme, H., & Distl, O. (2010b). Infertility and candidate gene markers for fertility in stallions: A review. *Veterinary Journal*, *185*, 265–271.

Gur, Y., & Breitbart, H. (2006). Mammalian sperm translate nuclear-encoded proteins by mitochondrial-type ribosomes. *Genes Development*, *20*, 411–416.

Gur, Y., & Breitbart, H. (2007). Protein translation in mammalian sperm. *Society of Reproductive Fertility*, *65*(Suppl.), 391–397.

Gur, Y., & Breitbart, H. (2008). Protein synthesis in sperm: Dialog between mitochondria and cytoplasm. *Molecular and Cell Endocrinology*, *282*, 45–55.

Hamann, H., Jude, R., Sieme, H., Mertens, U., Topfer-Petersen, E. et al. (2007). A polymorphism within the equine CRISP3 gene is associated with stallion fertility in Hanoverian warmblood horses. *Animal Genetics*, *38*, 259–264.

Harris, R. M., Finlayson, C., Weiss, J., Fisher, L., Hurley, L. et al. (2010). A missense mutation in LRR8 of RXFP2 is associated with cryptorchidism. *Mammalian Genome*, *21*, 442–449.

Hayes, H. M. (1986). Epidemiological features of 5009 cases of equine cryptorchism. *Equine Veterinary Journal*, *18*, 467–471.

Haynes, S. E., & Reisner, A. H. (1982). Cytogenetic and DNA analyses of equine abortion. *Cytogenetics and Cell Genetics*, *34*, 204–214.

Herzog, A., Höhn, H., & Hecht, W. C. (1989). Gonosomenmosaik beim maskulinen Pseudohermaphroditismus des Pferdes. *Tierärztliche Praxis*, *17*, 171–175.

Hunt, P. A., & Hassold, T. J. (2008). Human female meiosis: What makes a good egg go bad? *Trends in Genetics*, *24*, 86–93.

Ing, N. H., Laughlin, A. M., Varner, D. D., Welsh, Jr., T. H., Forrest, D. W. et al. (2004). Gene expression in the spermatogenically inactive "dark" and maturing "light" testicular tissues of the prepubertal colt. *Journal of Andrology*, *25*, 535–544.

Juengel, J. L., Bodensteiner, K. J., Heath, D. A., Hudson, N. L., Moeller, C. L. et al. (2004). Physiology of GDF9 and BMP15 signalling molecules. *Animal Reproduction Science*, *82–83*, 447–460.

Juras, R., Raudsepp, T., Das, P. J., Conant E., & Cothran, E. G. (2010). XX/XY blood lymphocyte chimerism in heterosexual dizygotic American Bashkir Curly horse twins. *Journal of Equine Veterinary Science*, *30*, 575–580.

Kakoi, H., Hirota, K., Gawahara, H., Kurosawa, M., & Kuwajima, M. (2005). Genetic diagnosis of sex chromosome aberrations in horses based on parentage test by microsatellite DNA and analysis of X- and Y-linked markers. *Equine Veterinary Journal*, *37*, 143–147.

King, S. S., Campbell, A. G., Dille, E. A., Roser, J. F., Murphy, L. L. et al. (2005). Dopamine receptors in equine ovarian tissues. *Domestic Animal Endocrinology*, *28*, 405–415.

King, W. A. (2008). Chromosome variation in the embryos of domestic animals. *Cytogenetic and Genome Research*, *120*, 81–90.

King, W. A., Bezard, J., Bousquet, D., Palmer, E., & Betteridge, K. J. (1987). The meiotic stage of preovulatory oocytes in mares. *Genome*, *29*, 679–682.

King, W. A., Desjardins, M. R., Xu, K. P., & Bousquet, D. (1990). Chromosome analysis of horse oocytes cutured *in vitro*. *Genetics and Selective Evolution*, *22*, 151–160.

Krawetz, S. A. (2005). Paternal contribution: New insights and future challenges. *National Review of Genetics*, *6*, 633–642.

Kubien, E. M., Pozor, M. A., & Tischner, M. (1993). Clinical, cytogenetic and endocrine evaluation of a horse with a 65,XXY karyotype. *Equine Veterinary Journal*, *25*, 333–335.

Laughlin, A. M., Forrest, D. W., Varner, D. D., Blanchard, T. L., Love, C. C. et al. (2002). DNA microarray analysis of gene expression in testicular tissue of stallions. *Theriogenology*, *58*, 413–415.

Lear, T. L., & Bailey, E. (2008). Equine clinical cytogenetics: The past and future. *Cytogenetic & Genome Research*, *120*, 42–49.

Lear, T. L., Lundquist, J., Zent, W. W., Fishback, Jr., W. D., & Clark, A. (2008). Three autosomal chromosome translocations associated with repeated early embryonic loss (REEL) in the domestic horse (Equus caballus). *Cytogenetic and Genome Research*, *120*, 117–122.

Lechniak, D., Wieczorek, M., & Sosnowski, J. (2002). Low incidence of diploidy among equine oocytes matured in vitro. *Equine Veterinary Journal*, *34*, 738–740.

Leeb, T. (2007). The horse genome project – sequence based insights into male reproductive mechanisms. *Reproduction of Domestic Animals*, *42*(Suppl 2), 45–50.

Leeb, T., Sieme, H., & Topfer-Petersen, E. (2005). Genetic markers for stallion fertility – lessons from humans and mice. *Animal Reproductive Science*, *89*, 21–29.

Lopes, A. M., Burgoyne, P. S., Ojarikre, A., Bauer, J., Sargent, C. A. et al. (2010). Transcriptional changes in response to X chromosome dosage in the mouse: Implications for X inactivation and the molecular basis of Turner Syndrome. *BMC Genomics*, *11*, 82.

Madill, S. (2002). Reproductive considerations: Mare and stallion. *Veterinary Clinical North American Equine Practice*, *18*, 591–619.

Makela, O., Gustavsson, I., & Hollmen, T. (1994). A 64,X,i(Xq) karyotype in a standardbred filly. *Equine Veterinary Journal*, *26*, 251–254.

Makinen, A., Hasegawa, T., Makila, M., & Katila, T. (1999). Infertility in two mares with XY and XXX sex chromosomes. *Equine Veterinary Journal*, *31*, 346–349.

Makinen, A., Katila, T., Andersson, M., & Gustavsson, I. (2000). Two sterile stallions with XXY-syndrome. *Equine Veterinary Journal*, *32*, 358–360.

Massip, K., Bonnet, N., Calgaro, A., Billoux, S., Baquie, V. et al. (2009). Male meiotic segregation analyses of peri- and paracentric inversions in the pig species. *Cytogenetic and Genome Research*, *125*, 117–124.

Matzuk, M. M., & Lamb, D. J. (2002). Genetic dissection of mammalian fertility pathways. *Nature & Cell Biology*, *4*(Suppl), S41–S49.

Matzuk, M. M., & Lamb, D. J. (2008). The biology of infertility: Research advances and clinical challenges. *Nature & Medicine*, *14*, 1197–1213.

McCue, M., Mickelson, J., Bannasch, D., Penedo, C., Bailey, E. et al. (2009). The horse gentrain project: initial evaluation of the Equine SNP50 BeadCips. Paper presented at The 8th Dorothy Russell Havemeyer Foundation International Equine Genome Mapping Workshop, July 22–25, Suffolk, UK.

Merkl, M., Otzdorff, C., Herbach, N., Wanke, R., Wolf, E. et al. (2010). Identification of differentially expressed genes in equine endometrium at day 12 of pregnancy. *Animal Reproduction Science*, *1215*, S288–S290.

Meyers-Wallen, V. N., Hurtgen, J., Schlafer, D., Tulleners, E., Cleland, W. R. et al. (1997). Sry-negative XX true hermaphroditism in a Pasa Fino horse. *Equine Veterinary Journal*, *29*, 404–408.

Montgomery, S. B., Sammeth, M., Gutierrez-Arcelus, M., Lach, R. P., Ingle, C. et al. (2010). Transcriptome genetics using second generation sequencing in a Caucasian population. *Nature*, *464*, 773–777.

Naz, R. K., & Catalano, B. (2010). Gene knockouts that affect female fertility: Novel targets for contraception. *Frontline Bioscience*, *2*(Schol ed.), 1092–1112.

Ogorevc, J., Dovc, P., & Kunej, T. (2011). Comparative genomics approach to identify candidate genetic loci for male fertility. *Reproduction of Domestic Animals*, *46*, 229–239.

Ostermeier, G. C., Dix, D. J., Miller, D., Khatri, P., & Krawetz, S. A. (2002). Spermatozoal RNA profiles of normal fertile men. *Lancet*, *360*, 772–777.

Ostermeier, G. C., Goodrich, R. J., Diamond, M. P., Dix, D. J., & Krawetz, S. A. (2005a). Toward using stable spermatozoal RNAs for prognostic assessment of male factor fertility. *Fertility and Sterility*, *83*, 1687–1694.

Ostermeier, G. C., Goodrich, R. J., Moldenhauer, J. S., Diamond, M. P., & Krawetz, S. A. (2005b). A suite of novel human spermatozoal RNAs. *Journal of Andrology*, *26*, 70–74.

Paget, S., Ducos, A., Mignotte, F., Raymond, I., Pinton, A. et al. (2001). 63,XO/65,XYY mosaicism in a case of equine male pseudohermaphroditism. *Veterinary Record*, *148*, 24–25.

Paria, N. (2009). *Discovery of candidate genes for stallion fertility*. College Station: Texas A&M University, College of Veterinary Medicine.

Paris, D. B. B. P., Kuijk, E. W., Roelen, B. A. J., & Stout, T. A. E. (2010). Stage specific changes in expression of *OCT4, CDX2, NANOC* and *GATA6* mRNA during blastocyst formation in the horse. *Animal Reproduction Science*, *1215*, S291–S292.

Pinton, A., Calgaro, A., Bonnet, N., Ferchaud, S., Billoux, S. et al. (2009). Influence of sex on the meiotic segregation of a t(13;17) Robertsonian translocation: A case study in the pig. *Human Reproduction*, *24*, 2034–2043.

Pinton, A., Ducos, A., & Yerle, M. (2004). Estimation of the proportion of genetically unbalanced spermatozoa in the semen of boars carrying chromosomal rearrangements using FISH on sperm nuclei. *Genetics, Selection, Evolution*, *36*, 123–137.

Pinton, A., Raymond Letron, I., Berland, H. M., Bonnet, N., Calgaro, A. et al. (2008). Meiotic studies in an azoospermic boar carrying a Y;14 translocation. *Cytogenetic and Genome Research*, *120*, 106–111.

Piprek, R. P. (2009). Genetic mechanisms underlying male sex determination in mammals. *Journal of Applied Genetics*, *50*, 347–360.

Platts, A. E., Dix, D. J., Chemes, H. E., Thompson, K. E., Goodrich, R. et al. (2007). Success and failure in human spermatogenesis as revealed by teratozoospermic RNAs. *Human Molecular Genetics*, *16*, 763–773.

Power, M. M., Gustavsson, I., Switonski, M., & Ploen, L. (1992). Synaptonemal complex analysis of an autosomal trisomy in the horse. *Cytogenetics and Cell Genetics*, *61*, 202–207.

Qu, Z., Chowdhary, B. P., Raudsepp, T., and Adelson, D. L. (2011). Design of an equine whole genome tiling array. Paper presented at Plant & Animal Genome XIX, January 15–19, San Diego, CA.

Rambags, B. P., Krijtenburg, P. J., Drie, H. F., Lazzari, G., Galli, C. et al. (2005). Numerical chromosomal abnormalities in equine embryos produced in vivo and in vitro. *Molecular Reproduction and Development*, *72*, 77–87.

Raudsepp, T., Durkin, K., Lear, T., Das, P. J., Avila, F. et al. (2010). Molecular heterogeneity of XY sex reversal in the horse. *Animal Genetics Special Issue on Horse Genetics and Genomics*.

Raudsepp, T., McCue, M. E., Das, P. J., Vishnoi, M., Fritz, K. L. et al. (2011). Genome-wide association studies and gene expression profiling in Thoroughbred stallions with acrosomal dysfunction. Paper presented at Plant & Animal Genome XIX, January 15–19, San Diego, CA.

Romagnano, A., Richer, C. L., King, W. A., & Betteridge, K. J. (1987). Analysis of X-chromosome inactivation in horse embryos. *Journal of Reproductive Fertility*, *35*(Suppl.), 353–361.

Rothfels, K. H., Alexrad, A. A., Siminovitch, L., & McCulloch Parker, R. C. (1959). The origin of altered cell lines from mouse, monkey and man, as indicated by chromosome and transplantation studies. *Proceedings of Canadian Cancer Research Conference*, *3*, 189–214.

Rubes, J., Vozdova, M., & Kubickova, S. (1999). Aneuploidy in pig sperm: Multicolor fluorescence in situ hybridization using probes for chromosomes 1, 10, and Y. *Cytogenetics and Cell Genetics*, *85*, 200–204.

Sabeur, K., Ball, B. A., Corbin, C. J., & Conley, A. (2008). Characterization of a novel, testis-specific equine serine/threonine kinase. *Molecular Reproduction and Development*, *75*, 867–873.

Safronova, L. D., & Pimentova, T. I. (1988). The karyotyping of bull (Bos taurus) and of horse (Equus caballus) on the basis of the synaptonemal complexes. *Genetika*, *24*, 708–714.

Scott, I. S., & Long, S. E. (1980). An examination of chromosomes in the stallion (Equus caballus) during meiosis. *Cytogenetics and Cell Genetics*, *26*, 7–13.

Shendure, J. (2008). The beginning of the end for microarrays? *Natural Methods*, *5*, 585–587.

Short, R. V., Chandley, A. C., Jones, R. C., & Allen, W. R. (1974). Meiosis in interspecific equine hybrids. II. The przewalski horse/domestic horse hybrid. *Cytogenetics and Cell Genetics*, *13*, 465–478.

Sudderth, A. K., Das, P. J., Varner, D. D., & Raudsepp, T. (2010). Determination of optimal semen processing methods for total RNA isolation and sperm genomic analysis. *Animal Reproduction Science*, *121*(Suppl.), S149–S150.

Templado, C., Donate, A., Giraldo, J., Bosch, M., & Estop, A. (2011). Advanced age increases chromosome structural abnormalities in human spermatozoa. *European Journal of Human Genetics*, *19*, 145–151.

Turner, R. M., & Casas-Dolz, R. (2002). Differential gene expression in stallions with idiopathic testicular degeneration. *Theriogenology*, *58*, 421–434.

Varner, D. D. (2008). Developments in stallion semen evaluation. *Theriogenology*, *70*, 448–462.

Varner, D. D., & Johnson, L. (2007). From a Sperm's eye view – revisiting our perception of this intriguing cell. *AAEP Proceedings*, *53*, 104–177.

Vaughan, L., Schofield, W., & Ennis, S. (2001). SRY-negative XX sex reversal in a pony: A case report. *Theriogenology*, *55*, 1051–1057.

Villagomez, D. A., Parma, P., Radi, O., Di Meo, G., Pinton, A. et al. (2009). Classical and molecular cytogenetics of disorders of sex development in domestic animals. *Cytogenetic and Genome Research*, *126*, 110–131.

Villagomez, D. A., & Pinton, A. (2008). Chromosomal abnormalities, meiotic behavior and fertility in domestic animals. *Cytogenetic and Genome Research*, *120*, 69–80.

von Nett, P., Jung, H. R., & Stranzinger, G. (1996). Neue Erkenntnisse über die Chromosomenpaarung in der Meiose. *Deutsche Tierärztl Wschr*, *103*, 369–448.

Wade, C. M., Giulotto, E., Sigurdsson, S., Zoli, M., Gnerre, S. et al. (2009). Genome sequence, comparative analysis, and population genetics of the domestic horse. *Science*, *326*, 865–867.

Wang, Z., Gerstein, M., & Snyder, M. (2009). RNA-Seq: A revolutionary tool for transcriptomics. *National Review of Genetics*, *10*, 57–63.

Watson, E. D., Bae, S. E., Steele, M., Thomassen, R., Pedersen, H. G. et al. (2004). Expression of messenger ribonucleic acid encoding for steroidogenic acute regulatory protein and enzymes, and luteinizing hormone receptor during the spring transitional season in equine follicles. *Domestic Animal Endocrinology*, *26*, 215–230.

Watson, E. D., Thomson, S. R., & Howie, A. F. (2000). Detection of steroidogenic acute regulatory protein in equine ovaries. *Journal of Reproduction & Fertility*, *119*, 187–192.

Wilhelm, D., Palmer, S., & Koopman, P. (2007). Sex determination and gonadal development in mammals. *Physiology Review*, *87*, 1–28.

Woods, J., Rigby, S., Brinsko, S., Stephens, R., Varner, D. et al. (2000). Effect of intrauterine treatment with prostaglandin E2 prior to insemination of mares in the uterine horn or body. *Theriogenology*, *53*, 1827–1836.

Zenteno-Ruiz, J. C., Kofman-Alfaro, S., & Mendez, J. P. (2001). 46,XX sex reversal. *Archives of Medical Research*, *32*, 559–566.

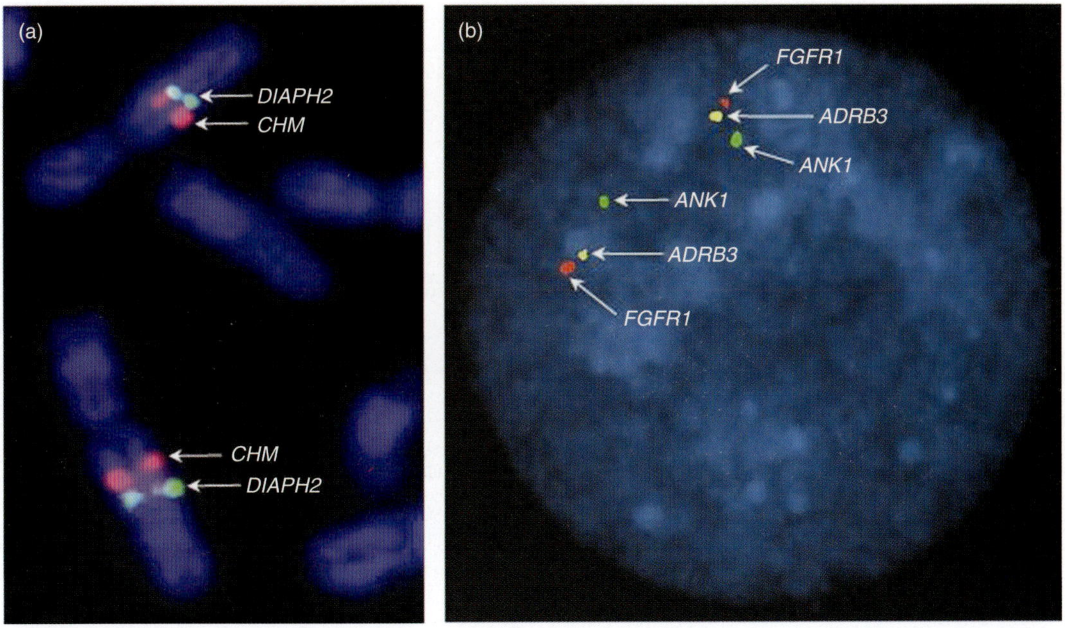

Figure 3.2 Multicolor FISH: (a) dual-color FISH to order two genes on ECAX in metaphase; (b) three-color FISH to order three closely located loci on ECA27 in interphase.

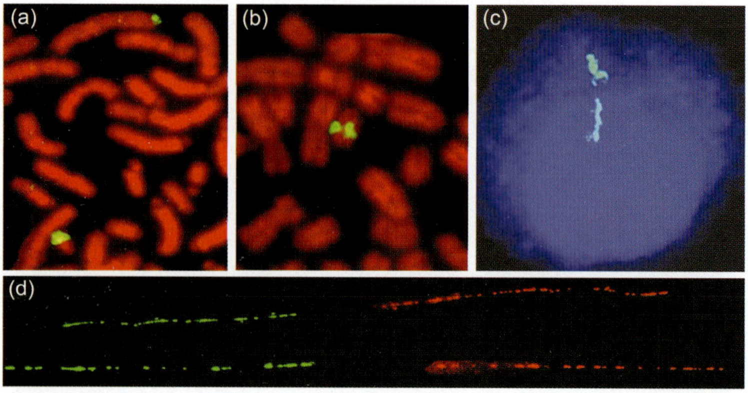

Figure 4.3 Examples of metaphase (a, b), interphase (c), and fiber FISH (d) to study the horse Y chromosome using (a) a pool of ECAY and PAR BAC clones; (b) BAC clone containing *RBMY, SRY, TSPY*, and *YIR2*; (c) BAC clone containing novel multicopy genes *ETY4* and *ETSTY5*; and (d) BAC clones containing *KDM5D* (green) and *SRY* (red).

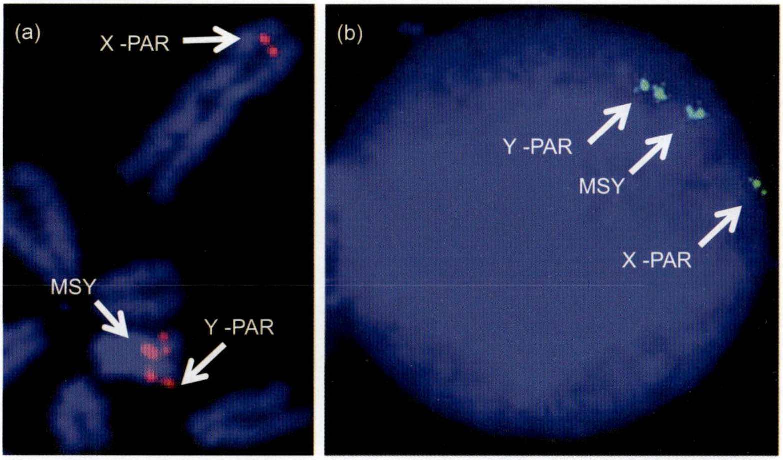

Figure 4.4 FISH on metaphase (a) and interphase (b) chromosomes showing the duplication between the proximal part of MSY and the PAR.

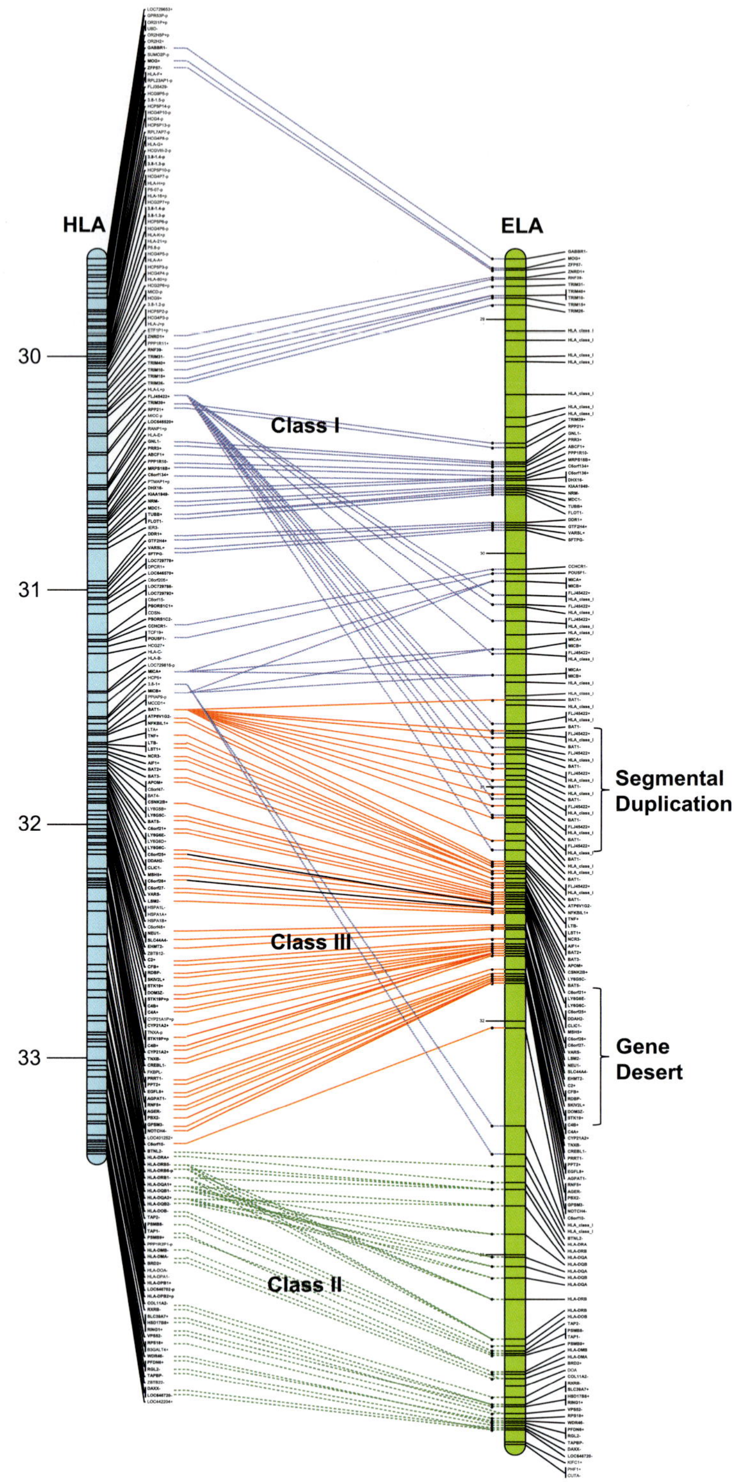

Figure 5.1 Alignment of genes between HLA and ELA based on best BLAST alignment of the horse sequence to the human sequence showing the locations of the Segmental Duplication and the Gene Desert. Class I and class II orthologues are difficult to identify across species, as evidenced by the several intersecting lines between genes in the class I and the class II regions.

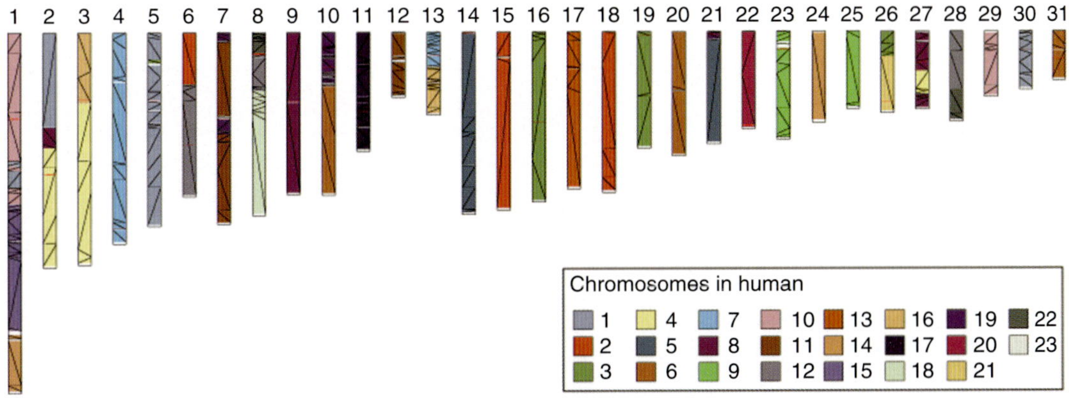

Figure 6.2 Conserved synteny between horse and human. Horse chromosomes are colored by corresponding human chromosomes.

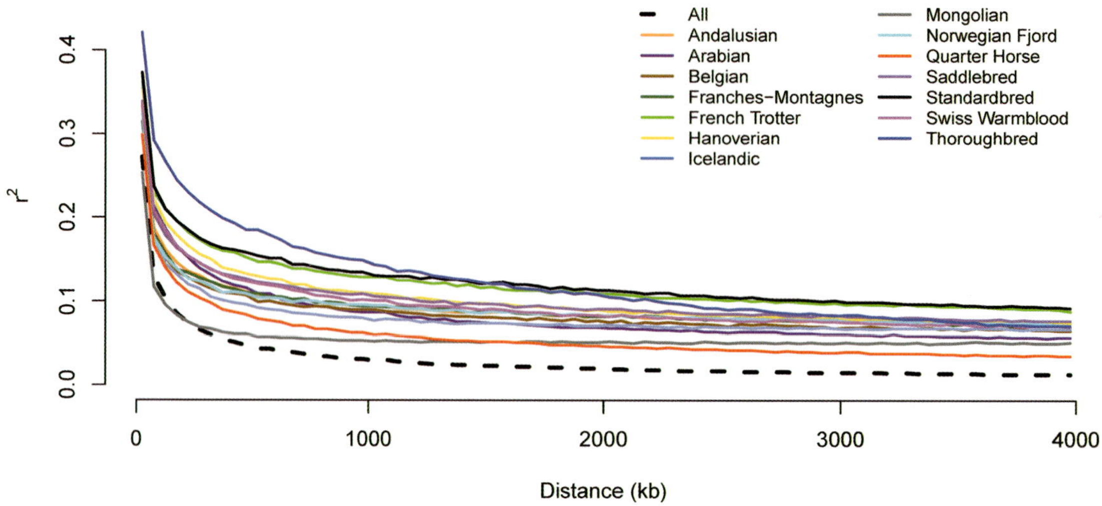

Figure 7.1 Decline in genome-wide linkage disequilibrium across and within breeds. Genome-wide linkage disequilibrium (LD) was estimated both within a given breed and across all breeds by calculating r^2 values between all pairs of SNPs with inter-SNP distances of less than 4 Mb.

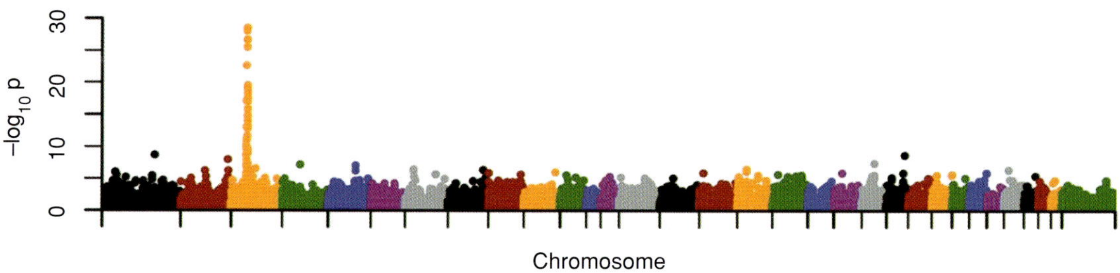

Figure 7.2 Allele association analysis for chestnut coat color loci. Mapping of the chestnut locus across breeds based on coat color phenotype inferred from 9 genotyped loci.

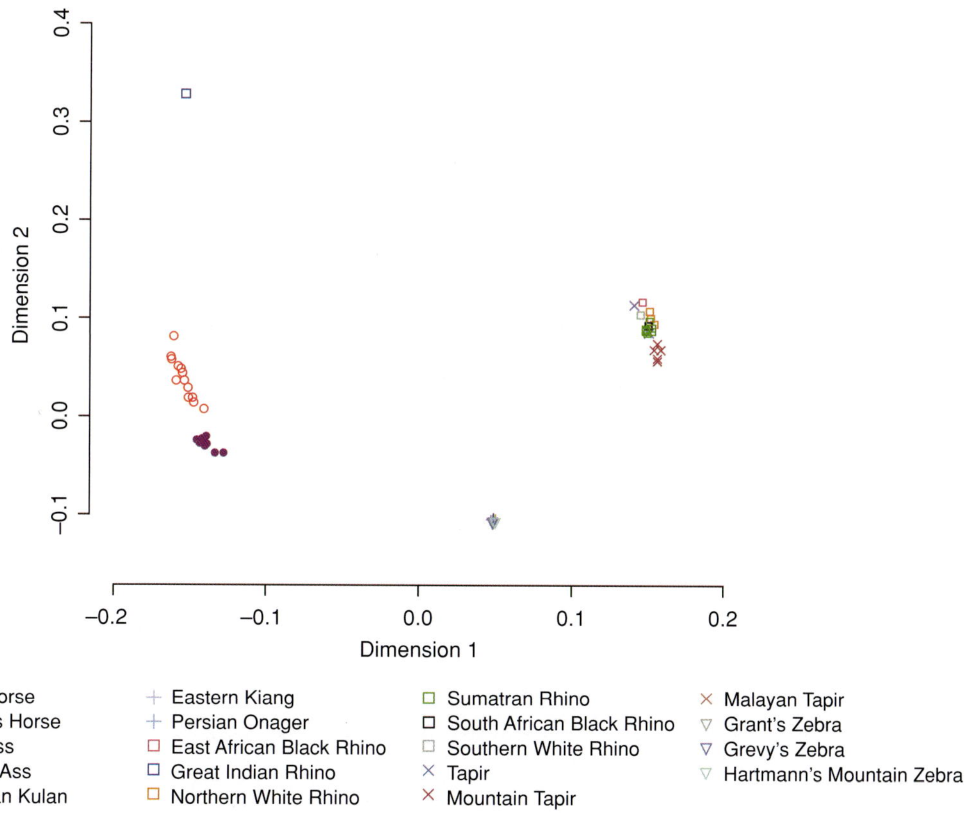

Figure 7.3 Multidimensional scaling with 14 domestic horses and all other extant *Perissodactyla*. Metric multidimensional scaling analysis of pair-wise genetic distance (up to 6 dimensions) was used as described in Materials and Methods to identify relationships between one individual drawn at random from each domestic horse breed and the individuals from each of the 17 other *Perissodactyl* species.

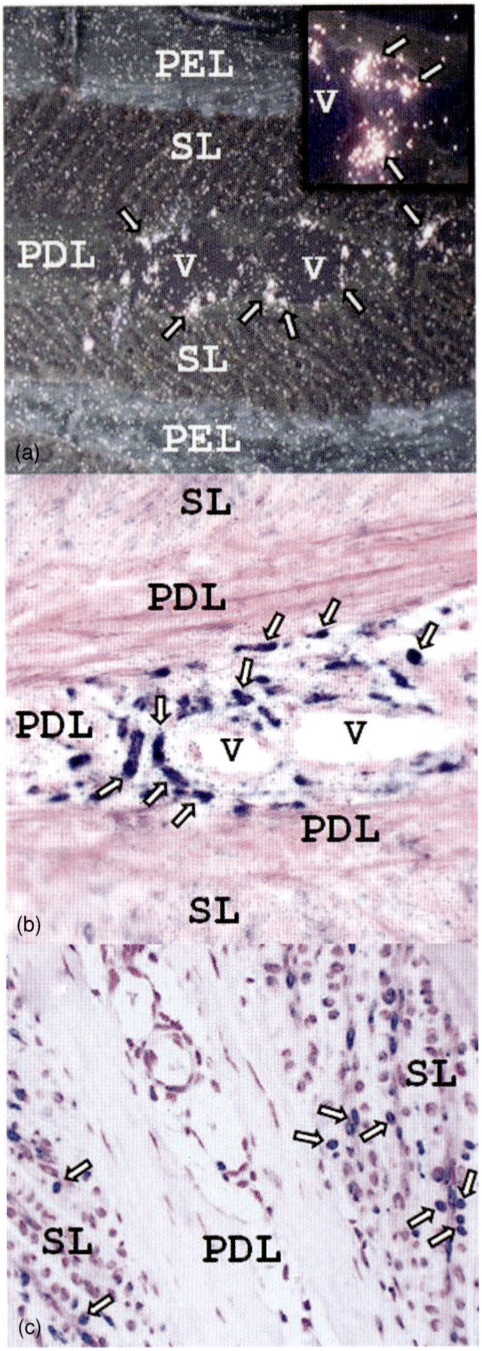

Figure 16.1 In situ hybridization of laminar sections for cytokines and chemokines in BWE-treated animals; (a) Note in the dark field micrograph (including magnified inset) the cells (white arrows) expressing IL-1β signal (white silver grains in photoemulsion exposed by ^{35}S-labeled probe) around venules (V) in the primary dermal laminae (PDL) at a developmental stage in the BWE model; (b) The same stage in the BWE model. Note in the light field micrograph of a non-isotopic ISH study the IL-6 positive cells (blue cells indicated by arrows) situated in the PDL around the venules (V); (c) Note the CXCL1 signal (blue stain) in the non-isotopic ISH study in the laminar basal epithelial cells (arrows); cells in the vascular wall (likely adhered leukocytes) in the primary dermal laminae and in secondary dermal laminae (with a morphology comparable to macrophages) are also positive. PEL = primary epidermal laminae, SL = interdigitating secondary dermal and epidermal laminae.

14 Genetics of equine neurologic disease

Carrie J. Finno and Monica Aleman

Introduction

Horses are used extensively for pleasure riding, competitive activities such as barrel racing, cutting, reining, dressage, and jumping, and for high-level performance activities such as racing, steeplechase, and upper-level eventing. The neurologic system plays an integral role in these functions, and subtle neurologic dysfunction may be readily noticed by riders and trainers of performance animals. In this chapter, we focus on diseases with known genetic mutations, including lethal white foal syndrome and lavender foal syndrome and other equine neurologic diseases with an underlying genetic etiology that are currently being investigated. As tools to perform whole genome interrogation become readily available in the field of equine genomics, the importance of accurate and complete phenotyping cannot be overestimated. For each disease, the clinical presentation including typical signalment, age of onset, and disease progression, diagnostic tools, treatment options, pathologic findings, and most recent information on the status of the genetic investigation is provided, in addition to information regarding comparative diseases in other species.

Neurologic Disorders with Known Mutations

Ileocolonic aganglionosis (lethal white foal syndrome)

Ileocolonic aganglionosis, or lethal white foal syndrome (LWFS), affects foals of both sexes born to American Paint horse, Quarter Horse, and Thoroughbred parents. Clinically, foals with LFWS are born with an all-white or nearly all-white coat and present with intestinal tract abnormalities that result in colic that is nonresponsive to analgesics between 5 and 20 hours of birth (Hultgren, 1982; Vonderfecht, Bowling, & Cohen, 1983). Fecal material is not passed in these foals despite enema administration, and the abdomen becomes progressively distended. The intestinal abnormalities are due to the complete absence of the intrinsic myenteric plexus in the distal small intestine, cecum, and entire colon, with the ileum most severely affected (Hultgren, 1982; Vonderfecht et al., 1983; McCabe et al., 1990). The white coat of affected foals is due to the absence of melanocytes from the skin; embryologically, both melanocytes and myenteric ganglia cells are of neural crest origin (Hultgren, 1982). If affected foals are not suffering from sepsis, hypovolemia, or any other concurrent disorder, the neurological evaluation in these foals is normal with the exception of lack

Equine Genomics, First Edition. Edited by Bhanu P. Chowdhary.

of behavioral response to auditory stimuli (Magdesian et al., 2009). Most foals have blue irides. In a recent study utilizing brainstem auditory-evoked response testing, three foals with confirmed LWFS had bilateral blue irides and complete bilateral sensorineural hearing loss (Magdesian et al., 2009).

Diagnosis of LWFS may include abdominal ultrasonography to assess intestinal motility, plain abdominal radiographs, and enteral or rectal contrast radiography to confirm ileus, reduced diameter of the small colon and rectum, and intestinal distension. Affected foals will be unable to defecate the rectally administered barium contrast material. These imaging alterations, in conjunction with the phenotype of these foals and their parents and progressive signs of abdominal discomfort, allow for a presumptive diagnosis of LWFS. It is important to mention that not all white foals with abdominal pain may have LWFS, and ruling out other causes of disease is essential. White foals born to Paint horse parents may not necessarily possess the mutation and should be genetically tested. There is no treatment available for LWFS and the prognosis is invariably grave. Post-mortem findings include constrictions of the distal intestine and aganglionosis of the distal small intestine and entire large and small colon. Ganglion cells and myenteric plexuses are present and normal in appearance from the mid-portion of the small intestine proximally. Sections of skin will have absence of pigmented cells.

LWFS is inherited as an autosomal recessive trait. In the heterozygous state, the gene for frame overo patterning yields the frame overo phenotype while in the homozygous state, the gene is associated with LWFS. Breeding between frame overos can produce foals that are all white or nearly all white and die shortly after birth of severe intestinal blockage. The highest incidence of heterozygotes (>94%) are frame overo, highly white calico overo, and frame blend overo, while the lowest incidence of heterozygotes (<21%) are tobiano, sabino, minimally blend overo, and breeding-stock solid (Santschi et al., 2001).

The genetic defect responsible for LWFS is a dinucleotide substitution that causes a missense mutation and results in a isoleucine/lysine substitution at codon 118 of the endothelin receptor B gene (*EDNRB*), which is located on chromosome 17 (Metallinos, Bowling, & Rine, 1998; Santschi et al., 1998). An affected foal is homozygous *Lys* (*Lys* 118 / *Lys* 118) while carriers are heterozygous (*Ile* 118 / *Lys* 118) (Metallinos et al., 1998; Santschi et al., 1998). As the heterozygous phenotype has variable penetrance, heterozygotes may not have the frame overo pattern (Metallinos et al., 1998). The equine *Ile* to *Lys EDNRB* substitution is located in transmembrane domain 1 of a seven transmembrane domain G-protein coupled receptor for the endothelins. A relationship has been found between the *EDNRB* mutation and deafness in the American Paint horse; however, there is no conclusive evidence at this time that the mutation is causative for deafness (see section on deafness in the American Paint horse).

Lethal white foal syndrome is an equine variant of Hirschsprung disease, where children are born with aganglionic megacolon from the absence of intramural ganglion cells in the myenteric and submucosal plexues within the distal intestine. There are two major signaling pathways involved in the formation of the enteric nervous system: the Rearranged during Transfection (RET)-cell line-derived neurotrophic factor and the endothelin 3 (EDN3)-EDNRB system (Amiel et al., 2008). Missense and nonsense mutations in *RET* and *EDNRB* have been found in patients with Hirschsprung disease (Auricchio et al., 1996; Chakravarti, 1996; Amiel et al., 2008). Recently, there is evidence that *EDN3* might be considered a common susceptibility gene for sporadic Hirschsprung disease (Sanchez-Mejias et al., 2010). Currently, there have been no cases of equine ileocolonic aganglionosis due to mutations in *EDN3* or *RET*.

A genetic test for LWFS syndrome is available through the Veterinary Genetics Laboratory at the University of California at Davis (www.vgl.ucdavis.edu). This test is useful in confirming suspect LWFS cases and determining carrier status, especially in horses that cannot be recognized by the overt phenotype (i.e., tobiano Paint horses, Thoroughbreds, Quarter Horses).

Lavender foal syndrome

Lavender foal syndrome (LFS) – also known as dilute lethal, lethal lavender foal syndrome, or coat color dilatation lethal – is a tetany syndrome of Egyptian lineage Arabian neonates. Foals have a silver, lavender, or pewter/chestnut hue to their coat, thereby giving the condition its name. From birth, foals demonstrate tetantic episodes with opisthotonus, paddling, and extensor rigidity. Affected foals remain in lateral recumbency and cannot become sternal or stand without assistance. Suckling may be strong or absent (Pascoe & Knottenbelt, 1999; Fanelli, 2005). Direct and indirect pupillary light responses are generally present. Ventral strabismus and nystagmus may be observed (Bowling, 1996; Page et al., 2006). Reflexes including the cervicofacial, cutaneous truni, and flexors may be present but result in an exaggerated response characterized by increased paddling and rigidity. It is not fully clear if the paddling episodes are due to seizure activity (Page et al., 2006) or attempts by the foal to stand (Fanelli, 2005), as electroencephalograms have not been performed. Diazepam and phenobarbital provided temporary relief in some cases; however, whether this was due to sedative or anticonvulsant effects of these drugs is unknown. Foals affected with LFS can defecate and urinate normally.

Diagnostic workup often reveals no specific abnormalities on routine clinicopathologic analysis for blood, serum, urine, or cerebrospinal fluid unless other conditions (i.e., sepsis) are present. Skull radiographs reveal no abnormalities. Cases of LFS need to be distinguished from cases of perinatal asphyxia syndrome (PAS), as stabilization of neurologic signs within 72 hours after birth followed by gradual improvement with up to an 80% recovery rate may be expected in foals affected by PAS (Vaala, 2003), whereas LFS is a fatal condition within 24–72 hours. Key features that distinguish LFS from PAS are typical signalment and coat color, severe neurological deficits at birth, and lack of response to therapy.

Treatment in affected cases has been aimed at controlling possible seizures through the use of anticonvulsant drugs, such as diazepam and phenobarbital, and providing supportive care with intravenous fluid therapy, prophylactic antibiotics, gastroprotectants, and enteral feeding through a nasogastric tube. The clinical condition of foals with LFS will continue to deteriorate despite treatment and necessitate euthanasia. Post-mortem examination and histopathological findings on affected foals may reveal no central nervous system lesions (Fanelli, 2005; Page et al., 2006). In other cases, an anomalous choroid plexus (Bowling, 1996) and central nervous system vacuolization of neurons (Madigan, 1997) have been described. Skin biopsies may reveal no notable pathology (Page et al., 2006) or abnormal clumping of melanin (Fanelli, 2005).

LFS is inherited as an autosomal recessive trait. LFS is the first equine disease to have the molecular genetic mutation mapped utilizing a single nucleotide polymorphism (SNP)-based genome-wide association study approach with the EquineSNP50 Beadchip (Illumina, San Diego, CA) (Brooks et al., 2010). Six LFS-affected and 30 healthy relatives were genotyped for the 56,402 SNP markers available on the array and 14 highly significant SNPs were identified on ECA1 spanning a region of 10.5 Mb. Homozygosity mapping further narrowed the region to a 1.6 Mb block that was homozygous in all six affected horses and heterozygous in obligate carriers. Ten genes fell into this region, including one of the two candidate genes for the disease, myosin Va (*MYO5A*). Sequencing of the 39 exons of *MYO5A* revealed a single base pair deletion in exon 30, causing the reading frame to shift and creating a premature stop codon in the translation of exon 30. Preliminary carrier frequencies were estimated to be 10.3% in Arabian (58 horses tested) and 1.8% in non-Arabians (56 tested). Although expression data is not available at this time, based on the known function of the gene in other species and high conservation across the mutated region, it is hypothesized that the frameshift mutation likely impairs binding of myosin Va to organelles with appropriate receptors, which leads to the loss of vesicle traffic and interferes with the function of melanocytes and neurons (Brooks et al., 2010).

A possible relationship between LFS and juvenile idiopathic epilepsy has been postulated as both conditions occur in Arabian foals of Egyptian breeding, and overlap between mares producing both LFS and juvenile idiopathic epilepsy has been reported (Fanelli, 2005). At this time, there have been no documented cases of juvenile idiopathic epilepsy that have been tested for the LFS mutation (Brooks et al., 2010).

LFS appears to be an equine variant of Griscelli syndrome, although the mutations of *MYO5A* that have been associated with Griscelli syndrome are often due to changes in a single amino acid rather than loss of a large portion of the transcript (Sanal et al., 2002). The clinical signs observed in foals with LFS are more severe than those observed in humans with mutations in the *MYO5A* gene. Humans with Griscelli disease may have mutations in *MYO5A* or *Rab27a*. Neurologic dysfunction, including hypotonicity and mental retardation from infancy, is observed in patients suffering from Griscelli disease due to mutations in *MYO5A*. Mutations in *Rab27a* lead to a different form of Griscelli disease, characterized by hemophagocytic syndrome, which is an uncontrolled process of lymphocyte and macrophage activation and proliferation, without evidence of neurologic disease (Sanal et al., 2002).

A genetic test for LFS is available through the Animal Health Diagnostic Center at Cornell University (http://ahdc.vet.cornell.edu/pdf/lavendarfoal.pdf). This test can be used to definitively diagnose LFS and should be used in Arabian horses of Egyptian lineage to determine carrier status.

Cerebellar abiotrophy

Well-defined syndromes of familial cerebellar abiotrophy (CA) have been reported in the Oldenburg (Koch & Fischer, 1950), Gottland pony (Bjorck et al., 1973), Eriskay pony (Hahn, Mayhew, & MacKay, 1999), and, most frequently, in the Arabian (Sponseller, 1967; Palmer et al., 1973; Turner Beatty et al., 1985; Blanco et al., 2006). In Arabians, there appears to be no sex predilection and clinical signs of cerebellar disease, including intention tremors of the head, ataxia, spasticity, dysmetria, and an absent or inconsistent menace response generally appear between six weeks and six months of age. Dysmetria is typically more pronounced in the thoracic limbs than in the pelvic limbs and conscious proprioceptive deficits are observed as a base-wide or base-narrow stance.

Definitive diagnosis of affected cases requires histopathological examination of the cerebellum; however, the presence of cerebellar ataxia and intention tremors in an Arabian under six months of age is considered pathognomonic but not exclusive for the disease (Palmer et al., 1973). There is currently no effective treatment for CA; however, affected cases may stabilize with time as there may be a learned accommodation as described in humans with cerebellar disease (DeBowes, Leipold, & Turner-Beatty, 1987). Gross post-mortem examination typically reveals a cerebellum of normal size. Histopathological evaluation of the cerebellum demonstrates degenerative Purkinje cells that have undergone apoptosis (Blanco et al., 2006) along with variable gliosis and thinning of the granular and molecular layers. Mineralized cell bodies have been reported in the thalamus of affected cases (Turner Beatty et al., 1985).

Cerebellar abiotrophy in Arabians is inherited as an autosomal recessive trait (Brault et al., 2011b). In 2010, using linkage analysis with 335 genome-wide microsatellite markers upon four paternal half-sibling families segregating for cerebellar abiotrophy, and subsequent homozygosity mapping, CA in Arabians was mapped to a region on equine chromosome 2 and a putative mutation was identified (Brault et al., 2011a). The mutation is an SNP (ECA2:13074277) located in exon 4 of

TOE1 (*Target of EGR1*) and approximately 1,200 bp upstream from *MUTYH (MutY Homolog)*. As the non-synonymous missense mutation does not appear to cause a deleterious amino acid change (arginine to histidine), it is unclear if the mutation affects the expression of *TOE1*, a gene expressed in the central nervous system and involved in cell cycle regulation by inducing the expression of TGFβ and p21 (De Belle et al., 2003; Sperandio et al., 2009). It is postulated that the mutation affects a binding site for the transcription factor, GATA2, which could affect *MUTYH* expression. *MUTYH* is highly expressed in the cerebellum. *MUTYH* encodes for a DNA glycosylase involved in post-replicative repair in the nuclei of rapidly proliferating Purkinje cells, as well as DNA repair due to oxidative damage of mitochondrial genomes (Gu et al., 2002; Lee et al., 2002). Preliminary work evaluating the expression of *MUTYH* in the cerebellar tissues of affected and control horses suggests that *MUTYH* is down-regulated in affected horses (Brault et al., 2011a). However, further studies are necessary utilizing additional affected cases and age-matched controls for the analysis, as *MUTYH* demonstrates age-dependent expression.

The available genetic test (www.vgl.ucdavis.edu) is based on the conserved haplotype (50 kb) identified in affected Arabians as compared to controls. This haplotype block contains 22 mutations, with only the SNP described above (ECA2:13074277) identified solely in the Arabian breed. A carrier rate of 19.7% was described based on this genetic test; however, these samples were most likely biased as they were submitted to diagnostic laboratory specifically for CA testing. The affected genotype was identified, to date, in nine apparently unaffected Arabian horses, which researchers attribute to variable expression of the disease due to varying amounts of DNA damage in the Purkinje cells of the cerebellum or the existence of a potential suppressor mutation (Brault et al., 2011b). Additional work is necessary to document that this missense mutation is causative for CA in Arabians and to investigate the underlying genetic basis for CA in other breeds.

Neurologic Disorders with Ongoing Mapping Efforts

Neuroaxonal dystrophy/Equine degenerative myeloencephalopathy

Neuroaxonal dystrophy (NAD) is a degenerative disease of selected neurons and their associated axonal processes. In horses, NAD is considered the underlying basis of equine degenerative myeloencephalopathy (EDM), and there is a high likelihood that the pathophysiology of the two diseases is similar (Miller & Collatos, 1997). The disease has been described in many different breeds, including Thoroughbreds, Standardbreds, Quarter Horses, Arabians, Haflingers, Appaloosa horses, Morgans, Paso Finos, Tennesse Walking Horses, Paints, a Welsh pony, Lusitanos, and various mixed breeds (Mayhew et al., 1978a; Liu et al., 1983; Beech & Haskins, 1987; Mayhew et al., 1987; Baumgartner, Frese, & Elmadfa, 1990; Dill et al., 1990; Blythe et al., 1991b; Cummings et al., 1995; Adams et al., 1996; Siso, Ferrer, & Pumarola, 2003; Gandini et al., 2004; Aleman et al., 2011). An inherited basis has been suggested in most of the cases reported and supported by pedigree analysis in Quarter Horses (Aleman et al., 2011) and prospective breeding trials in Morgans (Beech & Haskins, 1987) and Appaloosa horses (Blythe et al., 1991b). There is no sex predilection, and age of onset ranges from birth up to 36 months (Mayhew et al., 1978b; Beech & Haskins, 1987; Adams et al., 1996).

Clinical signs of NAD/EDM include a symmetric ataxia that is often more severe in the pelvic limbs than thoracic limbs, varying degrees of weakness, abnormal base-wide stance at rest (Figure 14.1a), and conscious proprioceptive deficits. In some reports, hyporeflexia of the cervicofacial and cutaneous trunci is described in addition to an absent laryngeal adductor reflex (Gandini

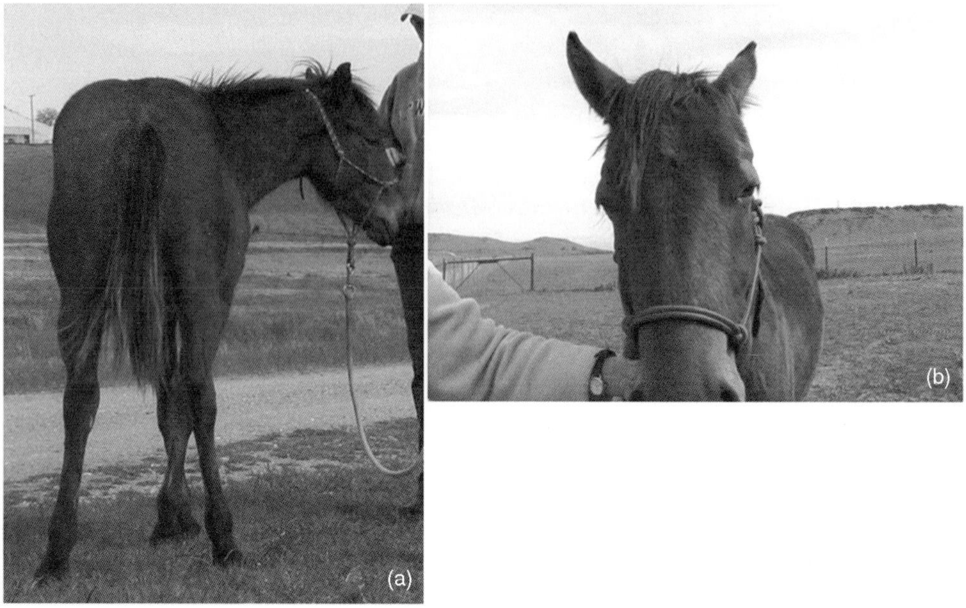

Figure 14.1 (a) Proprioceptive deficits and (b) dull mentation characteristic of the neuroaxonal dystrophy phenotype in Quarter Horses.

et al., 2004; Mayhew et al., 1987). In a group of Quarter Horses diagnosed with NAD/EDM, an inconsistent menace response with no apparent loss of vision was observed along with varying degrees of dullness and obtundation (Figure 14.1b) (Aleman et al., 2011). Clinical signs are often self-limiting, and horses that survive to 2–3 years of age commonly exhibit lifelong, stable neurological deficits. The disease rarely progresses to recumbency.

There is an association with a vitamin E deficiency in cases of NAD/EDM, although low vitamin E levels are not present consistently in all cases. It appears that vitamin E is a factor in the development of NAD/EDM in the first year of life in genetically predisposed foals (Blythe et al., 1991b). A study of risk factors associated with the development of EDM found that foals from dams that had previously produced an EDM-affected foal were at a significantly higher risk (25 times more likely) of developing EDM than were foals from other dams, whereas low serum vitamin E concentrations were not found to be a significant risk factor in disease development (Dill et al., 1990). In the family of Quarter Horses with NAD/EDM, we performed quantitative Real-Time PCR for alpha-tocopherol transfer protein (TTPA) mRNA to determine if the correlation between low vitamin E and NAD could be due to decreased or absent absorption of vitamin E, and there was no significant difference in expression between NAD-affected and unaffected horses (Aleman et al., 2011). In addition, we have performed serial measurements of serum, CSF and muscle vitamin E concentrations in affected horses before and after supplementation and found that vitamin E concentrations could be increased in serum, CSF, and muscle post-supplementation (unpublished results). Overall, there is very strong evidence that NAD/EDM is an inherited disorder, and it may be that vitamin E concentrations act as environmental modifiers to determine the overall severity of the phenotype of horses affected with NAD/EDM – a theory that has been supported by the research to date (Aleman et al., 2011; Blythe et al., 1991a).

An ante-mortem diagnosis of NAD/EDM is based solely on clinical signs, the elimination of other causes of neurologic disease, and possible association with a low serum vitamin E concentration. Definitive diagnosis requires histopathological evaluation of brainstem and spinal cord tissue at post-mortem. Histological lesions in NAD/EDM consist of dystrophic neurons and axons, vacuolization and spheroid formation (Beech, 1984). The distinction between NAD and EDM is made based on location of the lesions. Lesions in NAD are confined strictly and bilaterally to the lateral (accessory) cuneate, medial cuneate, and gracilis nuclei of the caudal myeloencephalon and the nucleus thoracicus (Clarke's nucleus) of the spinal cord from the first thoracic vertebrae (T1) to the third lumbar vertebrae (L3) (Beech, 1984; Miller & Collatos, 1997). With EDM, there is additional bilateral axonal degeneration of the dorsal spinocerebellar and ventromedial tracts. Transport disruption of axonal proteins has been recognized in affected nuclear regions (Siso et al., 2003). Currently, there is no treatment for NAD/EDM, although vitamin E supplementation may stabilize clinical cases in addition to preventing severe signs in predisposed individuals (Mayhew et al., 1987; Dill et al., 1990; Aleman et al., 2011).

NAD has been associated with neurologic disease in humans, sheep, cats, and dogs; however, the clinical and specific histopathological findings vary. In the majority of NAD cases in various species, dystrophic neurons and axons are most commonly found in specific nuclear areas of the gray matter in brain and spinal cord (Siso et al., 2006). In canine and feline NAD, there is involvement of the cerebellum, which is not observed in horses (Chrisman, Cork, & Gamble, 1984; Carmichael et al., 1983; Cork et al., 1983). In humans, an underlying genetic basis for the disease has been identified in two of the five NAD disorders: infantile NAD (mutations in *PLA2G6*) and Juvenile/Adult Hallervorden-Spatz disease (mutations in *PANK2* gene) (Zhou et al., 2001; Khateeb et al., 2006; Morgan et al., 2006). Equine NAD/EDM varies considerably from human NAD in both clinical signs and affected neuroanatomical pathways, and therefore a candidate gene approach is unlikely to identify a putative mutation for the disease in horses.

The prospective breeding trial in Morgan horses with NAD performed by Beech and Haskins (1987) demonstrated either an autosomal dominant mode of inheritance with variable expression or a polygenic mode of inheritance. Pedigree analysis performed on a group of Quarter Horses affected with NAD support an incompletely penetrant autosomal dominant mode of inheritance. Alpha-tocopherol transfer protein (TTPA) has been definitively excluded as a candidate gene through sequencing and expression studies (Finno, et al. 2012). At this time, genotyping of 133 horses has been performed using the EquineSNP50 Beadchip (Illumina, San Diego, CA); a genome-wide association study with family data (GWAF) is in progress.

Shivers

Shivers (shivering) is a chronic nervous or neuromuscular syndrome in horses that has been recognized since 1886 (Williams, 1886) and affects primarily draft-horse breeds, but has also been reported in Warmbloods, Warmblood crosses, and, less commonly, Quarter Horses and Thoroughbreds (Deen, 1984; Valentine et al., 1999; Davies, 2000; Firshman, Baird, & Valberg, 2005; Hunt et al., 2005). Although it has been reported that there is no sex predilection (Firshman et al., 2005), evidence from the University of Minnesota (UMN) Neuromuscular Laboratory reveals that shivers is predominantly diagnosed in male horses (80%) as compared to females (20%) (Valberg, 2009). There is a variable age of onset, with the mean age reported from UMN Neuromuscular laboratory of 6.4 years and a range from 1 to 18 years of age (Valberg, 2009). The disease is characterized by "shivering" and hyperflexion of the pelvic limbs and an elevated tail head caused by involuntary spasms of the muscles in the pelvic region and tail. The muscles of the thigh and quarters appear to quiver when the leg is hyperflexed. These signs are often most evident when the horse is moving backward (Baird, Firshman, & Valberg, 2006). Clinical signs are often exacerbated by excitement. Less commonly, there may be involvement of the muscles of the thoracic limbs, neck, or face. Pelvic limb weakness was evident in 58% (11/19) of Belgian horses with shivers (Firshman et al., 2005).

Diagnosis is primarily made based on clinical observation. There is no significant difference in muscle enzyme concentrations (creatine kinase and aspartate aminotransferase) or vitamin E concentrations in horses with shivers compared to control horses (Firshman et al., 2005). Although hypothesized for some time that shivers was a manifestation of polysaccharide storage myopathy (PSSM), no significant association was found between a diagnosis of PSSM and a diagnosis of shivers (Firshman et al., 2005). In most cases, shivers is progressive and gradually debilitating. Various treatments have been attempted, with no success, including acupuncture, chiropractic adjustment, and supplementation with magnesium or vitamin E. Recommendations include maintaining the horse in fit condition and providing frequent turnout in an effort to stabilize clinical signs (Valberg, 2009). Post-mortem examination is unrewarding in these cases, with no documented histopathological lesions observed despite extensive examination of the central and peripheral nervous systems (Mayhew, 2005; Valentine et al., 1999).

An underlying familial tendency to shivers is suspected, and it has been suggested that shivers is inherited or has a hereditary predisposition (Green, Davies, & Doucet, 1995). Shivering has been classified as a hereditary disease in the United Kingdom (Tutt, 1964). Currently, a genome-wide association study is being performed utilizing the EquineSNP50 (Illumina, San Diego, CA) (Valberg, personal communication).

Cervical vertebral malformation and malarticulation

Cervical vertebral compressive myelopathy (CVCM) – also known as cervical vertebral stenotic myelopathy, cervical vertebral malformation and malarticulation (CVM), spinal ataxia, cervical stenotic myelopathy, and "wobbler's" syndrome – is one of the most common causes of neurologic disease in horses worldwide. CVCM can be divided into two subsets of disease manifestation. The first (type I) is a developmental disease, caused by vertebral malformation or malarticulation that leads to a dynamic instability of the vertebral canal and is most common in young horses. There is a higher incidence of type I CVCM in Thoroughbred horses, and an inherited basis has been suggested. It has been demonstrated that young male horses are at risk and dietary factors, such as a high carbohydrate ration, low dietary copper and excess dietary zinc may predispose individuals to type I CVCM (Kronfeld DS & Donoghue, 1990). Type II CVCM occurs more commonly in older horses and results from osteoarthritic enlargement of the articular processes, typically within the caudal cervical vertebrae (C5, C6, C7). It does not appear to be a breed predisposition for type II CVCM, but it has been suggested that individuals are predisposed to develop type II CVCM due to a relatively narrow vertebral canal (Van Biervliet, 2007).

Clinical signs of CVCM include usually symmetric upper motor neuron dysfunction, leading to spastic paresis; and sensory dysfunction resulting in ataxia and proprioceptive deficits of all limbs. All limbs could be equally affected or more typically the pelvic limbs more affected than the thoracic limbs. Ante-mortem diagnosis has been supported by plain cervical radiography, myelography, computed tomography, and exclusion of other causes. Pathologic changes, including mild subluxation of the vertebrae, physeal enlargement and dorsal projection of the caudal physis of the vertebral body, osteoarthritis of the articular processes, osteochondrotic changes at the articular processes, and caudal extension of the dorsal vertebral arch over the cranial physis of the next caudal body may be evident (Mayhew et al., 1993). A definitive ante-mortem diagnosis of CVCM is challenging and therefore accurate phenotyping of affected cases can be difficult. In an attempt to determine whether cervical vertebral canal stenosis is present, radiographic measurements have been used as common denominators for correcting the absolute mean sagittal diameter measurements into correct value. Both intravertebral and intervertebral ratios have been established (Moore et al., 1994; Hahn et al., 2008a). Cervical myelography may be useful in confirming a diagnosis of CVCM and is required to define the site of spinal cord compression if surgical intervention is to be performed. There are no established diagnostic criteria for compression based on myelographic observations, and false-positive diagnoses may occur (van Biervliet et al., 2004). If available, contrast-enhanced computed tomography (CECT) can accurately diagnose the location and severity of compressive lesions in cases of CVCM (Reed, 2007). Although CECT is promising as a diagnostic option, most CT units at veterinary practices and universities are unable to image adult equine cervical vertebrae; therefore, in most cases, post-mortem examination remains the only way to achieve a definitive diagnosis of CVCM.

Medical therapy in horses with CVCM involves reducing inflammation and edema in an attempt to reduce the compression on the cervical spinal cord. Use of nonsteroidal anti-inflammatory agents and dimethyl sulfoxide are common. In young horses (less than 1 year of age), restricted exercise and diet are recommended, with particular attention to meeting the minimum requirements of essential nutrients. In adult horses with compressive lesions that are graded as 1 to 3 ataxic based on the scale proposed by Lunn and Mayhew (1989) and are compressed at a maximum of two sites, ventral interbody fusion can be used to stabilize the affected joint(s) (Reed, 2007).

It has been suggested that inheritance plays a role in the etiology of CVCM. In 1950, Dimock proposed, based on post-mortem findings and neurological examinations in 121 Thoroughbreds with CVCM and known pedigree information, that the disease was hereditary (Dimock, 1950). Dimock observed that foals that developed CVCM were frequently by the same or closely related sires, with 43% of the affected cases appearing in one line and some mares producing two or three cases. Conversely, in 1976, an investigation into the genetics of CVCM in Thoroughbreds was performed in Great Britain, but no conclusive evidence was found to support a direct role of inheritance in cases of CVCM (Falco, Whitwell, & Palmer, 1976). Supporting this finding, no evidence was found to suggest that CVCM is directly heritable by simple Mendelian dominant or recessive patterns (Wagner et al., 1985). Interestingly, in that same study, although the breeding of an affected CVCM stallion to an affected CVCM mare did not produce a foal with CVCM, there was a high incidence of osteochondrosis (OC) (10 of 22 foals) and physitis (9 of 22) found in the offspring of the crosses. Osteochondrosis has been postulated to have a substantial heritable component (Grondahl & Dolvik, 1993). Subsequent work supported an association between CVCM and osteochondrosis, with the frequency and severity of OC lesions in articular cartilage of the appendicular skeleton (patella, distal humerus, femur, third metacarpal/metatarsal bone) greater in CVCM horses as compared to normal controls (Stewart, Reed, & Weisbrode, 1991). As nutritional factors and rate of growth appear to be factors in both the development of CVCM and OC, it may be that there is a genetic predisposition to develop either or both conditions as underlying inabilities to develop normal mature bone and cartilage. Efforts to investigate the genetic contribution to CVCM are currently under way.

From a comparative perspective, cervical spinal cord compression has been described in Great Danes and Doberman pinchers, with males more frequently affected than females, in a similar ratio to that of horses (approximately 2:1) (Trotter, 1985). Clinical presentation is similar to CVCM in horses. Breeding affected dogs results in a higher incidence of the syndrome in the progeny (Selcer & Oliver, 1975); however, the determination of a genetic predisposition to cervical myelopathy due to an inherited spinal canal stenosis awaits confirmation.

Figure 14.2 Frame-splashed white overo phenotype in a deaf American Paint horse.

Sensorineural deafness in American Paint horses

There have been isolated case reports of American Paint horses suffering from sensorineural deafness, which was confirmed by brainstem auditory-evoked responses (BAER) (Harland et al., 2006). There is a small population of melanocytes within the inner ear that is essential for the development of the *stria vascularis*, a blood-vessel-rich zone of the cochlea that plays a role in modulating the chemical composition of endolymph and the resultant production of an endocochlear potential. In mice that have been genetically engineered to have no melanocytes in the stria (viable dominant spotting mice), the endocochlear potential is close to zero (wild type + 100 mV) at all stages of development and the mice are profoundly hearing impaired (Steel & Barkway, 1989). In many species, congenital deafness is associated with abnormal migration of these melanocytes from the neural crest and poor survival within the inner ears of some animals with coat and iris pigmentation alterations (Steel & Barkway, 1989; Price & Fisher, 2001).

Recently, a case-control series evaluating 14 confirmed deaf, 20 suspected deaf, and 13 non-deaf American Paint horses was performed to further evaluate deafness in the American Paint horse by phenotype, clinical findings, BAERs, and endothelin B receptor (*EDNBR*) genotype (Magdesian et al., 2009). Horses were confirmed to be bilaterally deaf through BAER testing. In both the confirmed and suspected deaf groups, the most common coat color patterns were splashed white overo and frame-splashed white overo blends (Figure 14.2). All confirmed deaf horses had extensive white facial markings, one or more blue irides, partially pigmented palpebral skin, and limbs with white markings. Otoscopic examination, routine clinicopathological analysis of blood and serum, skull radiographs, and neurologic examinations revealed no abnormalities in the confirmed deaf horses. There is currently no treatment available for equine deafness; however, it is important to recognize the condition in order to counsel owners about management and handling of deaf horses.

When evaluating the confirmed deaf horses for the *EDNRB* mutation that is causative for LWFS (I118K), 7/8 confirmed deaf horses were carriers for the mutation and 1/8 was not (Magdesian et al., 2009). Therefore, a relationship between the mutation and deafness was established; however, this could represent two independent mutations (i.e., *EDNRB* and the true causative mutation for deafness in the American Paint horse) that are prevalent in certain overo horses.

In humans, there are several conditions that cause hypopigmentation and deafness. Type II Waardenburg syndrome is limited to hypopigmentation and deafness without any craniofacial deficits and is caused by a mutation in the *Mitf* gene, which is a central mediator of melanocyte development (Price & Fisher, 2001). Type IV Waardenburg syndrome includes deafness and hypopigmentation in conjunction with Hirschsprung's disease and is associated with a mutation of the *SOX10*, *EDN3*, or *EDNRB* genes (Price & Fisher, 2001). It appears that the other genes involved in Waardenburg syndrome, namely *Pax3*, *Kit*, *Sox10*, *EDN3*, and *EDNRB*, mechanistically converge on *Mitf*, suggesting that *Mitf* plays a pivotal role during melanocytes development (Price & Fisher, 2001). Based on likely candidate genes, additional investigation into the genetics of deafness in horses is currently being pursued.

Equine inherited myoclonus

Suspected inherited myoclonus has been documented in Peruvian Paso horses. Three foals (aged 12 hours, 9 days, and 18 days at presentation) were evaluated at the University of California at Davis between 1988 and 1991. In the two older foals, weakness and lethargy were noted at birth and, in all three cases, the foals were hyper-responsive to auditory and tactile stimuli, demonstrating myoclonic contractions upon stimulation. Gait abnormalities were demonstrated, including a "rabbit hopping" of the pelvic limbs in one foal and an inability to stand without assistance in the second foal.

Diagnostics included a complete blood count, serum biochemistry, cerebrospinal fluid (CSF) analysis, pelvic radiographs, muscle biopsies, electrophysiological testing, and complete post-mortem examinations. Clinicopathological values for blood and CSF were within normal ranges, other than mild elevations in creatine kinase and aspartate aminotransferase. Pelvic radiographs, muscle biopsies and electrodiagnostic testing were unremarkable. On post-mortem examination, there were no gross or histopathological alterations noted in the central nervous system

Spinal tissue from two of these three cases was further evaluated by Gundlach et al. (1993). In calves with inherited myoclonus, there is a marked loss of spinal strychnine-sensitive glycine receptors (Gundlach et al., 1988; Gundlach, 1990). By using a [^{3}H] strychnine-binding assay, it was demonstrated that [^{3}H] binding was significantly reduced (by 40%) in cervical and lumbar spinal regions in the myoclonic horses when compared to control levels (Gundlach et al., 1993).

Glycine is a major inhibitory transmitter in the spinal cord and brainstem and acts by binding its receptor, transiently forming an anion-selective chloride transmembrane channel that increases chloride conductance and results in hyperpolarization, thereby inhibiting neuronal firing. In calves with myotonia, it has been demonstrated that there is an 80–90% loss of glycine receptors and the resultant phenotype is more severe than that observed in these two reported horses (Gundlach et al., 1993). It was confirmed in the two equine cases that there was only the loss of the glycine receptor and, in all other binding sites evaluated (muscarinic, β-adrenergic, $GABA_A$/benzodiazepine and peripheral benzodiazepine receptor binding sites, and cholecystokinin receptor binding site), densities were normal when compared to control levels.

Inherited startle syndromes are recognized in humans and mice and are caused by defects in inhibitory glycinergic pathways. In humans, missense mutations in the glycine receptor α-1 subunit gene have been identified as molecular causes of inherited startle syndromes (Shiang et al., 1993). In mice, two autosomal recessive mutations have been described that resemble the startle-disease phenotype in humans: *spasmodic* and *spastic*. The spasmodic mutation (missense mutation in the receptor α-1 subunit gene) is similar to the human mutation and impairs the function of the glycine receptors by reducing the receptor's sensitivity to agonists. The spastic mutation (intronic insertion of a LINE-1 transposable element (Kingsmore et al., 1994) or aberrant splicing of β-subunit mRNA (Mulhardt et al., 1994) prevents or diminishes the expression of glycine receptors, similar to what is observed in bovine and equine myoclonus. Both mechanisms result in reduced glycinergic inhibition in the spinal cord, resulting in a startle syndrome or myoclonus. Currently, there is ongoing research to definitively define the underlying molecular mechanism of equine myoclonus.

Juvenile idiopathic epilepsy

Juvenile idiopathic Epilepsy is a syndrome in Arabian foals of Egyptian lineage that is characterized by recurrent generalized seizures that are manageable and appear to be self-limiting, disappearing by 1 to 2 years of age (Mittel, 1987; Aleman et al., 2006). This form of epilepsy appears to be inherited; however, a proposed mode of inheritance had not been established. There appears to be no sex predilection (Aleman et al., 2006).

Clinical manifestations in affected cases include a history of multiple generalized or partial seizures, a decreased to absent menace response, blindness, and abnormal mentation ranging from lethargy to obtundation. In a review of 22 cases (Aleman et al., 2006), the median age at onset was 2 months (range of 2 days to 6 months). Clinical signs of seizure disorders can be divided into pre-ictal, ictal, and post-ictal signs. Pre-ictally, most foals exhibit either no clinical signs or mild behavioral changes. During the ictal phase, tonic seizures are observed followed by clonic motor activity in severe cases or focal head twitches and nystagmus in milder cases. These seizure episodes usually last for less than 1 minute and are responsive to treatment with benzodiazepines (diazepam, midazolam) (Mittel, 1987; Aleman et al., 2006). Post-ictally, blindness, lasting from a few minutes up to three weeks, was noted in 100% of the 22 cases reported by Aleman and colleagues. Other post-ictal signs included lethargy, disorientation, and obtundation. Injuries are common following seizure episodes, with corneal abrasions frequently noted (Mittel, 1987).

Diagnostics in suspect cases include routine clinicopathological evaluation of blood, serum, and CSF. An increase in serum creatine kinase and glucose may be noted on the serum biochemistry. The remainder of the clinicopathological evaluation is within reference ranges. Imaging of the head with plain radiographs and computed tomography reveal no abnormalities. The diagnosis of epilepsy is made primarily on clinical grounds and exclusion of other causes of seizures. Electroencephalography has been useful to document and characterize the presence of abnormal brain electrical activity in affected foals. In 9/13 foals, electrophysiological abnormalities, including spikes, sharp waves, spike and wave discharges, or multiple spike complexes were observed (Aleman et al., 2006). The majority of these changes were noted in the central region and the voltage maximum was often on midline. There were, however, times where the voltage maximum was localized to the left or right sides, indicating multiform events in some foals. In human cases of idiopathic generalized epilepsies, EEG findings should be abnormal even during the inter-ictal phase in untreated patients (Panayiotopoulos, 2005). If inter-ictal awake EEG findings in an untreated suspect equine juvenile idiopathic epilepsy case are unremarkable, it may be advisable to repeat the EEG while the patient is sleeping and during awakening as this may reveal abnormal activity. Intermittent photic stimulation has been demonstrated to induce seizures in humans with idiopathic generalized epilepsies (Panayiotopoulos, 2005); however, in the two foals where photic stimulation was performed during EEG readings, no epileptiform activity was noted (Aleman et al., 2006).

Therapeutic management of juvenile idiopathic epilepsy is generally successful. Phenobarbital is the main antiepileptic medication that has been used for the long-term control of seizures. Dose regimens vary among individuals and adjustments may be necessary. Phenobarbital concentrations should be considered therapeutic on an individual basis when seizures are controlled without causing excessive sedation (Aleman et al., 2006). Potassium bromide (KBr) may be added in some cases. Generally, affected foals will require anywhere from 2 to 9 months of treatment with antiepileptic drugs. Additional therapies include vitamin E and selenium supplementation and supportive care while adjusting the antiepileptic drug dosage as seizure-induced trauma may lead to abrasions, contusions, or fractures. Pneumonia was present in 50% of the cases diagnosed with juvenile idiopathic epilepsy and should be managed with appropriate antimicrobial therapy (Aleman et al., 2006). The prognosis for juvenile idiopathic epilepsy is excellent when the seizures are appropriately managed. The disease appears to be self-limiting, disappearing by 1 to 2 years of age.

There appears to be a strong hereditary component to juvenile idiopathic epilepsy, as all reported cases are in Arabian foals from Egyptian lineage. Many of the cases reported in the case series by Aleman et al. (2006) had siblings or parents that appeared to be affected with the same condition. In humans, many of the rare Mendelian idiopathic generalized epilepsies are ion voltage- and receptor-mediated channelpathies (Panayiotopoulos, 2005). There are three types of neonatal and infancy epilepsy syndromes in humans that are characterized by afebrile focal motor seizures that remit after weeks or months with administration of standard anti-epileptic drugs. All three syndromes have an autosomal dominant mode of inheritance with high penetrance (Baulac & Baulac, 2009).

(1) Benign familial neonatal convulsions, a rare autosomal-dominant idiopathic epilepsy that causes seizures in newborn infants from the second or third day of life that remit by six weeks, are due to mutations in *KCNQ2* or *KCNQ3*, both voltage-gated potassium channels that regulate neuronal excitability in the central nervous system by controlling the duration of the action potential.
(2) Benign familial neonatal-infantile seizures occur between 2 days to 7 months of age and are most frequently due to mutations in *SCN2A*, a subunit of the neuronal voltage-gated sodium channel.
(3) Benign familial infantile seizures occur between 4 and 8 months of age with spontaneous remission before the age of 3 years. No causative genes have been identified for this condition.

Other forms of epilepsy in humans have a later age of onset (between 6 months and 6 years) and generally have a particular provoking factor – fever. These are termed febrile seizures and generalized epilepsy with febrile seizures plus (GEFS+). At this time, specific genes that affect FS have not been identified; however, mutations in *SCN1A* (Escayg et al., 2000) and *SCN1B* (Wallace et al., 1996) and a subunit of the $GABA_A$ receptor (*GABRG2*) (Baulac et al., 2001) have been incriminated in GEFS+ families. Adolescence- and adult-onset epilepsy syndromes occur during the first two decades of life and are characterized by brief nocturnal motor seizures. Genes encoding for the nicotine acetylcholine receptors, *CHRNA4*, *CHRNB2*, and *CHRNA2*, have been implicated in these epileptic disorders (Baulac et al., 2008). With well-documented genes involved in epilepsy in humans, genetic investigation into the molecular cause of idiopathic epilepsy of Arabian foals appears promising.

Occipitoatlantoaxial malformation

Congenital malformations of the occiput, atlas, and axis (occipitoatlantoaxial malformations or OAAM) were initially classified into three disease groups by Mayhew et al. (1978b). Subsequently, three additional disease entities have been recognized, thus dividing these malformations into a total of six classes.

Familial occipitalization of the atlas with atlantalization of the axis in Arabian horses
These Arabian foals demonstrated one of three clinical syndromes: death at birth, tetraparesis at birth, or progressive ataxia as foals, characterized by symmetrical upper motor neuron signs and general proprioceptive deficits referable to a cervical lesion. Many of these foals had an extended neck posture and a "click" that could be heard when the neck moved, which was suggested to be due to movement of the dens over the body of the atlas as it would slip ventrally and luxate under the atlantal body upon neck extension and then move dorsal to its normal position upon neck flexion. Occipitalization of the atlas and atlantalization of the axis was evident on post-mortem examination, and bilateral symmetric compression of the cervical spinal cord was apparent. Pedigree analysis in cases described by Mayhew et al. (1978b) and Watson and Mayhew (1986) revealed common ancestry among affected foals.

Congenital asymmetrical occipitoatlantoaxial malformation
Two cases in non-Arabian breeds (Standardbred and Morgan) presented with cervical scoliosis and a head deviated from birth with no evidence of spinal cord disease. Fusion between the occiput and atlas (either symmetric or asymmetric) with absence of occipital condyles was noted. The mid-cervical scoliosis was due to a wedge-shaped piece of vertebrae fused to the caudal edge of the axis. There was no evidence of spinal cord compression. It was not possible to determine if the wedge-shaped piece was due to an addition or a loss of vertebral column material. Based on the occurrence in two distinct breeds, it was thought that these malformations were congenital but nonfamilial. An additional reported case of congenital asymmetrical occipitoatlantoaxial malformation was reported in a Miniature horse foal; however, mid-cervical scoliosis was absent (Rosenstein et al., 2000).

Asymmetric atlanto-occipital fusion (unknown breed)
This case (Mayhew et al., 1978b) was presented with a deviated head and cervical scoliosis centered at the atlantoaxial joint. There was asymmetric fusion between the occiput and atlas with absence of the occipital condyles, and the entire atlas was rotated on its vertical axis with no evidence of bony constriction of the spinal cord. To date, this appears to be an isolated case of this particular congenital defect.

Duplication of the axis and/or atlas
A half-Arabian foal that was recumbent since birth was found to have duplication of the atlas and partial duplication of the axis, leading to widespread cervical spinal cord compression (de Lahunta, Hatfield, & Dietz, 1989). Another case of atlas duplication was reported in an Arabian horse that developed a compressive myelopathy unrelated to that malformation (Watson, Quick, & de Lahunta, 1978). Pedigree evaluations have not been performed on these isolated cases; however, the reports have occurred in Arabian and half-Arabian horses.

Symmetrical OAAM in non-Arabian horses (Appaloosa, QH, Friesian, Miniature horse)
Symmetrical OAAM has been reported in non-Arabian horses. In two cases (Appaloosa and Quarter Horse), the foals had an abnormal head and neck carriage since birth. One case demonstrated hyperesthesia of the pelvic limbs that had developed at 2 years of age but no evidence of neurologic deficits (Appaloosa), whereas the Quarter Horse demonstrated signs of acute neurologic impairment (tetraparesis, conscious proprioceptive deficits) at 3 years of age (Wilson et al., 1985). In a 3-month-old

Friesian foal (Bell et al., 2007), the clinical presentation consisted of severe ataxia and an inability to rise since birth. Symmetric atlantooccipital fusion with modification of the atlas, atlantoaxial joint, and axis was found on radiographs and confirmed on post-mortem examination in all cases. Although it is suggested that these cases are random occurrences and therefore nonfamilial, the similarity to the features found in familial OAAM in Arabians is apparent. If a causative genetic mutation is discovered for OAAM in Arabians, it will be useful to screen cases of symmetrical OAAM in other breeds.

Subluxation of the atlantooccipital joint, fusion of the atlas and axis with lateral deviation of the atlantoaxial joint and 20° rotation of the atlas in a half-Arabian colt (Blikslager et al., 1991). This was an isolated case of a yearling half-Arabian colt that presented for a head tilt, and symmetrical weakness and ataxia of all four limbs. The head tilt was due to a malpositioned atlas and subluxation of the atlantooccipital joint with fusion of the atlas and axis was diagnosed on radiographs and confirmed at post-mortem examination.

Diagnosis in all cases can be made on plain cervical radiographs. Although it has been reported that, in certain cases, surgical stabilization by fusion of the axis to the atlas can be attempted, the prognosis for OAAM is poor.

The familial OAAM seen in Arabian horses is very similar to that described in *Hoxd-3*$^-$/*Hoxd-3*$^-$ mice, which are homozygous for a targeted disruption of the homeobox containing gene, *Hoxd-3* (Condie & Capecchi, 1993). Mice homozygous for this mutation demonstrate occipitilization of the atlas and atlantization of the axis, an aspect of the phenotype that is fully penetrant. Interestingly, different parts of the same vertebrae are differentially affected by the loss of *Hoxd-3* function, suggesting that this gene may contribute to the development of the vertebral cartilages by affecting mesenchymal proliferation rates. The extensive similarities between mouse and horse phenotypes make the *Hoxd3* gene a strong candidate gene for OAAM.

Narcolepsy

Narcolepsy in humans is characterized by a tetrad of clinical signs including excessive daytime sleepiness with sleep attacks, cataplexy, hypnagogic hallucinations, and sleep paralysis. In 2005, the International Classification of Sleep Disorders divided human narcolepsy into two types: narcolepsy with cataplexy and narcolepsy without cataplexy. Strict criteria exist in the human medical field for diagnosing narcolepsy. A multiple sleep latency test (MSLT), which consists of scheduled periods of sleep where continuous electroencephalogram (EEs), electrooculogram (EOG), electromyogram (EMG), and electrocardiogram (ECG) activities are recorded, documents the length of time to onset of sleep and stages of sleep. In true narcoleptic patients, there is an abnormal tendency to enter Rapid Eye Movement (REM) sleep immediately upon falling asleep ("sleep onset REM periods or SOREMP") instead of a period of slow wave sleep preceding this event as is found in non-narcoleptic individuals. In narcoleptic dogs, EEG findings have revealed disruption of the normal sleep-wake cycle, and MSLTs can be used to confirm the diagnosis (Foutz, Mitler, & Dement, 1980).

The primary complaint in human narcoleptics is excessive sleepiness. Cataplexy is typically observed in dogs with narcolepsy. In horses, excessive sleepiness with and without cataplexy can be seen. In humans and dogs, cataplexy is often elicited upon the occurrence of sudden emotion or exciting events. In dogs, playing or eating often incites episodes. Cataplexy-provocative tests can be used in dogs to support a diagnosis of narcolepsy (Foutz et al., 1980). Mildly affected cases may require more stringent testing such as a challenge with physostigmine salicylate, which, when administered intravenously, will increase the probability of a spontaneous or elicited cataplectic attack within 5 to 15 minutes, and a food-elicited or play-elicited cataplexy test is performed (Foutz et al., 1980).

Narcolepsy in horses has been broadly characterized into two age-dependent groups: onset at birth and an adult onset (generally older than 2 years of age). Currently, an established criterion for the diagnosis of narcolepsy in horses is lacking. In adult equids, there is evidence that some of the reported cases of narcolepsy may be idiopathic hypersomnia, or sleep deprivation (Aleman, Williams, & Holliday, 2008; Bertone, 2006), where orthopedic issues or social stresses within a herd prevents horses from lying down and obtaining the required 0.5h/day of REM sleep (Aleman et al., 2008). In these cases, treatment with anti-inflammatory drugs or changing herd mates may provide the horse with relief, and the episodes of sleepiness will resolve (Bertone, 2006). Although the strict criteria used to diagnose human narcolepsy has not been documented in the horse, cases of adult-onset narcolepsy do appear to exist (Sweeney et al., 1983) and can be triggered by repeatable precipitating factors. In these rare adult-onset cases, episodes may be inducible with physostigmine salicylate and may be prevented with imipramine.

There does appear to be a true familial form of narcolepsy with cataplexy present in Miniature horses (Lunn et al., 1993), Shetland ponies (Foutz et al., 1980), Lippizan foals (Aleman et al., 2008), and Suffolk foals (Sheather, 1924). These foals demonstrate signs of daytime sleepiness and episodes of partial to complete cataplexy at several weeks of age. These episodes may be induced by some form of stimulation. Although these cases appear to be cases of true familial narcolepsy, attempts to document SOREM in cases have not been successful (Dreifuss & Flynn, 1984; Lunn et al., 1993). Treatment with imipramine, a tricyclic antidepressant that blocks uptake of serotonin and noradrenaline, was attempted in two cases, and only partial resolution of signs was achieved when the drug was administered at 0.55 mg/kg twice daily for two weeks (Lunn et al., 1993). Pharmacokinetics of imipramine administration in horses has recently been investigated (Peck et al., 2001), and appropriate dosing information will allow for future studies to determine efficacy of the drug in appropriately diagnosed equine narcoleptic patients.

In humans, *HLA-DQB1**0602 appears to be the main susceptibility allele for narcolepsy with cataplexy across many ethnicities, with 90% to 100% of patients testing positive in cases of narcolepsy with cataplexy (Mignot et al., 2002). Narcolepsy is inherited as an autosomal recessive trait in Dobermans and Labrador Retrievers (Foutz et al., 1980). Initial linkage studies in dogs with narcolepsy found genetic linkage between canarc-1, which was found to be an element that cross-reacted with the μ switch-like immunoglobulin switch element (Mignot et al., 1991; Mignot et al., 1994). Through positional cloning, a mutation in the hypocretin (Orexin) receptor 2 (*Hcrtr2*) gene was identified in dogs (Lin et al., 1999). Subsequently, mouse knockout models of the hypocretin receptor 1 (*Hctr1*) or *Hctr2* have been demonstrated to display symptoms of narcolepsy as adults (Chemelli et al., 1999). Hypocretin-containing neurons are located in the hypothalamus where they project widely with numerous brain nuclei including those responsible for the regulation of sleep and alertness. The *Hcrtr2* mutation appears to alter the normal developmental course of CSF hypocretin levels, with the onset of cataplexy occurring at a time when CSF hypocretin concentrations are lower compared to control animals. Interestingly, cataplectic activities continue in these dogs despite subsequent increases in CSF hypocretin after 4 weeks of age, suggesting that damage to other systems or reduced hypocretin input to other systems may produce symptoms of narcolepsy without a reduction of CSF hypocretin levels (John et al., 2004).

Although human narcolepsy is generally not due to mutations in *Hctr2*, hypocretin neurotransmission is impaired and the condition is associated with low (<110 pg/mL) hypocretin-1 levels in the cerebrospinal fluid (Mignot et al., 2002). Many subjects that have low hypocretin-1 levels also are positive for *HLA-DQB1**0602 allele (Mignot et al., 2002). In humans, narcolepsy is hypothesized to be due to autoimmune destruction of hypocretin neurons in the brain (Mignot et al., 2002; Ahmed & Thorpy, 2010). Treating *Hctr2* mutated narcoleptic dogs with immunosuppressive drugs within 3 weeks after birth resulted in an increased time to the onset of cataplexy and reduced time in cataplexy, again supporting that the immune system and its interaction with hypocretin neurons plays a role in the development of narcolepsy (Boehmer et al., 2004). Recently, a case of idiopathic narcolepsy without cataplexy was described in a 3-month-old Icelandic foal, and hypocretin concentrations were measured and found to be comparable to control levels (Bathen-Nothen et al., 2009). Further work is necessary in the field of equine narcolepsy to provide accurate phenotyping and classification of affected cases and to investigate the potential role of the hypocretin receptor genes in this species.

Recurrent laryngeal neuropathy

Recurrent laryngeal neuropathy (RLN) – also known as idiopathic laryngeal paralysis, "roaring," "whistling," and laryngeal hemiplegia – is a common cause of respiratory compromise at exercise in larger breeds of horses due to impaired arytenoid abduction, secondary to lesions in the recurrent laryngeal nerves. A variety of breeds are affected with RLN and age of onset is variable.

Clinical signs of the disease include airway obstruction and exercise intolerance, especially in horses exercising at maximal efforts. Diagnosis is frequently made upon laryngoscopy, and the grading system proposed in 2003 (Anon, 2003) has been shown to correlate with histopathologic changes in the laryngeal muscles of affected horses (Collins et al., 2009). Resting endoscopy may misdiagnose cases of RLN that require maximal exercise to experience arytenoid collapse, and dynamic respiratory endoscopy is currently being utilized to simulate racing conditions and allow for a more accurate diagnosis of RLN. Treatment for RLN encompasses a variety of surgical techniques, including laryngoplasty alone or combined with a unilateral or bilateral ventriculectomy or ipsilateral vocalcordectomy, neuromuscular pedicle grafts, or partial arytenoidectomy with bilateral ventriculectomy.

RLN is more prevalent in some families of horses (Cook, 1988; Poncet et al., 1989; Duncan, 1992; Harrison, Duncan, & Clayton, 1992) and studies have demonstrated that offspring of RLN-affected stallions are more likely to be affected than offspring of unaffected stallions (Poncet et al., 1989; Ohnesorge et al., 1993). Although RLN is found in a variety of equine breeds, subclinical disease is most prevalent in taller (>170 cm) horses. In horses of this stature, the recurrent laryngeal nerves are about 250 cm long, with the left recurrent laryngeal nerve about 30 cm longer than the right due to a longer course, looping around the aorta (Hahn et al., 2008b). A recent study by Hahn et al. (2008b), upon examining both the right and left recurrent laryngeal nerves and distal median, peroneal, and phrenic nerves in horses affected with RLN, conclusively demonstrated that RLN is a bilateral mononeuropathy limited to recurrent laryngeal nerves.

There are no reports of an inherited bilateral mononeuropathy in humans or other mammals. The gold standard for phenotyping relies on histopathology of laryngeal muscles and this becomes extremely important when selecting control cases, as many cases of RLN are subclinical. Efforts to investigate the inheritance and potential molecular mechanism of RLN are currently underway.

Photic headshaking

Headshaking in horses can be attributed to many causes, such as middle ear disorders, ear mites, cranial nerve dysfunction, ocular disease, guttural pouch mycosis, dental disorders, and rhinitis, among others (Lane & Mair, 1987). Photic headshaking, trigeminal neuralgia, or idiopathic headshaking is a disorder that affects more commonly middle-aged geldings. The mean age of onset of signs is 9 years old (Madigan & Bell, 1998). Although a breed predisposition has not been observed, Thoroughbred, Quarter Horse and Warmblood breeds are commonly reported breeds. In most horses, the disorder appears to be seasonal, manifesting in the early spring or fall (Madigan & Bell, 1998). Photic headshaking appears to be a subset of the "idiopathic headshakers," and many cases improve or resolve with blindfolding or avoiding bright light (Madigan et al., 1995). It has been suggested that optic-trigeminal summation causes facial neuropathic pain in response to light stimulation in affected horses (Madigan et al., 1995), based on a similar postulated mechanism underlying photic sneezing in humans (Eckardt, MacLean, & Goodell, 1943).

The photic sneeze reflex has autosomal dominant heritability in humans (McKusick, 1990). It may be that photic sneezing is a "threshold trait" manifested as the variable presence of the reflex in addition to the differential response to varying light intensities and photic frequencies among individuals demonstrating the reflex (Whitman & Packer, 1993). The seasonal occurrence of photic headshaking is particularly interesting and it has been suggested that the seasonal onset of headshaking might be related to photoperiod, which is supported by the improvement of some clinical cases of photic headshaking when treated with melatonin (Madigan & Bell, 1998). Seasonal affective disorder in humans is characterized by depression in the fall and winter, and the depression has been suggested to be a heightened vulnerability to fluctuations in environmental light and an increased sensitivity of a central muscarinic mechanism (Dilsaver & Majchrzak, 1987). A link between the photic sneeze response and seasonality as it relates to photoperiod and seasonal affective disorder in humans has been suggested (Pies, 1990). It may be that equine photic headshaking, similar to the photic sneeze reflex in humans, is heritable; however, additional studies within families of affected horses are required.

Conclusion

A variety of neurologic disorders in horses have demonstrated to be heritable. With the sequencing of the equine genome and molecular tools currently available to perform whole genome interrogation, the identification of quantitative trait loci for multi-factorial traits in addition to mapping of simple Mendelian genetic traits is possible. Equine genetic research will continue to advance and the molecular discoveries will provide insight into disease pathophysiology and allow veterinarians the ability to definitively diagnose particular conditions, while advising breeders in order to decrease the overall prevalence of certain inherited diseases in the equine population.

References

Adams, A. P., Collatos, C., Fuentealba, C., Illanes, O., & Blanchard, R. (1996). Neuroaxonal dystrophy in a two-year-old quarter horse filly. *Canadian Veterinary Journal, 37*, 43–44.

Ahmed, I., & Thorpy, M. (2010). Clinical features, diagnosis and treatment of narcolepsy. *Clinical Chest Medicine, 31*, 371–381.

Aleman, M., Finno, C. J., Higgins, R. J., Puschner, B., Gericota, B. et al. (2011). Evaluation of epidemiological, clinical, and pathological features of neuroaxonal dystrophy in Quarter Horses. *Journal of American Medical Veterinary Association, 239*, 823–833.

Aleman, M., Gray, L. C., Williams, D. C., Holliday, T. A., Madigan, J. E. et al. (2006). Juvenile idiopathic epilepsy in Egyptian Arabian foals: 22 cases (1985–2005). *Journal of Veterinary Internal Medicine, 20*, 1443–1449.

Aleman, M., Williams, D. C., & Holliday, T. (2008). Sleep and sleep disorders in horses. In the Proceedings of the 54th Annual Convention o fthe American Association of Equine Practitioners, Orlando, FL, pp. 180–185.

Amiel, J., Sproat-Emison, E., Garcia-Barcelo, M., Lantieri, F., Burzynski, G. et al. (2008). Hirschsprung disease, associated syndromes and genetics: A review. *Journal of Medical Genetics*, *45*, 1–14.

Anon. (2003). Havemeyer consensus statement on equine laryngeal neuropathy. In *Havemeyer Monograph Series* No. 11, Newmarket, pp. 93–97.

Auricchio, A., Casari, G., Staiano, A., & Ballabio, A. (1996). Endothelin-B receptor mutations in patients with isolated Hirschsprung disease from a non-inbred population. *Human Molecular Genetics*, *5*, 351–354.

Baird, J. D., Firshman, A. M., & Valberg, S. J. (2006). Shivers (Shivering) in the horse: A review. In the Proceedings of the Annual Convention of the American Asssociation of Equine Practitioners, San Antonia, TX, pp. 359–364.

Bathen-Nothen, A., Heider, C., Fernandez, A. J., Beineke, A., Sewell, A. C. et al. (2009). Hypocretin measurement in an Icelandic foal with narcolepsy. *Journal of Veterinary Internal Medicine*, *23*, 1299–1302.

Baulac, S., & Baulac, M. (2009). Advances on the genetics of mendelian idiopathic epilepsies. *Neurology Clinician*, *27*, 1041–1061.

Baulac, S., Gourfinkel-An, I., Couarch, P., Depienne, C., Kaminska, A. et al. (2008). A novel locus for generalized epilepsy with febrile seizures plus in French families. *Archives of Neurology*, *65*, 943–951.

Baulac, S., Huberfeld, G., Gourfinkel-An, I., Mitropoulou, G., Beranger, A. et al. (2001). First genetic evidence of GABA(A) receptor dysfunction in epilepsy: A mutation in the gamma2-subunit gene. *Nature Genetics*, *28*, 46–48.

Baumgartner, W., Frese, K., & Elmadfa, I. (1990). Neuroaxonal dystrophy associated with vitamin E deficiency in two Haflinger horses. *Journal of Comparative Pathology*, *103*, 114–119.

Beech, J. (1984). Neuroaxonal dystrophy of the accessory cuneate nucleus in horses. *Veterinary Pathology*, *21*, 384–393.

Beech, J., & Haskins, M. (1987). Genetic studies of neuraxonal dystrophy in the Morgan. *American Journal of Veterinary Research*, *48*, 109–113.

Bell, S., Detweiler, D., Benak, J., & Pusterla, N. (2007). What is your diagnosis? Occipitoatlantoaxial malformation. *Journal of the American Veterinary Medical Association*, *231*, 1033–1034.

Bertone, J. J. (2006). Sleep deprivation is not narcolepsy in horses. In the Proceedings from the 24th Annual American College of Veterinary Internal Medicine Forum, Louisville, KY, pp. 167–169.

Bjorck, G., Everz, K. E., Hansen, H. J., & Henricson, B. (1973) Congenital cerebellar ataxia in the gotland pony breed. *Zentralbl Veterinarmed A*, *20*, 341–354.

Blanco, A., Moyano, R., Vivo, J., Flores-Acuna, R., Molina, A. et al. (2006). Purkinje cell apoptosis in arabian horses with cerebellar abiotrophy. *Journal of Veterinary Medicine and Physiological and Pathological Clinical Medicine*, *53*, 286–287.

Blikslager, A. T., Wilson, D. A., Constantinescu, G. M., Miller, M. A., & Corwin, Jr., L. A. (1991). Atlantoaxial malformation in a half-Arabian colt. *Cornell Veterinary*, *81*, 67–75.

Blythe, L. L., Craig, A. M., Lassen, E. D., Rowe, K. E., & Appell, L. H. (1991a). Serially determined plasma alpha-tocopherol concentrations and results of the oral vitamin E absorption test in clinically normal horses and in horses with degenerative myeloencephalopathy. *American Journal of Veterinary Research*, *52*, 908–911.

Blythe, L. L., Hultgren, B. D., Craig, A. M., Appell, L. H., Lassen, E. D. et al. (1991b). Clinical, viral, and genetic evaluation of equine degenerative myeloencephalopathy in a family of Appaloosas. *Journal of the American Veterinary Medical Association*, *198*, 1005–1013.

Boehmer, L. N., Wu, M. F., John, J., & Siegel, J. M. (2004). Treatment with immunosuppressive and anti-inflammatory agents delays onset of canine genetic narcolepsy and reduces symptom severity. *Experimental Neurology*, *188*, 292–299.

Bowling, A. T. (1996). Medical genetics. In A. T. Bowling (ed.), *Horse genetics* (pp. 105–106). Wallingford: CAB International.

Brault, L. S., Cooper, C. A., Famula, T. R., Murray, J. D., & Penedo, M. C. T. (2011a). Mapping of equine cerebellar abiotrophy to ECA2 and identification of a potential causative mutation affecting expression of MUTYH. *Genomics*, *97*, 121–129.

Brault, L. S., Famula, T. R., & Penedo, M. C. T. (2011b). Inheritance of equine cerellar abiotrophy in Arabian horses. *American Journal of Veterinary*, *72*(7), 940–944.

Brooks, S. A., Gabreski, N., Miller, D., Brisbin, A., Brown, H. E. et al. (2010). Whole-genome SNP association in the horse: identification of a deletion in myosin Va responsible for lavender foal syndrome. *PLoS Genetics*, *6*, e1000909.

Carmichael, K. P., Howerth, E. W., Oliver, Jr., J. E., & Klappenbach, K. (1993). Neuroaxonal dystrophy in a group of related cats. *Journal of Veterinary Diagnostic Investigation*, *5*, 585–590.

Chakravarti, A. (1996). Endothelin receptor-mediated signaling in hirschsprung disease. *Human Molecular Genetics*, *5*, 303–307.

Chemelli, R. M., Willie, J. T., Sinton, C. M., Elmquist, J. K., Scammell, T. et al. (1999). Narcolepsy in orexin knockout mice: Molecular genetics of sleep regulation. *Cell*, *98*, 437–451.

Chrisman, C. L., Cork, L. C., & Gamble, D. A. (1984). Neuroaxonal dystrophy of Rottweiler dogs. *Journal of the American Veterinary Medical Association*, *184*, 464–467.

Collins, N., Milne, E. M., Hahn, C. N., & Dixon, P. M. (2009). Correlation of the Havemeyer endoscopic laryngeal grading system with histopathological changes in equine cricoarytenoideus dorsalis muscles. *Irish Veterinary Journal*, *62*, 334–338.

Condie, B. G., & Capecchi, M. R. (1993). Mice homozygous for a targeted disruption of Hoxd-3 (Hox-4.1) exhibit anterior transformations of the first and second cervical vertebrae, the atlas and the axis. *Development, 119*, 579–595.

Cook, W. R. (1988). Diagnosis and grading of hereditary recurrent laryngeal neuropathy in the horse. *Journal of Equine Veterinary Science, 8*, 432–455.

Cork, L. C., Troncoso, J. C., Price, D. L., Stanley, E. F., & Griffin, J. W. (1983). Canine neuroaxonal dystrophy. *Journal of Neuropathology and Exploratory Neurology, 42*, 286–296.

Cummings, J. F., de Lahunta, A., Mohammed, H. O., Divers, T. J., Summers, B. A. et al. (1995). Endothelial lipopigment as an indicator of alpha-tocopherol deficiency in two equine neurodegenerative diseases. *Acta Neuropathologia 90*, 266–272.

Davies, P. C. (2000). Shivering in a thoroughbred mare. *Canadian Veterinary Journal, 41*, 128–129.

De Belle, I., Wu, J. X., Sperandio, S., Mercola, D., & Adamson, E. D. (2003). In vivo cloning and characterization of a new growth suppressor protein TOE1 as a direct target gene of Egr1. *Journal of Biology and Chemistry, 278*, 14306–14312.

de Lahunta, A., Hatfield, C., & Dietz, A. (1989). Occipitoatlantoaxial malformation with duplication of the atlas and axis in a half Arabian foal. *Cornell Veterinary, 79*, 185–193.

DeBowes, R. M., Leipold, H. W., & Turner-Beatty, M. (1987). Cerebellar abiotrophy. *Veterinary Clinicial North American Equine Practice, 3*, 345–352.

Deen, T. (1984). Shivering, a rare equine lameness. *Equine Practice, 6*, 19–21.

Dill, S. G., Correa, M. T., Erb, H. N., deLahunta, A., Kallfelz, F.A., & Waldron, C. (1990). Factors associated with the development of equine degenerative myeloencephalopathy. *American Journal of Veterinary Research, 51*, 1300–1305.

Dilsaver, S. C., & Majchrzak, M. J. (1987). Bright artificial light subsensitizes a central muscarinic mechanism. *Life Science, 41*, 2607–2614.

Dimock, W. W. (1950). "Wobbles" – an hereditary disease in horses. *Journal of Heredity, 41*, 318–323.

Dreifuss, F. E., & Flynn, D. V. (1984). Narcolepsy in a horse. *Journal of the American Veterinary Medical Association, 184*, 131–132.

Duncan, I. D. (1992). Determination of the early age of onset of equine recurrent laryngeal neuropathy. 2. Nerve pathology. *Acta Neuropathologica, 84*, 316–321.

Eckardt, L. B., McLean, J. M., & Goodell, H. (1943). Experimental studies on headache: The genesis of pain from the eye. In the Proceedings from the Association for Research in Nervous and Mental Disease, pp. 209–227.

Escayg, A., MacDonald, B. T., Meisler, M. H., Baulac, S., Huberfeld, G. et al. (2000). Mutations of SCN1A, encoding a neuronal sodium channel, in two families with GEFS+2. *Nature Genetics, 24*, 343–345.

Falco, M. J., Whitwell, K., & Palmer, A. C. (1976). An investigation into the genetics of 'wobbler disease' in thoroughbred horses in Britain. *Equine Veterinary Journal, 8*, 165–169.

Fanelli, H. H. (2005). Coat color dilution lethal ('lavender foal syndrome'): A tetany syndrome of Arabian foals. *Equine Veterinary Education, 17*, 260–263.

Finno, C. J., Famula, T., Aleman, M., Higgins, R. J., Madigan, J. E., & Bannasch, D. L. (2012). Pedigree analysis and exclusion of alpha-tocopherol associated transfer protein (TTPA) as a candidate gene for neuroaxonal dystrophy in the American Quarter horse. *Journal of Veterinary Internal Medicine.* doi: 10.1111/jvim.12015. [Epub ahead of print].

Firshman, A. M., Baird, J. D., & Valberg, S. J. (2005). Prevalences and clinical signs of polysaccharide storage myopathy and shivers in Belgian draft horses. *Journal of the American Veterinary Medical Association, 227*, 1958–1964.

Foutz, A. S., Mitler, M. M., & Dement, W. C. (1980). Narcolepsy. *Veterinary Clinical North American Small Animal Practice, 10*, 65–80.

Gandini, G., Fatzer, R., Mariscoli, M., Spadari, A., Cipone, M., & Jaggy, A. (2004). Equine degenerative myeloencephalopathy in five Quarter Horses: Clinical and neuropathological findings. *Equine Veterinary Journal, 36*, 83–85.

Green, S. L., Davies, M., & Doucet, M. (1995). The nervous system and behaviour. In A. J. H. & I. M. Wright (eds.), *The equine manual* (pp. 967–1055). London: W. B. Saunders.

Grondahl, A. M., & Dolvik, N. I. (1993). Heritability estimations of osteochondrosis in the tibiotarsal joint and of bony fragments in the palmar/plantar portion of the metacarpo- and metatarsophalangeal joints of horses. *Journal of the American Veterinary Medical Association, 203*, 101–104.

Gu, Y., Parker, A., Wilson, T. M., Bai, H., Chang, D. Y., & Lu, A. L. (2002). Human MutY homolog, a DNA glycosylase involved in base excision repair, physically and functionally interacts with mismatch repair proteins human MutS homolog 2/human MutS homolog 6. *Journal of Biology and Chemistry, 277*, 11135–11142.

Gundlach, A. L. (1990). Disorder of the inhibitory glycine receptor: Inherited myoclonus in Poll Hereford calves. *FASEB Journal, 4*, 2761–2766.

Gundlach, A. L., Dodd, P. R., Grabara, C. S., Watson, W. E., Johnston, G. A. et al. (1988). Deficit of spinal cord glycine/strychnine receptors in inherited myoclonus of Poll Hereford calves. *Science, 241*, 1807–1810.

Gundlach, A. L., Kortz, G., Burazin, T. C., Madigan, J., & Higgins, R. J. (1993). Deficit of inhibitory glycine receptors in spinal cord from Peruvian Pasos: Evidence for an equine form of inherited myoclonus. *Brain Research, 628*, 263–270.

Hahn, C. N., Handel, I., Green, S. L., Bronsvoort, M. B., & Mayhew, I. G. (2008a). Assessment of the utility of using intra- and intervertebral minimum sagittal diameter ratios in the diagnosis of cervical vertebral malformation in horses. *Veterinary Radiology and Ultrasound*, *49*, 1–6.

Hahn, C. N., Matiasek, K., Dixon, P. M., Molony, V., Rodenacker, K., & Mayhew, I. G. (2008b). Histological and ultrastructural evidence that recurrent laryngeal neuropathy is a bilateral mononeuropathy limited to recurrent laryngeal nerves. *Equine Veterinary Journal*, *40*, 666–672.

Hahn, C. N., Mayhew, I. G., & MacKay, R. J. (1999). Nervous system. In P. T. Calahan, A. M. Merritt, J. N. Moore, & I. G. Mayhew (eds.), *Equine medicine and surgery* (pp. 904–996). St. Louis, MO: Mosby.

Harland, M. M., Stewart, A. J., Marshall, A. E., & Belknap, E. B. (2006). Diagnosis of deafness in a horse by brainstem auditory evoked potential. *Canadian Veterinary Journal*, *47*, 151–154.

Harrison, G. D., Duncan, I. D., & Clayton, M. K. (1992). Determination of the early age of onset of equine recurrent laryngeal neuropathy. 1. Muscle pathology. *Acta Neuropathologica*, *84*, 307–315.

Hultgren, B. D. (1982). Ileocolonic aganglionosis in white progeny of over spotted horses. *Journal of the American Veterinary Medical Association*, *180*, 289–292.

Hunt, L. M., Valberg, S. J., Steffenhagen, K. et al. (2005) A retrospective study of myopathies and associated gait abnormalities in 65 warmblood horses. *Journal of Veterinary Internal Medicine*, *19*, 428–429.

John, J., Wu, M. F., Maidment, N. T., Lam, H. A., Boehmer, L. N. et al. (2004). Developmental changes in CSF hypocretin-1 (orexin-A) levels in normal and genetically narcoleptic Doberman pinschers. *Journal of Physiology*, *560*, 587–592.

Khateeb, S., Flusser, H., Ofir, R., Shelef, I., Narkis, G. et al. (2006). PLA2G6 mutation underlies infantile neuroaxonal dystrophy. *American Journal of Human Genetics*, *79*, 942–948.

Koch, P., & Fischer, H. (1950). Die Oldenburger Fohlen Taxie als Erbkrankeit. *Tierarztl Umschau*, *5*, 317–320.

Kronfeld DS, M. T., & Donoghue, S. (1990). Dietary aspects of developmental orthopedic disease. *Veterinary Clinicial North American Equine Practice*, *6*, 451–465.

Lane, J. G., & Mair, T. S. (1987). Observations on headshaking in the horse. *Equine Veterinary Journal*, *19*, 331–336.

Lee, H. M., Wang, C., Hu, Z., Greeley, G. H., Makalowski, W., Hellmich, H. L., & Englander, E. W. (2002). Hypoxia induces mitochondrial DNA damage and stimulates expression of a DNA repair enzyme, the Escherichia coli MutY DNA glycosylase homolog (MYH), in vivo, in the rat brain. *Journal of Neurochemistry*, *80*, 928–937.

Lin, L., Faraco, J., Li, R., Kadotani, H., Rogers, W. et al. (1999). The sleep disorder canine narcolepsy is caused by a mutation in the hypocretin (orexin) receptor 2 gene. *Cell*, *98*, 365–376.

Liu, S. K., Dolensek, E. P., Adams, C. R., & Tappe, J. P. (1983). Myelopathy and vitamin E deficiency in six Mongolian wild horses. *Journal of the American Veterinary Medical Association*, *183*, 1266–1268.

Lunn, D. P., Cuddon, P. A., Shaftoe, S., & Archer, R. M. (1993). Familial occurrence of narcolepsy in miniature horses. *Equine Veterinary Journal*, *25*, 483–487.

Lunn, D. P., & Mayhew, I. G. (1989). The neurologic evaluation of horses. *Equine Veterinary Education*, *1*, 94–101.

Madigan, J. (1997). Congenital anomalies and genetic disorders, In J. Madigan (ed.), *Manual of equine neonatal medicine* (pp. 148–149). Woodland, CA: Saunders.

Madigan, J. E., & Bell, S. A. (1998). Characterisation of headshaking syndrome – 31 cases. *Equine Veterinary Journal Supplement*, 28–29.

Madigan, J. E., Kortz, G., Murphy, C., & Rodger, L. (1995). Photic headshaking in the horse: 7 cases. *Equine Veterinary Journal*, *27*, 306–311.

Magdesian, K. G., Williams, D. C., Aleman, M., Lecouteur, R. A., & Madigan, J. E. (2009). Evaluation of deafness in American Paint Horses by phenotype, brainstem auditory-evoked responses, and endothelin receptor B genotype. *Journal of the American Veterinary Medical Association*, *235*, 1204–1211.

Mayhew, I. G. (2005). Does the stiff horse syndrome exist? In the 23rd American College of Veterinary Internal Medicine Forum, Baltimore, MD, pp. 151–153.

Mayhew, I. G., Brown, C. M., Stowe, H. D., Trapp, A. L., Derksen, F. J., & Clement, S. F. (1987). Equine degenerative myeloencephalopathy: A vitamin E deficiency that may be familial. *Journal of Veterinary Internal Medicine*, *1*, 45–50.

Mayhew, I. G., deLahunta, A., Whitlock, R. H., Krook, L., & Tasker, J. B. (1978a). Spinal cord disease in the horse. *Cornell Veterinary*, *68*(Suppl. 6), 1–207.

Mayhew, I. G., Donawick, W. J., Green, S. L., Galligan, D. T., Stanley, E. K., & Osborne, J. (1993). Diagnosis and prediction of cervical vertebral malformation in thoroughbred foals based on semi-quantitative radiographic indicators. *Equine Veterinary Journal*, *25*, 435–440.

Mayhew, I. G., Watson, A. G., & Heissan, J. A. (1978b). Congenital occipitoatlantoaxial malformations in the horse. *Equine Veterinary Journal*, *10*, 103–113.

McCabe, L., Griffin, L. D., Kinzer, A., Chandler, M., Beckwith, J. B., & McCabe, R. B. (1990). Ovetro Lethal White Foal Syndrome: Equine model of aganglionic megacolon (Hirschsprung disease). *American Journal of Medical Genetics*, *36*, 336–340.

McKusick, V. (1990). Autosomal dominant phenotypes, ACHOO syndrome, 100820. In V. McKusick (ed.), *Mendelian inheritance in man* (p. 11). Baltimore: The John Hopkins University Press.

Metallinos, D. L., Bowling, A. T., & Rine, J. (1998). A missense mutation in the endothelin-B receptor gene is associated with Lethal White Foal Syndrome: An equine version of Hirschsprung disease. *Mammalian Genome*, *9*, 426–431.

Mignot, E., Bell, R. A., Rattazzi, C., Lovett, M., Grumet, F. C., & Dement, W. C. (1994). An immunoglobulin switchlike sequence is linked with canine narcolepsy. *Sleep*, *17*, S68–S76.

Mignot, E., Lammers, G. J., Ripley, B., Okun, M., Nevsimalova, S. et al. (2002). The role of cerebrospinal fluid hypocretin measurement in the diagnosis of narcolepsy and other hypersomnias. *Archives of Neurology*, *59*, 1553–1562.

Mignot, E., Wang, C., Rattazzi, C., Gaiser, C., Lovett, M. et al. (1991). Genetic linkage of autosomal recessive canine narcolepsy with a mu immunoglobulin heavy-chain switch-like segment. *Proceedings of the National Academy of Sciences of the United States of America*, *88*, 3475–3478.

Miller, M. M., & Collatos, C. (1997). Equine degenerative myeloencephalopathy. *Veterinary Clinical North American Equine Practice*, *13*, 43–52.

Mittel, L. (1987). Seizures in the horse. *Veterinary Clinical North American Equine Practice*, *3*, 323–332.

Moore, B. R., Reed, S. M., Biller, D. S., Kohn, C. W., & Weisbrode, S. E. (1994). Assessment of vertebral canal diameter and bony malformations of the cervical part of the spine in horses with cervical stenotic myelopathy. *American Journal of Veterinary Research*, *55*, 5–13.

Morgan, N. V., Westaway, S. K., Morton, J. E., Gregory, A., Gissen, P. et al. (2006). PLA2G6, encoding a phospholipase A2, is mutated in neurodegenerative disorders with high brain iron. *Nature Genetics*, *38*, 752–754.

Mulhardt, C., Fischer, M., Gass, P., Simon-Chazottes, D., Guenet, J. L. et al. (1994). The spastic mouse: Aberrant splicing of glycine receptor beta subunit mRNA caused by intronic insertion of L1 element. *Neuron*, *13*, 1003–1015.

Ohnesorge, B., Deegen, E., Miesner, K., & Geldermann, H. (1993). Laryngeal hemiplegia in warmblood horses – a study of stallions, mares and their offspring. *Zentralbl Veterinarmed A*, *40*, 134–154.

Page, P., Parker, R., Harper, C., Guthrie, A., & Neser, J. (2006). Clinical, clinicopathologic, postmortem examination findings and familial history of 3 Arabians with lavender foal syndrome. *Journal of Veterinary Internal Medicine*, *20*, 1491–1494.

Palmer, A.C., Blakemore, W. F., Cook, W. R., Platt, H., & Whitwell, K. E. (1973). Cerebellar hypoplasia and degeneration in the young Arab horse: Clinical and neuropathological features. *Veterinary Record*, *93*, 62–66.

Panayiotopoulos, C. P. (2005). Idiopathic generalized epilepsies: A review and modern approach. *Epilepsia*, *46*(Suppl. 9), 1–6.

Pascoe, R. R., & Knottenbelt, D. C. (1999). Congenital/developmental diseases. In R. R. Pascoe & D. C. Knottenbelt (eds.), *Manual of equine dermatology* (pp. 148–149). London: W. B. Saunders.

Peck, K. E., Hines, M. T., Mealey, K. L., & Mealey, R. H. (2001). Pharmacokinetics of imipramine in narcoleptic horses. *American Journal of Veterinary Research*, *62*, 783–786.

Pies, R. (1990). Seasonal affective disorder and the photic sneeze response. *American Journal of Psychiatry*, *147*, 1094.

Poncet, P. A., Montavon, S., Gaillard, C., Barrelet, F., Straub, R., & Gerber, H. (1989) A preliminary report on the possible genetic basis of laryngeal hemiplegia. *Equine Veterinary Journal*, *21*, 137–138.

Price, E. R., & Fisher, D. E. (2001). Sensorineural deafness and pigmentation genes: Melanocytes and the Mitf transcriptional network. *Neuron*, *30*, 15–18.

Reed, S. M. (2007). Equine cervical myelopathy – surgical, medical, and dietary management of affected horses and outcome. In the North American Veterinary Conference, Orlando, FL, pp. 160–162.

Rosenstein, D. S., Schott, II, H. C., & Stickle, R. L. (2000). Imaging diagnosis – occipitoatlantoaxial malformation in a miniature horse foal. *Veterinary Radiology and Ultrasound*, *41*, 218–219.

Sanal, O., Ersoy, F., Tezcan, I., Metin, A., Yel, L. et al. (2002). Griscelli disease: Genotype-phenotype correlation in an array of clinical heterogeneity. *Journal of Clinical Immunology*, *22*, 237–243.

Sanchez-Mejias, A., Fernandez, R. M., Lopez-Alonso, M., Antinolo, G., & Borrego, S. (2010). New roles of EDNRB and EDN3 in the pathogenesis of Hirschsprung disease. *Genetics & Medicine*, *12*, 39–43.

Santschi, E. M., Purdy, A. K., Valberg, S. J., Vrotsos, P. D., Kaese, H., & Mickelson, J. R. (1998) Endothelin receptor B polymorphism associated with lethal white foal syndrome in horses. *Mammalian Genome*, *9*, 306–309.

Santschi, E. M., Vrotsos, P. D., Purdy, A. K., & Mickelson, J. R. (2001). Incidence of the endothelin receptor B mutation that causes lethal white foal syndrome in white-patterned horses. *American Journal of Veterinary Research*, *62*, 97–103.

Selcer, R. R., & Oliver, J. E. (1975). Cervical spondylopathy: Wobbler syndrome in dogs. *Journal of the American Animal Hospitals Associaton*, *11*, 175–177.

Sheather, A. L. (1924). Fainting in foals. *Journal of Comparative Pathology & Therapy*, *37*, 106–113.

Shiang, R., Ryan, S. G., Zhu, Y. Z., Hahn, A. F., O'Connell, P., & Wasmuth, J. J. (1993). Mutations in the alpha 1 subunit of the inhibitory glycine receptor cause the dominant neurologic disorder, hyperekplexia. *Nature Genetics*, *5*, 351–358.

Siso, S., Ferrer, I., & Pumarola, M. (2003). Abnormal synaptic protein expression in two Arabian horses with equine degenerative myeloencephalopathy. *Veterinary Journal*, *166*, 238–243.

Siso, S., Hanzlicek, D., Fluehmann, G., Kathmann, I., Tomek, A., Papa, V., & Vandevelde, M. (2006). Neurodegenerative diseases in domestic animals: A comparative review. *Veterinary Journal*, *171*, 20–38.

Sperandio, S., Tardito, S., Surzycki, A., Latterich, M., & de Belle, I. (2009). TOE1 interacts with p53 to modulate its transactivation potential. *FEBS Letters*, *583*, 2165–2170.

Sponseller, M. L. (1967). Equine cerebellar hypoplasia and degeneration. In the 13th Annual Convention of the American Association of Equine Practitioners, New Orleans, LA.

Steel, K. P., & Barkway, C. (1989). Another role for melanocytes: Their importance for normal stria vascularis development in the mammalian inner ear. *Development*, *107*, 453–463.

Stewart, R. H., Reed, S. M., & Weisbrode, S. E. (1991). Frequency and severity of osteochondrosis in horses with cervical stenotic myelopathy. *American Journal of Veterinary Research*, *52*, 873–879.

Sweeney, C. R., Hendricks, J. C., Beech, J., & Morrison, A. R. (1983). Narcolepsy in a horse. *Journal of the American Veterinary Medical Association*, *183*, 126–128.

Trotter, E. J. J. (1985). Canine wobbler syndrome. In C. D. Newton & D. M. Nunamaker (eds.), *Textbook of small animal orthopaedics*. Ithaca, NY: International Veterinary Information Service.

Turner Beatty, M., Leipold, H. W., Cash, W., & DeBowes, R. (1985). Cerebellar disease in Arabian horses. In the 31st Annual Proceedings of the American Association of Equine Practitioners, Toronto, Canada, pp. 241–255.

Tutt, J. F. D. (1964). Diseases of the nervous system. In M. H. Hayhes (ed.), *Veterinary notes for horse owners: A manual of horse medicine and surgery* (pp. 137–155). London: Stanley Paul.

Vaala, W. E. (2003). Perinatal asphyxia syndrome in foals, In N. E. Robinson (ed.), *Current therapy in equine medicine* (pp. 644–649). Philadelphia: Saunders.

Valberg, S. J. (2009). Applications for Equine Genetics – Shivers. *My Horse University*. Available online at: http://www.myhorseuniversity.com/resources/webcasts/genetics_oct_09

Valentine, B. A., de Lahunta, A., Divers, T. J., Ducharme, N. G., & Orcutt, R. S. (1999). Clinical and pathologic findings in two draft horses with progressive muscle atrophy, neuromuscular weakness, and abnormal gait characteristic of shivers syndrome. *Journal of the American Veterinary Medical Association*, *215*, 1661–1665.

Van Biervliet, J. (2007). An evidence-based approach to clinical questions in the practice of equine neurology. *Veterinary Clinical North American Equine Practice*, *23*, 317–328.

van Biervliet, J., Scrivani, P. V., Divers, T. J., Erb, H. N., de Lahunta, A., & Nixon, A. (2004). Evaluation of decision criteria for detection of spinal cord compression based on cervical myelography in horses: 38 cases (1981–2001). *Equine Veterinary Journal*, *36*, 14–20.

Vonderfecht, S. L., Bowling, A. T., & Cohen, M. (1983) Congenital intestinal aganglionosis in white foals. *Veterinary Pathology*, *20*, 65–70.

Wagner, P. C., Grant, B. D., Watrous, B. J., & Appell, L. H. B. (1985). A study of the heritability of cervical vertebral malformation in horses. In the 31st Annual Convetion of the American Association of Equine Practitioners, Toronto, Canada, pp. 43–50.

Wallace, R. H., Berkovic, S. F., Howell, R. A., Sutherland, G. R., & Mulley, J. C. (1996). Suggestion of a major gene for familial febrile convulsions mapping to 8q13-21. *Journal of Medical Genetics*, *33*, 308–312.

Watson, A. G., & Mayhew, I. G. (1986) Familial congenital occipitoatlantoaxial malformation (OAAM) in the Arabian horse. *Spine*, *11*, 334–339.

Watson, A. G., Quick, C. B., & de Lahunta, A. (1978). Congenital occipitoatlantoaxial malformation (OAAM) with two atlases in a young Arabian horse. *Anatomy & Histological Embryology*, *7*, 354.

Whitman, B. W., & Packer, R. J. (1993). The photic sneeze reflex: Literature review and discussion. *Neurology*, *43*, 868–871.

Williams, W. (1886). Immobilite, shivering, spained back, etc. In W. Williams (ed.), *The principles and practice of veterinary surgery* (pp. 247–249). New York: William R. Jenkins.

Wilson, W. D., Hughes, S. J., Ghoshal, N. G., & McNeel, S. V. (1985). Occipitoatlantoaxial malformation in two non-Arabian horses. *Journal of the American Veterinary Medical Association*, *187*, 36–40.

Zhou, B., Westaway, S. K., Levinson, B., Johnson, M. A., Gitschier, J., & Hayflick, S. J. (2001). A novel pantothenate kinase gene (PANK2) is defective in Hallervorden-Spatz syndrome. *Nature Genetics*, *28*, 345–349.

15 Molecular genetic testing and karyotyping in the horse

M. C. T. Penedo and Terje Raudsepp

Introduction

The goal of the horse genome project has been to develop a suite of tools that enable research on the molecular basis underlying disease and other phenotypes with simple and complex modes of inheritance (Chowdhary & Bailey, 2003; Spencer & Davis, 2007; Wade et al., 2009). The accomplishments to date in resource development, gene mapping, and mutation discovery are reviewed in other chapters of this book. The discoveries in the development of genetic markers or in the identification of causative mutations for disease or coat color phenotypes have found prompt application as molecular genetic tests that have and will continue to benefit the horse industry at large. The impact that horse genomic research has had in accelerating the development of molecular tests can be appreciated by the rate at which mutation discoveries are being reported in the scientific literature. Between 1990 and 1999, the molecular basis for three diseases and one coat color phenotype were published. Between 2000 and 2009, the numbers are six diseases and eight coat colors.

Molecular genetic tests employ methods based on polymerase chain reaction (PCR) for amplification of specific DNA targets and detection technologies that permit the unambiguous identification of variants present in the genome of individual animals (genotyping). The tests fall into two categories: identification/parentage testing and disease/trait diagnostic tests. Horse genotyping services are provided by private and institutional laboratories that specialize in the processing of biological samples submitted by individual owners, breed organizations, or veterinarians. Another important aspect of genetic testing concerns the detection of chromosomal abnormalities. Anomalies that affect sex chromosomes and result in infertility are not a rare occurrence in horses, with overall incidence of about 2% (Bugno, Slota, & Koscielny, 2007). Among mares presenting gonadal dysgenesis, 54–56% of the cases are explained by sex chromosome abnormalities (Bowling et al., 1987; Bugno et al., 2007).

This chapter provides an overview of current applications of genetic testing in horses for identification and parentage, phenotypic traits, and cytogenetic analyses.

Genetic Tests for Individual Identification and Parentage Analysis

Genetic markers for individual identification and parentage testing have been used by the horse industry since the early 1960s to validate pedigree records or to solve cases where parentage

Equine Genomics, First Edition. Edited by Bhanu P. Chowdhary.

is questioned. Before the advent of DNA-genotyping technologies, the procedures were based on blood group and biochemical polymorphisms assayed by serologic and electrophoretic methods (Sandberg & Cothran, 2000). The development of the PCR method and the discovery of polymorphic microsatellite DNA sequences in the horse genome allowed the implementation of robust and cost-effective procedures that became the standard for horse genotyping in the late 1990s (Bowling et al., 1997). Microsatellites or short tandem repeats (STRs) are repetitive sequence motifs, commonly of 2–4 bases, that are found throughout the genome and that display polymorphism in the number of repeats. The Mendelian codominant inheritance of STR variants, degree of polymorphism, and the flexibility to use sample tissues other than blood established STRs as the markers of choice for identification and parentage testing. The first horse STR panel for parentage testing was described in 1997 (Bowling et al., 1997).

The STR panels for horse genotyping currently in use typically contain 17–20 markers that are very polymorphic, stably inherited, have low mutation rates (1×10^{-4} or less), and are easy to score. The international nature of the horse industry, with movement of live animals, frozen embryos, and frozen semen, has required standardization of marker panels and allele nomenclature to facilitate inter-laboratory genotype record exchange. Sharing of DNA records eliminates the need of retesting animals and provides a means to verify horse identity if questions arise about animals transferred from one country to another. The standardization of horse STR genotyping occurs under the auspices of the International Society of Animal Genetics (ISAG) by means of biennial DNA comparison tests and workshops held since 1996. The International Studbook Committee also played a significant role in this process by requiring DNA genotyping standardization among laboratories providing services to Thoroughbred registries. These have been important activities that have secured seamless record exchange across laboratories and breed registries throughout the world. As an example of the global scale of horse genotyping, 80 laboratories from 31 countries participated in the 2009–2010 ISAG comparison test.

Genotyping methods for STRs are based on multiplex PCR amplification of DNA with fluorescence-tagged primers, separation by capillary or gel electrophoresis, amplicon detection by laser detection systems, and software-driven fragment size analysis. Commercially available kits with 17 markers, such as the Equine Genotypes™ Panel 1.1 (Finnzymes Diagnostics) and StockMarks® for Horses Equine 17-Plex Genotyping Kit (Applied Biosystems), provide a cost-effective source of standardized reagents, although high-throughput laboratories opt to maintain stocks of primers and reagents to prepare in-house master mixes.

Nomenclature of STR alleles follows the recommendations by the Horse Standard Committee of ISAG proposed in Tours, France, 1996. Alleles are designated with alphabetical symbols, from smallest to largest, based on a middle-sized allele having been assigned as *M* by the laboratory that characterized the STR. Comparison testing of reference samples allows each laboratory to make the appropriate conversions of allele sizes to the alphabetical equivalent. From these early workshop activities a set of 9 STRs, known as the ISAG International Panel or ISAG Panel, was selected to be included in any marker panel implemented by laboratories. Three additional STRs were added at the last Horse Standing Committee workshop of 2010 in Edinburgh, Scotland. The ISAG panel thus consists of the 12 dinucleotide markers shown in Table 15.1. The allele size ranges and *M* allele values provided are according to the given primers and dye labels, which are used by the Veterinary Genetics Laboratory (VGL), University of California at Davis. Description of the *M* alleles in terms of the number of repeat units is based on the work of van de Goor, Panneman, and van Haeringen (2010).

The ISAG Panel defines the minimum number of markers to be used for horse identification and parentage analysis. In practice, panels in current usage contain 17–20 markers. Such panels have a very high level of performance for individual identification and parentage testing,

Table 15.1 International ISAG STR panel for horse identification and parentage.

Locus	Chromosome location	Primer sequences	Label	Size range (bp)	Repeat # and (size) of M allele	Ref.
AHT4	24q14	F: AACCGCCTGAGCAAGGAAGT R: CCCAGAGAGTTTACCCT	FAM	145–165	30 (155)	a
AHT5	8	F: ACGGACACATCCCTGCCTGC R: GCAGGCTAAGGAGGCTCAGC	PET	127–151	19 (139)	a
ASB2	15q21.3–23	F: CACTAAGTGTCGTTTCAGAAGG R: (GTTT)CACAACTGAGTTCTCTGATAGG	FAM	224–260	20 (246)	b
ASB17	2p14–15	F: GGTGGCCAACTGCTAACCTC R: (GTTTCTT)TGGAGGGCGGTACCTTTGTA	PET	198–242	20 (218)	b
ASB23	3q22.1–22.3	F: GAGGGCAGCAGGTTGGGAAGG R: ACATCCTGGTCAAATCACAGTCC	VIC	181–215	21 (195)	c
HMS2	10	F: CAGTAAATCACAGGAACTAAT R: ACTGCCAAGGAAGCCACTACA	FAM	276–302	20 (288)	d
HMS3	9	F: ACATCAGTCAGAAGCTGCGAAC R: CCCCTCTTGCTCTAAAGCCCCA	VIC	257–283	25 (269)	d
HMS6	4	F: GAAGCTGCCAGTATTCAACCATTG R: CTCCATCTTGTGAAGTGTAACTCA	VIC	151–169	15 (163)	d
HMS7	1q25	F: CAGGAAACTCATGTTGATACCATC R: TGTTGTTGAAACATACCTTGACTGT	FAM	166–190	19 (178)	d
HTG4	9	F: CTATCTCAGTCTTGATTGCAGGAC R: CTCCCTCCCTCCCTCTGTTCTC	FAM	121–141	32 (141)	e
HTG10	21	F: CCTAATGTCATATGGAAAGCCTTG R: TGGGCTTTTTATTCTGATCTGTCACATTT	NED	135–165	21 (149)	f
VHL20	30	F: TCGATGGTGCTGTCAAGTCC R: (GTTTCTT)AACTCAGGGAGAATCTTCCTCA	NED	105–129	17 (115)	g

a: Binns et al. (1995); b: Breen et al. (1997); c: Irvin et al. (1998); d: Guérin et al. (1994); e: Ellegren et al. (1992); f: Marklund et al. (1994); g: van Haeringen et al. (1994).

based on conventional metrics that assume random mating. Because horse-breeding practices often involve mating of close relatives, information from additional markers may be needed for specific situations. By recommendation of the Horse Standing Committee in 2010, a set of 13 STRs (TKY287, TKY294, TKY297, TKY301, TKY312, TKY325, TKY333, TKY337, TKY341, TKY343, TKY344, TKY374, and TKY394) from the panel developed by Tozaki et al. (2001) has been designated as the secondary panel. Measures of genetic variability and exclusion probabilities for the two panels obtained from records of the VGL (available at http://www.vgl.ucdavis.edu/genomic/Supplemental%20Table%2018.S1.pdf) demonstrate the usefulness of the two panels across 26 breeds.

Horse racing and forensic analysis

Horse identification and sample matching is becoming a more common practice in the horse racing industry. Proper identification of horses competing in official races and maintenance of legal chain of custody of samples undergoing drug testing are important in doping control regulations. Positive results from drug tests can be challenged by claims of sample switching. The STR panels described above provide the necessary level of discrimination to match DNA profiles of a horse obtained from blood or hair root with those obtained from urine samples submitted to a drug-testing laboratory. A multiplex panel with 24 STRs with 21 horse and 3 human markers has also been developed and validated for Thoroughbreds and Standardbreds to address the specific requirements of doping control in racetracks, including detection of fraudulent contamination of horse samples with human urine (Chen et al., 2010). Horse genotyping laboratories are usually equipped to handle forensic cases.

Genetic Tests for Genetic Diseases

Since the molecular basis of a horse genetic disease was first described (hyperkalemic periodic paralysis or HYPP) (Rudolph et al., 1992), there has been a steady stream of discoveries of the causative mutations for other inherited defects. DNA tests offered by commercial laboratories are currently available for 10 genetic diseases – glycogen branching enzyme deficiency (GBED), hereditary equine regional dermal asthenia (HERDA), HYPP, lethal white overo foal syndrome (LWO), two forms of junctional epidermolysis bullosa (JEB), lavender foal syndrome (LFS), malignant hyperthermia (MH), polysaccharide storage myopathy type 1 (PSSM1), and severe combined immunodeficiency (SCID). Details about the specific mutations and horse breeds in which they occur are provided in Table 18.2. Reviews of the conditions and the supporting genomic research are covered in other chapters. None of the causative mutations are restricted to one breed. As reviewed by Finno, Spier, and Valberg (2009), genetic testing for each disease is applicable to related breeds and crossbred animals such as, for example, GBED, HERDA, and HYPP in Quarter Horses, Appaloosa, American Paint horse, and Quarter Horse crossbred animals; SCID in Arabians and part-Arabians; JEB (*LAMC2*) in Belgian, Breton, Comtois, Vlaams Paard, and Belgische Koudbloed Flander; and PSSM1, which has widespread occurrence among breeds.

Recent work by Bellone et al. (2010) identified *TRPM1* (*Transient Receptor Potential Cation Channel, Subfamily M, Member 1*) as the gene responsible for the white spotting pattern known as leopard complex (LP), or Appaloosa spotting. Homozygosity for the *LP* allele is associated with the eye defect known as congenital stationary night blindness (CSNB). A non-coding SNP (ECA1:108249293C>T), has been found to be in complete association with *LP* and CSNB in Appaloosas and Knabstruppers, and can be used as proxy in a diagnostic assay for zygosity of *LP* and CSNB. The causative LP mutation has since been identified and will soon become available from service laboratories.

PCR methods for detection of normal (wild-type) and disease-causing alleles are described in references listed in Table 15.2 or elsewhere as indicated next for specific diseases. The methods are based on restriction fragment length polymorphism (RFLP) procedures for GBED (Wagner et al., 2006), HERDA, HYPP, LFS, MH, and PSSM1; fragment length differences for both types of JEB and for SCID (Bernoco & Bailey, 1998), or allele-specific amplification for LWO (Metallinos, Bowling, & Rine, 1998). Laboratories providing genetic disease testing are found in many countries (e.g., Australia, Brazil, Canada, France, Germany, Italy, South Africa, Sweden, The Netherlands, the United Kingdom, and the United States). The specific tests offered by each laboratory vary according to local demand or licensing restrictions for patent protected tests. GBED, HERDA, HYPP, JEB, LWO, and PSSM1 are patented in the United States; SCID is covered by international patent in the United States, Canada, and the United Kingdom. In the United States, LFS is available only from Cornell University's Animal Health Diagnostic Center, PSSM1 from the University of Minnesota's Veterinary Diagnostic Laboratory, and SCID from VetGen (Ann Arbor, MI).

Genetic disease testing is not subject to oversight for quality control of service laboratories or for standardization of allele nomenclature. As the primary users of the tests are breeders, owners, and veterinarians, a uniform designation of normal and disease alleles facilitates the understanding of

Table 15.2 Genetic disease and allelic variants identified by DNA testing.

Disease	Gene	Alleles	Breeds[a]	Mutation Description	Ref.
GBED	*GBE1*	*N, G*	QH and related	ECA26 (see ref.)	Ward et al. (2004)
HERDA	*PPIB*	*N, HRD*	QH and related	ECA1g.128056148G>A	Tryon et al. (2007)
HYPP	*SCN4A*	*N, H*	QH and related	ECA11g.15500439C>G	Rudolph et al. (1992)
LWO	*EDNRB3*	*N, O*	Several[b]	ECA17g.50624681-50624682TC>AG	*
JEB	*LAMC2*	*N, J*	BE and related	ECA5g.20256789-20256790insC	Spirito et al. (2002)
JEB	*LAMA3*	*N, Js*	AS	ECA8g.3724_10312del6589	Graves et al. (2009)
LFS	*MYO5A*	*N, L*	AR and related	ECA1g.138235715del	Brooks et al. (2010)
MH	*RyR1*	*N, MH*	QH, PT	ECA10g.9554699C>G	Aleman et al. (2004)
PSSM	*GYS1*	*N, P*	Several[c]	ECA10g.18940346G>A	McCue et al. (2008)
SCID	*PRKDC*	*N, S*	AR and related	ECA9g.35528429-35528433del	Shin et al. (1997)

a: QH (Quarter Horse), BE (Belgian Draft), AS (American Saddlebred), AR (Arabian), PT (American Paint Horse); b: breeds with overo white spotting; c: widespread including QH and related, Warmbloods and Draft breeds. *: Metallinos et al. (1998), Santschi et al. (1998), Yang et al. (1998).

results reported by different test providers. In Table 15.2 we propose a standard nomenclature based on assigning the letter N to the normal allele and a letter symbol for the disease allele that is easily associated with each condition.

Mandatory testing for genetic diseases is not standard practice among registries. Exceptions to this are the registration regulations of the American Quarter Horse Association (AQHA) for HYPP and of the U.S. and Canada Belgian Draft registries for JEB. In 1996, AQHA instituted the requirement of HYPP testing and disclosure of status on registration certificates for all foals that descended from the stallion Impressive and that were born on or after January 1, 1998. The rule was amended in 2004 to require parentage verification and HYPP testing of all of Impressive's descendants born on or after January 1, 2007, and to make ineligible for registration HYPP homozygous foals. JEB became a mandatory test for Belgian stallions coming into service after November 1, 2002, for mares inseminated with frozen semen or used as embryo donors, and for the resulting foals.

HYPP, JEB, and SCID have the longest history of genetic testing. A survey of 12 age cohorts (1998 to 2009) of Belgians and Quarter Horses tested at the VGL (Table 15.3; Penedo,

Table 15.3 Survey of JEB and HYPP incidence and disease allele frequency in 12 year of birth cohorts of Belgians and Quarter Horses.

	JEB			HYPP			
Year of Birth	N	Freq (*N/J*)	Freq (*J*)	N	Freq (*H/H*)	Freq (*N/H*)	Freq (*H*)
1998	79	0.14	0.07	179	0	0.24	0.12
1999	121	0.10	0.05	179	0	0.29	0.15
2000	200	0.13	0.06	214	0.005	0.21	0.11
2001	274	0.14	0.07	273	0	0.30	0.15
2002	215	0.13	0.07	292	0.007	0.29	0.15
2003	209	0.11	0.05	378	0.003	0.26	0.13
2004	217	0.10	0.05	403	0.007	0.37	0.19
2005	187	0.11	0.06	455	0.004	0.35	0.18
2006	159	0.10	0.05	1060	0.016	0.37	0.20
2007	121	0.12	0.06	4530	0.010	0.39	0.20
2008	99	0.12	0.06	3679	0.006	0.41	0.21
2009	45	0.11	0.06	1787	0.011	0.47	0.24

personal communcation) shows no change in disease allele frequency for JEB and an increase for HYPP of disease incidence and allele frequency. Information about SCID testing from the Arabian Horse Association (http://www.arabianhorses.org/education/genetic/docs/10Genetic_SCID_2010.pdf) suggests a pattern similar to JEB.

Although the estimates are not based on random sampling within each year, the data for Quarter Horses suggests continued human-driven positive selection for HYPP in the halter subpopulation (Naylor, 1994; Tryon et al., 2009). The situation with JEB and SCID, both fatal at an early age, shows that carriers remain reproductively active. If other diseases follow the same trend of JEB and SCID, the defective alleles will remain in the gene pool of breeds and testing for these mutations will be relevant for a long time.

Genetic Tests for Phenotypic Traits

Horse coat color phenotypes have been collected and propagated by breeders since the early days of domestication. Analyses of ancient horse DNA samples provided evidence for rapid accumulation of color variants in the fossil record of early domesticates dating 5,000 to 2,600 years before present (Ludwig et al., 2009). Most modern breeds display variation in coat color resulting from action of several genes, with rare exceptions in which a specific phenotype is fixed (e.g., all black Friesian, all flaxen Haflinger). Coat color has long been the object of interest to breeders, intent on producing animals with particular colors that enhance the aesthetic and economic value of animals.

Since the discovery of the first coat color mutation in horses (Marklund et al., 1996), advances in horse genomic resources have accelerated such that the genes and mutations for most major color variants have now been identified. The genomic research of coat color traits is reviewed in depth in Chapter 11. We provide here information about diagnostic tests available to the horse industry based on published work. These tests determine genotypes that help with coat color classification of animals and that allow breeders to devise mating schemes for consistent production of foals with desired phenotypes.

The known genes and causative mutations associated with coat color variants are summarized in Table 15.4. The allele nomenclature for Extension and Agouti is in common usage among laboratories. For the dilution and white-spotting genes, the allele designations follow the same scheme used with genetic disease alleles, with *N* indicating absence of the mutation that defines the color. The 11 Dominant White *KIT* variants, all presumed to be homozygous-lethal (Haase et al., 2007; Haase et al., 2009), have not found wide application in coat color testing yet because of their restricted occurrence to specific lineages within breeds, very recent origin in most cases, and rare incidence. However, testing for presence of the W4 mutation (Haase et al., 2007) is now one of the requirements for registration with the Camarillo White Horse Association in the United States, which represents a small historic breed founded by one white stallion. In breeds with overo white spotting pattern, testing of breeding stock for presence of the mutation associated with Lethal White Overo Foal Syndrome is common practice among owners to reduce the risk of producing homozygous foals that are afflicted with ileocolonic aganglionosis and die shortly after birth because of functional intestinal obstruction.

As mentioned in the previous section, a non-coding SNP in *TRPM1,* associated with leopard complex spotting (Bellone et al., 2010) in Appaloosas and Knabstruppers, can be used as a genetic marker for the pattern in these breeds and should soon become available from service laboratories. Other coat color phenotypes (e.g., dun dilution, roan, splash white) are currently under investigation and diagnostic tests for the specific mutations should become available in the near future. Multi-locus allele frequencies of coat color variants in different breeds are rarely documented in the literature.

Table 15.4 Coat color genes and allelic variants identified by DNA testing.

Effect/Locus	Gene	Alleles	Mutation description	Ref.
Base Color				
Extension	*MC1R*	*E* (black), *e* (red)	ECA3g.36259552C>T	a
		e^a (red)	ECA3g.36259552T, 36259555G>A	b
Agouti	*ASIP*	*A* (bay), *a* (black)	*a*: ECA22g.25168579-25168589del	c
Color Dilution				
Cream	*SLC45A2*	*N*, Cr	ECA21g.30666626G>A	d
Champagne	*SLC36A1*	*N, Ch*	ECA14g.26701114C>G	e
Silver	*PMEL17*	*N, Z*	ECA6g.73665315C>T (Exon 11)*	f
White Pattern				
White	*KIT*	*N, W*	Several variants (see ref.)	g
Gray	*STX17*	*N, G*	ECA15: large duplication (see ref.)	h
Tobiano	*KIT*	*N, TO*	ECA3, large inversion (see ref.)	i
Sabino-1	*KIT*	*N, SB1*	ECA3g.77735542T>A	j
Overo (LWO)	*EDNRB3*	*N, O*	ECA17g.50624681-50624682TC>AG	k

a: Marklund et al. (1996); b: Wagner and Reissmann (2000); c: Rieder et al. (2001); d: Mariat et al. (2003); e: Cook et al. (2008); f: Brunberg et al. (2006); g: Haase et al. (2007, 2009); h: Pielberg et al. (2008); i: Brooks et al. (2007); j: Brooks & Bailey (2005); k: Metallinos et al. (1998). *: A SNP in intron 9 is also associated with Silver dilution.

Data for 11 breeds in the United States obtained from the VGL (Table 15.5; Penedo, personal communication) illustrate the widespread occurrence of the variants in Extension, Agouti, Cream, and Gray loci, and the more limited distribution of Champagne and Silver dilutions, and white-spotting patterns, such as Sabino-1, Tobiano, and Lethal White Overo. We note that frequencies for *TO* and *O* in Paints may be overestimated because of a bias in testing for homozygosity and

Table 15.5 Allele frequencies of coat color variants in 11 horse breeds: Andalusian (AN), Appaloosa (AP), Gypsy Cob (GC), Morgan Horse (MH), Miniature Horse (MI), Paint (PT), Quarter Horse (QH), Rocky Mountain Horse (RM), Tennessee Walking Horse (TW), and Thoroughbred (TB).

Locus	Allele	AN (124)	AP (61)	AR (124)	GC (64)	MH (103)	MI (112)	PT (170)	QH (208)	RM (82)	TW (76)	TB (100)
Extension	*E*	0.83	0.59	0.57	0.73	0.47	0.51	0.28	0.44	0.67	0.48	0.61
	e	0.17	0.41	0.43	0.27	0.53	0.49	0.72	0.56	0.33	0.52	0.39
Agouti	*A*	0.40	0.35	0.43	0.09	0.44	0.25	0.51	0.44	0.16	0.30	0.69
	a	0.60	0.65	0.57	0.91	0.56	0.75	0.49	0.56	0.84	0.70	0.31
Cream	*N*	0.93	0.91	1.00	0.95	0.80	0.80	0.95	0.79	0.92	0.72	0.97
	Cr	0.07	0.09	0.00	0.05	0.20	0.20	0.05	0.21	0.08	0.28	0.03
Champagne	*N*	1.00	0.99	1.00	1.00	1.00	0.99	1.00	0.99	1.00	0.97	1.00
	Ch	0	0.01	0	0	0	0.01	0	0.01	0	0.03	0
Silver	*N*	1.00	0.98	1.00	0.96	0.98	0.85	1.00	1.00	0.56	0.95	1.00
	Z	0	0.02	0	0.04	0.02	0.15	0	0	0.44	0.05	0
Sabino-1	*N*	1.00	1.00	1.00	0.99	1.00	0.96	0.97	1.00	1.00	0.92	1.00
	SB1	0	0	0	0.01	0	0.04	0.03	0	0	0.08	0
Gray	*N*	0.81	0.98	0.88	0.98	0.97	0.95	0.99	0.68	0.99	0.95	0.97
	G	0.19	0.02	0.12	0.02	0.03	0.05	0.01	0.32	0.01	0.05	0.03
Tobiano	*N*	1.00	1.00	1.00	0.59	1.00	0.74	0.81	1.00	1.00	0.91	1.00
	TO	0	0	0	0.41	0	0.26	0.19	0	0	0.09	0
Overo (LWO)	*N*	1.00	1.00	1.00	1.00	1.00	0.95	0.72	1.00	1.00	0.99	1.00
	O	0	0	0	0	0	0.05	0.28	0*	0	0.01	0*

* Known to occur as rare variant in breed.

presence of *O* in overo-patterned horses, respectively. In a random sampling of Paint horses, the allele frequency of LWO was estimated to be 0.11 (Tryon et al., 2009).

Genotyping methods for individual loci are described in each of the original papers, but protocols for simultaneous genotyping of multiple coat color loci have been proposed based on SNAPshot™ (Kakoi et al., 2009) and real-time PCR (Royo et al., 2008) techniques. Other SNP-genotyping technologies and instruments available from companies such as Applied Biosystems, Fluidigm, KBioscience, Illumina, and Sequenom lend themselves to simultaneous genotyping for all known color variants, with the possible exception of Gray for which accurate genotyping relies on long-range PCR to detect presence and dosage of the 4.6 Kb tandem duplication.

Future Developments in Molecular Genetic Testing

DNA testing for identification and parentage verification is an integral part of the process to record pedigrees with breed organizations. The need for this application will continue to exist well into the future. The number of diagnostic tests stands to increase exponentially as a result of ongoing research on simple and complex traits related to health, coat color, reproduction, and performance. The recent report of a SNP associated with race distance aptitude in Thoroughbreds (Hill et al., 2010) provides an example of future trends in applied genomics; markers predictive of "performance" will be in demand and become an important component of genetic testing. As most trait-related mutations are simple base variations, use of SNP-genotyping technologies allow implementation of more streamlined and cost-effective procedures for simultaneous screening of multiple traits. SNP panels for identification and parentage have begun to emerge (see Hirota et al., 2010). Data from the EquineSNP50 chip are being mined and will contribute to the selection of a suitable standard SNP panel. Combining SNPs for parentage and economic traits in a single test is a reality in other livestock species, namely cattle. Future development in horse molecular genetic testing is poised to follow a similar path.

Karyotyping Services Available for the Horse

The basic chromosome analysis

The aim of clinical karyotyping or chromosome analysis is to determine whether or not a horse has a normal set of chromosomes. The analysis is usually carried out on animals with various reproductive and developmental disorders to ascertain the possible cause of the abnormal phenotype. Additionally, cytogenetic tests are ordered by insurance companies, horse breeders, and owners to evaluate potential breeding stallions, broodmares, or the horses to be purchased.

The analysis includes counting the diploid chromosome number – which for the normal horse is 2n = 64 – arranging homologous chromosomes into pairs and assessing their morphology, and ordering chromosomes into a karyotype according to the international standard (ISCNH, 1997). The analysis also involves identification of the sex chromosomes to count their number, check morphology and determine whether or not the genetic sex corresponds to the phenotypic sex. Female horses should have two X chromosomes and male horses should have one X and one Y chromosome.

Chromosome preparations are typically obtained from peripheral blood lymphocyte culture (see Raudsepp & Chowdhary, 2008; Durkin, Raudsepp, & Chowdhary, 2010). In case of suspected chimerism or mosaicism, chromosomes are also analyzed from cultured skin fibroblasts. For lymphocyte culture, 5–10 mL of blood is aseptically collected from the jugular vein in heparinized vacutainers and allowed to settle. Lymphocytes are obtained from the buffy coat and are cultured for 72 hours in medium supplemented with fetal calf serum, pokeweed mitogen, and antibiotics. The cultures are harvested with colcemid using standard approaches. The cells are subjected to hypotonic treatment and four rounds of fixation. Finally, the fixed cell suspension is dropped on a clean wet glass microscope slide to burst the cells and obtain metaphase spreads.

For basic chromosome analysis, the preparations are routinely stained with Giemsa, which is usually sufficient for counting the chromosome number and arranging chromosomes roughly into a karyotype. To further verify the sex chromosomes, the preparations are stained by C-banding (Arrighi & Hsu, 1971). The X chromosome shows a characteristic dark C-band at Xq21, which corresponds to facultative heterochromatin, while the predominantly heterochromatic Y chromosome stains overall dark (see Chapter 5 for details). An example of clinical cytogenetic analysis results is shown in Figure 15.1.

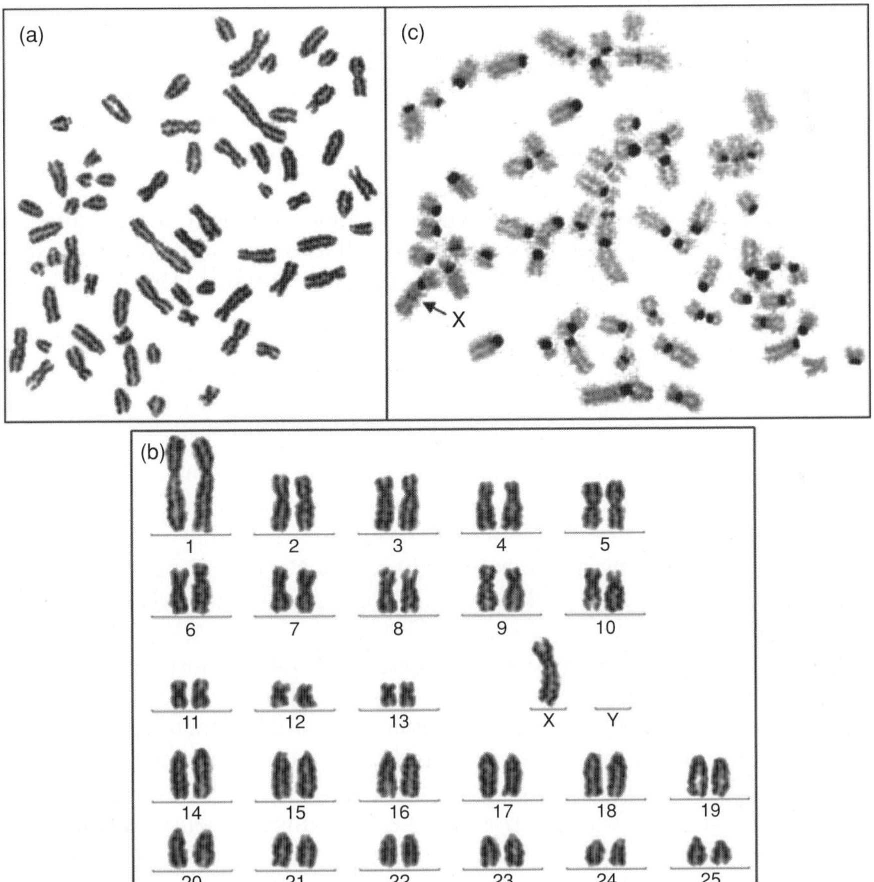

Figure 15.1 Cytogenetic analysis results for a mare with 63,XO karyotype: (a) a Giemsa stained metaphase spread; (b) the same metaphase spread arranged into a karyotype; and (c) C-banded metaphase spread showing the location of the single X chromosome (arrow).

These standard procedures are expanded if a structural abnormality is suspected or a numerical aberration is present. For this, the chromosomes are stained by the GTG-banding method, which includes trypsin digestion of chromosomal proteins followed by Giemsa staining (Seabright, 1971). The method produces distinct banding patterns for chromosome identification and the detection of structural rearrangements.

Normally, about 20–30 metaphase spreads are analyzed for each case and the results can be reported to the owners in about 10 working days. The procedures and the turnaround time for the basic chromosome analysis, however, vary between service laboratories and depend on the complexity of each individual case.

Molecular cytogenetic analysis

If an aberration is found, further analysis might involve molecular cytogenetic techniques, such as fluorescence in situ hybridization (FISH) with chromosome-specific molecular probes to precisely define losses or rearrangements. For example, chromosome-specific painting probes have been used to identify X chromosome abnormalities in infertile mares (Breen et al., 1997a) or to carry out large-scale cytogenetic surveys in horse populations (Bugno et al., 2007). Also, large insert genomic clones, such as BACs, are useful for the identification of small autosomes involved in aneuploidies (Brito et al., 2008) or for the characterization of complex structural rearrangements like translocations (Lear et al., 2008; Durkin et al., 2010) and inversions (Brooks et al., 2007). However, the use of molecular cytogenetic methods is costly, laborious, and sets high demands on the microscopy and imaging systems, as well as on the skills of the personnel. Therefore, molecular methods are mainly used for research and less frequently for basic chromosome analysis service.

Equine clinical cytogenetics in perspective

Chromosome analysis in horses and other domestic species started in the late 1960s – early 1970s, being directly influenced and inspired by the achievements in human cytogenetics. The most prominent among these were the development of improved cell culture and harvesting techniques and the introduction of differential banding methods (Arrighi & Hsu 1971; Seabright, 1971; for review, see Trask, 2002; Ducos et al., 2008; Lear & Bailey, 2008) that revolutionized chromosome studies both in humans and animals.

The first report of mare infertility associated with chromosome abnormality dates back to 1968 when the first case of X chromosome monosomy was described by a human cytogenetics group in California (Payne et al., 1968). Thereafter, equine clinical cytogenetics was initiated in many laboratories worldwide. Among these, the most productive and prominent groups in Europe were in England (S. Long, A. C. Chandley, R. V. Short, W. R. Allen), Ireland (M. Power), Sweden (I. Gustavsson), France (E. P. Cribiu), Switzerland (G. Stranzinger), Denmark (M. Ronne), Finland (A. Makinen), Poland (M. Switonski, E. M. Kubien), and in the United States in California (A. T. Bowling, L. V. Millon, K. Benirschke), Minnesota (L. C. Buoen, A. F. Weber), Kentucky (T. L. Lear), Tennessee (M. G. Kent), and Pennsylvania (L. R. Klunder, R. A. McFeely). Outstanding horse and animal cytogenetics studies were carried out also in Canada (K. Basrur), Australia (C. R. E. Halnan), and Japan (Y. Miyake). These laboratories were created almost exclusively within academic research institutions with a primary focus on basic research and less on service.

The "golden age" of equine cytogenetics was reached in 1980s when the majority of autosomal and sex chromosome abnormalities known to date were described (for review, see Chowdhary & Raudsepp, 2000; Lear & Bailey, 2008). The most important outcome of these studies was the establishment of a connection between chromosomal abnormalities and reduced fertility, making the veterinarians recognize the value of clinical cytogenetics. Yet, the beginning of 1990s marked a clear decline in equine and overall animal cytogenetics (for review, see Ducos et al., 2008). The primary reason was the shift of interest from chromosome analysis to physical gene mapping. This was facilitated by the inception of FISH (see Trask, 2002) and Zoo-FISH (Scherthan et al., 1994), construction of horse-chromosome-specific probes (Breen et al., 1997b; Chaudhary et al., 1998; Raudsepp & Chowdhary, 1999; Bugno & Slota, 2007), and the availability of equine BAC libraries (see Rubes et al., 2009). As a result, traditional cytogenetics turned into cytogenomics (Trask, 2002; Speicher & Carter 2005), improving the depth and quality of chromosome research in horses (see Ducos et al., 2008; Lear & Bailey 2008; Rubes et al., 2009), but introducing an unwanted downside: the number of cytogenetic service laboratories carrying out basic chromosome analysis started a steady decline. Currently, significant activities in equine clinical cytogenetics are limited to Europe, North America, and Japan involving less than 10 laboratories worldwide. Despite this, the need for chromosome analysis in horses continues. It is important for evaluating potential broodmares and stallions that are about to become studs, and for helping owners and breeders make informed decisions. Equine geneticists and veterinarians should learn lessons from humans: despite the "genomics era," traditional cytogenetics remains an essential part of clinical practice in prenatal, reproductive, and cancer medicine.

Molecular Genetic Testing and Equine Cytogenetic Resources

A comprehensive list of laboratories and tests provided is beyond the scope of this chapter. In the United States, services are available from: Animal Genetics, Inc. (http://www.animalgenetics.us/); Animal Genetics Laboratory, Texas A&M University (http://vetmed.tamu.edu/vibs/service-labs), Animal Genetic Testing and Research Laboratory, University of Kentucky, Lexington (http://www.ca.uky.edu/gluck/AGTRL.asp), VetGen (http://www.vetgen.com), and Veterinary Genetics Laboratory, University of California, Davis (http://www.vgl.ucdavis.edu/). All of these laboratories provide DNA genotyping for identification and parentage. Diagnostic testing for diseases and coat color are also available but specific tests offered vary. For horse forensic analyses, the Veterinary Genetics Laboratory, UC Davis, has a dedicated animal forensics unit accredited by the American Society of Crime Lab Directors Laboratory Accreditation Board (ASCLD/LAB). Horse karyotyping services are available at the Molecular Cytogenetics and Genomics Laboratory, Texas A&M University (http://vetmed.tamu.edu/labs/cytogenics-genomics), at Gluck Equine Research Center, the University of Kentucky (http://www.ca.uky.edu/gluck/leart_proj_clinical.asp), and Veterinary Genetics Laboratory, University of California, Davis (http://www.vgl.ucdavis.edu/). Resources in other countries can be requested from ISAG for laboratories that are institutional members of the Society.

References

Aleman, M., Riehl, J., Aldridge, B. M., Lecouteur, R. A., Stott, J. L., & Pessah, I. N. (2004). Association of a mutation in the ryanodine receptor 1 gene with equine malignant hyperthermia. *Muscle Nerve*, *30*(3), 356–365.

Arrighi, F. E., & Hsu, T. C. (1971). Localization of heterochromatin in human chromosomes. *Cytogenetics*, *10*(2), 81–86.

Bellone, R. R., Forsyth, G., Leeb, T., Archer, S., Sigurdsson, S. et al. (2010). Fine-mapping and mutation analysis of TRPM1: A candidate gene for leopard complex (LP) spotting and congenital stationary night blindness in horses. *Briefings in Functional Genomics*, *9*(3), 193–207.

Bernoco, D., & Bailey, E. (1998). Frequency of the SCID gene among Arabian horses in the USA. *Animal Genetics*, *29*(1), 41–42.

Binns, M. M., Holmes, N. G., Holliman, A., & Scott, A. M. (1995). The identification of polymorphic microsatellite loci in the horse and their use in thoroughbred parentage testing. *British Veterinary Journal*, *151*(1), 9–15.

Bowling, A. T., Eggleston-Stott, M. L., Byrns, G., Clark, R. S., Dileanis, S., & Wictum, E. (1997). Validation of microsatellite markers for routine horse parentage testing. *Animal Genetics*, *28*(4), 247–252.

Bowling, A. T., Millon, L., & Hughes, J. P. (1987). An update of chromosomal abnormalities in mares. *Journal of Reproduction and Fertility*, *35*(Suppl.), 149–155.

Breen, M., Langford, C. F., Carter, N. P., Fischer, P. E., Marti, E. et al. (1997a). Detection of equine X chromosome abnormalities in equids using a horse X whole chromosome paint probe (WCPP). *Veterinary Journal*, *153*(3), 235–238.

Breen, M., Lindgren, G., Binns, M. M., Norman, J., Irvin, Z. et al. (1997b). Genetical and physical assignments of equine microsatellites – first integration of anchored markers in horse genome mapping. *Mammalian Genome*, *8*(4), 267–273.

Brito, L. F. C., Sertich, P. L., Durkin, K., Chowdhary, B. P. Turner, R. M. et al. (2008). Autosomic 27 trisomy in a standardbred colt. *Journal of Equine Veterinary Science*, *45*, 431–436.

Brooks, S. A., & Bailey, E. (2005). Exon skipping in the KIT gene causes a Sabino spotting pattern in horses. *Mammalian Genome*, *16*(11), 893–902.

Brooks, S. A., Gabreski, N., Miller, D., Brisbin, A., Brown, H. E. et al. (2010). Whole-genome SNP association in the horse: Identification of a deletion in myosin Va responsible for lavender foal syndrome. *PLoS Genetics*, *6*(4), e1000909.

Brooks, S. A., Lear, T. L., Adelson, D. L., & Bailey, E. (2007). A chromosome inversion near the KIT gene and the Tobiano spotting pattern in horses. *Cytogenetic and Genome Research*, *119*(3–4), 225–230.

Brunberg, E., Andersson, L., Cothran, G., Sandberg, K., Mikko, S., & Lindgren, G. (2006). A missense mutation in PMEL17 is associated with the Silver coat color in the horse. *BMC Genetics*, *7*, 46.

Bugno, M., & Slota, E. (2007). Application of arm-specific painting probes of horse X chromosome for karyotype analysis in an infertile Hutsul mare with 64,XX/65,XX+Xp karyotype: Case report. *Acta Veterinaria Hungarica*, *55*(3), 309–314.

Bugno, M., Slota, E., & Koscielny, M. (2007). Karyotype evaluation among young horse populations in Poland. *Schweizer Archiv fur Tierheilkunde*, *149*(5), 227–232.

Bugno, M., Slota, E., Pienkowska-Schelling, A., & Schelling, C. (2009). Identification of chromosome abnormalities in the horse using a panel of chromosome-specific painting probes generated by microdissection. *Acta Veterinaria Hungarica*, *57*(3), 369–381.

Bugno, M., Zabek, T., Golonka, P., Pienkowska-Schelling, A., Schelling, C., & Slota, E. (2008). A case of an intersex horse with 63,X/64,XX/65,XX,del(Y)(q?) karyotype. *Cytogenetic and Genome Research*, *120*(1–2), 123–126.

Chaudhary, R., Raudsepp, T., Guan, X. Y., Zhang, H., & Chowdhary, B. P. (1998). Zoo-FISH with microdissected arm specific paints for HSA2, 5, 6, 16, and 19 refines known homology with pig and horse chromosomes. *Mammalian Genome*, *9*(1), 44–49.

Chen, J. W., Uboh, C. E., Soma, L. R., Li, X., Guan, F. et al. (2010). Identification of racehorse and sample contamination by novel 24-plex STR system. *Forensic Science International Genetics*, *4*(3), 158–167.

Chowdhary, B. P., & Bailey, E. (2003). Equine genomics: Galloping to new frontiers. *Cytogenetic and Genome Research*, *102*(1–4), 184–188.

Chowdhary, B. P., & Raudsepp, T. (2000). Cytogenetics and physical gene maps. In A. T. Bowling & A Ruvinsky (eds.), *The genetics of the horse*. Wallingford, Oxon: CABI.

Cook, D., Brooks, S., Bellone, R., & Bailey, E. (2008). Missense mutation in exon 2 of SLC36A1 responsible for champagne dilution in horses. *PLoS Genetics*, *4*(9), e1000195.

Ducos, A., Revay, T., Kovacs, A., Hidas, A., Pinton, A. et al. (2008). Cytogenetic screening of livestock populations in Europe: An overview. *Cytogenetic and Genome Research*, *120*(1–2), 26–41.

Durkin, K., Raudsepp, T., & Chowdhary, B. P. (2010). Cytogenetic evaluation of the stallion. In E. L. S., A. O. McKimmon, W. E. Vaala, & D. D. Varner (eds.), *Equine reproduction*. London: Wiley-Blackwell.

Ellegren, H., Johansson, M., Sandberg, K., & Andersson, L. (1992). Cloning of highly polymorphic microsatellites in the horse. *Animal Genetics*, *23*(2), 133–142.

Finno, C. J., Spier, S. J., & Valberg, S. J. (2009). Equine diseases caused by known genetic mutations. *Veterinary Journal*, *179*(3), 336–347.

Graves, K. T., Henney, P. J., & Ennis, R. B. (2009). Partial deletion of the LAMA3 gene is responsible for hereditary junctional epidermolysis bullosa in the American Saddlebred Horse. *Animal Genetics*, *40*(1), 35–41.

Guerin, G., Bertaud, M., & Amigues, Y. (1994). Characterization of seven new horse microsatellites: HMS1, HMS2, HMS3, HMS5, HMS6, HMS7, and HMS8. *Animal Genetics*, *25*(1), 62.

Haase, B., Brooks, S. A., Schlumbaum, A., Azor, P. J., Bailey, E. et al. (2007). Allelic heterogeneity at the equine KIT locus in dominant white (W) horses. *PLoS Genetics*, *3*(11), e195.

Haase, B., Brooks, S. A., Tozaki, T., Burger, D., Poncet, P. A. et al. (2009). Seven novel KIT mutations in horses with white coat colour phenotypes. *Animal Genetics*, *40*(5), 623–629.

Hill, E. W., McGivney, B. A., Gu, J., Whiston, R., & Machugh, D. E. (2010). A genome-wide SNP-association study confirms a sequence variant (g.66493737C>T) in the equine myostatin (MSTN) gene as the most powerful predictor of optimum racing distance for Thoroughbred racehorses. *BMC Genomics*, *11*, 552.

Hirota, K., Kakoi, H., Gawahara, H., Hasegawa, T., & Tozaki, T. (2010). Construction and validation of parentage testing for thoroughbred horses by 53 single nucleotide polymorphisms. *Journal Veterinary Medicine Science*, *72*(6), 719–726.

Irvin, Z., Giffard, J., Brandon, R., Breen, M., & Bell, K. (1998). Equine dinucleotide repeat polymorphisms at loci ASB 21, 23, 25 and 37–43. *Animal Genetics*, *29*(1), 67.

International system for cytogenetic nomenclature of the domestic horse (ISCNH), Bowling, A. Breen, T., Chowdhary, M. Hirota, B. P., Lear, K. et al. (1997). Report of the Third International Committee for the Standardization of the domestic horse karyotype, Davis, CA, 1996. *Chromosome Research*, *5*(7), 433–443.

Kakoi, H., Tozaki, T., Nagata, S., Gawahara, H., & Kijima-Suda, I. (2009). Development of a method for simultaneously genotyping multiple horse coat colour loci and genetic investigation of basic colour variation in Thoroughbred and Misaki horses in Japan. *Journal of Animal Breeding and Genetics*, *126*(6), 425–431.

Lear, T. L., & Bailey, E. (2008). Equine clinical cytogenetics: The past and future. *Cytogenetic and Genome Research*, *120*(1–2), 42–49.

Lear, T. L., Lundquist, J., Zent, W. W., Fishback, Jr., W. D., & Clark, A. (2008). Three autosomal chromosome translocations associated with repeated early embryonic loss (REEL) in the domestic horse (Equus caballus). *Cytogenetic and Genome Research*, *120*(1–2), 117–122.

Ludwig, A., Pruvost, M., Reissmann, M., Benecke, N., Brockmann, G. A. et al. (2009). Coat color variation at the beginning of horse domestication. *Science*, *324*(5926), 485.

Mariat, D., Taourit, S., & Guerin, G. (2003). A mutation in the MATP gene causes the cream coat colour in the horse. *Genetics, Selection and Evolution*, *35*(1), 119–133.

Marklund, L., Moller, M. J., Sandberg, K., & Andersson, L. (1996). A missense mutation in the gene for melanocyte-stimulating hormone receptor (MC1R) is associated with the chestnut coat color in horses. *Mammalian Genome*, *7*(12), 895–899.

Marklund, S., Ellegren, H., Eriksson, S., Sandberg, K., & Andersson, L. (1994). Parentage testing and linkage analysis in the horse using a set of highly polymorphic microsatellites. *Animal Genetics*, *25*(1), 19–23.

McCue, M. E., Valberg, S. J., Lucio, M., & Mickelson, J. R. (2008). Glycogen synthase 1 (GYS1) mutation in diverse breeds with polysaccharide storage myopathy. *Journal of Veterinary Internal Medicine*, *22*(5), 1228–1233.

Metallinos, D. L., Bowling, A. T., & Rine, J. (1998). A missense mutation in the endothelin-B receptor gene is associated with Lethal White Foal Syndrome: an equine version of Hirschsprung disease. *Mammalian Genome*, *9*(6), 426–431.

Naylor, J. M. (1994). Selection of quarter horses affected with hyperkalemic periodic paralysis by show judges. *Journal of the American Veterinary Medical Association*, *204*(6), 926–928.

Payne, H. W., Ellsworth, K., & DeGroot, A. (1968). Aneuploidy in an infertile mare. *Journal of the American Veterinary Medical Association*, *153*(10), 1293–1299.

Pielberg, G. R., Golovko, A., Sundstrom, E., Curik, I., Lennartsson, J. et al. (2008). A cis-acting regulatory mutation causes premature hair graying and susceptibility to melanoma in the horse. *Nature Genetics*, *40*(8), 1004–1009.

Raudsepp, T., & Chowdhary, B. P. (1999). Construction of chromosome-specific paints for meta- and submetacentric autosomes and the sex chromosomes in the horse and their use to detect homologous chromosomal segments in the donkey. *Chromosome Research*, *7*(2), 103–114.

Raudsepp, T., & Chowdhary, B. P. (2008). FISH for mapping single copy genes. In W. Murphy (ed.), *Phylogenomics: Methods in molecular biology*. New Brunswick, NJ: Humana Press.

Rieder, S., Taourit, S., Mariat, D., Langlois, B., & Guerin, G. (2001). Mutations in the agouti (ASIP), the extension (MC1R), and the brown (TYRP1) loci and their association to coat color phenotypes in horses (Equus caballus). *Mammalian Genome*, *12*(6), 450–455.

Royo, L. J., Fernandez, I., Azor, P. J., Alvarez, I., Perez-Pardal, L., & Goyache, F. (2008). Technical note: A novel method for routine genotyping of horse coat color gene polymorphisms. *Journal of Animal Science*, *86*(6), 1291–1295.

Rubes, J., Pinton, A., Bonnet-Garnier, A., Fillon, V., Musilova, P. et al. (2009). Fluorescence in situ hybridization applied to domestic animal cytogenetics. *Cytogenetic and Genome Research*, *126*, 34–48.

Rudolph, J. A., Spier, S. J., Byrns, G., Rojas, C. V., Bernoco, D., & Hoffman, E. P. (1992). Periodic paralysis in quarter horses: a sodium channel mutation disseminated by selective breeding. *Nature Genetics*, *2*(2), 144–147.

Sandberg, K., & Cothran, E. G. (2000). Blood groups and biochemical polymorphisms. In B. A. T. & A. Ruvinsky (eds.), *The genetics of the horse*. Wallingford: CAB International.

Santschi, E. M., Purdy, A. K., Valberg, S. J., Vrotsos, P. D., Kaese, H., & Mickelson, J. R. (1998). Endothelin receptor B polymorphism associated with lethal white foal syndrome in horses. *Mammalian Genome*, *9*(4), 306–309.

Scherthan, H., Cremer, T., Arnason, U., Weier, H. U., Lima-de-Faria, A., & Fronicke, L. (1994). Comparative chromosome painting discloses homologous segments in distantly related mammals. *Nature Genetics*, *6*(4), 342–347.

Seabright, M. (1971). A rapid banding technique for human chromosomes. *Lancet*, *2*(7731), 971–972.

Shin, E. K., Perryman, L. E., & Meek, K. (1997). A kinase-negative mutation of DNA-PK(CS) in equine SCID results in defective coding and signal joint formation. *Journal of Immunology*, *158*(8), 3565–3569.

Speicher, M. R., & Carter, N. P. (2005). The new cytogenetics: Blurring the boundaries with molecular biology. *Nature Reviews in Genetics*, *6*(10), 782–792.

Spencer, G., & Davis, N. (2007). *Horse genome assembled*. Available from http://www.genome.gov/20519480 (accessed July 15, 2008).

Spirito, F., Charlesworth, A., Linder, K., Ortonne, J. P., Baird, J., & Meneguzzi, G. (2002). Animal models for skin blistering conditions: Absence of laminin 5 causes hereditary junctional mechanobullous disease in the Belgian horse. *Journal of Investigative Dermatology*, *119*(3), 684–691.

Tozaki, T., Kakoi, H., Mashima, S., Hirota, K., Hasegawa, T. et al. (2001). Population study and validation of paternity testing for Thoroughbred horses by 15 microsatellite loci. *Journal of Veterinary Medical Science*, *63*(11), 1191–1197.

Trask, B. J. (2002). Human cytogenetics: 46 chromosomes, 46 years and counting. *National Review of Genetics*, *3*(10), 769–778.

Tryon, R. C., Penedo, M. C., McCue, M. E., Valberg, S. J., Mickelson, J. R. et al. (2009). Evaluation of allele frequencies of inherited disease genes in subgroups of American Quarter Horses. *Journal of the American Veterinary Medical Association*, *234*(1), 120–125.

Tryon, R. C., White, S. D., & Bannasch, D. L. (2007). Homozygosity mapping approach identifies a missense mutation in equine cyclophilin B (PPIB) associated with HERDA in the American Quarter Horse. *Genomics*, *90*(1), 93–102.

van de Goor, L. H., Panneman, H., & van Haeringen, W. A. (2010). A proposal for standardization in forensic equine DNA genotyping: allele nomenclature for 17 equine-specific STR loci. *Animal Genetics*, *41*(2), 122–127.

van Haeringen, H., Bowling, A. T., Stott, M. L., Lenstra, J. A., & Zwaagstra, K. A. (1994). A highly polymorphic horse microsatellite locus: VHL20. *Animal Genetics*, *25*(3), 207.

Wade, C. M., Giulotto, E., Sigurdsson, S., Zoli, M., Gnerre, S. et al. (2009). Genome sequence, comparative analysis, and population genetics of the domestic horse. *Science*, *326*(5954), 865–867.

Wagner, H. J., & Reissmann, M. (2000). New polymorphism detected in the horse MC1R gene. *Animal Genetics*, *31*(4), 289–290.

Wagner, M. L., Valberg, S. J., Ames, E. G., Bauer, M. M., Wiseman, J. A. et al. (2006). Allele frequency and likely impact of the glycogen branching enzyme deficiency gene in Quarter Horse and Paint Horse populations. *Journal of Veterinary Internal Medicine*, *20*(5), 1207–1211.

Ward, T. L., Valberg, S. J., Adelson, D. L., Abbey, C. A., Binns, M. M., & Mickelson, J. R. (2004). Glycogen branching enzyme (GBE1) mutation causing equine glycogen storage disease IV. *Mammalian Genome*, *15*(7), 570–577.

Yang, G. C., Croaker, D., Zhang, A. L., Manglick, P., Cartmill, T., & Cass, D. (1998). A dinucleotide mutation in the endothelin-B receptor gene is associated with lethal white foal syndrome (LWFS): A horse variant of Hirschsprung disease. *Human Molecular Genetics*, *7*(6), 1047–1052.

16 Genomics of laminitis

Jim K. Belknap

Introduction

Equine laminitis is a devastating disease in which there is a structural failure of the adhesion between the epidermal laminae (attached to the hoof wall) and the dermal laminae (attached to the distal phalanx), resulting in a loss of suspension of the distal phalanx within the hoof wall, and therefore a crippling distal displacement of the phalanx due to the weight of the horse. Equine laminitis most commonly occurs as a sequel to primary disease processes involving sepsis (i.e., enterocolitis due to infectious agents or carbohydrate overload), acute metritis, equine metabolic syndrome, or Cushing's syndrome/pituitary adenoma. The majority of research has been performed on models of sepsis related to laminitis, including the carbohydrate overload models (traditional wood flour/corn starch or oligofructose models) (Garner et al., 1978; van Eps & Pollitt, 2006) and the black walnut extract (BWE) model. The BWE model, based on the fact that horses bedded on black walnut shavings develop laminitis, induces a similar (but transient) systemic inflammation as observed in sepsis (Belknap, 2010). Due to the devastating nature of this disease to the horse, client, and industry, laminitis has been listed as the top priority for equine research by members of the world's largest professional organization dedicated to equine veterinary medicine, the American Association of Equine Practitioners (AAEP). As discussed below, functional genomics has been a key factor in the progress of laminitis research, which otherwise has been plagued for more than two decades by dogmas and false extrapolations from human studies, and by the lack of knowledge about biochemical processes in equine tissues affecting the studies of laminar physiology and histology.

Functional Genomics of Laminitis

Functional genomics has played a central role in the study of equine laminitis for the past 15 years, primarily due to the fact that there are only a few antibodies from human biomedical research that react with the equine mediators of interest. This is especially a problem with inflammatory mediators, because the majority of antibodies against pro-inflammatory cytokines used in human and laboratory rodent tissues do not recognize equine epitopes. An important advantage to using functional genomics for the research of pro-inflammatory cytokines/chemokines is that the majority

Equine Genomics, First Edition. Edited by Bhanu P. Chowdhary.

of the regulation takes place at mRNA level (both gene expression and post-transcriptional control, including mRNA stability) (Anderson, 2008; Khabar, 2010), with minimal post-translational modifications. Prior to the sequencing of the equine genome (Wade et al., 2009), PCR with oligomers designed from alignments of published mRNA sequences from other species was used to obtain equine-specific cDNA fragments. Cloning these fragments into plasmid vectors provided equine-specific probes for techniques such as in situ hybridization (ISH) and Northern hybridization (see Figure 16.1) (Rodgerson et al., 2000; Fontaine et al., 2001). In the last decade, quantitative real-time PCR (qRT-PCR) has replaced other types of quantitative and semi-quantitative assays as the method of choice for assessing gene expression of molecules of interest (Rodgerson et al., 2001; Belknap et al., 2007; Loftus et al., 2007; Kyaw-Tanner et al., 2008; Coyne et al., 2009). The majority of these studies have employed methods for relative quantitation of mRNA concentrations of genes of interest, and normalizing the individual data for analyses by either the use of a normalization factor obtained from more than one housekeeping gene (Waguespack et al., 2004a; Waguespack et al., 2004b; Loftus et al., 2007), or the use of the $\Delta\Delta C_T$ method using a single housekeeping gene (Noschka et al., 2009). More recently, primer design has been much more successful using the published equine mRNA sequences, which are available as a result of the successful sequencing of the equine genome.

Molecular biology has been crucial in refuting a central dogma in equine laminitis and in organ injury in human sepsis, stating that laminar/organ injury was due to decreased blood flow, and not due to inflammatory injury. Many physiology studies indicated a decreased vascular supply or blood flow to the digital laminae at different clinical stages of the disease. Because routine histology (i.e., H&E staining) showed the lack of leukocytes, it was concluded that laminar inflammation was not playing a role in the etiology of laminitis. The first study to challenge this dogma involved ISH, using equine-specific ^{35}S-labeled riboprobes for interleukin-1 beta (IL-1β) in paraffin-embedded tissues from a model of laminitis (the black walnut extract [BWE] model), in which the digits were perfused with 10% formalin immediately following euthanasia. The ISH study demonstrated IL-1β-positive cells in the laminar dermis located outside of the laminar venules which were markedly positive, whereas no IL-1β signal was present in the laminae from control animals (Figure 16.1a). Further, studies using CD13 (a marker of myeloid leukocytes) demonstrated a similar cellular pattern for emigrating leukocytes, indicating that the IL-1β–positive cells are emigrating neutrophils and possibly monocytes (Black et al., 2006). More recently, we have used non-isotopic ISH methodology for IL-6 to demonstrate a similar pattern indicating that leukocytes are the source of laminar IL-6 (Figure 16.1b), and to demonstrate that an important chemokine, CXCL1, is expressed in laminar epithelium (discussed below, Figure 16.1c).

There are limited studies of laminar gene expression prior to the invention of qRT-PCR. This was partially because only a few equine investigators had training in molecular biology, but also because of the limited amount of RNA that could be obtained from laminar samples for Northern hybridization (Rodgerson et al., 2001; Kyaw-Tanner & Pollitt, 2004). Quantitative RT-PCR was rapidly incorporated into laminitis research starting in 2004, when it was used both for the assessment of laminar mRNA concentrations of inflammatory proteins (Waguespack et al., 2004b), and for the assessment of laminar mRNA concentrations for a protease proposed to be involved in breakdown of the laminar adhesion, matrix metalloprotease-2 (MMP-2) (Kyaw-Tanner & Pollitt, 2004). In the first study, mRNA differential display, a genomic screening technique commonly used prior to the introduction of microarray technology, was employed to assess differential gene expression in affected laminae of BWE-treated horses compared to control laminae (Waguespack et al., 2004b). In this study, the investigators did not screen pooled samples due to the variability of response in individual horses in the laminitis models, but screened 3 pairs each of laminar samples from

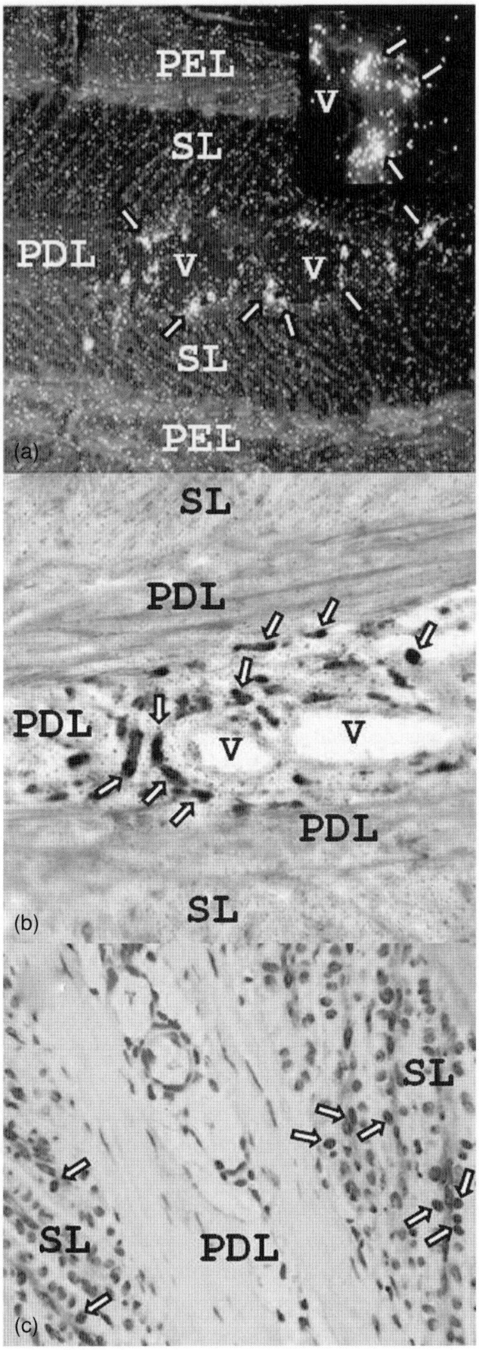

Figure 16.1 In situ hybridization of laminar sections for cytokines and chemokines in BWE-treated animals; (a) Note in the dark field micrograph (including magnified inset) the cells (white arrows) expressing IL-1β signal (white silver grains in photoemulsion exposed by ^{35}S-labeled probe) around venules (V) in the primary dermal laminae (PDL) at a developmental stage in the BWE model; (b) The same stage in the BWE model. Note in the light field micrograph of a non-isotopic ISH study the IL-6 positive cells (blue cells indicated by arrows) situated in the PDL around the venules (V); (c) Note the CXCL1 signal (blue stain) in the non-isotopic ISH study in the laminar basal epithelial cells (arrows); cells in the vascular wall (likely adhered leukocytes) in the primary dermal laminae and in secondary dermal laminae (with a morphology comparable to macrophages) are also positive. PEL = primary epidermal laminae, SL = interdigitating secondary dermal and epidermal laminae. (*For color details, see color plate section.*)

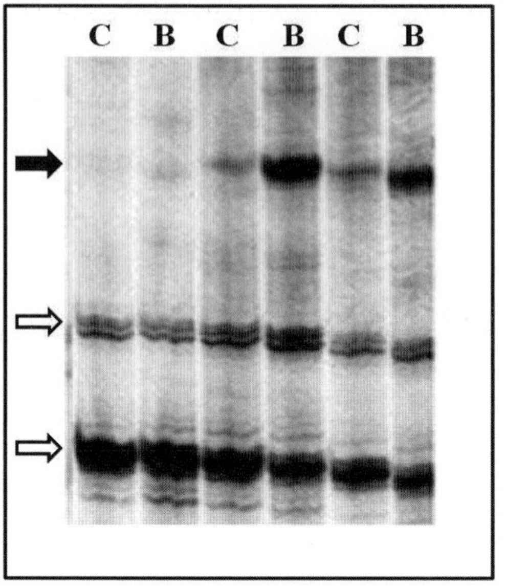

Figure 16.2 Differential mRNA display of laminar samples from control and BWE-treated horses. Note the band (black arrow) that is differentially regulated in two of the three pairs of laminae from control (C) and BWE-treated horses (B) compared to the non-regulated bands (open arrows). The regulated band was cut out from the polyacrylamide gel, purified, and sequenced; it was found to be the equine homologue of the inflammatory gene MAIL/ IκB-ξ.

BWE-treated and control animals (Figure 16.2). Due to the variability in using the outbred equine population, the investigators assessed genes that were differentially regulated in ≥2 of the 3 pairs of samples (Figure 16.2) (Waguespack et al., 2004b). Interestingly, although this was an unbiased screening (i.e., that could have resulted in cloning of an equine homologue of a gene important for toenail growth), the first differentially expressed mRNA that was cloned from the gel and sequenced was MAIL (molecule with ankyrin repeats induced by lipopolysaccharide, LPS) (Waguespack et al., 2004b). This molecule is directly related to inflammatory signaling and was originally discovered in a survey of differentially regulated genes in the brain of mice administered with LPS (Haruta et al., 2001). MAIL is now known as one of the IκB proteins, IκB zeta (IκB-ξ). This protein is induced by numerous bacterial products/TLR ligands and by IL-1β, but not by TNF-α (Yamamoto et al., 2004; Motoyama et al., 2005). IκB-ξ is essential for the up-regulation of specific cytokines including IL-6 and IL-12. In the same study in which MAIL/IκB-ξ was found by differential mRNA display in affected equine laminae, qRT-PCR was used to demonstrate a 4-fold increase in IκB-ξ, 30-fold increase in IL-1β, and 160-fold increase in IL-6 mRNA concentrations in the laminae of BWE-treated animals at a developmental stage in the BWE model of laminitis versus controls (Waguespack et al., 2004b). In later studies, qRT-PCR has documented marked increases in laminar mRNA concentrations of cytokines (IL-1β, IL-6, IL-12) and chemokines (CXCL8/IL-8, CXCL1/Groα) in the developmental and lameness stages of both the BWE model and the carbohydrate overload (CHO) model of laminitis (Belknap et al., 2007; Loftus et al., 207; Leise et al., 2011). Interestingly, TNF-α has never been reported to be increased in any model of laminitis (Belknap et al., 2007; Loftus et al., 2007; Leise et al., 2011). Laminar COX-2 was also demonstrated to be up-regulated, originally by qRT-PCR (Waguespack et al., 2004a; Blikslager et al., 2006) and later by immunochemical techniques (Blikslager et al., 2006). These studies contrasted the differences between the BWE and CHO models: cytokine mRNA concentrations increase at early developmental time points (as early as 1.5 hours post BWE administration) in the BWE model (Loftus et al., 2007), whereas cytokine gene expression did not increase until the onset of clinical signs of lameness in the CHO model (Leise et al., 2011). Also, qRT-PCR was recently used to demonstrate that the increased incidence of laminitis in the forelimbs is not due to any difference between the front and hind feet regarding physiological or cellular events affecting cell signaling, but rather due to differences in weight bearing (Leise et al., 2009). Because digital laminae appears to be the "target organ" in the horse with sepsis, versus visceral organs like liver and lung in the human, we used qRT-PCR to compare inflammatory events between visceral organs and the laminae in horses. Although we did find increases in pulmonary and hepatic cytokine and chemokine mRNA concentrations (up to 83-fold increase in pulmonary IL-6 mRNA concentrations), the increases were not as high as those observed in the laminae in the same animals (up to 553-fold increase in laminar IL-6 mRNA concentration) (Loftus et al., 2007). The results are consistent with the clinical reality that the septic equine patient is much more likely to undergo clinically apparent laminar injury than significant pulmonary or hepatic dysfunction. However, the increased inflammatory gene expression in the liver and lungs suggests that, if we were able to maintain horses with severe state of sepsis in intensive care like it is done with human patients (equine patients have usually been euthanized at this point due to the inability to maintain a recumbent adult horse with severe sepsis or due to severity of laminitis), we would probably observe a similar visceral organ dysfunction as occurs in the septic human patient. There are no reports regarding the use of functional genomics in the study of laminar injury related to equine metabolic syndrome.

Similarly to sepsis-related organ injury in other species (Belknap et al., 2009), leukocyte emigration appears to be the source of inflammatory cytokines in laminar injury in laminitis. Therefore, two events essential for leukocyte adhesion and extravasation were assessed by qRT-PCR: endothelial adhesion molecule expression and laminar chemokine (cytokines chemotactic for leukocytes) expression. It was found that two important adhesion molecules, ICAM-1 and E-selectin, had a similar pattern of expression at the onset of leukocyte extravasation in the two laminitis models. The two molecules undergo peak expression at the 1.5 hour time point in the BWE model (Loftus et al., 2007), but not until the onset of lameness (approximately 24–36 hours) in the CHO model (Leise et al., 2009). Chemokines, reported to be essential for leukocyte activation and migration, demonstrated a somewhat similar pattern peaking very early in the BWE model (Loftus et al., 2007) and peaking at the onset of lameness in the CHO model, but undergoing some increases at the developmental stage (approximately 12 hours post CHO administration) in the CHO model. The chemokines that have been examined include CXCL1/Gro-α and CXCL8/IL-8. Both were found to undergo a marked increase (140–160-fold) at the 1.5 hour time point in the BWE model, decreasing to more moderate increases (approximately 20-fold) at later developmental stages (Loftus et al., 2007). As leukocytes are thought to migrate on a chemokine gradient (toward higher chemokine concentrations), we were interested to determine if laminar epithelial cells express CXCL1, a chemokine commonly expressed by epithelial cells in different organs. Using, non-isotopic ISH, we found CXCL1 expression in the laminar basal epithelial cells of the epidermal laminae, in endothelium, and also in cells that appear morphologically identical to those we stain with the macrophage marker CD163 in the secondary dermal laminae (Figure 16.1c) (Faleiros et al., 2009). This work demonstrates that there might be a chemokine gradient that induces leukocytes to migrate toward the point of failure in laminitis – the laminar dermal/epidermal interface.

The possibility that dysadhesion of the basal epithelial cells from the underlying basement membrane and dermis may occur due to degradation of matrix proteins by proteases, including matrix metalloproteases, led to the assessment of laminar mRNA concentrations for MMP-2,

MMP-9, and, more recently, for aggrecanase ADAMTS-4. Quantitative RT-PCR results for laminar MMP-2 and MMP-9 expression in laminitis models are not consistent with reports of either minor increases or no change (Kyaw-Tanner and Pollitt, 2004; Loftus et al., 2006; Loftus et al., 2007). Most likely, this is because a great deal of the regulation of MMPs is post-translational. There are zymography data showing increases in MMP-2 activation in the CHO laminitis model, thus indicating that this MMP may play a role in laminar injury (Loftus et al., 2009). Laminar mRNA concentrations of the aggrecanase ADAMTS-4 undergo consistent increases in both BWE and CHO models, and in acute clinical cases of laminitis (Coyne et al., 2009). This increase correlates with an increase at the laminar protein level (personal communication, Samuel J. Black, University of Massachusetts, Amherst).

Whole Genome Transcriptional Profiling

Whole-genome transcriptional profiling techniques in the past included incredibly work-intensive techniques including mRNA differential display (discussed above) and subtractive cloning techniques that required a great deal of effort to determine a small number of differentially regulated genes. More recently, microarray analyses have come to the forefront of human and veterinary research due to optimized formats that allow the screening of thousands of genes at one time. Although these genome-wide screening techniques have been derided at times as "fishing expeditions" or "non-hypothesis driven research", they are extremely important in complex disease processes, such as laminitis in which the small number of laboratories throughout the world investigating this disease are unlikely to discover all of the different signaling mechanisms important to the disease process. Presently, only two publications on laminitis include the use of microarray technology. These studies were performed using a bovine oligonucleotide microarray chip (over 15,000 transcripts) and a custom equine cDNA microarray (over 3,000 genes) to assess gene expression in the early stages of laminitis in the BWE and CHO models of laminitis (Budak et al., 2009; Noschka et al., 2009). Using the bovine microarray chip, the investigators reported up-regulation of 155 genes (none down-regulated), with genes associated with "pro-inflammatory biochemical or cellular processes" and "protein degradation/turnover" being most prevalent (Budak et al., 2009). However, because a large number of genes were considered as up-regulated at less than a 2-fold increase, and the microarray had not been validated on equine tissues, these results require further validation by qRT-PCR. The investigators did perform qRT-PCR on 3 genes that showed a 4–6-fold increase in microarray signal intensity: the mRNA concentrations for all three had a 1.7–11.6-fold increase by qRT-PCR.

The custom equine cDNA microarray in the other study represents approximately 10% of the equine transcriptome and was designed using ESTs obtained from several equine cDNA libraries of leukocyte origin, and cDNA fragments of inflammatory genes provided by other investigators (Noschka et al., 2009). Thus, this microarray is largely dedicated to inflammatory signaling. The investigators assessed laminar signaling at three time points in the BWE model including two developmental time points (1.5 and approximately 3 hours [onset of leucopenia] post BWE administration), and the onset of first signs of laminitis (Obel Grade 1 defined as weight shifting between forelimbs and bounding digital pulses [approx 10–12 hours post BWE administration). The number of differentially regulated (DR) transcripts detected by the microarray increased temporally: with only 14 DR genes at 1.5 hours (8 genes up-regulated and 6 genes down-regulated), 22 DR transcripts at the 3 hour time point (17 up- and 5 down-regulated at each time point), and 62 DR transcripts at the onset of Obel Grade 1 laminitis (55 up- and 7 down-regulated). Four genes consisting of chemokines (CCL7/MCP-3, MCP-1, CXCL10/IP-10) and an inflammation-related acute phase protein (serum amyloid A protein, SAA) were up-regulated at all three time points in the laminar tissue. The discovery of up-regulation of SAA proteins, acute-phase proteins induced by IL-6 mediated activation on STAT3 in other tissues (Nerstedt et al., 2010), is in good agreement with other recent findings showing that IL-6 is the most highly up-regulated inflammatory gene in equine laminitis (Loftus et al., 2007; Leise et al., 2011), and that laminar STAT3 is activated in the two laminitis models (Leise et al., 2010). Quantitative assessment by qRT-PCR of the expression of SOD2 and MCP-3 showed that the two transcripts undergo a much greater increase (up to 68-fold for SOD2 and 404-fold for MCP-3) than calculated from the microarray data (Noschka et al., 2009). Furthermore, the authors stated that they would have detected DR in genes previously reported to undergo significant up-regulation in BWE laminitis, such as IL-1β (up to 50-fold increase) (Loftus et al., 2007) and COX-2 (32-fold increase) (Blikslager et al., 2006), if they lowered the stringency of analysis. These results indicate that the microarray may not be the most sensitive method of detecting differentially regulated genes, and that more attention needs to be paid to genes with small increases in the expression (i.e., need to be assessed quantitatively by qRT-PCR). In addition to the increased number of inflammatory transcripts at the onset of OG1 laminitis, there is also an increase in the number of transcripts encoding anti-inflammatory proteins including Mn-superoxide disumutase (Mn-SOD/SOD2), elafin, and TIMP-1, which may play a protective role to limit laminar injury.

It is anticipated that future microarray studies will take advantage of the availability of higher fidelity equine-specific microarrays, advanced software allowing to organize the DR genes into different signaling cascades, and cutting-edge techniques, such as laser capture microdissection (LCM), where investigators can dissect individual cell types from cryosections for qRT-PCR and microarray analysis. There is an ongoing study in the author's laboratory where the laminar basal epithelial cell, the cell that dysadheres from the underlying laminar dermis, resulting in laminar failure in laminitis, is being obtained by LCM from serial laminar biopsies taken prior to induction of laminitis (control), and at developmental and lameness time points. The cells will be subjected for microarrary analysis combined with gene networking to assess the signaling mechanisms occurring early in the disease process.

Contribution of Genomics to the Therapy of Laminitis

Functional genomics has recently been used to assess the efficacy of two different therapies being used in laminitis to block inflammatory signaling (Van Eps et al., 2010; Williams et al., 2010). In one, a constant-rate infusion of intravenous lidocaine, which has been studied and used for its purported anti-inflammatory properties, was assessed in the BWE model of laminitis. Quantitative RT-PCR analysis demonstrated that a lidocaine CRI did not change expression of cytokines or COX-2, whereas it actually induced an increase in the expression of the adhesion molecule E selectin indicating endothelial activation in the laminar vasculature by the lidocaine infusion (Williams et al., 2010). In distinction to these results, qRT-PCR was recently used to assess the effect of local digital hypothermia (termed "cryotherapy") on inflammatory signaling in the oligofructose model of laminitis. Local hypothermia/cryotherapy has come to the forefront of therapies for laminitis due to recent studies indicating clinical efficacy of the treatment when the horse's limbs are maintained in ice water following intragastric administration of a laminitis-inducing solution of oligofructose

(Van Eps & Pollitt, 2009). In the study, animals were administered oligofructose and then had one forelimb placed and maintained in ice water (treatment foot) while the other foot was maintained at ambient temperature (control foot); this model is very powerful for statistical analysis of qRT-PCR data due to the paired nature of the comparisons. In that work, local hypothermia was demonstrated to markedly decrease laminar mRNA concentrations of inflammatory mediators including cytokines, chemokines, COX-2, and endothelial adhesion molecules (Van Eps et al., 2010). Due to the ability of qRT-PCR to easily provide data from the paired samples from individual horses (vs. the usual use of pooled samples in former techniques such as Northern hybridization), we were able to discover the efficacy of hypothermia in effectively decreasing inflammatory mediator gene expression as much as 1,000-fold in individual horses (Figure 16.3).

Functional genomics approaches have only recently been incorporated into the study of equine metabolic syndrome of which laminitis is usually the most important complication. One study reported a correlation between circulating leukocyte cytokine expression levels (assessed by qRT-PCR) and insulin sensitivity in obese horses (Vick, 2007), whereas another study used qRT-PCR to demonstrate that, unlike the human where the visceral fat depots are the primary sources of inflammatory mediators important in the disease process, nuchal ligament fat appears to be the primary source on inflammatory cytokine gene expression in the obese horse (Burns et al., 2010).

Despite increasing availability of protein biochemistry reagents for equine studies, functional genomics is likely to play a central role in laminitis research in the future. This is thanks to constant advances in the field improving the ability to accurately assess tissue and cellular mRNA concentrations of multiple genes of interest with techniques such as qRT-PCR, in situ hybridization (including PCR in situ hybridization), microarray analyses, and ancillary techniques such as LCM.

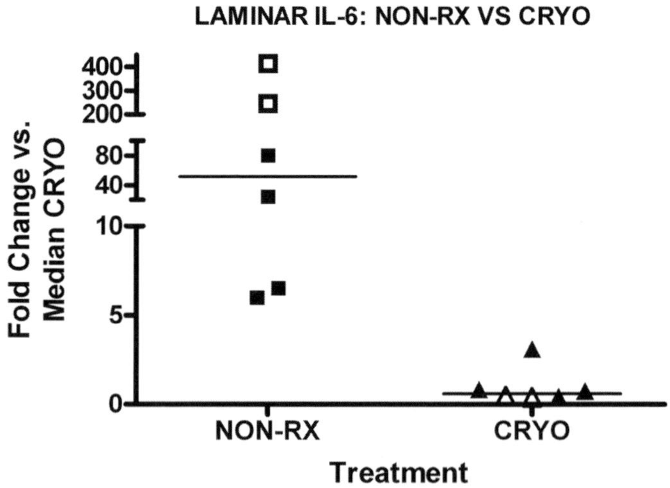

Figure 16.3 Quantitative RT-PCR of laminar IL-6 in oligofructose-administered animals treated with digital hypothermia/cryotherapy on the left forefoot. Note the efficacy of digital hypothermia of the left forefoot of these horses (CRYO) in inhibiting the up-regulation of laminar IL-6 observed in the foot kept at ambient temperature (NON-RX) in horse administered oligofructose. The two open squares in NON-RX group (representing IL-6 mRNA concentrations of the untreated/ambient temperature feet of the two horses with the greatest increase in laminar IL-6 when compared to the median value of the hypothermic feet) are the same horses represented by the two open triangles (CRYO group, the opposite forefoot maintained in ice water), demonstrating the efficacy of local hypothermia in blocking the inflammatory response even in the animals that undergo up to an 800-fold increase in IL-6 gene expression when not treated with digital hypothermia/cryotherapy.

References

Anderson, P. (2008). Post-transcriptional control of cytokine production. *Nature Immunology*, *9*, 353–359.

Belknap, J. K. (2010). Black walnut extract: An inflammatory model. *Veterinary Clinical North American Equine Practice*, *26*, 95–101.

Belknap, J. K., Giguere, S., Pettigrew, A., Cochran, A. M., Van Eps, A.W. et al. (2007). Lamellar pro-inflammatory cytokine expression patterns in laminitis at the developmental stage and at the onset of lameness: Innate vs. adaptive immune response. *Equine Veterinary Journal*, *39*, 42–47.

Belknap, J. K., Moore, J. N., & Crouser, E. C. (2009). Sepsis – from human organ failure to laminar failure. *Veterinary Immunology and Immunopathology*, *129*, 155–157.

Black, S. J., Lunn, D. P., Yin, C., Hwang, M., Lenz, S. D. et al. (2006). Leukocyte emigration in the early stages of laminitis. *Veterinary Immunology and Immunopathology*, *109*, 161–166.

Blikslager, A. T., Yin, C., Cochran, A. M., Wooten, J. G., Pettigrew, A. et al. (2006). Cyclooxygenase expression in the early stages of equine laminitis: A cytologic study. *Journal of Veterinary Internal Medicine*, *20*, 1191–1196.

Budak, M. T., Orsini, J. A., Pollitt, C. C., & Rubinstein, N. A. (2009). Gene expression in the lamellar dermis-epidermis during the developmental phase of carbohydrate overload-induced laminitis in the horse. *Veterinary Immunology and Immunopathology*, *131*, 86–96.

Burns, T. A., Geor, R. J., Mudge, M. C., McCutcheon, L. J., Hinchcliff, K. W. et al. (2010). Proinflammatory cytokine and chemokine gene expression profiles in subcutaneous and visceral adipose tissue depots of insulin-resistant and insulin-sensitive light breed horses. *Journal of Veterinary Internal Medicine*, *24*, 932–939.

Coyne, M. J., Cousin, H., Loftus, J. P., Johnson, P. J., Belknap, J. K. et al. (2009). Cloning and expression of ADAM-related metalloproteases in equine laminitis. *Veterinary Immunology and Immunopathology*, *129*, 231–241.

Faleiros, R. R., Leise, B. B., Westerman, T., Yin, C., Nuovo, G. J. et al. (2009). In vivo and in vitro evidence of the involvement of CXCL1, a keratinocyte-derived chemokine, in equine laminitis. *Journal of Veterinary Internal Medicine*, *23*, 1086–1096.

Fontaine, G. L., Belknap, J. K., Allen, D., Moore, J. N., & Kroll, D. L. (2001). Expression of interleukin-1beta in the digital laminae of horses in the prodromal stage of experimentally induced laminitis. *American Journal of Veterinary Research*, *62*, 714–720.

Garner, H. E., Moore, J. N., Johnson, J. H., Clark, L., Amend, J. F. et al. (1978). Changes in the caecal flora associated with the onset of laminitis. *Equine Veterinary Journal*, *10*, 249–252.

Haruta, H., Kato, A., & Todokoro, K. (2001). Isolation of a novel interleukin-1-inducible nuclear protein bearing ankyrin-repeat motifs. *Journal of Biology and Chemistry*, *276*, 12485–12488.

Khabar, K. S. (2010). Post-transcriptional control during chronic inflammation and cancer: A focus on AU-rich elements. *Cellular and Molecular Life Sciences*, *67*, 2937–2955.

Kyaw-Tanner, M., & Pollitt, C. C. (2004). Equine laminitis: Increased transcription of matrix metalloproteinase-2 (MMP-2) occurs during the developmental phase. *Equine Veterinary Journal*, *36*, 221–225.

Kyaw-Tanner, M. T., Wattle, O., van Eps, A. W., & Pollitt, C. C. (2008). Equine laminitis: Membrane type matrix metalloproteinase-1 (MMP-14) is involved in acute phase onset. *Equine Veterinary Journal*, *40*, 482–487.

Leise, B. S., Faleiros, R. R., Watts, M. R., Johnson, P. J., Black, S. J. et al. (2011). Laminar inflammatory gene expression in the carbohydrate overload model of laminitis. *Equine Veterinary Journal*, *43*, 54–61.

Leise, B. S., Johnson, P. J., Faleiros, R. R., Black, S. J., & Belknap, J. K. (2010). Laminar regulation of STAT1 and STAT3 in the black walnut extract and carbohydrate induced models of lamintis. *Journal of Equine Veterinary Science*, *30*, 109–110.

Leise, B. B., Johnson, P. J., Faleiros, R. R., Black, S. J., Watts, M. R. et al. (2009). Hind laminar proinflammatory response is present after carbohydrate overload. *2nd AAEP Foundation Equine Laminitis Research Workshop.*

Loftus, J. P., Belknap, J. K., & Black, S. J. (2006). Matrix metalloproteinase-9 in laminae of black walnut extract treated horses correlates with neutrophil abundance. *Veterinary Immunology and Immunopathology*, *113*, 267–276.

Loftus, J. P., Black, S. J., Pettigrew, A., Abrahamsen, E. J., & Belknap, J. K. (2007). Early laminar events involving endothelial activation in horses with black walnut- induced laminitis. *American Journal of Veterinary Research*, *68*, 1205–1211.

Loftus, J. P., Johnson, P. J., Belknap, J. K., Pettigrew, A., & Black, S. J. (2009). Leukocyte-derived and endogenous matrix metalloproteinases in the lamellae of horses with naturally acquired and experimentally induced laminitis. *Veterinary Immunology and Immunopathology*, *129*, 221–230.

Motoyama, M., Yamazaki, S., Eto-Kimura, A., Takeshige, K., & Muta, T. (2005). Positive and negative regulation of nuclear factor-kappaB-mediated transcription by IkappaB-zeta, an inducible nuclear protein. *Journal of Biology and Chemistry*, *280*, 7444–7451.

Nerstedt, A., Johansson, A., Andersson, C. X., Cansby, E., Smith, U. et al. (2010). AMP-activated protein kinase inhibits IL-6-stimulated inflammatory response in human liver cells by suppressing phosphorylation of signal transducer and activator of transcription 3 (STAT3). *Diabetologia*, *53*, 2406–2416.

Noschka, E., Vandenplas, M. L., Hurley, D. J., & Moore, J. N. (2009). Temporal aspects of laminar gene expression during the developmental stages of equine laminitis. *Veterinary Immunology and Immunopathology*, *129*, 242–253.

Rodgerson, D. H., Belknap, J. K., Fontaine, G. L., & Kroll, D. L. (2000). Characterization of cultured smooth muscle cells obtained from the palmar digital arteries of horses. *American Journal of Veterinary Research*, *61*, 1602–1608.

Rodgerson, D. H., Belknap, J. K., Moore, J. N., & Fontaine, G. L. (2001). Investigation of mRNA expression of tumor necrosis factor-alpha, interleukin-1beta, and cyclooxygenase-2 in cultured equine digital artery smooth muscle cells after exposure to endotoxin. *American Journal of Veterinary Research*, *62*, 1957–1963.

Van Eps, A. W., Leise, B. S., Watts, M. R., Pollitt, C. C., & Belknap, J. K. (2010). Digital hypothermia inhibits early laminary inflammatory signaling in the oligofrucose laminitis model. *Journal of Veterinary Internal Medicine*, *24*, 690.

Van Eps, A. W., & Pollitt, C. C. (2006). Equine laminitis induced with oligofructose. *Equine Veterinary Journal*, *38*, 203–208.

Van Eps, A. W., & Pollitt, C. C. (2009). Equine laminitis model: Cryotherapy reduces the severity of lesions evaluated seven days after induction with oligofructose. *Equine Veterinary Journal*, *41*, 741–746.

Vick, M. M., Adams, A. A., Murphy, B. A., Sessions, D. R., Horohov, D. W. et al. (2007). Relationships among inflammatory cytokines, obesity, and insulin sensitivity in the horse. *Journal of Animal Science*, *85*, 1144–1155.

Wade, C. M., Giulotto, E., Sigurdsson, S., Zoli, M., Gnerre, S. et al. (2009). Genome sequence, comparative analysis, and population genetics of the domestic horse. *Science*, *326*, 865–867.

Waguespack, R. W., Cochran, A., & Belknap, J. K. (2004a). Expression of the cyclooxygenase isoforms in the prodromal stage of black walnut-induced laminitis in horses. *American Journal of Veterinary Reseach*, *65*, 1724–1729.

Waguespack, R. W., Kemppainen, R. J., Cochran, A., Lin, H. C., & Belknap, J. K. (2004b). Increased expression of MAIL, a cytokine-associated nuclear protein, in the prodromal stage of black walnut-induced laminitis. *Equine Veterinary Journal*, *36*, 285–291.

Williams, J. M., Lin, Y. J., Loftus, J. P., Faleiros, R. R., Peroni, J. F. et al. (2010). Effect of intravenous lidocaine administration on laminar inflammation in the black walnut extract model of laminitis. *Equine Veterinary Journal*, *42*, 261–269.

Yamamoto, M., Yamazaki, S., Uematsu, S., Sato, S., Hemmi, H. et al. (2004). Regulation of Toll/IL-1-receptor-mediated gene expression by the inducible nuclear protein IkappaBzeta. *Nature*, *430*, 218–222.

17 Genomics of performance

Emmeline W. Hill, Lisa M. Katz, and David E. MacHugh

Introduction

As flight animals and grazers, the wild ancestors of modern horses were naturally selected for speed and the ability to traverse long distances. Since horses were domesticated on the Eurasian steppe some 6,000 years ago (Levine, 1999), they have been selected for strength, speed, and endurance-exercise traits. More recently the development of specific breeds has resulted in selection for athletic phenotypes that enable the use of the horse for riding, recreation, sport, and racing. Some breeds, most notably the Thoroughbred, Standardbred, French Trotter, Swedish Trotter, and Quarter Horse, are particularly suited to racing, while others have been strongly selected for jumping and other athletic phenotypes (Table 17.1).

Horses are natural athletes and have evolved many structural and functional adaptations that contribute to an athletic phenotype. In some cases these adaptations are not influenced by training, while in other instances training has a marked effect on function. Briefly, the superior athletic ability of the horse may be attributed to a high maximal aerobic capacity, large intramuscular stores of energy substrates, high skeletal mitochondrial density and oxidative enzyme activity, and an increased oxygen carrying capacity. As a result, horses are considerably superior in physical performance than other species of similar size.

Selection for superior racing ability has been particularly pronounced in the Thoroughbred, which for three centuries has been subject to intense artificial selection for the system-wide structural and functional adaptations that have significantly enhanced the physiological characteristics that enable elite athletic performance (Constantinopol et al., 1989; Jones et al., 1989; Evans et al., 1993). In particular, the Thoroughbred has a large muscle mass (approximately 55%) to body weight ratio, while other, less athletic horse breeds have a lower proportion of muscle (~42%). This contrasts with most other mammalian species, in which muscle mass accounts for only 30–40% of body weight (Gunn, 1987). A large lung volume and high maximum haemoglobin concentration and cardiac output allow for efficient transport of oxygen to exercising muscle where it is utilized to generate ATP in the mitochondria via oxidative phosphorylation. Thoroughbreds contain approximately twice the number of mitochondria in contrast to similar sized mammals (Kayar et al., 1989), and in general VO_{2max} is restricted by oxygen supply to the mitochondria rather than by mitochondrial oxidative capacity (Katz et al., 2005). As well as a high skeletal muscle mitochondrial density and oxidative enzyme activity, Thoroughbreds have considerable intramuscular stores of energy substrates

Equine Genomics, First Edition. Edited by Bhanu P. Chowdhary.

Table 17.1 World Breeding Federation for Sport Horses (WBFSH) Rankings 2009 for Eventing, Show Jumping, and Dressage; Stud Book name is provided.

Eventing	Show Jumping	Dressage
1. Irish Sport Horse	1. KWPN	1. KWPN
2. Hannoverian	2. Selle Francais	2. Hannoverian
3. Selle Francais	3. Holstein	3. Danish Warmblood
4. Baden-Wurttemberg	4. Belgian Warmblood	4. Oldenburger
5. KWPN	5. Westfalian	5. Swedish Warmblood
6. Holstein	6. Hannoverian	6. Westfalian
7. Trakehner	7. Oldenburger	7. Rheinisches
8. Westfalian	8. Zangersheide	8. Belgian Warmblood
9. Oldenburger	9. Swedish Warmblood	9. Polish Horse breeders Assoc
10. Sport Horse Breeding of GB	10. Danish Warmblood	10. Trakhener
11. Anglo-Arab (Francais)	11. Irish Sport Horse	11. Baden-Wurttemberg
12. Belgian Warmblood	12. SBS (Belgian Sport Horse)	12. P.R.E.

(primarily glycogen). Together, these physiological adaptations allow for enhanced oxidative capacity and set the Thoroughbred horse apart from other horse breeds for superior athletic ability.

Heritability of Athletic Performance

Athletic phenotypes are influenced markedly by environment, management, and training; however, it has long been accepted that there are underlying genetic factors that influence a horse's athletic performance capabilities (Gaffney & Cunningham, 1988). Indeed, selection and breeding of racehorses is predicated on the belief that racing performance is inherited. Each time a pedigree is consulted by a breeder, it represents an attempt to capture and evaluate the genetic contributions from ancestors within that pedigree.

Many studies have tried to quantify the genetic contribution or heritability of various equine performance traits, which may be categorised broadly into work, riding, and racing traits (Hintz, 1980). Estimates for the heritability of riding performance vary widely from 0.00 ± 0.12 for $\log_{10}$ of total earnings and $\log_{10}$ of average annual earnings for three-day eventing (Bade et al., 1975b) to 0.71 ± 0.41 for points for jumping (Bade et al., 1975a). Mean values for $\log_{10}$ of earnings from jumping, three-day eventing, and dressage have been estimated as 0.18, 0.19, and 0.17, respectively (Hintz, 1980). Athleticism, scored as free-jumping ability, has a heritability of approximately 0.24–0.56 (Huizinga et al., 1991; Brockmann, 1998; Hascher, 1998; Von Velsen-Zerweck, 1999).

The majority of studies investigating heritability of performance traits have been performed in racing breeds, and in particular Thoroughbreds (Hintz, 1980; Langlois, 1980; Tolley et al., 1985; Oki et al., 1995; Williamson & Beilharz, 1996, 1998; Bakhtiari & Kashan, 2009; Thiruvenkadan et al., 2009). Estimates vary depending on the analytical model and the performance measure; in most instances earnings, handicap ratings, and race times are used. The estimates for heritability of ($\log_{10}$) earnings range from 0.23 to 0.56 among three-year-old Thoroughbreds (Tolley et al., 1985). For lifetime earnings and lifetime rank, heritability has been estimated as 0.12 and 0.11 among Japanese Thoroughbreds (Tozaki et al., 2010). Estimates for heritability of handicap rating range from 0.24 to 0.76 (Tolley et al., 1985; Gaffney & Cunningham, 1988) and heritability of race time ranges from 0.09 to 0.78 (Tolley et al., 1985). The heritability of race time has been observed to decrease as the race length increases (Oki et al., 1995). The most strongly inherited performance attribute reported is best race distance, which in a population of Australian Thoroughbreds has been estimated as 0.94 (Williamson & Beilharz, 1998).

Athletic Performance Genes Encoded by the Nuclear Genome

Genetic differences that may influence fitness or performance in humans were first examined in the 1990s, and at the turn of the millennium, the first review of the genes contributing to human performance and health-related fitness phenotypes reported just 24 genes (Rankinen et al., 2001). During the subsequent decade, six human gene maps for performance and health-related fitness traits have been published (Rankinen et al., 2002; Perusse et al., 2003; Rankinen et al., 2004; Wolfarth et al., 2005; Rankinen et al., 2006; Bray et al., 2009). The most recent catalog of human performance and health-related fitness traits identified 221 genes – a 10-fold increase since 2000 (Bray et al., 2009). In particular, studies have focused on genes influencing exercise behaviour (i.e., variation between sedentary and active cohorts), genes relevant to muscle strength and power, cardiorespiratory genes, adiposity, glucose metabolism, lipid metabolism, and cardiovascular phenotypes (Bray et al., 2009). In particular, polymorphisms in the angiotensin I converting enzyme (peptidyl-dipeptidase A) 1 gene (*ACE*) and the actinin, alpha 3 gene (*ACTN3*) have been most extensively described (for recent reviews, see MacArthur & North, 2007; Ostrander et al., 2009; Woods, 2009; Lek et al., 2010; Rankinen et al., 2010).

The availability of the equine genome sequence (Wade et al., 2009) and the parallel development of high-throughput genomic assay platforms for the horse have rapidly enabled the identification of genomic sequence variation associated with athletic performance phenotypes in Thoroughbreds (Gu et al., 2009; Hill et al., 2010b; Hill et al., 2010d; Tozaki et al., 2010). Functional groups of genes that contribute to the control of substrate utilization, insulin signaling, and muscle strength seem to be of greatest importance to the athletic phenotype of the Thoroughbred. For instance, the first three reported sequence variants located within genes associated with performance traits are at the myostatin gene (*MSTN*) (Hill et al., 2010b; Hill et al., 2010d; Tozaki et al., 2010), the cytochrome c oxidase, subunit 4, isoform 2 gene (*COX4I2*) (Gu et al., 2010), and the pyruvate dehydrogenase kinase isozyme 4, mitochondrial gene (*PDK4*) (Hill et al., 2010c) loci.

Over the coming decade, the list of genes contributing to performance in the Thoroughbred is likely to lengthen dramatically as increasingly sophisticated equine genomics resources and tools become available (Figure 17.1) that will significantly improve our knowledge of the genes and molecular mechanisms underpinning athleticism and exercise adaptation.

Mitochondrial DNA and Athletic Performance in Thoroughbreds

In 2006, Harrison and Turrion-Gomez reported a relationship between mitochondrial DNA (mtDNA) haplotypes and athletic performance in the Thoroughbred (Harrison & Turrion-Gomez, 2006). Mammalian mitochondrial DNA haplotypes are inherited exclusively from the female parent and therefore solely represent female contributions to the phenotype. The mtDNA data presented by Harrison and Turrion-Gomez purported to support the traditional belief that particular Thoroughbred female lines were superior to others (Lowe & Allison, 1913; Bobinski & Zamoyski, 1960) and that mtDNA sequence variation was associated with a considerable female contribution to stamina

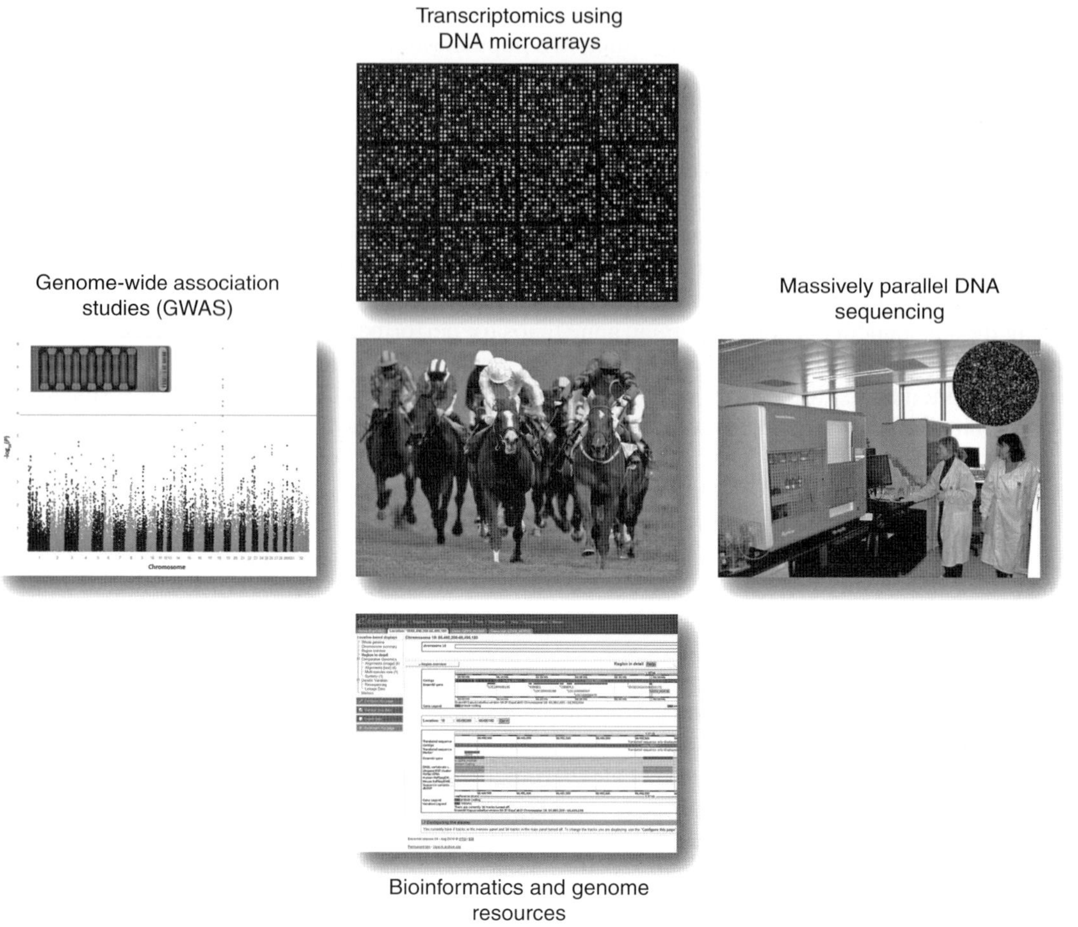

Figure 17.1 Genomics technologies and resources that may be used to study exercise adaptation and athletic performance in Thoroughbred horses.

potential. In this regard, previous work by one of the authors of this chapter demonstrated that a number of Thoroughbred female lineages had been found to contain mitochondrial haplotypes that were unexpected in the family lines, which suggested, therefore, that the prediction of performance aptitude required verification of mtDNA haplotype (Hill et al., 2002). However, more recently, it has been shown that female family sublineages are a more accurate reflection of mtDNA ancestry and that many of the errors in the stud book occurred early in the foundation stages (Bower et al., 2011). Mitochondrial DNA haplotypes may therefore be directly inferred from pedigree information with reasonable confidence if subfamily lineages are considered.

Some breeders emphasize the importance of the mitochondrial genome to performance in the Thoroughbred. Indeed, the eukaryotic mitochondrion is essential for cellular metabolism and its principal function is to support aerobic respiration. Several metabolic pathways operate within the mitochondrion, including the Krebs cycle (citric acid cycle), β-oxidation of fatty acids, and lipid and cholesterol synthesis. Mitochondria possess a discrete circular DNA genome, which in mammals is approximately 17 kilobases (kb) in size. The mitochondrial genome encodes 13 proteins that are subunits of the respiratory chain and oxidative phosphorylation system (OXPHOS) (Scheffler, 2008). Although these 13 mtDNA-encoded protein subunits are essential for respiration, 72 of the 85 subunits of the OXPHOS system are actually encoded by genes within the nuclear genome that are subject to conventional biparental Mendelian patterns of genetic transmission and inheritance. However, it is important to note that recent work has described statistical associations between mitochondrial haplotypes and endurance in human athletes, but small sample sizes, ethnic heterogeneity, and conflicting results from different surveys currently obscure the significance of these observations, and more detailed studies will be required to clarify the exact nature of any mitochondrial genetic contributions to human athletic performance (Bray et al., 2009; Ostrander et al., 2009; Rankinen et al., 2010). Significantly, in this context, more than 1,500 proteins – encoded by genes on nuclear chromosomes – function within the mitochondrion (Scheffler, 2008). Therefore, it is most likely that mitochondrial functions relating to athletic performance are influenced primarily by genetic variation in the nuclear genome.

Detection of Genomic Regions under Selection in the Thoroughbred Genome

Natural selection for athletic traits among the wild ancestors of the domestic horse has been uniquely augmented in Thoroughbreds through recent and strong artificial selection. Domestic animal species provide unique opportunities to identify genes underlying specific phenotypes that have been strongly selected because discrete breeds have arisen relatively recently from a small number of founder animals (Georges, 2007; Sellner et al., 2007; Goddard & Hayes, 2009). The Thoroughbred breed is a closed population established in the sixteenth and seventeenth centuries from crosses between local Galloway and Irish hobby horses with imported Eastern stock (Willett, 1975). As with many domestic breeds, the Thoroughbred originates from a small number of founders; just one founder stallion contributes to 95% of paternal lineages and ten founder mares account for 72% of maternal lineages (Cunningham et al., 2001). However, despite a limited number of founders and strong selection for racetrack performance, some 35–55% of variation in racing performance is heritable (Gaffney & Cunningham, 1988; Mota et al., 2005). Consequently, the demographic history of the Thoroughbred coupled with the intense recent selection for athleticism offer a unique opportunity to understand genomic contributions to exercise-related traits.

The first report of genes and genomic regions contributing to athletic potential in the horse was published in 2009, describing regions of the genome that have been under selection during the 300-year development of the Thoroughbred (Gu et al., 2009). This work involved a population genetics-based genome scan of genetic variation at 394 autosomal and X chromosome microsatellite loci in four geographically diverse horse populations (Connemara, Akhal-Teke, Tuva, and Thoroughbred). Positively selected loci were identified in the extreme tail-ends of the distributions for population genetic parameters and test statistics (F_{ST} and the Ewens-Watterson test) that identify departures from patterns of genetic variation expected under neutral genetic drift (Gu et al., 2009). Deviations from expected heterozygosity in Thoroughbred and global differentiation among horse populations identified outlier loci that are indicative of selection. Such outlier approaches have led to a deeper understanding of the selective forces that have shaped the recent evolution of human populations and also domestic dog breeds (Akey, 2009; Novembre & Di Rienzo, 2009; Pickrell et al., 2009; Akey et al., 2010; Grossman et al., 2010; Oleksyk et al., 2010; Vonholdt et al., 2010).

The positively selected genomic regions in Thoroughbred identified by Gu et al. (2009) are enriched for genes involved in phosphatidylinositol 3-kinase (PI3K) mediated signaling, insulin receptor signaling, and lipid transport – biochemical pathways with well-characterized roles in adaptation to exercise. Insulin stimulates glucose transport to maintain glucose homeostasis via a range of different transcriptionally active signaling pathways (O'Brien & Granner, 1996). Among these, the PI3K pathway plays a key role in insulin-stimulated glucose transport in skeletal muscle (Hayashi et al., 1998; Shepherd et al., 1998; Roques & Vidal, 1999) via its interaction with IRS-1 (insulin receptor substrate 1) (Andreelli et al., 1999) and its regulation by insulin of phosphoinositide-3-kinase, regulatory subunit 1 (alpha) gene (*PIK3R1*) expression (Roques & Vidal, 1999). In this regard, it is noteworthy that regulation by insulin of genes via the PI3K pathway is disrupted in type 2 diabetes (T2DM) (Ducluzeau et al., 2001). Among the regions in the Thoroughbred genome that displayed clear signatures of strong selection were the *IRS1*, *PIK3R1*, and phosphoinositide 3-kinase, class 3 (*PIK3C3*), genes, transcripts of which are dysregulated in skeletal muscle from T2DM patients following stimulation with insulin (Andreelli et al., 1999; Tsuchida et al., 2002). Other genes identified by Gu et al. (2009) in positively selected regions of the Thoroughbred genome include the insulin-receptor signaling pathway genes *FOXO1* (forkhead box O1), *GRB2* (growth factor receptor-bound protein 2), *PTPN1* (protein tyrosine phosphatase, non-receptor type 1), *SOCS3* (suppressor of cytokine signalling 3), *SOCS7* (suppressor of cytokine signalling 7), and *STXBP4* (syntaxin-binding protein 4).

The importance of muscle function in the recent evolution of the Thoroughbred population was also highlighted by a significant overrepresentation of sarcoglycan complex and focal adhesion pathway genes located within the selected regions (Gu et al., 2009). The sarcoglycan complex is found associated with the dystophin-glycoprotein complex, which is located at the sarcolemma of cardiac and skeletal muscle cells and links the contractile apparatus of the muscle with the lamina, thus providing a mechano-signaling role (Pardo et al., 1983). Mutations in any one of the sarcoglycan genes destabilizes the entire sarcoglycan complex (Ozawa et al., 2005), which may lead to progressive loss of skeletal myofibres or cardiomyocytes (Mizuno et al., 1994). The dystrophin-glycoprotein complex contributes to the integrity and stability of skeletal muscle by its association with laminin receptors and the integrin-associated complex in the costamere (Pardo et al., 1983). Focal adhesion complexes form part of the costamere and genes in genomic regions that displayed signatures of recent selection included *TNC* (tenascin C), which functions in the focal adhesion pathway and may be particularly important for muscle integrity because of its role in protection against mechanically induced damage (Fluck et al., 2008).

Muscle-related genes within positively selected regions of the Thoroughbred genome also included the *ACTA1* (actin, alpha 1, skeletal muscle) and *ACTN2* (actinin, alpha 2) genes. The alpha actin protein is found principally in muscle and is a major constituent and regulator of the contractile apparatus (Tobacman, 1996; Gordon et al., 2000). In skeletal muscle α-actinin is responsible for cross-linking actin filaments between adjacent sarcomeres and is known to interact with PI3K signaling pathways (Shibasaki et al., 1994). Polymorphisms in the gene encoding α-actinin 3 (*ACTN3*) are among the best-characterized athletic-performance-associated variants in human endurance athletes (Yang et al., 2003; Chan et al., 2008; MacArthur et al., 2008), and evidence for positive selection in the genomic region surrounding *ACTN3* has been reported in humans (MacArthur et al., 2007). While *ACTN3* is expressed principally in fast muscle fibres, *ACTN2* is more widely expressed in skeletal and cardiac muscle. The ACTN2 protein is structurally and functionally similar to ACTN3 and it has been suggested that ACTN2 has a compensatory functional role in the absence of ACTN3 (Mills et al., 2001).

In summary, the identification of genomic regions that have been influenced by selection for athletic phenotypes has enabled the identification of the first candidate athletic performance genes in the Thoroughbred (Table 17.2). Based on these analyzes, it is apparent that recent selection in the ancestors of the present-day Thoroughbred population principally targeted genes associated with

Table 17.2 Functional groups of genes significantly overrepresented in selected genomic regions in the Thoroughbred. Functional categories are separated into GO Biological Processes, GO Cellular Components, GO Molecular Functions, and KEGG Pathways. The *P*-value, fold-change, and statistical parameter by which each functional category was identified are given, where F_{ST} is indicative of among population differentiation and Dh/sd indicates a reduction in heterozygosity.

Functional group	*P*-value	Fold enrichment	Population genetics parameter
GO Biological Process			
Actin filament organization	<0.01	3.9	F_{ST}
Anatomical structure formation	<0.05	2.3	F_{ST}
Cellular calcium ion homeostasis	<0.01	3.1	F_{ST}
Detection of temperature stimulus	<0.01	20.8	F_{ST}
G-protein signaling, coupled to IP3 second messenger (phospholipase C activating)	<0.01	3.4	Dh/sd
Hydrogen peroxide catabolic/metabolic process	<0.05	9.7	F_{ST}
Insulin receptor signaling pathway	<0.01	5.0	F_{ST} & Dh/sd
Lipid transport	<0.05	2.2	Dh/sd
Mating	<0.05	8.9	F_{ST}
Muscle development	<0.01	2.0	F_{ST} & Dh/sd
Peptidyl-histidine phosphorylation	<0.05	9.7	Dh/sd
Phosphoinositide-mediated signalling	<0.05	3.2	Dh/sd
Porphyrin biosynthetic process	<0.05	7.2	F_{ST} & Dh/sd
Regulation of catalytic activity	<0.05	1.6	F_{ST}
Regulation of osteoblast differentiation	<0.05	9.6	F_{ST}
Response to hydrogen peroxide	<0.05	5.2	Dh/sd
Response to oxidative stress	<0.01	2.6	F_{ST}
Sexual reproduction	<0.01	2.0	F_{ST}
Skeletal development	<0.05	2.0	
Spermatogenesis	<0.01	2.3	F_{ST} & Dh/sd
GO Cellular Component			
Cortical actin cytoskeleton	<0.01	6.7	F_{ST}
Mitochondrion	<0.05	1.5	F_{ST} & Dh/sd
Sarcoglycan complex	<0.05	11.1	Dh/sd
GO Molecular Function			Dh/sd
Diacylglycerol kinase activity	<0.05	9.1	Dh/sd
G-protein-coupled receptor binding	<0.01	4.0	F_{ST}
Pyruvate dehydrogenase (acetyl-transferring) kinase activity	<0.01	21.2	F_{ST} & Dh/sd
KEGG Pathway			F_{ST}
Focal adhesion	<0.01	1.8	F_{ST}

fatty acid oxidation, increased insulin sensitivity, and muscle strength – highlighting the central role for muscle function and integrity in the Thoroughbred athletic phenotype.

DNA Sequence Variation and Athletic Performance Traits in the Thoroughbred

Natural selection and recent artificial selection giving rise to adaptations associated with exercise and athletic performance have resulted in changes in the frequencies of advantageous sequence variants in genes that contribute to athletic phenotypes among successful subgroups of the Thoroughbred population. A number of approaches may be taken to identify genes underlying phenotypic adaptations; these include the candidate gene approach, which requires *a priori* knowledge of gene function and linkage mapping, which requires information about familial relationships as well as access to samples from large numbers of relatives. The most powerful strategies, however, have been population-based approaches using microsatellite marker panels (Gu et al., 2009; Tozaki et al., 2010) or the pan-genomic single nucleotide polymorphism (SNP) assay platform available from Illumina® Corp. (Equine SNP50 BeadChip genotyping array) (Hill et al., 2010d), which have facilitated hypothesis-free, genome-wide discovery to detect nuclear-encoded performance gene variants.

For candidate gene and genome-wide association studies, the study population has generally been segregated into cohorts of individuals with divergent racing phenotypes. In the case of studies undertaken by our group, horses were categorized based on retrospective racecourse performance records as follows: elite Thoroughbreds or non-elite Thoroughbreds, and short-distance elite winners or long-distance elite winners. In our studies, elite Thoroughbreds are considered Flat racehorses that have won at least one Group race (Group 1, Group 2, or Group 3) or a Listed race. Such Group (or Stakes) races are the most prestigious and highest grade of race and have the greatest prize money. The international standards are set for these races by the International Federation of Horseracing Authorities.

To date, three genes with molecular functions relevant to physiological processes important for exercise have been reported to be associated with racing performance, including the myostatin gene (*MSTN*) (Binns et al., 2010; Hill et al., 2010b; Hill et al., 2010d; Tozaki et al., 2010), the cytochrome c oxidase, subunit 4, isoform 2 gene (*COX4I2*) (Gu et al., 2010), and the pyruvate dehydrogenase kinase isozyme 4, mitochondrial gene [*PDK4*] (Hill et al., 2010c). A variant in the genomic sequence for *PDK4* is the first example of a statistically significant association of a SNP with elite race winning performance. The association was detected by investigating sequence variation in 20 candidate exercise-relevant genes that were selected for the study on the basis of gene ontology and their presence in one of the top ranked regions with a signature of selection ascertained from a genome scan (Gu et al., 2009). The expression of *PDK4* is coordinated by the transcriptional co-activator PGC-1α (Wende et al., 2005), a key regulator of energy metabolism that regulates insulin sensitivity by controlling glucose transport, drives the formation of oxidative muscle fibres and co-ordinates mitochondrial biogenesis via its interaction with nuclear encoded mitochondrial protein genes (Scarpulla, 2008). Furthermore, the oxidation of fatty acids, which is highly efficient in the generation of ATP, is controlled by the expression of PDK4 in skeletal muscle during and after exercise (Pilegaard & Neufer, 2004). In the horse, *PDK4* gene expression has been observed to increase almost 7-fold following a bout of moderate intensity treadmill exercise in untrained Thoroughbreds (Eivers et al., 2010) and is differentially regulated after sprint exercise in trained Thoroughbreds (Hill et al., 2010a). In addition, the *PDK4* gene is located in the region of the equine genome that displayed the strongest selection signature, emphasizing its role as a key target of selection for exercise adaptation.

In a cohort of 148 Thoroughbreds, three non-coding SNPs in the *PDK4* genomic sequence (*PDK4*_38968139, *PDK4*_38969307, and *PDK4*_38973231) were found to be significantly associated with elite race winning performance; *PDK4*_38973231 had the strongest association ($P = 0.0017$; odds ratio = 2.20). When handicap rating was considered as a quantitative phenotype, the association was confirmed (*PDK4*_38973231, $P = 0.0252$). The associations were validated in an independent sample set ($n = 130$) (elite vs. non-elite, *PDK4*_38973231, $P = 0.0150$; handicap rating, *PDK4*_38973231, $P = 0.0252$) and when all samples ($n = 278$) were considered the significance of association was stronger (*PDK4*_38973231, $P = 0.0004$, odds ratio = 1.97, C.I. (95) = 1.35–2.87). A dominant model in which the A:A and A:G genotypes were favorable provided the best explanation for the data ($P = 0.0003$), with the A:A and A:G genotypes more common among elite (70%) than non-elite (47%) racehorses. When all individuals with a RPR handicap rating[1] ($n =$ 228) were considered the A:A and A:G genotypes (*PDK4*_38973231) had on average a 16.2–16.6 lb handicap advantage over G:G horses. Additional preliminary associations between candidate gene loci and racing performance have been reported. For example, weak but significant associations with racing performance were observed for the *COX4I2* and *CKM* (creatine kinase, muscle) genes, but only the *COX4I2* association was validated in an extended sample set (Gu et al., 2010).

In order for equine genomic information to have real applicability, it is imperative that all preliminary genetic associations are validated in adequate and appropriate cohorts of animals. Equine genomics is a new and emerging field and should adhere to the rigorous standards for experimental design, data integrity, statistical replication, and validation that have been established for human genomics research (Hughes, 2009; Igl et al., 2009; Ioannidis et al., 2009; Jorgensen et al., 2009; Little et al., 2009; Singer, 2009). This will ensure that equine genomic information has real value to owners, breeders, and trainers, and that it can be exploited and implemented for maximum benefit throughout the Thoroughbred industry.

Identification of the Myostatin Gene (*MSTN*) – the "Speed" Gene – as a Major Locus Affecting Race Distance Aptitude

It is widely recognized among horse breeders that inherited variation in physical and physiological characteristics is responsible for variation in individual aptitude for race distance, and that muscle phenotypes in particular are important. Similar to their human counterparts, sprint-racing Thoroughbreds are observed to be generally more compact and muscular than horses suited to longer-distance races. The International Federation of Horseracing Authorities recognizes five race distance categories: Sprint (5–6.5 furlongs[2] (f), ≤1, 300 m); Mile (6.51–9.49 f, 1,301–1,900 m); Intermediate (9.5–10.5 f, 1,901–2,112 m); Long (10.51–13.5 f, 2,114–2,716 m); and Extended (>13.51 f, >2,717 m) races. Horses that compete within these race categories are generally termed "sprinters" (<6 furlongs), "middle distance" or "milers" (7–8 f), or "stayers" (>8 f).

The most extensively studied locus that has been associated with a performance trait in the Thoroughbred population contains the gene encoding myostatin (*MSTN*). Myostatin is a growth and differentiation factor that functions as a negative regulator of skeletal muscle mass development. In several mammalian species, including cattle, sheep, dogs, and mice, muscle hypertrophy phenotypes are associated with sequence variants in the *MSTN* gene (Grobet et al., 1997; McPherron et al., 1997; McPherron & Lee, 1997; Schuelke et al., 2004; Mosher et al., 2007).

In horses, sequence and structural variation in the intergenic and proximal upstream and downstream sequences of the *MSTN* gene has been identified and associated with optimum racing distance in Thoroughbreds (Binns et al., 2010; Hill et al., 2010b; Hill et al., 2010d; Tozaki et al., 2010), and increased muscle phenotypes among heavy draft horse breeds (Dall'Olio et al., 2010). The equine *MSTN* gene contains three exons and spans ~6 kb on chromosome 18 (reverse strand nucleotides: 66,489,608–66,495,780; EquCab2.0 genome assembly). No exonic sequence variants have been described in Thoroughbreds but the following polymorphisms displaying a minor allele frequency

[1] Handicap ratings are an expression in pounds of the merit of the horse's form. The scale on the Flat runs from around 20 for the poorest animals to 140 or more for the very best performers. For this study, handicap ratings designated by the Racing Post (RPR, Racing Post Ratings) were used, as these were available for the greatest number of horses.

[2] 1 furlong = $^1/_8$ mile = 201.2 metres.

(MAF) greater than 0.05 have been identified: two SNPs in intron 1; a 227 bp SINE insertion polymorphism located 145 bp upstream of the transcriptional start site; and four 3′UTR SNPs (Hill et al., 2010b; Hill et al., 2010d). Re-sequencing the *MSTN* locus in a panel of 12 horses from 10 diverse horse breeds (Bardigiano, Haflinger, Italian Saddle, Italian Trotter, Noric, Rapid Heavy Draft, Salernitano, Throroughbred, and Ventasso) identified 6 further SNPs in the promoter region and intron 1 and 2, though none of these polymorphisms displayed a MAF > 0.05 in the Thoroughbred sample (Dall'Olio et al., 2010).

Our research group has carried out a series of population-based case-control investigations of variation at the *MSTN* gene, where a sample of Thoroughbreds was separated on the basis of retrospective racecourse performance into discrete cohorts containing unrelated animals. Considering the relative contribution of muscle power to sprint and longer-distance racing, elite Group race winning animals were subdivided into those that had won their best (most valuable or highest grade) race over distances $\leq$8 f and those that had won their best race over distances >8 f. Among the two distance cohorts a highly significant ($P = 3.70 \times 10^{-5}$) association with a SNP in intron 1 (g.66493737C>T) was observed (Hill et al., 2010b). In an increased sample set, which was not restricted by excluding relatives, a very strong association was observed ($n = 197$, $P = 3.28 \times 10^{-13}$) and this association became stronger when the long-distance cohort was compared with individuals that had won their best race over $\leq$7 f (n = 167, $P = 8.55 \times 10^{-14}$). Two alleles were observed at this biallelic SNP: a 'C' allele and a 'T' allele, with the 'C' allele more than twice as frequent in the short-distance ($\leq$7 f) cohort of animals compared to the long-distance (>8 f) group (0.75 and 0.34 respectively), corresponding to an odds ratio of 5.81 (Hill et al., 2010d). Considering best race distance (BRD) as a quantitative trait, the data for the elite cohort was analyzed using the distance (furlongs) of the highest grade or most valuable Group race won as the phenotype. BRD was highly significantly associated ($n = 197$, $P = 1.47 \times 10^{-19}$) with the g.66493737C>T SNP (Hill et al., 2010b; Hill et al., 2010d).

A genetic test is now available to horse breeders and trainers for this polymorphism. The *Equinome Speed Gene Test* may be used to make a prediction about the type of horse an individual is most likely to be and may be used to improve decision making in selection, breeding and training. At this locus, homozygotes for the 'C' allele (i.e., C:C) have been found to compete preferentially in faster, shorter-distance races (mean BRD = 6.5 $\pm$ 1.5 f). The C:T horses, on the other hand, are best suited to middle-distance races (mean BRD = 9.1 $\pm$ 2.3 f), while T:T horses have greater stamina and tend to excel in longer-distance races (mean BRD = 11.0 $\pm$ 2.1 f) (Hill et al., 2010b). A distribution of the genotypes according to BRD is shown in Figure 17.2. The physical phenotypes of genetically different horses also vary significantly; among males, C:C horses were found to have ~7% greater muscle mass at two years old than T:T horses, suggesting a more precocious development of the skeletal musculature. In terms of racing performance, C:C individuals earned up to 13 times more in prize money than T:T horses at two years old, when race distances are primarily limited to $\leq$8 f.

Evaluating the genotypes of horses that have competed in races in Europe, Australia, New Zealand, and North America, individual genotypes at the g.66493737C>T SNP were not observed to be more common among elite Group race winning Thoroughbreds than horses that had never won a race. Also, no association was detected when handicap ratings – reflecting retrospective racing ability – were evaluated as a quantitative phenotype. However, in a study investigating genotype associations among Japanese Thoroughbreds, a significant association with performance rank and lifetime earnings has been observed (Tozaki et al., 2010); however, this may be explained by the distribution of race distances and the opportunities to achieve rank and win high-valued races.

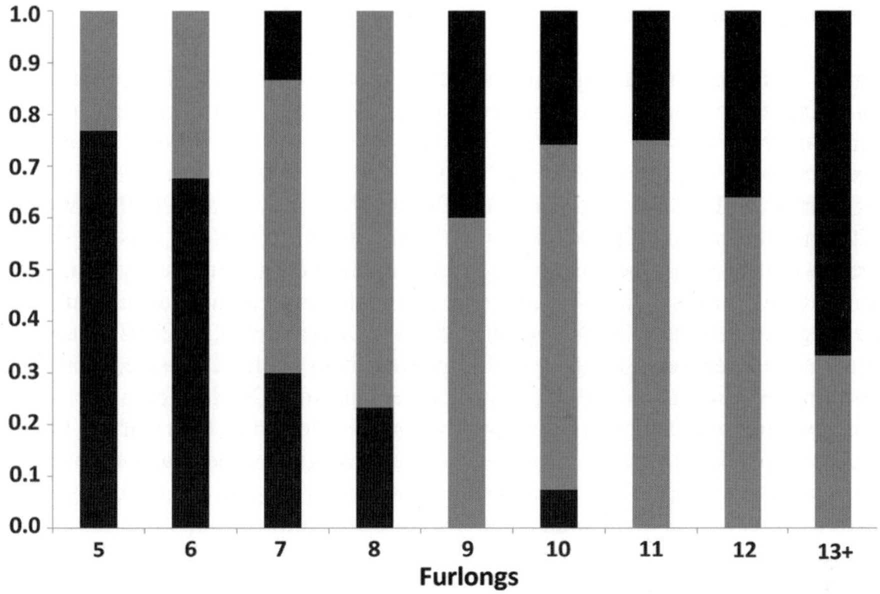

Figure 17.2 Optimum genetic profile for distance determined by genotypes for the *Equinome Speed Gene Test*, founded in variation at the *MSTN* g.66493737C>T locus; the frequency of C:C – maroon, C:T – aqua, and T:T – brown individuals are shown for each race distance category (f, furlongs). As the race distances increase, the frequency of C:C individuals decreases and is mirrored by an increase in the proportion of T:T individuals that have won their best race over distances >1 mile (8 furlongs).

A Genome-Wide Association Study (GWAS) for Optimum Race Distance in the Thoroughbred

While the studies described above focusing on the *MSTN* locus demonstrated a strong effect of *MSTN* genotype on distance, they were constrained by a lack of comparative data to evaluate the effects of additional nuclear gene variants contributing to distance aptitude. Therefore, a genome-wide association study (GWAS) using the Illumina® EquineSNP50 BeadChip, which assays 54,602 pan-genomic SNPs in parallel, was performed using a cohort of 118 elite Thoroughbred racehorses divergent for race distance aptitude (Hill et al., 2010d). This study identified the genomic region on chromosome 18 containing the *MSTN* gene as the highest-ranked region for best racing distance and a set of seven SNPs within a 1.7 Mb region that reached genome-wide significance (Hill et al., 2010d). The most predictive SNP (BIEC2-417495) and the second-most predictive SNP (BIEC2-417372) were 692 kb and 28 kb from the *MSTN* gene, respectively. Evaluation of genotypes for the g.66493737C>T SNP allowed a comparative assessment of predictive power of each sequence variant. A comparison of trait association in the same set of samples confirmed the superior power of the g.66493737C>T SNP ($P_{\text{unadj.}} = 1.02 \times 10^{-10}$) for the prediction of best race distance when compared with BIEC2-417495 ($P_{\text{unadj.}} = 1.61 \times 10^{-9}$). While LD was high between g.66493737C>T and BIEC2-417495, no inter-locus epistatic effects between g.66493737C>T and any chromosome 18 SNP represented on the array was observed to influence best race distance (P >0.0001 for all interactions). Therefore, the effect of genotype on racing phenotype is highly likely a result of the reported variation in the *MSTN* gene at SNP g.66493737C>T.

Functional Genomics and Proteomics

The integration of complementary genomics technologies is a powerful approach to understand the biology of the exercising athlete. In recent years there has been extensive research into the functional genomics of human and rodent muscle physiology in response to exercise, and the number of reports using the equine model is growing rapidly (Eizema et al., 2005; Jose-Cunilleras et al., 2005; Barrey et al., 2006; Mucher et al., 2006; McGivney et al., 2009; Eivers et al., 2010; Hill et al., 2010a; Martin et al., 2010; McGivney et al., 2010). In skeletal muscle, exercise induces multiple stresses that result in increased energy demand and oxygen consumption, and the muscle tissue responds to alterations in such bioenergetic demands through changes in gene expression (for reviews, see Pilegaard & Neufer, 2004; Joseph et al., 2006; Hoppeler et al., 2007; Olesen et al., 2010). A single bout of acute exercise induces multiple stresses in skeletal muscle, including increased oxygen consumption (Sahlin et al., 1998; Martineau & Gardiner, 2001; McGivney et al., 2009); the responses to these stressors can be divided into two broad categories: the return to homeostasis and the adaptive response. The principle processes associated with homeostatic recovery are glucose sparing, elevated fat oxidation, glycogen resynthesis, and free radical quenching, as well as the repairing of free-radical-mediated damage and restoration of intracellular electrolyte concentrations and pH (Pilegaard et al., 2000; Richter et al., 2001; Ji, 2002; Jeukendrup, 2003; Green, 2004). The adaptive response is the process whereby skeletal muscle responds to repeated exercise bouts (conditioning or training) in ways that cumulatively lead to an enhanced ability to maintain muscle homeostasis during exercise.

The products of many genes are likely to influence system-wide physiological responses. However, the protein products of two genes have been identified as key regulators of the adaptive response to exercise in humans and model species. These are the hypoxia inducible factor 1, alpha subunit (basic helix-loop-helix transcription factor) gene (*HIF1A*), which encodes HIF-1α; and the peroxisome proliferator-activated receptor gamma, coactivator 1 alpha gene (*PPARGC1A*), which encodes PGC-1α (for reviews, see Bonen, 2009; Gibala, 2009; Lundby et al., 2009; Yan, 2009; Lira et al., 2010; Olesen et al., 2010). There is a large requirement of the musculature for oxygen during exercise in the Thoroughbred. Eivers et al. (2010) have reported the gene expression responses of a panel of HIF-responsive genes in skeletal muscle biopsies collected before and after a standardized incremental-step treadmill exercise test in untrained Thoroughbred horses. The genes examined included those encoding important glycolytic enzymes (Semenza, 1999), genes encoding key proteins for oxidative metabolism (Fukuda et al., 2007), and *PPARGC1A*. Analyses of mRNA profiles revealed no significant transcriptomic differences immediately post-exercise (T_0 *vs* T_1). However, four hours post-exercise (T_2) the *CKM*, *COX4I1*, *COX4I2*, *PDK4*, *PPARGC1A*, and *SLC2A4* genes had significant differences relative to pre-exercise levels. The observed relationships with measured physiological variables (VHR_{max} and [La] peak) indicated that local transcriptional microadaptations influence the overall athletic phenotype.

Proteomics studies offer a complementary approach to understanding the key molecular contributions to exercise adaptation and training. Using 2D gel electrophoresis, Bouwman et al. (2010) cataloged the skeletal muscle proteome in a cohort of Standardbred horses. Twenty spots representing 16 proteins differed between untrained and trained samples, and their functions suggested structural alterations toward a higher oxidative capacity, an increased capacity to take up long-chain fatty acids, and energy storage in the form of glycogen as central to the adaptive response to training. Interestingly, by further intensifying training, additional differential expression was detected. For example, alpha-1-antitrypsin increased after intensified training but not after normal training, and it has been suggested therefore that this protein may be considered a marker for overtraining.

Global Gene Expression Changes: Microarrays and Other Highly Parallel Gene Expression Approaches

During the last ten years the availability of microarray platforms has enabled hypothesis-free approaches in the evaluation of gene expression and exercise (Fehrenbach et al., 2003; Mahoney & Tarnopolsky, 2005; Virtanen & Takahashi, 2008). Prior to the availability of equine-specific microarrays, studies reported gene expression profiles in peripheral blood mononuclear cells (PBMC) in Arabian endurance horses using a mouse-specific cDNA array (Barrey et al., 2006). This study investigated associations with poor performance among competing horses and identified functional groups of genes that were differentially expressed between successful performers and individuals that were disqualified due to a metabolic disorder. Significant relationships between gene expression and creatine kinase (CK) and aspartate aminotransferase (AST) activity – indicators of metabolic stress – were identified, and these may be good candidates to explain underlying molecular contributions to poor performance in Arabian endurance horses.

Following on from the first microarray-based study, analysis of PBMC using a cDNA-AFLP technique identified a set of genes with altered expression in Arabian endurance horses after exercise (Cappelli et al., 2007). Four of the 49 identified mRNA transcript-derived fragments had sequence similarity with genes involved in an exercise-induced stress response: the interleukin 8 gene (*IL8*); the retinoblastoma binding protein 6 gene (*RBBP6*); the eukaryotic translation initiation factor 4 gamma, 3 gene (*EIF4G3*); and the heat shock protein 90kDa alpha (cytosolic), class A member 1 gene (*HSP90AA1*). The expression of these genes was subsequently confirmed by real-time qRT-PCR, suggesting their involvement in an exercise-induced stress response.

The recent availability of equine-specific gene expression microarrays has allowed our group to identify novel genes and key regulatory pathways responsible for exercise adaptation in Thoroughbred skeletal muscle (McGivney et al., 2009). Functional analysis of genes differentially expressed four hours post-exercise revealed an overrepresentation of genes localized to the actin cytoskeleton and with functions in the MAPK signaling, focal adhesion, insulin signaling, mTOR signaling, p53 signaling, and T2DM pathways, suggesting that protein synthesis, mechanosensation, and muscle remodelling contribute to skeletal muscle adaptation toward improved integrity and hypertrophy. These data indicate that the response to cumulative bouts of exercise (i.e., training) involves both morphological changes, such as hypertrophy, and metabolic responses such as an increased capacity for oxidative substrate metabolism in mitochondria, and a shift toward oxidizing proportionately more fats and less glucose during exercise (Carter et al., 2001; Adhihetty et al., 2003); they also suggest that these adaptations lead to an enhanced ability to maintain muscle homeostasis during exercise.

Our group has also exploited a novel high-throughput, massively parallel gene expression approach as an alternative to microarrays for examining the set of genes expressed in skeletal muscle and the subset of these that are differentially expressed following training (McGivney et al., 2010). This new method is digital gene expression (DGE) mRNA tag profiling, which is based on the high-throughput sequencing by synthesis technology available from Illumina Inc. (Mardis, 2008; Fox et al., 2009; Morozova et al., 2009). Using DGE mRNA tag profiling, we showed that the most abundant mRNA transcripts in the Thoroughbred muscle transcriptome were those involved in muscle contraction, aerobic respiration, and mitochondrial function. Following training, 16 genes increased in expression, including the genes *ACADVL*, *MRPS21*, and *SLC25A29* that function in the mitochondria, and 58 genes displayed decreased expression. Notably, transcripts representing the *MSTN* gene had the greatest decrease (–4.2-fold, $P = 0.0043$), indicative of the key role of the myostatin protein in the adaptive response and its function as a negative regulator of muscle development. Functional groups displaying highly significant altered expression were mitochondrion, oxidative phosphorylation and fatty acid metabolism, and structural genes, including those linked to the sarcoplasm, laminin complex, and cytoskeleton.

Genetic Testing for Thoroughbred Performance Potential

Thoroughbred horse racing is highly competitive and commands large financial investments. The evaluation of prospective racing success relies mostly on pedigree evaluation and observation of the physical characteristics of the horse. A range of scientific approaches including endoscopic evaluation (Garrett et al., 2010), radiographic evaluation (Spike-Pierce & Bramlage, 2003), assessment of heart size (Young et al., 2005), muscle fibre type (Rivero et al., 1993; Rivero et al., 1995; Barrey et al., 1999; Young et al., 2005), post-exercise lactate concentration (Evans et al., 1993), speed at maximal heart rate (Gramkow & Evans, 2006), hematological (Revington, 1983), and other physiological variables (Harkins et al., 1993) have been investigated as possible tools for predicting racing aptitude. However, currently there is no consensus among horse trainers, breeders, and owners about the most appropriate and informative scientific methods. The stage, therefore, is now set for the integration of molecular genetic know-how in the breeding and training of racehorses. The establishment of commercial companies offering genetic tests for performance, and the eager acceptance by the Thoroughbred industry of such technologies, is promising. Genomics studies are rapidly increasing our knowledge of the molecular mechanisms underpinning exercise adaptation and contributing to athletic performance in racehorses. This growing field of research will lend to improved breeding strategies and to training regimes that may be optimized for each individual horse, thereby minimizing risk and maximizing potential.

Acknowledgments

The authors thank Dr. Suzanne Eivers, Dr. Rita Fonseca, Dr. Jingjing Gu, and Dr. Beatrice McGivney for their valuable contributions to the understanding of genomics of exercise adaptation and athletic performance in the Thoroughbred horse.

References

Adhihetty, P. J. et al. (2003). Plasticity of skeletal muscle mitochondria in response to contractile activity. *Exploratory Physiology*, *88*, 99–107.

Akey, J. M. (2009). Constructing genomic maps of positive selection in humans: Where do we go from here? *Genome Research*, *19*, 711–722.

Akey, J. M. et al. (2010). Tracking footprints of artificial selection in the dog genome. *Proceedings of the National Academy of Sciences of the United States of America*, *107*, 1160–1165.

Andreelli, F. et al. (1999). Defective regulation of phosphatidylinositol-3-kinase gene expression in skeletal muscle and adipose tissue of non-insulin-dependent diabetes mellitus patients. *Diabetologia*, *42*, 358–364.

Bade, B. et al. (1975a). Die entwicklung von selektionskriterien für die reitpferdezucht. I. genetische parameter für kriterien der eigenleistungsprüfung von junghengsten auf station. *Züchtungsk*, *47*, 67–77.

Bade, B. et al. (1975b). Die entwicklung von selektionskriterien für die reitpferdezucht. II. genetische parameter für kriterien der nachkommenprüfung von hengsten im feld. *Züchtungsk*, *47*, 154–163.

Bakhtiari, J., & Kashan, N. E. J. (2009). Estimation of genetic parameters of racing performance in Iranian Thoroughbred horses. *Livestock Science*, *120*, 151–157.

Barrey, E. et al. (2006). Gene expression profiling in blood cells of endurance horses completing competition or disqualified due to metabolic disorder. *Equine Veterinary Journal Supplement*, 43–49.

Barrey, E. et al. (1999). Heritability of percentage of fast myosin heavy chains in skeletal muscles and relationship with performance. *Equine Veterinary Journal*, *30*(Suppl.), 289–292.

Binns, M. M. et al. (2010). Identification of the myostatin locus (MSTN) as having a major effect on optimum racing distance in the Thoroughbred horse in the USA. *Animal Genetics*, *41*(Suppl. 2), 154–158.

Bobinski, K., & Zamoyski, S. (1960). *Family tables of racehorses*. London: J. A. Allen.

Bonen, A. (2009). PGC-1alpha-induced improvements in skeletal muscle metabolism and insulin sensitivity. *Applied Physiology, Nutrition and Metabolism*, *34*, 307–314.

Bouwman, F. G. et al. (2010). Differential expression of equine muscle biopsy proteins during normal training and intensified training in young standardbred horses using proteomics technology. *Comparative Biochemistry and Physiology Part D Genomics Proteomics*, *5*, 55–64.

Bower, M. A., Campana, M. G., Whitten, M., Edwards, C. J., Jones, H., et al. (2011). The cosmopolitan maternal heritage of the Thoroughbread racehorse breed shows a significant contribution from British and Irish native mares. *Biology Letters*, *7*, 316–320.

Bray, M. S. et al. (2009). The human gene map for performance and health-related fitness phenotypes: The 2006–2007 update. *Medical Science of Sports and Exercise*, *41*, 35–73.

Brockmann, A. (1998). *Entwicklung einer eigenleistungsprüfung im feld für hengste unter berücksichtigung der turniersportprüfung*. Göttingen University, Göttingen, Göttingen.

Cappelli, K. et al. (2007). Analysis of peripheral blood mononuclear cells gene expression in endurance horses by cDNA-AFLP technique. *Research in Veterinary Science*, *82*, 335–343.

Carter, S. L. et al. (2001). Substrate utilization during endurance exercise in men and women after endurance training. *American Journal of Physiology, Endocrinology and Metabolism*, *280*, E898–E907.

Chan, S. et al. (2008). A gene for speed: Contractile properties of isolated whole EDL muscle from an alpha-actinin-3 knockout mouse. *American Journal of Physiology and Cell Physiology*, *295*, C897–C904.

Constantinopol, M. et al. (1989) Oxygen transport during exercise in large mammals. II. Oxygen uptake by the pulmonary gas exchanger. *Journal of Applied Physiology*, *67*, 871–878.

Cunningham, E. P. et al. (2001). Microsatellite diversity, pedigree relatedness and the contributions of founder lineages to thoroughbred horses. *Animal Genetics*, *32*, 360–364.

Dall'Olio, S. et al. (2010). Analysis of horse myostatin gene and identification of single nucleotide polymorphisms in breeds of different morphological types. *Journal of Biomedicine and Biotechnology*, *2010,* 2010.

Ducluzeau, P. H. et al. (2001). Regulation by insulin of gene expression in human skeletal muscle and adipose tissue. Evidence for specific defects in type 2 diabetes. *Diabetes*, *50*, 1134–1142.

Eivers, S. S. et al. (2010). Alterations in oxidative gene expression in equine skeletal muscle following exercise and training. *Physiology and Genomics*, *40*, 83–93.

Eizema, K. et al. (2005). Myosin heavy chain isoforms in equine gluteus medius muscle: comparison of mRNA and protein expression profiles. *Journal of Histochemistry and Cytochemistry*, *53*, 1383–1390.

Evans, D. L. et al. (1993). Correlation of racing performance with blood lactate and heart rate after exercise in thoroughbred horses. *Equine Veterinary Journal*, *25*, 441–445.

Fehrenbach, E. et al. (2003). Microarray technology – the future analyses tool in exercise physiology? *Exercise and Immunology Review*, *9*, 58–69.

Fluck, M. et al. (2008). Mechano-regulated tenascin-C orchestrates muscle repair. *Proceedings of the National Academy of Sciences of the United States of America*, *105*, 13662–13667.

Fox, S. et al. (2009). Applications of ultra-high-throughput sequencing. *Methods in Molecular Biology*, *553*, 79–108.

Fukuda, R. et al. (2007). HIF-1 regulates cytochrome oxidase subunits to optimize efficiency of respiration in hypoxic cells. *Cell*, *129*, 111–122.

Gaffney, B., & Cunningham, E. P. (1988). Estimation of genetic trend in racing performance of thoroughbred horses. *Nature*, *332*, 722–724.

Garrett, K. S. et al. (2010). Endoscopic evaluation of arytenoid function and epiglottic structure in Thoroughbred yearlings and association with racing performance at two to four years of age: 2,954 cases (1998–2001). *Journal of the American Veterinary Medical Association*, *236*, 669–673.

Georges, M. (2007). Mapping, fine mapping, and molecular dissection of quantitative trait loci in domestic animals. *Annual Review of Genomics and Human Genetics*, *8*, 131–162.

Gibala, M. (2009). Molecular responses to high-intensity interval exercise. *Applied Physiology, Nutrition and Metabolism*, *34*, 428–432.

Goddard, M. E., & Hayes, B. J. (2009). Mapping genes for complex traits in domestic animals and their use in breeding programmes. *National Review of Genetics*, *10*, 381–391.

Gordon, A. M. et al. (2000). Regulation of contraction in striated muscle. *Physiology Review*, *80*, 853–924.

Gramkow, H. L., & Evans, D. L. (2006). Correlation of race earnings with velocity at maximal heart rate during a field exercise test in thoroughbred racehorses. *Equine Veterinary Journal Supplement*, 118–122.

Green, H. J. (2004). Membrane excitability, weakness, and fatigue. *Canadian Journal of Applied Physiology*, *29*, 291–307.

Grobet, L. et al. (1997). A deletion in the bovine myostatin gene causes the double-muscled phenotype in cattle. *Nature Genetics*, *17*, 71–74.

Grossman, S. R. et al. (2010). A composite of multiple signals distinguishes causal variants in regions of positive selection. *Science*, *327*, 883–886.

Gu, J. et al. (2010). Association of sequence variants in *CKM* (creatine kinase, muscle) and *COX4I2* (cytochrome c oxidase, subunit 4, isoform 2) genes with racing performance in Thoroughbred horses. *Equine Veterinary Journal*, *38*, 569–575.

Gu, J. et al. (2009). A genome scan for positive selection in thoroughbred horses. *PLoS One*, *4*, e5767.

Gunn, H. M. (1987). Muscle, bone and fat proportions and muscle distribution of Thoroughbreds and other horses, In J. R. Gillespie & N. E. Robinson (eds.), *Equine exercise physiology 2* (pp. 253–264). Davis, CA: ICEEP Publications.

Harkins, J. D. et al. (1993). The correlation of running ability and physiological variables in thoroughbred racehorses. *Equine Veterinary Journal*, *25*, 53–60.

Harrison, S. P., & Turrion-Gomez, J. L. (2006). Mitochondrial DNA: An important female contribution to thoroughbred racehorse performance. *Mitochondrion*, *6*, 53–63.

Hascher, H. (1998). Schätzung von populationsparametern mittels exterieur- und leistungsdaten für Haflinger, Freiberger und CH Warmblut, Institute of Animal Production, Swiss Federal Institute of Technology (ETH), Zurich, Zurich.

Hayashi, K. et al. (1998). Differentiated phenotype of smooth muscle cells depends on signaling pathways through insulin-like growth factors and phosphatidylinositol 3-kinase. *Journal of Biology and Chemistry*, *273*, 28860–28867.

Hill, E. W. et al. (2002). History and integrity of thoroughbred dam lines revealed in equine mtDNA variation. *Animal Genetics*, *33*, 287–294.

Hill, E. W., Eivers, S. S., McGivney, B. A., Fonseca, R. G., Gu, J., et al. (2010a). Moderate and high intensity sprint exercise induce differential responses in *COX4I2* and *PDK4* gene expression in Thoroughbred horse skeletal muscle. *Equine Veterinary Journal*, *38*(Suppl.), 576–581.

Hill, E. W. et al. (2010b). A sequence polymorphism in MSTN predicts sprinting ability and racing stamina in thoroughbred horses. *PLoS One*, *5*, e8645.

Hill, E. W., Gu, J., MacGivney, B. A., & MacHugh, D. E. (2010c). Targets of selection in the Thoroughbred genome contain exercise-relevant gene SNPs associated with elite racecourse performance. *Animal Genetics*, *41*(Suppl. 2), 56–63.

Hill, E. W. et al. (2010d). A genome-wide SNP-association study confirms a sequence variant (g.66493737C>T) in the equine myostatin (*MSTN*) gene as the most powerful predictor of optimum racing distance for Thoroughbred racehorses. *BMC Genomics*, *11*, 552.

Hintz, R. L. (1980). Genetics of performance in the horse. *Journal of Animal Science*, *51*, 582–594.

Hoppeler, H. et al. (2007). Gene expression in working skeletal muscle. *Advances in Experimental Medicine and Biology*, *618*, 245–254.

Hughes, T. R. (2009). 'Validation' in genome-scale research. *Journal of Biology*, *8*, 3.

Huizinga, H. A. et al. (1991). Stationary performance testing of stallions from the Dutch Warmblood riding horse population .1. estimated genetic-parameters of scored traits and the genetic relation with dressage and jumping competition from offspring of breeding stallions. *Livestock Production Science*, *27*, 231–244.

Igl, B. W. et al. (2009). What do we mean by 'replication' and 'validation' in genome-wide association studies? *Human Heredity*, *67*, 66–68.

Ioannidis, J. P. et al. (2009). Validating, augmenting and refining genome-wide association signals. *National Review of Genetics*, *10*, 318–329.

Jeukendrup, A. E. (2003). Modulation of carbohydrate and fat utilization by diet, exercise and environment. *Biochemical Society Transcripts*, *31*, 1270–1273.

Ji, L. L. (2002). Exercise-induced modulation of antioxidant defense. *Annals of the New York Academy of Science*, *959*, 82–92.

Jones, J. H. et al. (1980). Oxygen transport during exercise in large mammals. I. Adaptive variation in oxygen demand. *Journal of Applied Physiology*, *67*, 862–870.

Jorgensen, T. J. et al. (2009). Hypothesis-driven candidate gene association studies: Practical design and analytical considerations. *American Journal of Epidemiology*, *170*, 986–993.

Jose-Cunilleras, E. et al. (2005). Expression of equine glucose transporter type 4 in skeletal muscle after glycogen-depleting exercise. *American Journal of Veterinary Research*, *66*, 379–385.

Joseph, A. M. et al. (2006). Control of gene expression and mitochondrial biogenesis in the muscular adaptation to endurance exercise. *Essays in Biochemistry*, *42*, 13–29.

Katz, L. M. et al. (2005). Ventilatory responses of ponies and horses to exercise. *Equine and Comparative Exercise Physiology*, *2*, 229–240.

Kayar, S. R. et al. (1989). Total muscle mitochondrial volume in relation to aerobic capacity of horses and steers. *Pflugers Archif*, *413*, 343–347.

Langlois, B. (1980). Heritability of racing ability in thoroughbreds – A review. *Livestock Production Science*, *7*, 591–605.

Lek, M. et al. (2010). The evolution of skeletal muscle performance: Gene duplication and divergence of human sarcomeric alpha-actinins. *Bioessays*, *32*, 17–25.

Levine, M. A. (1999). The origins of horse husbandry on the Eurasian Steppe. In M. A. Levine, Y. Y. Rassamakin, A. M. Kislenko, & N. S. Tatarintseva (eds.), *Late prehistoric exploitation of the Eurasian steppe* (pp. 5–58). Cambridge: McDonald Institute for Archaeological Research.

Lira, V. A. et al. (2010). PGC-1alpha regulation by exercise training and its influences on muscle function and insulin sensitivity. *American Journal of Physiology, Endocrinology and Metabolism*, *299*, E145–E161.

Little, J. et al. (2009). STrengthening the REporting of Genetic Association Studies (STREGA): An extension of the STROBE statement. *PLoS Medicine*, *6*, e22.

Lowe, C. B., & Allison, W. (1913). *Breeding racehorses by the figure system*. London: The Field & Queen (H. Cox) ltd.

Lundby, C. et al. (2009). The response of human skeletal muscle tissue to hypoxia. *Cellular and Molecular Life Science*, *66*, 3615–3623.

MacArthur, D. G., & North, K. N. (2007). *ACTN3*: A genetic influence on muscle function and athletic performance. *Exercise and Sports Science Review*, *35*, 30–34.

MacArthur, D. G. et al. (2007). Loss of *ACTN3* gene function alters mouse muscle metabolism and shows evidence of positive selection in humans. *Nature Genetics*, *39*, 1261–1265.

MacArthur, D. G. et al. (2008). An *Actn3* knockout mouse provides mechanistic insights into the association between alpha-actinin-3 deficiency and human athletic performance. *Human Molecular Genetics*, *17*, 1076–1086.

Mahoney, D. J., & Tarnopolsky, M. A. (2005). Understanding skeletal muscle adaptation to exercise training in humans: contributions from microarray studies. *Physical and Medical Rehabilitation Clinics in North America*, *16*, 859–873.

Mardis, E. R. (2008). Next-generation DNA sequencing methods. *Annual Review of Genomics and Human Genetics*, *9*, 387–402.

Martin, A. M. et al. (2010). Circadian regulation of locomotor activity and skeletal muscle gene expression in the horse. *Journal of Applied Physiology*, *109*, 1328–1336.

Martineau, L. C., & Gardiner, P. F. (2001). Insight into skeletal muscle mechanotransduction: MAPK activation is quantitatively related to tension. *Journal of Applied Physiology*, *91*, 693–702.

McGivney, B. A. et al. (2009). Transcriptional adaptations following exercise in thoroughbred horse skeletal muscle highlights molecular mechanisms that lead to muscle hypertrophy. *BMC Genomics*, *10*, 638.

McGivney, B. A. et al. (2010). Characterization of the equine skeletal muscle transcriptome identifies novel functional responses to exercise training. *BMC Genomics*, *11*: 398.

McPherron, A. C. et al. (1997). Regulation of skeletal muscle mass in mice by a new TGF-beta superfamily member. *Nature*, *387*: 83–90.

McPherron, A. C., & Lee, S. J. (1997). Double muscling in cattle due to mutations in the myostatin gene. *Proceedings of the National Academy of Sciences of the United States of America*, *94*, 12457–12461.

Mills, M. et al. (2001). Differential expression of the actin-binding proteins, alpha-actinin-2 and -3, in different species: implications for the evolution of functional redundancy. *Human Molecular Genetics*, *10*, 1335–1346.

Mizuno, Y. et al. (1994). Expression of utrophin (dystrophin-related protein) and dystrophin-associated glycoproteins in muscles from patients with Duchenne muscular dystrophy. *Muscle and Nerve*, *17*, 206–216.

Morozova, O. et al. (2009). Applications of new sequencing technologies for transcriptome analysis. *Annual Review of Genomics and Human Genetics*, *10*, 135–151.

Mosher, D. S. et al. (2007). A mutation in the myostatin gene increases muscle mass and enhances racing performance in heterozygote dogs. *PLoS Genetics*, *3*, e79.

Mota, M. D. et al. (2005). Genetic and environmental parameters for racing time at different distances in Brazilian Thoroughbreds. *Journal of Animal Breeding Genetics*, *122*, 393–399.

Mucher, E. et al. (2006). Gene expression profiling in equine muscle tissues using mouse cDNA microarrays. *Equine Veterinary Journal Supplement*, 359–364.

Novembre, J., & Di Rienzo, A. (2009). Spatial patterns of variation due to natural selection in humans. *National Review of Genetics*, *10*, 745–755.

O'Brien, R. M., & Granner, D. K. (1996). Regulation of gene expression by insulin. *Physiology Review*, *76*, 1109–1161.

Oki, H. et al. (1995). Genetic parameter estimates for racing time by restricted maximum-likelihood in the Thoroughbred horse of Japan. *Journal of Animal Breeding Genetics*, *112*, 146–150.

Oleksyk, T. K. et al. (2010). Genome-wide scans for footprints of natural selection. *Philosophical Transcripts of the Royal Society of London B Biological Science*, *365*, 185–205.

Olesen, J. et al. (2010). PGC-1alpha-mediated adaptations in skeletal muscle. *Pflugers Archif*, *460*, 153–162.

Ostrander, E. A. et al. (2009). Genetics of athletic performance. *Annual Review of Genomics and Human Genetics*, *10*, 407–429.

Ozawa, E. et al. (2005). Molecular and cell biology of the sarcoglycan complex. *Muscle and Nerve*, *32*, 563–576.

Pardo, J. V. et al. (1983). A vinculin-containing cortical lattice in skeletal muscle: Transverse lattice elements ("costameres") mark sites of attachment between myofibrils and sarcolemma. *Proceedings of the National Academy of Sciences of the United States of America*, *80*, 1008–1012.

Perusse, L. et al. (2003). The human gene map for performance and health-related fitness phenotypes: The 2002 update. *Medical Science of Sports and Exercise*, *35*, 1248–1264.

Pickrell, J. K. et al. (2009). Signals of recent positive selection in a worldwide sample of human populations. *Genome Research*, *19*, 826–837.

Pilegaard, H., & Neufer, P. D. (2004). Transcriptional regulation of pyruvate dehydrogenase kinase 4 in skeletal muscle during and after exercise. *Proceedings of Nutrition Society*, *63*, 221–226.

Pilegaard, H. et al. (2000). Transcriptional regulation of gene expression in human skeletal muscle during recovery from exercise. *American Journal of Physiology, Endocrinology and Metabolism*, *279*, E806–E814.

Rankinen, T. et al. (2006). The human gene map for performance and health-related fitness phenotypes: The 2005 update. *Medical Science of Sports and Exercise*, *38*, 1863–1888.

Rankinen, T. et al. (2001). The human gene map for performance and health-related fitness phenotypes. *Medical Science of Sports and Exercise*, *33*, 855–867.

Rankinen, T. et al. (2002). The human gene map for performance and health-related fitness phenotypes: The 2001 update. *Medical Science of Sports and Exercise*, *34*, 1219–1233.

Rankinen, T. et al. (2004). The human gene map for performance and health-related fitness phenotypes: The 2003 update. *Medical Science of Sports and Exercise*, *36*, 1451–1469.

Rankinen, T. et al. (2010). Advances in exercise, fitness, and performance genomics. *Medical Science of Sports and Exercise*, *42*, 835–846.

Revington, M. (1983). Haematology of the racing Thoroughbred in Australia 2: Haematological values compared to performance. *Equine Veterinary Journal*, *15*, 145–148.

Richter, E. A. et al. (2001). Glucose, exercise and insulin: Emerging concepts. *Journal of Physiology*, *535*, 313–322.

Rivero, J. L. et al. (1995). Effects of a 3 month endurance training programme on skeletal muscle histochemistry in Andalusian, Arabian and Anglo-Arabian horses. *Equine Veterinary Journal*, *27*, 51–59.

Rivero, J. L. et al. (1993). Muscle fiber type composition and fiber size in successfully and unsuccessfully endurance-raced horses. *Journal of Applied Physiology*, *75*, 1758–1766.

Roques, M., & Vidal, H. (1999). A phosphatidylinositol 3-Kinase/p70 ribosomal S6 protein kinase pathway is required for the regulation by insulin of the p85alpha regulatory subunit of phosphatidylinositol 3-kinase gene expression in human muscle cells. *Journal of Biology and Chemistry*, *274*, 34005–34010.

Sahlin, K. et al. (1998). Energy supply and muscle fatigue in humans. *Acta Physiologica Scandinavia*, *162*, 261–266.

Scarpulla, R. C. (2008). Transcriptional paradigms in mammalian mitochondrial biogenesis and function. *Physiology Review*, *88*, 611–638.

Scheffler, I. E. (2008). *Mitochondria*. Hoboken, NJ: Wiley-Liss.

Schuelke, M. et al. (2004). Myostatin mutation associated with gross muscle hypertrophy in a child. *New England Journal of Medicine*, *350*, 2682–2688.

Sellner, E. M. et al. (2007). Board-invited review: Applications of genomic information in livestock. *Journal of Animal Science*, *85*, 3148–3158.

Semenza, G. L. (1999). Regulation of mammalian O_2 homeostasis by hypoxia-inducible factor 1. *Annual Review of Cellular Development and Biology*, *15*, 551–578.

Shepherd, P. R. et al. (1998). Phosphoinositide 3-kinase: the key switch mechanism in insulin signalling. *Biochemistry Journal*, *333*(Pt. 3), 471–490.

Shibasaki, F. et al. (1994). Phosphatidylinositol 3-kinase binds to alpha-actinin through the p85 subunit. *Biochemistry Journal*, *302*(Pt. 2), 551–557.

Singer, J. B. (2009). Candidate gene association analysis. *Methods in Molecular Biology*, *573*, 223–230.

Spike-Pierce, D. L., & Bramlage, L. R. (2003). Correlation of racing performance with radiographic changes in the proximal sesamoid bones of 487 Thoroughbred yearlings. *Equine Veterinary Journal*, *35*, 350–353.

Thiruvenkadan, A. K. et al. (2009). Inheritance of racing performance of Thoroughbred horses. *Livestock Science*, *121*, 308–326.

Tobacman, L. S. (1996). Thin filament-mediated regulation of cardiac contraction. *Annual Review of Physiology*, *58*, 447–481.

Tolley, E. S. et al. (1985). A review of the inheritance of racing performance in horses. *Animal Breeding Abstracts*, *53*, 163–185.

Tozaki, T. et al. (2010). A genome-wide association study for racing performances in Thoroughbreds clarified a candidate region nearby *MSTN* gene. *Animal Genetics*, *41*(Suppl. 2), 28–35.

Tsuchida, H. et al. (2002). Gene expression of the p85alpha regulatory subunit of phosphatidylinositol 3-kinase in skeletal muscle from type 2 diabetic subjects. *Pflugers Archif*, *445*, 25–31.

Virtanen, C., & Takahashi, M. (2008). Muscling in on microarrays. *Applied Physiology of Nutrition and Metabolism*, *33*, 124–129.

Von Velsen-Zerweck, A. (1999). Integrierte zuchtwertschätzung für zuchtpferde. Göttingen University, Göttingen, Göttingen.

Vonholdt, B. M. et al. (2010). Genome-wide SNP and haplotype analyses reveal a rich history underlying dog domestication. *Nature*, *464*, 898–902.

Wade, C. M. et al. (2009). Genome sequence, comparative analysis, and population genetics of the domestic horse. *Science*, *326*, 865–867.

Wende, A. R. et al. (2005). PGC-1alpha coactivates *PDK4* gene expression via the orphan nuclear receptor ERRalpha: A mechanism for transcriptional control of muscle glucose metabolism. *Molecular and Cellular Biology*, *25*, 10684–10694.

Willett, P. (1975). *An introduction to the thoroughbred*. London: Paul.

Williamson, S. A., & Beilharz, R. G. (1996). Heritabilities of racing performance in Thoroughbreds: A study of Australian data. *Journal of Animal Breeding Genetics*, *113*, 505–524.

Williamson, S. A., & Beilharz, R. G. (1998). The inheritance of speed, stamina and other racing performance characters in the Australian Thoroughbred. *Journal of Animal Breeding Genetics*, *115*, 1–16.

Wolfarth, B. et al. (2005). The human gene map for performance and health-related fitness phenotypes: The 2004 update. *Medical Science of Sports and Exercise*, *37*, 881–903.

Woods, D. (2009). Angiotensin-converting enzyme, renin-angiotensin system and human performance. *Medical and Sports Science*, *54*, 72–87.

Yan, Z. (2009). Exercise, PGC-1alpha, and metabolic adaptation in skeletal muscle. *Applied Physiology of Nutrition and Metabolism*, *34*, 424–427.

Yang, N. et al. (2003). *ACTN3* genotype is associated with human elite athletic performance. *American Journal of Human Genetics*, *73*, 627–631.

Young, L. E. et al. (2005). Left ventricular size and systolic function in Thoroughbred racehorses and their relationships to race performance. *Journal of Applied Physiology*, *99*, 1278–1285.

18 Genomics of the circadian clock

Barbara A. Murphy

Introduction

The circadian system provides animals with a means to adapt their internal physiology to the constantly changing environmental stimuli that exists on a rotating planet. Light information is translated into molecular timing mechanisms within pacemaker cells of the mammalian hypothalamic suprachiasmatic nucleus (SCN) via transcriptional-translational feedback loops. Humoral and neural outputs from this "master" clock result in circadian rhythms of physiology and behavior. The larger circadian system involves SCN synchronization of cellular clocks throughout the organism such that individual organs can adapt their specific function to the correct time of day. In the short history of this scientific field, the vast majority of mammalian chronobiological research has been conducted using small laboratory animals. This chapter examines what these studies have revealed, how recent chronobiological findings in the horse compare to what is known, and highlight how the principles of circadian biology are applicable to veterinary and husbandry concerns of horses.

Introduction to Circadian Biology

Presence of the sun and the continuous rotation of our planet on its axis result in constantly changing 24-hour cycles of light and dark. Over millennia, natural selection pressures gave way to the development of self-sustained cellular clock mechanisms that were responsive to light. The information inherent in the daily day-night cycles was used to adjust their timing adaptively to best meet the challenges of the many daily and annual environmental cycles of the planet, such as temperature, food availability, predation pressure, and so forth. The circadian clock system that subsequently developed provides organisms with the ability to anticipate periods of activity and to time their behavior and internal physiology in ways that optimize survival.

A biological clock mechanism first evolved in single-celled algae and fungi that are directly exposed and responsive to sunlight (Hastings & Sweeney, 1960; Sommer et al., 1989; Lee et al., 2000), as well as in higher-order animals, such as fruit flies and zebrafish, where light can still directly penetrate "clock-containing cells" in many parts of the body (Plautz et al., 1997; Delaunay et al., 2000; Whitmore et al., 2000). However, as the complexity of the nervous system increased, the light-receptive elements of the circadian clock became more centralized in the process of evolution.

Equine Genomics, First Edition. Edited by Bhanu P. Chowdhary.

In mammals, a "master" circadian pacemaker developed in which the dominant circadian clock containing cells are concentrated in the light-translating region of the nervous system known as the suprachiasmatic nucleus (SCN).

The term "biological clock" was originally coined to describe the physiological entity responsible for navigation in migratory birds (Kramer, 1952). This was a pivotal point in the emerging field of chronobiology. Two scientists – Jürgen Aschoff, whose research included biological rhythms in mice, birds and humans, and Colin S. Pittendrigh, who studied circadian organization in *Drosophila* spp. – are considered the founders of modern biological rhythm research, and are credited with developing the unifying concepts and key principles of biological clocks (Daan & Gwinner, 1998; Daan, 2000). It was not until one of the first international symposia on biological clocks in 1960, and a lecture given by Colin Pittendrigh entitled "Circadian Rhythms and the Circadian Organisation of Living Systems" (Pittendrigh, 1960), that the principles of circadian biology began to be accepted as applicable to human societal and medical concerns (DeCoursey, 2004). "Suddenly, and on a world scale, endogenous timing was seen by scientists and lay people alike as a cardinal feature of living organisms on a rhythmically revolving planet. Biological clocks were core features of life" (DeCoursey, 2004, p. 13).

As a specialized scientific field, chronobiology has its own set of terms and definitions. The word "circadian" originates from the Latin *circa* (around) and *dies* (a day) and refers to the (approximate) 24-hour period of free-running rhythms in the absence of environmental stimuli. Light is the primary environmental time cue serving to entrain circadian rhythms to the 24-hour period of the Earth's rotation (Pittendrigh & Minis, 1964), while additional non-photic cues, such as exercise patterns (Mrosovsky et al., 1989; Reebs & Mrosovsky, 1989; Turek, 1989; Wickland & Turek, 1991) and feeding times (Damiola et al., 2000; Stokkan et al., 2001), generally play a secondary role. Combined, these environmental cues are called *entraining signals* as they provide the periodic stimuli that enable the endogenous internal clock system to entrain or "lock onto" the driving oscillation of the environment.

A *circadian rhythm* is defined as an endogenous rhythm with a period length close to 24 hours, which persists under constant conditions, is unaffected by changes in temperature, and can be entrained by external time cues (Pittendrigh, 1960, 1993). It is necessary to differentiate between the terms "circadian" and "diurnal." A diurnal rhythm simply implies that it occurs during the day and may or may not have a circadian basis. In summary, the circadian system can be conceptualized as having three linked components, namely (1) an input or stimulus from the external environment necessary to synchronize (2) a central clock or pacemaker that drives (3) an output in the form of gene expression changes that influence physiological function and behavior. Outputs from the circadian clock are often termed *overt rhythms* as they are the physiological and/or behavioral endpoints that act as indirect markers of the internal circadian timing system. Examples of the behavioral and physiological outputs of the human circadian system are illustrated in Figure 18.1.

The Mammalian Clockwork Mechanism

Multiple gene-protein-gene feedback loops support the mammalian molecular clock mechanism that exists in almost all mammalian cell types (Reppert & Weaver, 2002). The genes in question consist of a group of highly conserved core components termed clock genes. The only known tissues that show constant, rather than cyclic, expression of clock genes are cells from the thymus

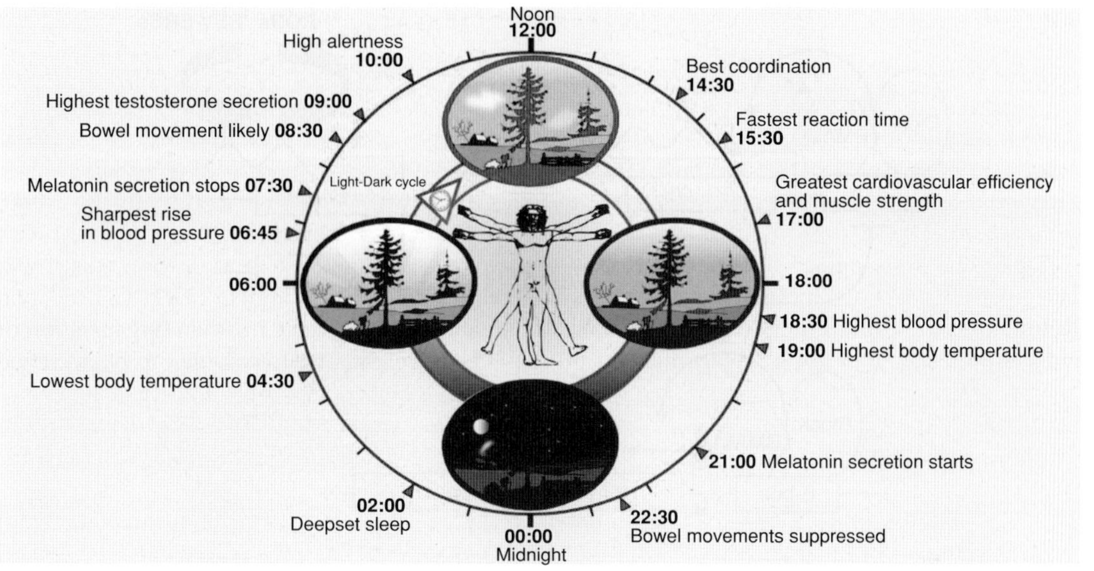

Figure 18.1 Overview of outputs from the circadian clock in humans. The circadian system influences the daily rhythm of many physiological processes. This diagram depicts the circadian patterns typical of a person who rises early in the morning, eats lunch around noon, and sleeps at around 10 p.m. While the primary synchronizer of circadian rhythms is the light/dark cycle, non-photic entrainment cues such as ambient temperature, meal times, stress, and exercise can also influence circadian timing of physiological processes.

and testis, and this is thought to be due to the immature, differentiating nature of cells from these tissues (Alvarez & Sehgal, 2005). Core clock components have been defined as genes whose protein products are necessary for the generation and/or regulation of circadian rhythms within individual cells (Takahashi, 2004). The core components can be assembled into a diagram of interconnecting loops involving multiple transcriptional feedback circuits, which are, in turn, regulated by post-translational modification processes (Figure 18.2).

The primary feedback loop consists of three *Period* genes (*Per1*, *Per2*, and *Per3*), two cryptochrome genes (*Cry1* and *Cry2*), a *Clock* gene, and the gene encoding brain-muscle-Arnt-like protein 1 (*Bmal1*[1]) (Dunlap, 1999). With the exception of *Clock*, which is constitutively expressed in most tissues, all transcripts exhibit oscillatory expression, with *Per* and *Cry* transcripts peaking in reverse phase to those of *Bmal1* (Morse & Sassone-Corsi, 2002). Positive regulation is provided by CLOCK and BMAL1, both members of the basic helix-loop-helix (bHLH)-PAS (Period-Arnt-Single-minded) transcription factor family. Dimerized CLOCK-BMAL1 complexes induce the expression of a large number of output genes, as well as binding to E-box enhancer motifs upstream of their own repressor genes, *Cry1* and *Cry2*, and *Per1* and *Per2* to initiate their transcription (Gekakis et al., 1998; Kume et al., 1999; Bunger et al., 2000; Zheng et al., 2001). Over the course of the day, the PER and CRY proteins accumulate and form multimers in the cytoplasm, where they are targets for phosphorylation by casein kinase Iε (*CKIε*) and glycogen synthase kinase-3

[1] *Bmal1* is more correctly now known as *ARNTL*, according to HUGO nomenclature. However, *Bmal1* and *ARNTL* are used interchangeably in this text due to it being more consistently recognized in chronobiological literature as *Bmal1*.

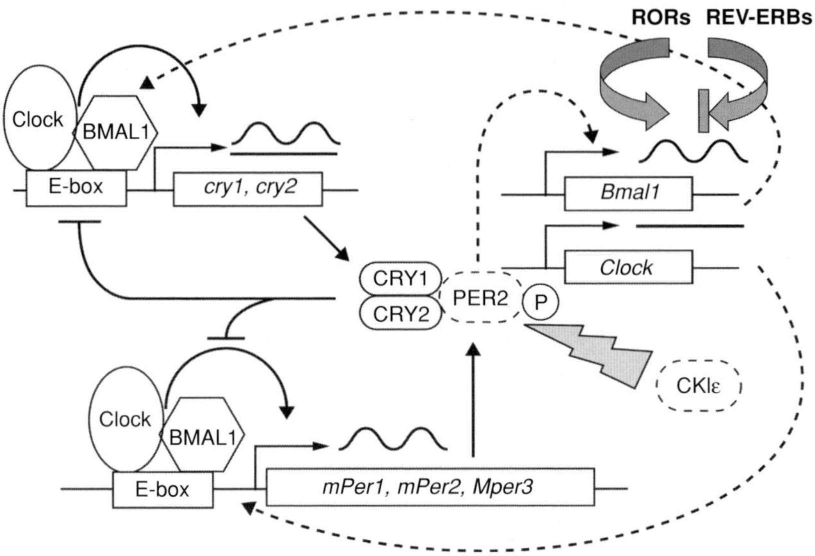

Figure 18.2 Model of mammalian molecular clock. Transcription of *Period* (*mPer*) and *Cryptochrome* (*Cry*) genes is initiated by CLOCK–BMAL1 heterodimers. The CRY and PER2 proteins heterodimerise and enter into the nucleus where they shut off their own synthesis. PER2 has an additional role in the activation of the *Bmal1* gene. The kinase CKIε may affect nuclear translocation and half-life of mPER proteins via phosphorylation. A second feedback loop comprising the opposing activities of the ROR and REVERB orphan nuclear receptors regulate *Bmal1* transcription. The next cycle begins when the concentration of the repressors decreases. This interplay of genes and their protein products give rise to the temporal clock gene expression patterns observed in Figure 18.3. Adapted from Ripperger & Schibler (2001), with permission from Elsevier.

(*GSK3*) (Iitaka et al., 2005). This facilitates translocation to the nucleus where they interact with the CLOCK-BMAL1 complexes to repress their own activation in a negative feedback manner (Kume et al., 1999; Okamura et al., 1999; Shearman et al., 2000; Lee et al., 2001; Sato et al., 2006). For each circadian cycle to end, the PER and CRY proteins are degraded by further CKIε phosphorylation and degradation, which releases the repression of the CLOCK-BMAL1 transcription and allows the next cycle to start (Gallego & Virshup, 2007).

CLOCK-BMAL1 induction of output genes contribute to rhythmical biological processes outside of the clockwork mechanism. Microarray studies have revealed that up to 10% of the transcriptome is under circadian regulation and that unique subsets of genes oscillate with 24-hour profiles in each individual tissue (Akhtar et al., 2002; Kita et al., 2002; Panda et al., 2002; Storch et al., 2002; Ueda et al., 2002; Zambon et al., 2003; Desai et al., 2004; Yamamoto et al., 2004). Recent studies have identified an additional stabilizing feedback loop within the molecular clock mechanism comprising the opposing activities of the RORα and REV-ERBα orphan nuclear receptors (Sato et al., 2004). RORα acts as a transcriptional activator of *Bmal1* while REV-ERBα inhibits its expression (Sato et al., 2004). This combination of loops, in conjunction with important post-translational mechanisms contributing to the time delays needed for the 24-hour period of the clock (Reppert & Weaver, 2002), ensures perpetuation of the self-sustaining nature of the molecular clock. Increasing complexity within this molecular mechanism continues to be revealed with recent demonstrations of crucial roles for new clock genes (Honma et al., 2002; Godinho et al., 2007; Siepka et al., 2007).

Hierarchy of Master and Peripheral Clocks

The mammalian circadian system is organized as a hierarchy of oscillators. The "master" clock in the SCN resides at the top of this hierarchy and is responsible for receiving and transducing the light information from the retina to directly drive many rhythms throughout the organism via neural and hormonal pathways. In this manner, the SCN synchronizes peripheral tissue clocks much as a conductor might conduct an orchestra, thereby achieving harmony in the many physiological and biochemical rhythms of the body.

The SCN regulates diverse physiological processes, such as blood pressure, heart rate (Arraj & Lemmer, 2006), activity cycles (Aston-Jones et al., 2001), hormone secretion (Weibel & Brandenberger, 2002), metabolism (Kita et al., 2002), immune function (Born et al., 1997; Petrovsky et al., 1998; Arjona & Sarkar, 2006), and body temperature (Moore & Danchenko, 2002). Robust diurnal variations in many physiological parameters have recently been reported in horses and are discussed later in the chapter. Importantly, the molecular components and temporal relationships of clock gene mRNA expression rhythms within rodent peripheral oscillators appear identical to those of the SCN, as exemplified by the antiphase oscillations of *PER2* and *BMAL1* (Oishi et al., 1998; Yagita et al., 2001; Muhlbauer et al., 2004) (Figure 18.3). A recent exception to this rule is the *Clock* transcript, which expression has been shown to be rhythmic, as opposed to constitutive, in some tissues outside of the SCN (Lowrey & Takahashi, 2004).

Timing signals from the SCN reach peripheral clocks to ensure that each tissue can then adapt its specific function to the correct time of day by means of tissue-specific circadian regulation of transcription. This phenomenon has been revealed by microarray studies in rodents, which identify unique subsets of genes under circadian regulation in different peripheral tissues (Kita et al., 2002; Panda et al., 2002; Storch et al., 2002; Zambon et al., 2003; Desai et al., 2004; Yamamoto et al., 2004). For example, clock-controlled genes relating to metabolism and detoxification are found to be differentially expressed in microarray analyses of the liver (Akhtar et al., 2002; Kita et al., 2002). Moreover, very little overlap was found between groups of oscillating circadian genes identified in the heart and the liver in one study (Storch et al., 2002), and between the liver and the SCN in another (Panda et al., 2002). These findings support a specialized and local role for circadian clocks in each tissue.

An important differentiation between the SCN clock and peripheral clocks was identified using transgenic rats in which luciferase was rhythmically expressed under the control of a *Per1* promoter (Yamazaki et al., 2000). It was shown that cultured SCN explants are capable of sustaining synchronous 24-hour rhythms of bioluminescence for weeks, whereas rhythmical clock gene expression from explanted peripheral tissues quickly lose amplitude and dampen in the absence of resetting stimuli from the SCN. Dampening of peripheral circadian rhythms ex vivo is now understood to reflect a gradual desynchronizing of many independent cellular oscillators (Welsh et al., 2004), thus defining the role of the SCN as a conductor rather than a driver of the circadian orchestra. Further important studies using cultured rat fibroblast cell lines revealed that the component oscillators within individual cells could be temporarily resynchronized by a number of different methods, most commonly by a change of culture medium to one containing high serum concentration (Balsalobre et al., 1998; Balsalobre et al., 2000). It is therefore very important that researchers involved in cell culture experiments are cognizant of the fact that each change of media likely resets a phase for the cultured cells and, depending on the tissue they are derived from, will influence specific circadian regulated pathways downstream from the clock.

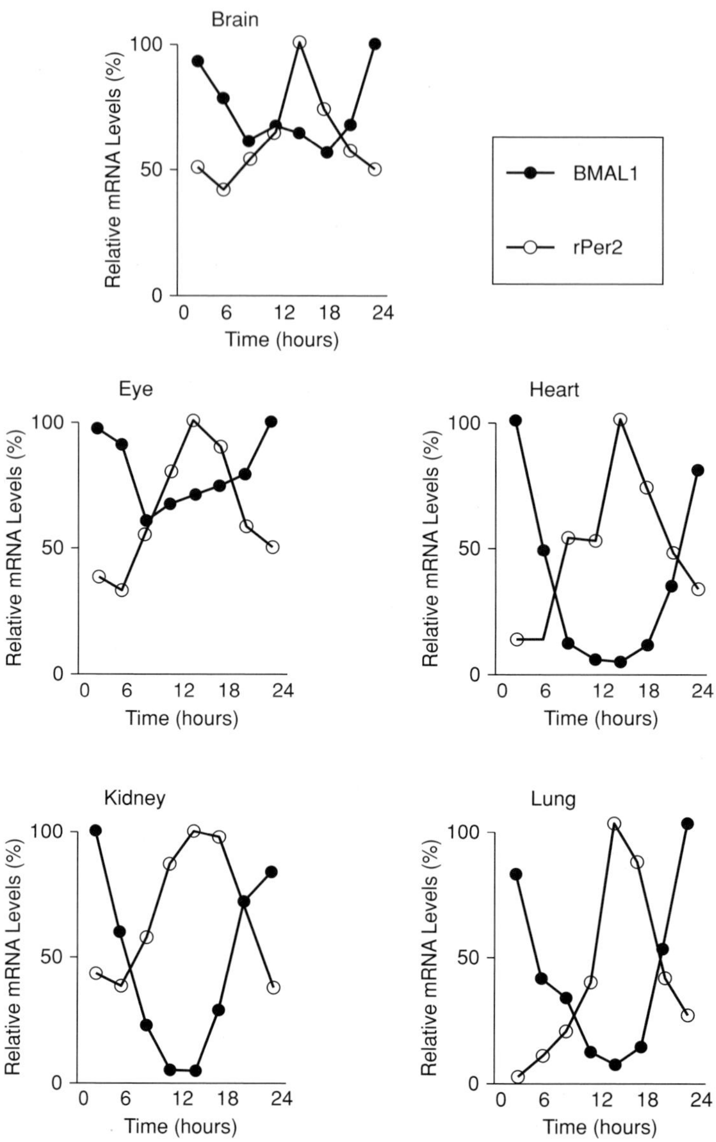

Figure 18.3 Antiphase daily expressions of *BMAL1* and *rPer2* mRNA in peripheral tissues. Rats were housed in a 12-hour light–12-hour dark (LD) cycle (lights on at ZT 0). The graphs depict a comparison of the expression patterns of *BMAL1* and *rPer2* mRNA in each tissue. The maximum value was expressed as 100% in each gene as determined by Northern blot analysis of total RNA. Reprinted from Oishi et al. (1998), with permission from Elsevier.

Equine Peripheral Clocks

The equine core clock genes were identified, sequenced, and their chromosomal locations determined in 2006 (Murphy et al., 2007b). This was quickly followed by the first investigation of clock gene expression in the horse by examining temporal profiles of equine *Per2*, *Bmal1*, *Cry1*, and *Clock* in equine fibroblasts, peripheral blood, and adipose tissue (Murphy et al., 2006) using real-time

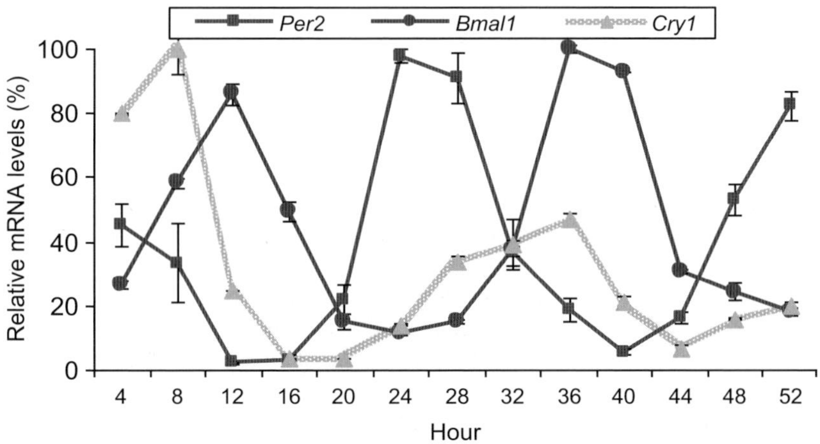

Figure 18.4 mRNA levels of *Per2*, *Bmal1*, and *Cry1* in equine fibroblast cells over a 52-hour period as determined by Real Time quantitative PCR. Expression levels of clock genes are reported as the number of transcripts relative to the number of molecules of housekeeping gene product *β-glucuronidase* (*GUS*). Each time point represents the mean ± SEM for three separate experiments ($n = 3$). Evidence of antiphase expression profiles of *Per2* and *Bmal1* suggests a similar temporal profile of clock gene expression to that previously observed in the SCN and peripheral tissues of rodents. From data presented in Murphy et al. (2006) and reviewed in Murphy (2009), with permission from Elsevier.

quantitative PCR detection methods. Blood and adipose tissue were chosen as they permitted minimally invasive tissue collection methods and multiple collection times. Equine fibroblasts of dermal origin were synchronized by shocking the cells with a high-percent serum media for two hours before transfer into a serum-free media. The robust rhythmic oscillations of equine *Per2*, *BMAL1*, and *Cry1* are clearly evident in Figure 18.4 in response to this treatment. The temporal relationships between the genes closely mimicked those observed in the SCN and peripheral tissues of other species. As fibroblasts are thought to serve as a valid model for investigation of core circadian clock function in mammals (Rosbash, 1998; Yagita et al., 2001), it was hypothesized that a similar molecular clockwork mechanism functioned in the horse. Surprisingly, in vivo results yielded unexpected findings. While low-amplitude clock gene rhythms were revealed in equine adipose tissue, there was no discernible oscillation in equine peripheral blood (Murphy et al., 2006).

The physiological advantage of a blood clock could be to temporally regulate transcriptional output from circulating leukocyte populations. As a heterogeneous tissue, it might be argued that failure to detect a rhythmic peripheral clock in whole blood is due to different temporal patterns of expression, contributed by a number of differentially synchronized cell types, resulting in a dampened overall rhythm. This hypothesis is supported by a recent finding that more robust clock gene temporal variation is observed in bovine neutrophils than in lymphocytes (Nebzydoski et al., 2010). Clock gene oscillations in human (Takata et al., 2002; Boivin et al., 2003) and rat (Oishi et al., 1998) peripheral blood cells have also been reported. However, Kusanagi et al. (2004) demonstrated that *Per1* rhythms in both human mononuclear and polymorphonuclear cell types oscillate in phase with each other, supporting the opposing view that different cell types within a tissue are entrained at the same phase angle.

To make matters more complicated, highly variable inter-individual clock gene expression profiles have recently been documented in human subjects (Teboul et al., 2005), leading to the suggestion that the circadian oscillator in peripheral blood may be regulated differently from other known peripheral clocks. In addition, the absence of a neural communication pathway between the SCN and peripheral blood lends further support to this assumption, as communication between the SCN and peripheral tissues is thought to occur via both neural and humoral mechanisms (Allen et al., 2001; Terazono et al., 2003; Guo et al., 2005). Future studies investigating clock gene expression in specific blood cell subpopulations may shed more light on the absence of oscillating clock genes in equine whole blood.

The evidence that equine adipose tissue possesses an oscillating peripheral clock was the first of its kind in a large mammal (Murphy et al., 2006). Adipose tissue is known to secrete a variety of biologically active molecules including leptin, resistin, and adiponectin (Matsuzawa et al., 2004), many of which have been shown to exhibit diurnal rhythms in humans (Gavrila et al., 2003) and horses (Piccione et al., 2004b; Gordon & McKeever, 2005). *Bmal1* is known to regulate adipogenesis in the mouse (Shimba et al., 2005). Furthermore, evidence of an obese phenotype in the *Clock* mutant mouse (Turek et al., 2005) strongly supports a regulatory role for clock genes in the production of adipocytokines (Ando et al., 2005). It is therefore clear that the circadian regulation of adipose tissue has significant metabolic implications. Further characterization of its role in the horse may provide new therapeutic possibilities with respect to the pathogenesis and treatment of diseases such as laminitis and hyperlipidemia.

Circadian Regulation of Performance

The recent demonstration that a large subset of genes undergo circadian regulation in mouse skeletal muscle (McCarthy et al., 2007b) corroborates numerous reports of daily variations in athletic performance parameters such as muscle force, strength, and power in humans (Zhang et al., 2009). A circadian rhythm in human athletic performance was recently clearly demonstrated in professional swimmers (Kline et al., 2007). It is considered likely that circadian variation in muscle transcription may contribute to this rhythm in performance (Zhang et al., 2009), in addition to 24-hour rhythmicity in many other cardio-respiratory factors (Millar-Craig et al., 1978; Giacomoni et al., 1999; Spengler et al., 2000).

Secondary to light stimuli, exercise acts as another important synchronizer of circadian rhythms (Edgar & Dement, 1991; Atkinson et al., 2007), supporting a hypothesis that enhanced performance may occur when times of training and competition coincide (Hill et al., 1989). This theory is especially important for equine athletes, particularly racehorses that are trained in the early morning hours and are then expected to perform optimally in the late afternoon. It could be postulated that there is an increased risk of musculoskeletal injury on racetracks if strenuous activity occurs at times that conflict with entrained rhythms. This is supported by evidence that some equine diurnal rhythms, such as those of platelet aggregation, shift in response to an exercise regime (Piccione et al., 2008b) (Figure 18.5). The nadir of the platelet aggregation rhythm was shown to shift to the time of day closest to the time of training, potentially functioning to reduce clotting capacity at a time associated with microvascular bleeding.

In order to provide a foundation for future studies investigating circadian regulation of performance in the equine athlete, an initial set of experiments investigating circadian gene expression profiles in equine skeletal muscle and their relationship to activity patterns in the horse were conducted (Martin et al., 2010). Up until this point there had been some confusion in the literature with regard to the diurnal (day-active) or ultradian (multiple activity bouts <24 hours throughout day and night) nature of horse activity rhythms. Studies from housed animals demonstrated diurnal patterns (Piccione et al., 2005), whereas observations of feral herds reported a predominantly

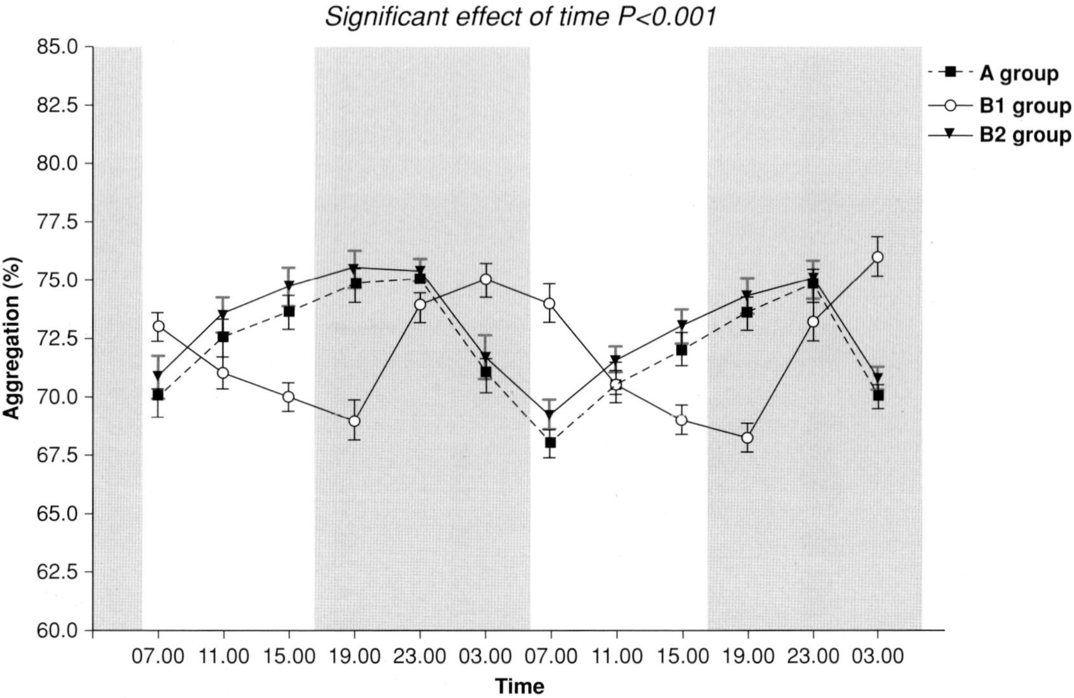

Figure 18.5 Daily rhythms of platelet aggregation (%) in horses. Platelet aggregation was measured every 4 hour for 48 hours in groups of A) sedentary horses, B1) athletic horses following a 60-day training program, and B2) the same athletic horses following two weeks of inactivity. Each point represents the mean ($\pm$ SEM) ($n = 6$) of parameters. Grey bars indicate the dark phase of the 48-hour LD cycle. Adapted from Piccione et al. (2008a), with permission from Elsevier.

ultradian behavioral pattern (Berger, 1999). Six healthy, untrained, sedentary mares were studied to determine whether locomotor activity behavior and skeletal muscle gene expression reflect endogenous circadian regulation. Activity was recorded for three consecutive 48-hour periods using halter-mounted Actiwatch-L® data-loggers, as a group at pasture (P), individually stabled under a light-dark (LD) cycle, and in constant darkness (DD). Animals had *ad libitum* access to hay and water while housed indoors. Figure 18.6 shows representative actograms displaying temporal patterns of activity (counts/min, Actiwatch L) and light exposure for two representative mares from the study. Visual inspection of the raw activity data supports the subjective summary that in P, the temporal variation in activity was predominantly ultradian, with multiple bouts of elevated activity distributed rather equally over day and night; by contrast, in LD and DD, there is a substantial decrease in overall activity levels along with the emergence of diurnality, exemplified by increased activity during daytime hours. Quantitative time series analysis of circadian cosinor parameters corroborated the predominantly ultradian (8.9 $\pm$ 0.7 bouts/24 hours) and weakly circadian pattern of activity in all three conditions (P, LD, DD). A more robust circadian pattern was observed during LD and DD (Figure 18.7). The cosinor method calculated estimates of four rhythm parameters: acrophase (time of peak value of the fitted cosine function), mesor (middle value of the fitted cosine curve representing the rhythm adjusted mean), amplitude (difference between maximum and mesor of the fitted cosine function), and Q value (goodness of fit, i.e., how well the rhythm reflected a circadian waveform).

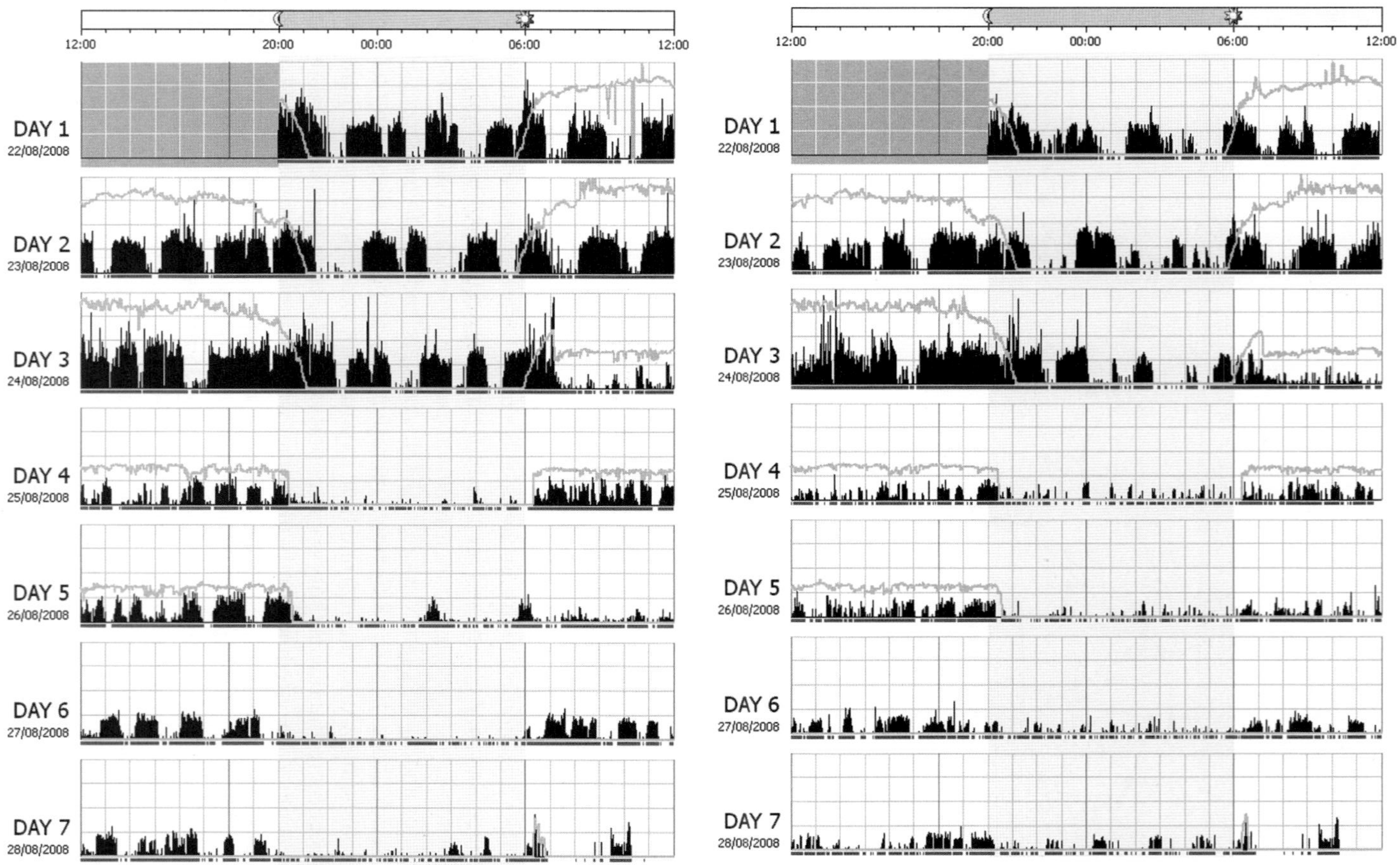

Figure 18.6 Actograms representing recorded activity from two representative mares. Black vertical lines represent activity (counts/min) and superimposed curves indicate light intensity. Days 1–7 (y-axis label) represent successive 24-hour periods (noon to noon). Mares were at pasture (P) on Days 1–3 and moved into the barn on the morning of Day 3 (0700), where they remained on an artificial light-dark (LD) cycle (14 hours – 10 hours) until lights out on Day 5. Thereafter (Days 6–7), they remained in continuous darkness (DD). White and grey bars above each actograph represent light and dark periods, respectively, of the environmental LD cycle present naturally at P and artificially in Barn LD conditions. Note the prominence of ultradian activity bouts (multiple peaks per 24 hours) when horses are outdoors and the subsequent emergence of a 24-hour rhythm when mares are stabled both in LD and DD. From Martin et al. (2010), with permission from the American Physiological Society.

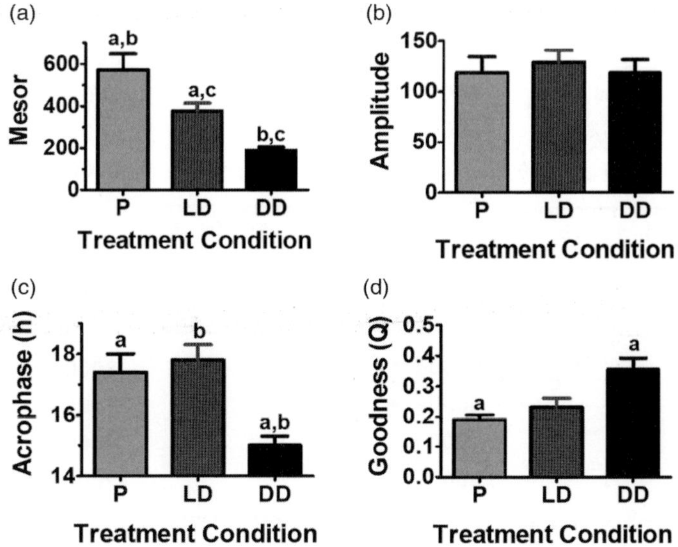

Figure 18.7 Cosinor analysis of activity data for circadian rhythmicity. Panels A–D plot bar graphs illustrating differences in cosine analysis parameters (mean ± SE). The cosinor method gave estimates of four rhythm parameters: acrophase (time of peak value of the fitted cosine function), mesor (middle value of the fitted cosine curve representing the rhythm adjusted mean), amplitude (difference between maximum, and mesor of the fitted cosine function), and Q value (goodness of fit – a value that relates to the degree of robustness of the circadian rhythm). Activity data was compared for horses observed sequentially in three contrasting environments: at Pasture (P), and while stabled in light-controlled barn, first in a light cycle (LD) and second in continuous darkness (DD). Shared letters (a, b, c) indicate group means that differ from each other ($P < 0.05$). From Martin et al. (2010), with permission from the American Physiological Society.

Muscle biopsies were then obtained from the horses' middle gluteal muscle every 4 hours for 24 hours under DD. Quantitative RT-PCR results from isolated total RNA confirmed the circadian expression ($P < 0.05$) of five core clock genes (*Arntl*, *Per1*, *Per2*, *Nr1d1*, *Nr1d2*), the clock-controlled gene *Dbp*, and the muscle-specific transcript *Myf6* (Figure 18.8). Additional genes, *Ucp3*, *MyoD1*, and *Vegfa*, while not significant, did clearly display circadian-like waveform expression profiles (Figure 18.8). *Myf6* is a member of the myogenic regulatory transcription factor (MRF) family, along with *Myf5* (myogenic factor 5), *Myod1* (myogenic differentiation 1) and myogenin (Megeney & Rudnicki, 1995). The identification of *Myf6* as a circadianly regulated transcript suggests that this MRF plays a role in the normal daily functioning of equine skeletal muscle. *Myf6* is the most abundantly expressed gene of the MRF family in adult muscle, and is therefore purported to play a role in the maintenance of skeletal muscle phenotype (Wyszynska-Koko et al., 2006). Recent studies have identified *Myf6* in newly developed myotubes of regenerating muscle in the amphibian *Xenopus* (Becker et al., 2003) and the rat (Zhou & Bornemann, 2001). These observations are consistent with numerous reports of the role of this gene in myogenesis (Montarras et al., 1991). Furthermore, elevated levels of *Myf6* mRNA have been detected in human skeletal muscle following heavy-resistance training, indicating that this gene may also play a role in skeletal muscle hypertrophy (Psilander et al., 2003). Lowe et al. (1998) reported elevated levels of *Myf6* mRNA in stretch-overloaded muscles of adult quails (Lowe et al., 1998), further supporting this theory. In untrained, sedentary horses, *Myf6* mRNA peaked between 0300 and 0700 (local time), exactly opposite the timing of the circadian peak in the observed activity rhythm (see Figure 18.7). This supports the

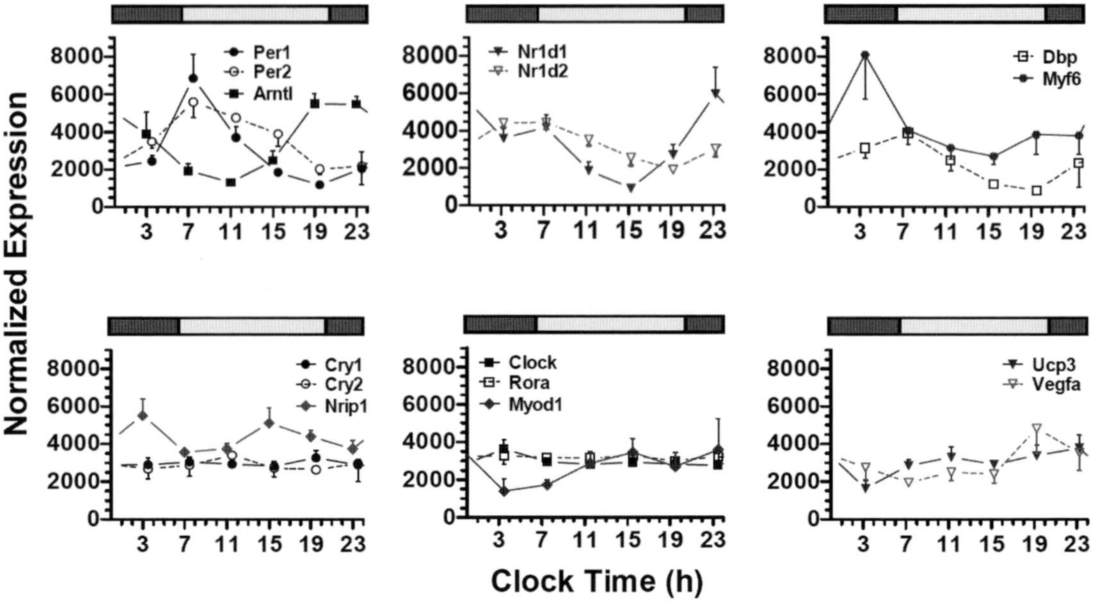

Figure 18.8 Twenty-four-hour profiles of skeletal muscle gene expression. Plotted are mRNA levels of candidate genes relative to the internal control gene *Ttn*, in equine skeletal muscle over 24 hours in constant darkness. Top: Genes that displayed a significant variation over time: *Per1*, *Per2*, *Arntl*, *Nr1d1*, *Nr1d2*, *Dbp*, and *Myf6* ($P < 0.05$). Bottom: Non-significant core clock genes: *Cry1*, *Cry2*, *Clock*, and *Rora* ($P > 0.05$); and potential clock-controlled genes; *Nrip1*, *Myod1*, *Ucp3*, and *Vegfa* ($P > 0.05$). Each time point represents the mean ± SE ($n = 6$). The barn light cycle in effect prior to entry into constant darkness (DD) is depicted above each graph with the dark grey shading representing subjective night (~CT14-CT24) and light grey shading representing subjective day (~CT0-CT14), corresponding to times of natural or simulated night and day existing prior to DD. From Martin et al. (2010), with permission from the American Physiological Society.

known function of Myf6 in muscle regeneration and repair during the rest phase of the activity cycle (Goetsch et al., 2003).

Ucp3 is purported to play a role in the protection of muscle from reactive oxygen species (ROS) damage during oxidative stress (MacLellan et al., 2005). ROS are normal by-products of mitochondrial respiration (MacLellan et al., 2005) that rise during physical exercise and may result in oxidative stress – a state in which ROS production exceeds the body's antioxidant defence mechanisms and subsequently induces lipid, protein, and DNA damage (Kinnunen et al., 2005). Exercise-induced oxidative stress is associated with muscle damage and decreased muscle performance (Kinnunen et al., 2005), an important consideration for horse trainers, as musculoskeletal injury is the most common reason for wastage in racehorses (Rose et al., 1983). Jiang et al. (2009) reported that *Ucp3* expression increased dramatically in response to intense exercise in rat skeletal muscle and coincided with a reduction in ROS levels (Jiang et al., 2009). These findings led the authors to suggest that *Ucp3* may promote uncoupling respiration during prolonged exercise and thus reduce ROS generation. This implies that *Ucp3* upregulation could act as an antioxidant defense mechanism to protect skeletal muscle mitochondria from exercise-induced oxidative insults (Jiang et al., 2009). It will be interesting to determine via future studies when this gene peaks in skeletal muscle from trained horses.

Vegfa regulates angiogenesis (Ferrara, 1999a) and has been proposed to play an important role in the maintenance of adult skeletal muscle microvasculature (Olfert et al., 2009). *Vegfa* stimulates vascular endothelial cell growth, survival, and proliferation and in addition promotes vascular permeability (Ferrara, 1999b). Exercise-induced increases in *Vegfa* expression are thus associated with the formation of new capillaries within skeletal muscle (Amaral et al., 2001). Expanded capillary network formation in response to aerobic exercise training serves to promote O_2 transport between the microcirculation and mitochondria by increasing the surface area available for diffusion of O_2 and decreasing the diffusional distance of O_2 to the mitochondria (Kraus et al., 2004). As a result, increases in the level of this growth factor likely contribute to improvements in skeletal muscle oxidative capacity and performance. Furthering our knowledge of when the potential for angiogenesis is highest in equine athletes will have important implications for trainers.

While *Myod1* did not display significant circadian regulation, this gene clearly demonstrated a 24-hour waveform (Figure 18.8). A central player in skeletal myogenesis (Weintraub, 1993), *Myod1* specifies skeletal muscle lineage in mice (Rudnicki et al., 1993; Tapscott, 2005) and is required for proliferation of muscle satellite cells (Yoshida et al., 1998). Furthermore, it was proposed that *Myod1* may function in regulating skeletal muscle hypertrophy and/or fiber-type transitions, due to up-regulation of this gene following heavy-resistance training (Psilander et al., 2003). It was also suggested that *Myod1* acts as an important clock-controlled gene and thus is regulated directly by the skeletal muscle molecular clock rather than by neural or humoral circadian signals from the SCN (McCarthy et al., 2007a; Zhang et al., 2009). The same authors also propose that the cellular clock contributes to the maintenance of muscle structure via its direct effects on *Myod1* and consequent effects on *Myod1*-regulated genes. Our findings further highlight the importance of elucidating the role of this gene in daily regulation and maintenance of muscle tissue, especially in routinely exercised performance horses.

These findings demonstrate the diurnal nature of horse activity rhythms, evidenced by the presence of a circadian molecular clock in the skeletal muscle that regulates muscle function. Because of the greater circadianicity of activity rhythms under DD, it is clear that human management regimes may strengthen, or unmask, equine circadian behavioral outputs (Martin et al., 2010). As exercise acts as a known synchronizer of circadian rhythms, these findings provide a basis for future work determining peak times for training and competing horses, to reduce injury and to achieve optimal performance.

Immune-Circadian Interaction

The immune system functions to protect and defend an organism's physiological status quo and thus represents an integral component of homeostasis. Homeostasis encompasses the mechanisms that react to maintain a constant, fixed set point of a physiological variable (reactive homeostasis), but also incorporates those mechanisms that are active in advance to maintain a set point that in itself is rhythmic (predictive homeostasis) (Moore-Ede, 1986). Hence, the ability to adaptively anticipate predictable changes in the environment, as conferred by the circadian system, is an important and often overlooked component of homeostasis. Consequently, interaction between these two systems is fundamental to survival.

Extensive evidence exists for circadian regulation of immune parameters exemplified by rhythmic secretion of the neuroendocrine hormone cortisol and the pineal hormone melatonin, both of which exhibit diurnal variation in the horse (Hoffsis et al., 1970; Bottoms et al., 1972; Larsson et al., 1979;

Piccione et al., 2005; Murphy et al., 2006). Glucocorticoids act as potent inhibitors of inflammatory mediators (Russo-Marie, 1992), and the ability of cortisol to suppress the pro-inflammatory cytokines, such as interferon (IFN)-γ, interleukin (IL)-12, tumour necrosis factor (TNF)-α, IL-1 and, to a lesser extent, IL-6 and IL-10 production has been described in humans (Petrovsky et al., 1998). The finding that cytokine production is negatively entrained by cortisol explains why the symptoms of immuno-inflammatory disorders, such as rheumatoid arthritis and asthma (Reinberg et al., 1963; Harkness et al., 1982; Bush, 1991; Martin et al., 1991), are exacerbated at the time of the early-morning nadir in plasma concentrations (Petrovsky et al., 1998). For this reason the importance of chronotherapeutics (circadian-time-specified drug administration or treatment) is of increasing importance in human medicine (Smolensky & Peppas, 2007) and should soon become increasingly relevant to veterinary practitioners as further advances are made in large animal chronobiology.

The immunomodulatory action of the pineal hormone melatonin is also well established (Colombo et al., 1992; Morrey et al., 1994; Carrillo-Vico et al., 2005), and numerous studies have described the interaction between photoperiod and the immune system (Nelson, 2004). It has been known for some time that exogenously administered melatonin can improve the outcome of acute and chronic inflammation (Maestroni, 1996). This immunosuppressant effect is elicited partly via the hormone's ability to inhibit TNFα levels (Wu et al., 2001) and reduce the levels of *IL6* (Sullivan et al., 1996) in mouse models of endotoxin-induced inflammation. Studies that investigate this therapeutic application of melatonin during inflammatory conditions (such as septic shock) in larger mammals are warranted.

Furthermore, the cytokine environment in which T lymphocytes are initially activated determines whether an immune response develops in a Type 1 (cellular) or Type 2 (humoral) direction (Petrovsky & Harrison, 1997). As IFN-γ and IL10 are markers of Type 1 and Type 2 immune activation, respectively, and under the opposing regulation of melatonin and cortisol, findings from the Petrovsky and Harrison's (1997) study strongly suggest that the time of day of antigen presentation will determine the direction of the immune response.

It has been proposed that responses to vaccination may be significantly modulated by time of vaccine administration relevant to the light/dark (LD) cycle, and that therapeutic manipulation using cortisol or melatonin may improve vaccine efficiency (Petrovsky & Harrison, 1997). An initial investigation of diurnal variation in circadian clock and immune mediator response to antigenic challenge in the horse was recently conducted (McGlynn et al., 2010). Blood samples were collected from young healthy animals at 4-hour intervals for 24 hours and immediately incubated for 6 hours in the presence or absence of lipopolysaccharide – a gram negative bacterial endotoxin that elicits strong immune cell activation. Quantitative RT-PCR analysis of total RNA harvested from cells post incubation revealed a significant effect of time on expression of the clock genes *Per2, Cry1, Arntl, Nr1d2* ($p<.001$, $p<.05$, $p<.05$, $p<.01$; respectively) and the immunomodulatory cytokine *interleukin (IL)6* ($p<.0001$) (Figure 18.9). These results confirmed that equine peripheral blood differentially responds to antigenic challenge over the 24-hour cycle, impacting upon our understanding of the pathophysiology of inflammatory responses. *IL6* was found to be up-regulated midway through the dark phase of the 24-hour photoperiod in equine circulation, in direct contrast to the temporal pattern observed in mice (Marpegan et al., 2009). This is likely indicative of a contrasting temporal immune surveillance regulation between diurnal and nocturnal species. The results of this preliminary study suggest that equine Th1 humoral responses may be favored when antigen exposure occurs in the evening as the involvement of *IL6* in the transition from innate to acquired immunity is known. This has clear implications regarding the potential optimal time of day for vaccination in the horse, emphasizing the importance of further research in this area.

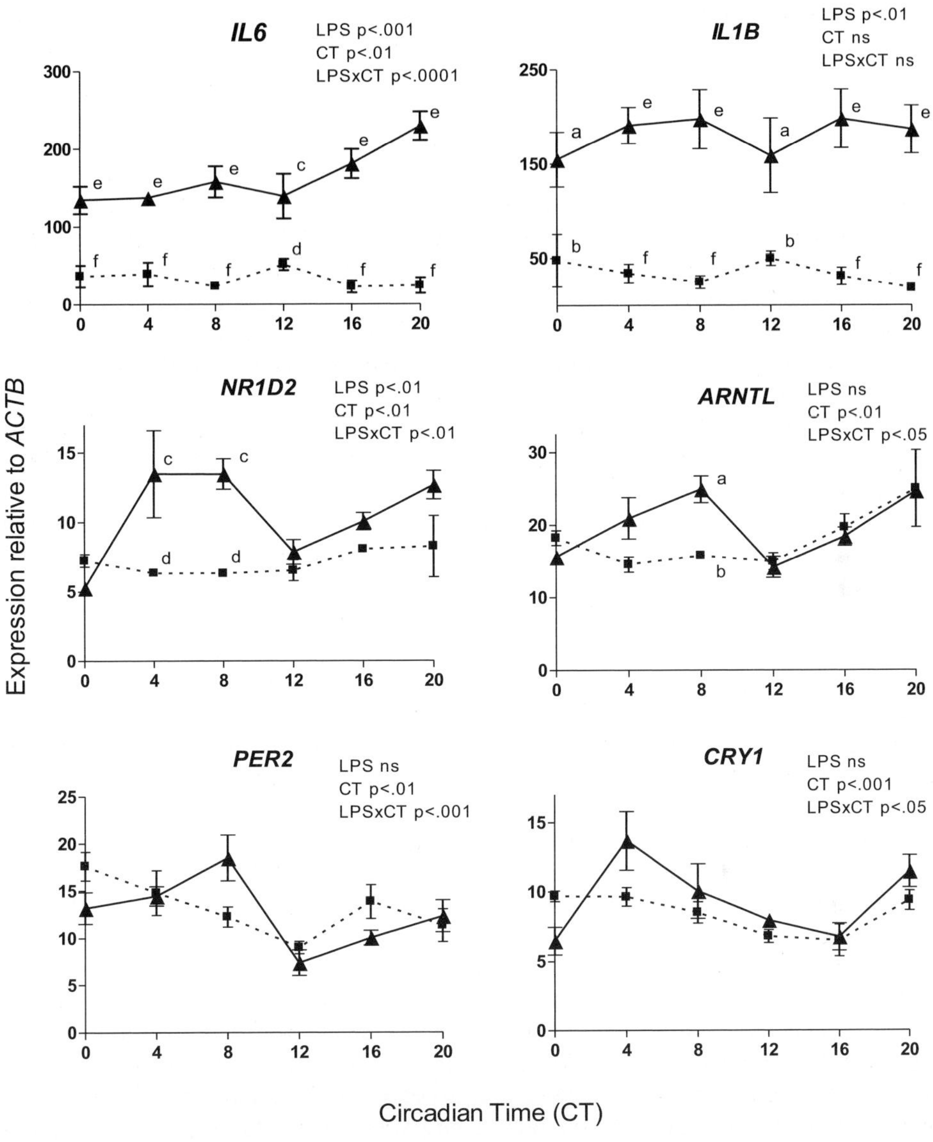

Figure 18.9 Time of day influences cytokine and clock gene response to immune stimulation in equine whole blood. Diurnal variation in gene expression from equine whole blood collected at 4-hour intervals over the 24-hour LD cycle and cultured for 6 hours with (solid line) or without (dotted line) LPS. Data are presented as means ± SE ($n = 4$ per time point). Within time point significant Bonferoni statistical post hoc differences are indicated by superscript lettering; a, b = $P < 0.05$; c, d = $P < .01$; e, f = $P < .001$. From McGlynn et al. (in press), *Animal Genetics*.

Circadian Desynchrony

Life on a rotating planet ensured that only the fittest survived by developing a circadian system capable of dealing with gradual seasonal changes in day length. However, this system is not adequately equipped to cope with today's 24-hour society and the demands of shift work and jet travel. The circadian disruption that arises from the mismatch between the previously entrained program of the internal clock and the new environmental LD cycle in response to rapid transition across multiple time zones is known as jet lag.

Significant decreases in reaction times, track speed, cardiorespiratory functions, and muscle strength have been reported following transmeridian travel in humans (Wright et al., 1983; Loat, 1989; Manfredini et al., 1998; Lemmer et al., 2002; Reilly et al., 2005). The same applies for the horse due to the frequency of air travel to international equestrian competitions now imposed on equine athletes. The rate of adjustment to a new time zone depends on the circadian output rhythm being measured, the number of time zones crossed, the flight direction (eastward flights cause prolonged disturbance), and the strength of the entraining cues (e.g., light intensity) in the new time zone (Gander et al., 1985; Loat, 1989).

In order to understand circadian readjustment to acute time zone transitions, analysis of both the molecular and cellular events that occur during resetting within the pacemaker in the brain and peripheral tissues is required. Dissociation of the cycles of transcript expression of *Cry1* from *Per2* recorded within the SCN of mice exposed to a 6-hour advance of the LD cycle was suggested as a possible origin of the temporal physiological disorder known in humans as jet lag (Reddy et al., 2002). Furthermore, clock gene rhythms in different regions of the rat SCN were found to re-entrain at different rates, providing another potential source of physiological desychrony (Nagano et al., 2003). Finally, it has been clearly demonstrated that clock gene rhythms in peripheral tissues such as skeletal muscle, liver, and lung shift slower than the SCN following both light cycle advances and delays (Yamazaki et al., 2000), further contributing to the physical malaise associated with rapid transmeridian travel.

Because it is typically not possible in humans or horses to directly monitor the molecular timing of the endogenous circadian clock in the brain, several key marker rhythms are frequently used to measure clock output and assess circadian phase position. Two commonly used and physiologically important markers of circadian phase are the rhythms of blood melatonin and core body temperature (Lewy & Sack, 1989; Klerman et al., 2002; Benloucif et al., 2005), both of which have been well characterized in the horse (Guerin et al., 1995; Piccione, 2002; Piccione et al., 2005; Murphy et al., 2006), and are widely used in humans to provide reliable estimates of circadian adaptation to phase shifts (Akerstedt et al., 1979; Van Cauter et al., 1998; Boivin & James, 2002).

The first experiment to investigate the effects of simulated jet lag on resynchronization of circadian markers in the horse demonstrated a rapid advance of melatonin in response to an abrupt 6-hour advance of the LD cycle (Murphy et al., 2007a) (Figure 18.10). Body temperature, however, was slower to re-entrain, requiring 3 days for the acrophase of the rhythm to shift. In stark contrast to previous human (Feve-Montange et al., 1981; Barnes et al., 1998) and rodent (Illnerova et al., 1989; Kennaway, 1994; Drijfhout et al., 1997) studies that demonstrate gradual adaptation of the melatonin rhythm (taking up to 14 days) to an advanced photoperiod, the equine melatonin rhythm appeared to complete the 6-hour phase advance essentially on the first post-shift day (Murphy et al., 2007a). These results raise the question of potential species differences in the ability to re-entrain melatonin production or produce melatonin endogenously. As the true circadian rhythmicity of melatonin has never been examined under constant conditions in the horse, it is possible that production of the hormone could be driven by the light/dark cycle rather than endogenously generated by the biological clock. The implications are that melatonin may be a poor candidate for monitoring internal clock phase in horses.

Re-entrainment of the equine body temperature rhythm was slower and involved disturbances in rhythm waveform that persisted on Day 11, although re-entrainment of the rhythm peak was still considerably faster than observed in rodents (Goel & Lee, 1996; Satoh et al., 2006), monkeys

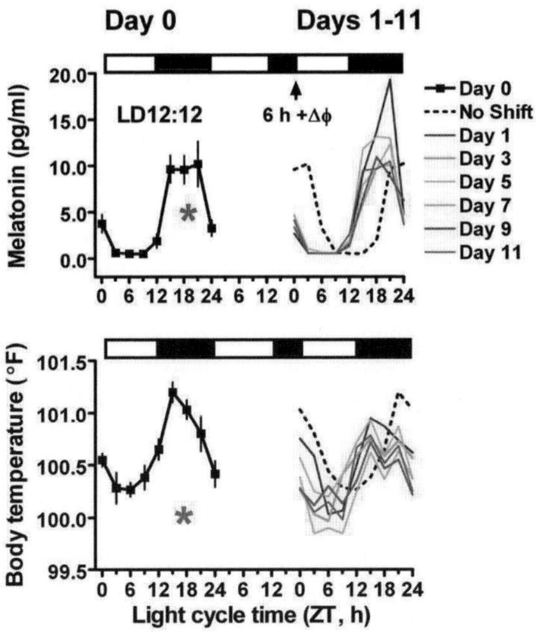

Figure 18.10 Response of equine melatonin and body temperature rhythms to a 6-hour phase advance of light/dark (LD) cycle. The LD cycle with 6-hour phase advance is depicted above each graph with white bars representing light and black bars representing times of darkness. The abscissa represents light cycle time (ZT) in hours, where ZT 0 corresponds to lights on and ZT 12 to lights off in a 12-hour photoperiod. Through Day 0 (curves at left) and for one additional day, lights were on 0730 to 1930 hours (ZT0-12). As indicated by the arrow, the photoperiod was advanced 6 hours on Day 1 (from ZT18 to ZT 0) to give new lights on from 0130 to 1330 hours (Days 1–11). Baseline curves (mean + /-SEM) for melatonin (top) and body temperature (bottom) are plotted on the left from ZT 0 to ZT 24. Curves for Day 1 through 11 are plotted on the right with point symbols and SEM removed for clarity. The dotted lines retrace the Day 0 curve assuming no phase shift. Asterisks represent Day 0 mean acrophase times. From Murphy et al. (2007a).

(Moore-Ede, 1986), and humans (Wever, 1980; Boivin and James, 2002). Direct neural connections exist between the SCN and the regions of the hypothalamus responsible for regulating melatonin (Teclemariam-Mesbah et al., 1999) and body temperature rhythms (Lu et al., 2001), whereas messages from the SCN to the periphery are thought to occur via more circuitous routes, resulting in delayed re-adaptation of clock gene rhythms in peripheral organs (Yamazaki et al., 2000). Therefore, determining the true extent to which horses are affected by jet lag will require further examination of peripheral tissue adaptation at the level of gene expression. Based on the recent findings of robust clock gene rhythms in equine skeletal muscle (Martin et al., 2010), this would appear to present a useful candidate tissue for future studies evaluating jet lag effects in the equine athlete.

Discussion

It is clear that the circadian system plays an important role in equine physiological homeostasis from the robust clock gene rhythms in equine skeletal muscle, immune cells, and in important regulatory genes specific to the function of these tissues (e.g., *Myf6, IL6*). However, the observed differences between horses and other species in terms of rhythmicity of peripheral blood and adaptability to abrupt shifts in the LD cycle remain to be fully understood. It is intriguing that an animal that is dependent on gradual seasonal changes in the duration of nocturnal melatonin for seasonal breeding is capable of rapidly shifting the 24-hour profile of this hormone immediately in response to an abrupt 6-hour advance of the photoperiod.

The following are characteristic differences between the horse and more commonly studied species that may explain the reduced robustness in activity rhyhms and the absence of clock gene rhythmicity in peripheral blood. Nocturnal rodents consume around 80% of their food during the hours of darkness, which likely explains the food-induced phase-resetting of peripheral clocks that have been observed in the liver, kidney, heart, and pancreas of mice (Challet & Pevet, 2003). This contrasts with large grazers like the horse, which, in a feral environment, devote some 15 hours per day to feeding, and disperses this behavior throughout the 24-hour period (Boyd, 1988). Ruminants, on the other hand, alternate their day between periods of foraging and ruminating, and are therefore probably less dependent on feeding cues for peripheral circadian entrainment (Andersson et al., 2005). It is feasible that the same holds true for non-ruminant equids, who require a constant slow supply of nutrients to facilitate their smaller stomach capacity and hindgut fermentation. Indeed, the recent finding that the daily rhythm of glycaemia is abolished in the horse when regular feeding cues are removed supports this (Piccione et al., 2008a).

Another important distinction between the horse and other species is their sleep-wake patterns. Rodents sleep an average of 12–13 hours a day while most healthy adult humans allocate about 8 hours out of every 24 hours for consolidated sleep (Campbell & Tobler, 1984; Challet & Pevet, 2003). The horse only sleeps in short 15-minutes bursts for an average of 2.9 hours a day, and rest periods are not confined to the hours of darkness (Dallaire, 1986; Martin et al., 2010). These differences likely reflect the different evolutionary pressures of survival experienced by small versus large prey animals. Ancient horses, as large, migratory mammals, were highly visible to predators when at rest, resulting in an increased requirement for constant wakefulness and herd vigilance. Finally, it is important to bear in mind that many laboratory species that have been used for studies investigating circadian phenomenon were selected and bred based on the strength of their circadian phenotype (Takahashi et al., 2008), as determined by the confinement of wheel-running activity rhythms to hours of darkness. It is assumed that wild-type animals, particularly horses, may experience greater masking of overt physiological rhythms due to the influence of social interactions and food availability in a feral environment. Importantly, a domestic setting removes much environmental variability, thus potentially permitting endogenous periodicities to emerge more robustly, as was clearly observed in the significantly greater circadianicity of activity profiles from horses under barn conditions and in the absence of light cues (Martin et al., 2010).

Robust rhythmicity of many physiological variables has been documented in the horse (Piccione et al., 2004a; Piccione et al., 2004b; Gordon & McKeever, 2005; Piccione et al., 2005; Bertolucci, 2008). However, closer inspection of the literature suggests that experimental animals are often exposed to management regimes that reflect human diurnality, typified by feeding and exercise times that are confined to daylight hours. The equine cortisol rhythm, for example, only emerges where horses are accustomed to a management routine, comprising stabling, feeding, and often exercise. This was evident in a study that examined diurnal variation of the hormone in horses maintained in environments that varied in their degree of human intervention (Irvine & Alexander, 1994). The fact that the most stable rhythm was observed in stabled racehorses involved in regimental training led the authors to suggest that the greater robustness of the cortisol rhythm may have been a product of the daily routine of the horses' environment. Similarly, housing conditions that better mimic the horses' natural environment, such as stalls with adjacent paddocks, have been shown to reduce the diurnal nature of the activity rhythm (Piccione, 2007). This is in agreement with past studies of Przewalski horse populations that reported both diurnal and ultradian (<24-hour cycles) behavioral patterns (Berger, 1999).

Secondary to light and similar to feeding cues, exercise is a known synchronizer of circadian clocks (Reebs & Mrosovsky, 1989; Turek, 1989; Atkinson et al., 2007). It is possible that, through domestic management, humans impose on horses timing cues that induce or enhance the expression of diurnal rhythms in physiology and behavior. Therefore, it will be important to understand the mechanisms of that influence on performance by conducting studies that determine the impact on the circadian muscle clock of set training times and performance capacity at different times of day.

It is clear that research in equine chronobiology has only begun to scratch the surface and its future is bright with potential and possibility. Areas of particular interest will include immune-circadian interaction, peripheral clocks in multiple tissues, and the influence of circadian disruption on parameters of performance in equine athletes.

Bibliography

Akerstedt, T., Froberg, J. E., Friberg, Y., & Wetterberg, L. (1979). Melatonin excretion, body temperature and subjective arousal during 64 hours of sleep deprivation. *Psychoneuroendocrinology*, *4*, 219–225.

Akhtar, R. A., Reddy, A. B., Maywood, E. S., Clayton, J. D., King, V. M. et al. (2002). Circadian cycling of the mouse liver transcriptome, as revealed by cDNA microarray, is driven by the suprachiasmatic nucleus. *Current Biology*, *12*, 540–550.

Allen, G., Rappe, J., Earnest, D. J., & Cassone, V. M. (2001). Oscillating on borrowed time: diffusible signals from immortalized suprachiasmatic nucleus cells regulate circadian rhythmicity in cultured fibroblasts. *Journal of Neuroscience*, *21*, 7937–7943.

Alvarez, J. D., & Sehgal, A. (2005). The thymus is similar to the testis in its pattern of circadian clock gene expression. *Journal of Biological Rhythms*, *20*, 111–121.

Amaral, S. L., Papanek, P. E., & Greene, A. S. (2001). Angiotensin II and VEGF are involved in angiogenesis induced by short-term exercise training. *American Journal of Physiology and Heart Circulation Physiology*, *281*, H1163–H1169.

Andersson, H., Johnston, J. D., Messager, S., Hazlerigg, D., & Lincoln, G. (2005). Photoperiod regulates clock gene rhythms in the ovine liver. *General and Comparative Endocrinology*, *142*, 357–363.

Ando, H., Yanagihara, H., Hayashi, Y., Obi, Y., Tsuruoka, S. et al. (2005). Rhythmic messenger ribonucleic acid expression of clock genes and adipocytokines in mouse visceral adipose tissue. *Endocrinology*, *146*, 5631–5636.

Arjona, A., & Sarkar, D. K. (2006). Evidence supporting a circadian control of natural killer cell function. *Brain Behavior and Immunology*, *20*, 469–476.

Arraj, M., & Lemmer, B. (2006). Circadian rhythms in heart rate, motility, and body temperature of wild-type C57 and eNOS knock-out mice under light-dark, free-run, and after time zone transition. *Chronobiology International*, *23*, 795–812.

Aston-Jones, G., Chen, S., Zhu, Y., & Oshinsky, M. L. (2001). A neural circuit for circadian regulation of arousal. *Nature Neuroscience*, *4*, 732–738.

Atkinson, G., Edwards, B., Reilly, T., & Waterhouse, J. (2007). Exercise as a synchroniser of human circadian rhythms: an update and discussion of the methodological problems. *European Journal of Applied Physiology*, *99*, 331–341.

Balsalobre, A., Brown, S. A., Marcacci, L., Tronche, F., Kellendonk, C. et al. (2000). Resetting of circadian time in peripheral tissues by glucocorticoid signaling. *Science*, *289*, 2344–2347.

Balsalobre, A., Damiola, F., & Schibler, U. (1998). A serum shock induces circadian gene expression in mammalian tissue culture cells. *Cell*, *93*, 929–937.

Barnes, R. G., Deacon, S. J., Forbes, M. J., & Arendt, J. (1998). Adaptation of the 6-sulphatoxymelatonin rhythm in shiftworkers on offshore oil installations during a 2-week 12-h night shift. *Neuroscience Letters*, *241*, 9–12.

Becker, C., Della Gaspera, B., Guyot, M., Donsez, E., Armand, A. S. et al. (2003). Expression of MRF4 protein in adult and in regenerating muscles in Xenopus. *Developmental Dynamics*, *227*, 445–449.

Benloucif, S., Guico, M. J., Reid, K. J., Wolfe, L. F., L'Hermite-Baleriaux, M., & Zee, P. C. (2005). Stability of melatonin and temperature as circadian phase markers and their relation to sleep times in humans. *Journal of Biological Rhythms*, *20*, 178–188.

Berger, A. S., Eichhorn, K., Scheibe, A., & Streich, J. (1999). Diurnal and ultradian rhythms of behaviour in a mare group of Przewalski horse (Equus ferus przewalskii), measured through one year under semi-reserve conditions. *Applied Animal Behaviour Science*, *64*, 1–17.

Bertolucci, C., Giannetto, C., Fazio, F., & Piccione, G. (2008). Seasonal variations in daily rhythms of activity in athletic horses. *Animal, 2*, 1055–1060.

Boivin, D. B., & James, F. O. (2002). Phase-dependent effect of room light exposure in a 5-h advance of the sleep-wake cycle: Implications for jet lag. *Journal of Biological Rhythms, 17*, 266–276.

Boivin, D. B., James, F. O., Wu, A., Cho-Park, P. F., Xiong, H., & Sun, Z. S. (2003). Circadian clock genes oscillate in human peripheral blood mononuclear cells. *Blood, 102*, 4143–4145.

Born, J., Lange, T., Hansen, K., Molle, M., & Fehm, H. L. (1997). Effects of sleep and circadian rhythm on human circulating immune cells. *Journal of Immunology, 158*, 4454–4464.

Bottoms, G. D., Roesel, O. F., Rausch, F. D., & Akins, E. L. (1972). Circadian variation in plasma cortisol and corticosterone in pigs and mares. *American Journal of Veterinary Research, 33*, 785–790.

Boyd, L. C., & Houpt, K. A. (1988). The 24-hour time budget of Przewalski horses. *Applied Animal Behaviour Science, 21*, 5–17.

Bunger, M. K., Wilsbacher, L. D., Moran, S. M., Clendenin, C., Radcliffe, L. A. et al. (2000). Mop3 is an essential component of the master circadian pacemaker in mammals. *Cell, 103*, 1009–1017.

Bush, R. K. (1991). Nocturnal asthma: Mechanisms and the role of theophylline in treatment. *Postgraduate Medical Journal, 67*(Suppl. 4), S20–S24.

Campbell, S. S., & Tobler, I. (1984). Animal sleep: A review of sleep duration across phylogeny. *Neuroscience and Biobehavior Review, 8*, 269–300.

Carrillo-Vico, A., Guerrero, J. M., Lardone, P. J., & Reiter, R. J. (2005). A review of the multiple actions of melatonin on the immune system. *Endocrine, 27*, 189–200.

Challet, E., & Pevet, P. (2003). Interactions between photic and nonphotic stimuli to synchronize the master circadian clock in mammals. *Frontiers of Bioscience, 8*, s246–s257.

Colombo, L. L., Chen, G. J., Lopez, M. C., & Watson, R. R. (1992). Melatonin induced increase in gamma-interferon production by murine splenocytes. *Immunology Letters, 33*, 123–126.

Daan, S. (2000). The Colin S. Pittendrigh Lecture. Colin Pittendrigh, Jurgen Aschoff, and the natural entrainment of circadian systems. *Journal of Biological Rhythms, 15*, 195–207.

Daan, S., & Gwinner, E. (1998). Jurgen Aschoff (1913–98). *Nature, 396*, 418.

Dallaire, A. (1986). Rest behavior. *Veterinary Clinical North American Equine Practice, 2*, 591–607.

Damiola, F., Le Minh, N., Preitner, N., Kornmann, B., Fleury-Olela, F., & Schibler, U. (2000). Restricted feeding uncouples circadian oscillators in peripheral tissues from the central pacemaker in the suprachiasmatic nucleus. *Genes Development, 14*, 2950–2961.

DeCoursey, P. J. (2004). Overview of biological timing from unicells to humans. In J. C. Dunlap, J. J. Loros, & P. J. DeCoursey (eds.), *Chronobiology* (pp. 3–24). Sunderland, MA: Sinauer Associates.

Delaunay, F., Thisse, C., Marchand, O., Laudet, V., & Thisse, B. (2000). An inherited functional circadian clock in zebrafish embryos. *Science, 289*, 297–300.

Desai, V. G., Moland, C. L., Branham, W. S., Delongchamp, R. R., Fang, H. et al. (2004). Changes in expression level of genes as a function of time of day in the liver of rats. *Mutation Research, 549*, 115–129.

Drijfhout, W. J., Brons, H. F., Oakley, N., Hagan, R. M., Grol, C. J., & Westerink, B. H. (1997). A microdialysis study on pineal melatonin rhythms in rats after an 8-h phase advance: New characteristics of the underlying pacemaker. *Neuroscience, 80*, 233–239.

Dunlap, J. C. (1999). Molecular bases for circadian clocks. *Cell, 96*, 271–290.

Edgar, D. M., & Dement, W. C. (1991). Regularly scheduled voluntary exercise synchronizes the mouse circadian clock. *American Journal of Physiology, 261*, R928–R933.

Ferrara, N. (1999a). Molecular and biological properties of vascular endothelial growth factor. *Journal of Molecular Medicine, 77*, 527–543.

Ferrara, N. (1999b). Vascular endothelial growth factor: Molecular and biological aspects. *Current Topics in Microbiology and Immunology, 237*, 1–30.

Feve-Montange, M., Van Cauter, E., Refetoff, S., Desir, D., Tourniaire, J., & Copinschi, G. (1981). Effects of "jet lag" on hormonal patterns. II. Adaptation of melatonin circadian periodicity. *Journal of Clinical Endocrinology and Metabolism, 52*, 642–649.

Gallego, M., & Virshup, D. M. (2007). Post-translational modifications regulate the ticking of the circadian clock. *National Review of Molecular and Cellular Biology, 8*, 139–148.

Gander, P. H., Kronauer, R. E., & Graeber, R. C. (1985). Phase shifting two coupled circadian pacemakers: implications for jet lag. *American Journal of Physiology, 249*, R704–R719.

Gavrila, A., Peng, C. K., Chan, J. L., Mietus, J. E., Goldberger, A. L., & Mantzoros, C. S. (2003). Diurnal and ultradian dynamics of serum adiponectin in healthy men: Comparison with leptin, circulating soluble leptin receptor, and cortisol patterns. *Journal of Clinical Endocrinology and Metabolism, 88*, 2838–2843.

Gekakis, N., Staknis, D., Nguyen, H. B., Davis, F. C., Wilsbacher, L. D. et al. (1998). Role of the CLOCK protein in the mammalian circadian mechanism. *Science*, *280*, 1564–1569.

Giacomoni, M., Bernard, T., Gavarry, O., Altare, S., & Falgairette, G. (1999). Diurnal variations in ventilatory and cardiorespiratory responses to submaximal treadmill exercise in females. *European Journal of Applied Physiology and Occupational Physiology*, *80*, 591–597.

Godinho, S. I., Maywood, E. S., Shaw, L., Tucci, V., Barnard, A. R. et al. (2007). The after-hours mutant reveals a role for Fbxl3 in determining mammalian circadian period. *Science*, *316*, 897–900.

Goel, N., & Lee, T. M. (1996). Relationship of circadian activity and social behaviors to reentrainment rates in diurnal Octodon degus (Rodentia). *Physiology of Behavior*, *59*, 817–826.

Goetsch, S. C., Hawke, T. J., Gallardo, T. D., Richardson, J. A., & Garry, D. J. (2003). Transcriptional profiling and regulation of the extracellular matrix during muscle regeneration. *Physiology of Genomics*, *14*, 261–271.

Gordon, M. E., & McKeever, K. H. (2005). Diurnal variation of ghrelin, leptin, and adiponectin in Standardbred mares. *Journal of Animal Science*, *83*, 2365–2371.

Guerin, M. V., Deed, J. R., Kennaway, D. J., & Matthews, C. D. (1995). Plasma melatonin in the horse: Measurements in natural photoperiod and in acutely extended darkness throughout the year. *Journal of Pinealogical Research*, *19*, 7–15.

Guo, H., Brewer, J. M., Champhekar, A., Harris, R. B., & Bittman, E. L. (2005). Differential control of peripheral circadian rhythms by suprachiasmatic-dependent neural signals. *Proceedings of the National Academy of Sciences of the United States of America*, *102*, 3111–3116.

Halberg, F. (1969). Chronobiology. *Annual Review of Physiology*, *31*, 675–725.

Halberg, F., Carandente, F., Cornelissen, G., & Katinas, G. S. (1977). Glossary of chronobiology. *Chronobiologia*, *4*(Suppl. 1), 1–189.

Harkness, J. A., Richter, M. B., Panayi, G. S., Van de Pette, K., Unger, A. et al. (1982). Circadian variation in disease activity in rheumatoid arthritis. *British Medical Journal (Clinical Research Edition)*, *284*, 551–554.

Hastings, J. W., & Sweeney, B. M. (1960). The action spectrum for shifting the phase of the rhythm of luminescence in Gonyaulax polyedra. *Journal of General Physiology*, *43*, 697–706.

Hill, D. W., Cureton, K. J., & Collins, M. A. (1989). Circadian specificity in exercise training. *Ergonomics*, *32*, 79–92.

Hoffsis, G. F., Murdick, P. W., Tharp, V. L., & Ault, K. (1970). Plasma concentrations of cortisol and corticosterone in the normal horse. *American Journal of Veterinary Research*, *31*, 1379–1387.

Honma, S., Kawamoto, T., Takagi, Y., Fujimoto, K., Sato, F. et al. (2002). Dec1 and Dec2 are regulators of the mammalian molecular clock. *Nature*, *419*, 841–844.

Iitaka, C., Miyazaki, K., Akaike, T., & Ishida, N. (2005). A role for glycogen synthase kinase-3beta in the mammalian circadian clock. *Journal of Biology and Chemistry*, *280*, 29397–29402.

Illnerova, H., Trentini, G. P., & Maslova, L. (1989). Melatonin accelerates reentrainment of the circadian rhythm of its own production after an 8-h advance of the light-dark cycle. *Journal of Comparative Physiology A*, *166*, 97–102.

Irvine, C. H., & Alexander, S. L. (1994). Factors affecting the circadian rhythm in plasma cortisol concentrations in the horse. *Domestic Animal Endocrinology*, *11*, 227–238.

Jiang, N., Zhang, G., Bo, H., Qu, J., Ma, G. et al. (2009). Upregulation of uncoupling protein-3 in skeletal muscle during exercise: A potential antioxidant function. *Free Radicals, Biology and Medicine*, *46*, 138–145.

Kennaway, D. J. (1994). Effect of a phase advance of the light/dark cycle on pineal function and circadian running activity in individual rats. *Brain Research Bulletin*, *33*, 639–644.

Kinnunen, S., Hyyppa, S., Lappalainen, J., Oksala, N., Venojarvi, M. et al. (2005). Exercise-induced oxidative stress and muscle stress protein responses in trotters. *European Journal of Applied Physiology*, *93*, 496–501.

Kita, Y., Shiozawa, M., Jin, W., Majewski, R. R., Besharse, J. C. et al. (2002). Implications of circadian gene expression in kidney, liver and the effects of fasting on pharmacogenomic studies. *Pharmacogenetics*, *12*, 55–65.

Klerman, E. B., Gershengorn, H. B., Duffy, J. F., & Kronauer, R. E. (2002). Comparisons of the variability of three markers of the human circadian pacemaker. *Journal of Biological Rhythms*, *17*, 181–193.

Kline, C. E., Durstine, J. L., Davis, J. M., Moore, T. A., Devlin, T. M. et al. (2007). Circadian variation in swim performance. *Journal of Applied Physiology*, *102*, 641–649.

Kramer, G. (1952). Experiments on bird orientation. *Ibis*, *94*, 265–268.

Kraus, R. M., Stallings, III, H. W., Yeager, R. C., & Gavin, T. P. (2004). Circulating plasma VEGF response to exercise in sedentary and endurance-trained men. *Journal of Applied Physiology*, *96*, 1445–1450.

Kume, K., Zylka, M. J., Sriram, S., Shearman, L. P., Weaver, D. R. et al. (1999). mCRY1 and mCRY2 are essential components of the negative limb of the circadian clock feedback loop. *Cell*, *98*, 193–205.

Larsson, M., Edqvist, L. E., Ekman, L., & Persson, S. (1979). Plasma cortisol in the horse, diurnal rhythm and effects of exogenous ACTH. *Acta Veterinaria Scandinavia*, *20*, 16–24.

Lee, C., Etchegaray, J. P., Cagampang, F. R., Loudon, A. S., & Reppert, S. M. (2001). Posttranslational mechanisms regulate the mammalian circadian clock. *Cell, 107*, 855–867.

Lee, K., Loros, J. J., & Dunlap, J. C. (2000). Interconnected feedback loops in the Neurospora circadian system. *Science, 289*, 107–110.

Lemmer, B., Kern, R. I., Nold, G., & Lohrer, H. (2002). Jet lag in athletes after eastward and westward time-zone transition. *Chronobiology International, 19*, 743–764.

Lewy, A. J., & Sack, R. L. (1989). The dim light melatonin onset as a marker for circadian phase position. *Chronobiology International, 6*, 93–102.

Loat, C. E., & Rhodes, E. C. (1989). Jet lag and human performance. *Sports Medicine, 8*, 226–238.

Lowe, D. A., Lund, T., & Alway, S. E. (1998). Hypertrophy-stimulated myogenic regulatory factor mRNA increases are attenuated in fast muscle of aged quails. *American Journal of Physiology and Cellular Physiology, 275*, C155–C162.

Lowrey, P. L., & Takahashi, J. S. (2004). Mammalian circadian biology: Elucidating genome-wide levels of temporal organization. *Annual Review of Genomics and Human Genetics, 5*, 407–441.

Lu, J., Zhang, Y. H., Chou, T. C., Gaus, S. E., Elmquist, J. K. et al. (2001). Contrasting effects of ibotenate lesions of the paraventricular nucleus and subparaventricular zone on sleep-wake cycle and temperature regulation. *Journal of Neuroscience, 21*, 4864–4874.

MacLellan, J. D., Gerrits, M. F., Gowing, A., Smith, P. J., Wheeler, M. B., & Harper, M. E. (2005). Physiological increases in uncoupling protein 3 augment fatty acid oxidation and decrease reactive oxygen species production without uncoupling respiration in muscle cells. *Diabetes, 54*, 2343–2350.

Maestroni, G. J. (1996). Melatonin as a therapeutic agent in experimental endotoxic shock. *Journal of Pinealogy Research, 20*, 84–89.

Manfredini, R., Manfredini, F., Fersini, C., & Conconi, F. (1998). Circadian rhythms, athletic performance, and jet lag. *British Journal of Sports Medicine, 32*, 101–106.

Marpegan, L., Leone, M. J., Katz, M. E., Sobrero, P. M., Bekinstein, T. A., & Golombek, D. A. (2009). Diurnal variation in endotoxin-induced mortality in mice: correlation with proinflammatory factors. *Chronobiology International, 26*, 1430–1442.

Martin, A. M., Elliott, J. A., Duffy, P., Blake, C. M., Ben Attia, S. et al. (2010). Circadian regulation of locomotor activity and skeletal muscle gene expression in the horse. *Journal of Applied Physiology, 109*, 1328–1336.

Martin, R. J., Cicutto, L. C., Smith, H. R., Ballard, R. D., & Szefler, S. J. (1991). Airways inflammation in nocturnal asthma. *American Review of Respiratory Diseases, 143*, 351–357.

Matsuzawa, Y., Funahashi, T., Kihara, S., & Shimomura, I. (2004). Adiponectin and metabolic syndrome. *Arteriosclerosis, Thrombosis and Vascular Biology, 24*, 29–33.

McCarthy, J. J., Andrews, J. L., McDearmon, E. L., Campbell, K. S., Barber, B. K. et al. (2007a). Identification of the circadian transcriptome in adult mouse skeletal muscle. *Physiology and Genomics, 31*, 86–95.

McCarthy, J. J., Andrews, J. L., McDearmon, E. L., Campbell, K. S., Barber, B. K. et al. (2007b). Identification of the circadian transcriptome in adult mouse skeletal muscle. *Physiology and Genomics, 31*, 86–95.

McGlynn, O. F., Browne, J. A., Blake, C. M., & Murphy, B. A. (2010). Time of day influences cytokine and clock gene response to immune stimulation in equine whole blood. *Animal Genetics*.

Megeney, L. A., & Rudnicki, M. A. (1995). Determination versus differentiation and the MyoD family of transcription factors. *Biochemistry and Cellular Biology, 73*, 723–732.

Millar-Craig, M. W., Bishop, C. N., & Raftery, E. B. (1978). Circadian variation of blood-pressure. *Lancet, 1*, 795–797.

Montarras, D., Chelly, J., Bober, E., Arnold, H., Ott, M. O. et al. (1991). Developmental patterns in the expression of Myf5, MyoD, myogenin, and MRF4 during myogenesis. *New Biology, 3*, 592–600.

Moore-Ede, M. C. (1986). Physiology of the circadian timing system: predictive versus reactive homeostasis. *American Journal of Physiology, 250*, R737–R752.

Moore, R. Y., & Danchenko, R. L. (2002). Paraventricular-subparaventricular hypothalamic lesions selectively affect circadian function. *Chronobiology International, 19*, 345–360.

Morrey, K. M., McLachlan, J. A., Serkin, C. D., & Bakouche, O. (1994). Activation of human monocytes by the pineal hormone melatonin. *Journal of Immunology, 153*, 2671–2680.

Morse, D., & Sassone-Corsi, P. (2002). Time after time: Inputs to and outputs from the mammalian circadian oscillators. *Trends in Neuroscience, 25*, 632–637.

Mrosovsky, N., Reebs, S. G., Honrado, G. I., & Salmon, P. A. (1989). Behavioural entrainment of circadian rhythms. *Experientia, 45*, 696–702.

Muhlbauer, E., Wolgast, S., Finckh, U., Peschke, D., & Peschke, E. (2004). Indication of circadian oscillations in the rat pancreas. *FEBS Letters, 564*, 91–96.

Murphy, B. A. (2009). Chronobiology and the horse: Recent revelations and future directions. *Veterinary Journal*, doi:10.1016/j.tvjl.2009.1004.1013

Murphy, B. A., Elliott, J. A., Sessions, D. R., Vick, M. M., Kennedy, E. L., & Fitzgerald, B. P. (2007a). Rapid phase adjustment of melatonin and core body temperature rhythms following a 6-h advance of the light/dark cycle in the horse. *Journal of Circadian Rhythms*, *5*, 5.

Murphy, B. A., Lear, T. L., Adelson, D. L., & Fitzgerald, B. P. (2007b). Chromosomal assignments and sequences for the equine core circadian clock genes. *Animal Genetics*, *38*, 84–85.

Murphy, B. A., Vick, M. M., Sessions, D. R., Cook, R. F., & Fitzgerald, B. P. (2006). Evidence of an oscillating peripheral clock in an equine fibroblast cell line and adipose tissue but not in peripheral blood. *Journal of Comparative Physiology A Neuroethology and Sensory Neural Behavioral Physiology*, *192*, 743–751.

Nagano, M., Adachi, A., Nakahama, K., Nakamura, T., Tamada, M. et al. (2003). An abrupt shift in the day/night cycle causes desynchrony in the mammalian circadian center. *Journal of Neuroscience*, *23*, 6141–6151.

Nebzydoski, S. J., Pozzo, S., Nemec, L., Rankin, M. K., Gressley, T. F. (2010). The effect of dexamethasone on clock gene mRNA levels in bovine neutrophils and lymphocytes. *Veterinary Immunology and Immunopathology*, *138*, 183–192.

Nelson, R. J. (2004). Seasonal immune function and sickness responses. *Trends in Immunology*, *25*, 187–192.

Oishi, K., Sakamoto, K., Okada, T., Nagase, T., & Ishida, N. (1998). Antiphase circadian expression between BMAL1 and period homologue mRNA in the suprachiasmatic nucleus and peripheral tissues of rats. *Biochemistry and Biophysics Research Community*, *253*, 199–203.

Okamura, H., Miyake, S., Sumi, Y., Yamaguchi, S., Yasui, A. et al. (1999). Photic induction of mPer1 and mPer2 in cry-deficient mice lacking a biological clock. *Science*, *286*, 2531–2534.

Olfert, I. M., Howlett, R. A., Tang, K., Dalton, N. D., Gu, Y. et al. (2009). Muscle-specific VEGF deficiency greatly reduces exercise endurance in mice. *Journal of Physiology*, *587*, 1755–1767.

Panda, S., Antoch, M. P., Miller, B. H., Su, A. I., Schook, A. B. et al. (2002). Coordinated transcription of key pathways in the mouse by the circadian clock. *Cell*, *109*, 307–320.

Petrovsky, N., & Harrison, L. C. (1997). Diurnal rhythmicity of human cytokine production: a dynamic disequilibrium in T helper cell type 1/T helper cell type 2 balance? *Journal of Immunology*, *158*, 5163–5168.

Petrovsky, N., McNair, P., & Harrison, L. C. (1998). Diurnal rhythms of pro-inflammatory cytokines: regulation by plasma cortisol and therapeutic implications. *Cytokine*, *10*, 307–312.

Piccione, G., Assenza, A., Grasso, F., & Caola, G. (2004a). Daily rhythm of circulating fat soluble vitamin concentration (A, D, E and K) in the horse. *Journal of Circadian Rhythms*, *2*, 3.

Piccione, G., Bertolucci, C., Foa, A., & Caola, G. (2004b). Influence of fasting and exercise on the daily rhythm of serum leptin in the horse. *Chronobiology International*, *21*, 405–417.

Piccione, G., Caola, G., & Refinetti, R. (2002). The circadian rhythm of body temperature of the horse. *Biological Rhythm Research*, *33*, 113–119.

Piccione, G., Caola, G., & Refinetti, R. (2005). Temporal relationships of 21 physiological variables in horse and sheep. *Comparative Biochemical Physiology A Molecular Integrative Physiology*, *142*, 389–396.

Piccione, G., Costa, A., Gianetto, C., & Caola, G. (2007). Daily rhythms of activity in horses housed in different stabling conditions. *Biological Rhythm Research*, *39*, 79–84.

Piccione, G., Fazio, F., Caola, G., & Refinetti, R. (2008a). Daily rhythmicity of glycemia in four species of domestic animals under various feeding regimes. *Journal of Physiological Science*, *58*, 271–275.

Piccione, G., Grasso, F., Fazio, F., & Giudice, E. (2008b). The effect of physical exercise on the daily rhythm of platelet aggregation and body temperature in horses. *Veterinary Journal*, *176*, 216–220.

Pittendrigh, C. (1960). Circadian rhythms and the circadian organisation of living things. *Cold Spring Harbor Symposium on Quantitative Biology*, *25*, 159–184.

Pittendrigh, C. S. (1993). Temporal organization: Reflections of a Darwinian clock-watcher. *Annual Review of Physiology*, *55*, 16–54.

Pittendrigh, C. S., & Minis, D. H. (1964). The entrainment of circadian oscillations by light and their role as photoperiodic clocks. *The American Naturalist*, *118*, 261–285.

Plautz, J. D., Kaneko, M., Hall, J. C., & Kay, S. A. (1997). Independent photoreceptive circadian clocks throughout Drosophila. *Science*, *278*, 1632–1635.

Psilander, N., Damsgaard, R., & Pilegaard, H. (2003). Resistance exercise alters MRF and IGF-I mRNA content in human skeletal muscle. *Journal of Applied Physiology*, *95*, 1038–1044.

Reddy, A. B., Field, M. D., Maywood, E. S., & Hastings, M. H. (2002). Differential resynchronisation of circadian clock gene expression within the suprachiasmatic nuclei of mice subjected to experimental jet lag. *Journal of Neuroscience*, *22*, 7326–7330.

Reebs, S. G., & Mrosovsky, N. (1989). Effects of induced wheel running on the circadian activity rhythms of Syrian hamsters: Entrainment and phase response curve. *Journal of Biological Rhythms*, *4*, 39–48.

Reilly, T., Waterhouse, J., & Edwards, B. (2005). Jet lag and air travel: Implications for performance. *Clinical and Sports Medicine*, *24*, 367–380.

Reinberg, A., Ghata, J., & Sidi, E. (1963). Nocturnal asthma attacks: Their relationship to the circadian adrenal cycle. *Journal of Allergy and Clinical Immunology*, *34*, 323–330.

Reppert, S. M., & Weaver, D. R. (2002). Coordination of circadian timing in mammals. *Nature*, *418*, 935–941.

Rosbash, M. (1998). Why the rat-1 fibroblast should replace the SCN as the in vitro model of choice. *Cell*, *93*, 917–919.

Rose, R. J., Hodgson, D. R., Sampson, D., & Chan, W. (1983). Changes in plasma biochemistry in horses competing in a 160 km endurance ride. *Australian Veterinary Journal*, *60*, 101–105.

Rudnicki, M. A., Schnegelsberg, P. N., Stead, R. H., Braun, T., Arnold, H. H., & Jaenisch, R. (1993). MyoD or Myf-5 is required for the formation of skeletal muscle. *Cell*, *75*, 1351–1359.

Russo-Marie, F. (1992). Macrophages and the glucocorticoids. *Journal of Neuroimmunology*, *40*, 281–286.

Sato, T. K., Panda, S., Miraglia, L. J., Reyes, T. M., Rudic, R. D. et al. (2004). A functional genomics strategy reveals Rora as a component of the mammalian circadian clock. *Neuron*, *43*, 527–537.

Sato, T. K., Yamada, R. G., Ukai, H., Baggs, J. E., Miraglia, L. J. et al. (2006). Feedback repression is required for mammalian circadian clock function. *Nature Genetics*, *38*, 312–319.

Satoh, Y., Kawai, H., Kudo, N., Kawashima, Y., & Mitsumoto, A. (2006). Temperature rhythm reentrains faster than locomotor rhythm after a light phase shift. *Physiology and Behavior*, *88*, 404–410.

Shearman, L. P., Sriram, S., Weaver, D. R., Maywood, E. S., Chaves, I. et al. (2000). Interacting molecular loops in the mammalian circadian clock. *Science*, *288*, 1013–1019.

Shimba, S., Ishii, N., Ohta, Y., Ohno, T., Watabe, Y. et al. (2005). Brain and muscle Arnt-like protein-1 (BMAL1), a component of the molecular clock, regulates adipogenesis. *Proceedings of the National Academy of Sciences of the United States of America*, *102*, 12071–12076.

Siepka, S. M., Yoo, S. H., Park, J., Song, W., Kumar, V. et al. (2007). Circadian mutant Overtime reveals F-box protein FBXL3 regulation of cryptochrome and period gene expression. *Cell*, *129*, 1011–1023.

Smolensky, M. H., & Peppas, N. A. (2007). Chronobiology, drug delivery, and chronotherapeutics. *Advances in Drug Delivery Review*, *59*, 828–851.

Sommer, T., Chambers, J. A., Eberle, J., Lauter, F. R., & Russo, V. E. (1989). Fast light-regulated genes of Neurospora crassa. *Nucleic Acids Research*, *17*, 5713–5723.

Spengler, C. M., Czeisler, C. A., & Shea, S. A. (2000). An endogenous circadian rhythm of respiratory control in humans. *Journal of Physiology*, *526*(Pt. 3), 683–694.

Stokkan, K. A., Yamazaki, S., Tei, H., Sakaki, Y., Menaker, M. (2001). Entrainment of the circadian clock in the liver by feeding. *Science*, *291*, 490–493.

Storch, K. F., Lipan, O., Leykin, I., Viswanathan, N., Davis, F. C. et al. (2002). Extensive and divergent circadian gene expression in liver and heart. *Nature*, *417*, 78–83.

Sullivan, D. J., Shelby, J., Shao, Y., Affleck, D. G., Hinson, D. M., & Barton, R. G. (1996). Melatonin and a 21-aminosteroid attenuate shock after hemorrhage but differentially affect serum cytokines. *Journal of Surgical Research*, *64*, 13–18.

Takahashi, J. S. (2004). Finding new clock components: Past and future. *Journal of Biological Rhythms*, *19*, 339–347.

Takahashi, J. S., Shimomura, K., & Kumar, V. (2008). Searching for genes underlying behavior: Lessons from circadian rhythms. *Science*, *322*, 909–912.

Takata, M., Burioka, N., Ohdo, S., Takane, H., Terazono, H. et al. (2002). Daily expression of mRNAs for the mammalian Clock genes Per2 and clock in mouse suprachiasmatic nuclei and liver and human peripheral blood mononuclear cells. *Japanese Journal of Pharmacology*, *90*, 263–269.

Tapscott, S. J. (2005). The circuitry of a master switch: Myod and the regulation of skeletal muscle gene transcription. *Development*, *132*, 2685–2695.

Teboul, M., Barrat-Petit, M.A., Li, X. M., Claustrat, B., Formento, J. L. et al. (2005). Atypical patterns of circadian clock gene expression in human peripheral blood mononuclear cells. *Journal of Molecular Medicine*, *83*, 693–699.

Teclemariam-Mesbah, R., Ter Horst, G. J., Postema, F., Wortel, J., & Buijs, R. M. (1999). Anatomical demonstration of the suprachiasmatic nucleus-pineal pathway. *Journal of Comparative Neurology*, *406*, 171–182.

Terazono, H., Mutoh, T., Yamaguchi, S., Kobayashi, M., Akiyama, M. et al. (2003). Adrenergic regulation of clock gene expression in mouse liver. *Proceedings of the National Academy of Sciences of the United States of America*, *100*, 6795–6800.

Turek, F. W. (1989). Effects of stimulated physical activity on the circadian pacemaker of vertebrates. *Journal of Biological Rhythms*, *4*, 135–147.

Turek, F. W., Joshu, C., Kohsaka, A., Lin, E., Ivanova, G. et al. (2005). Obesity and metabolic syndrome in circadian Clock mutant mice. *Science*, *308*, 1043–1045.

Ueda, H. R., Chen, W., Adachi, A., Wakamatsu, H., Hayashi, S. et al. (2002). A transcription factor response element for gene expression during circadian night. *Nature*, *418*, 534–539.

Van Cauter, E., Moreno-Reyes, R., Akseki, E., L'Hermite-Baleriaux, M., Hirschfeld, U. et al. (1998). Rapid phase advance of the 24-h melatonin profile in response to afternoon dark exposure. *American Journal of Physiology*, *275*, E48–E54.

Weibel, L., & Brandenberger, G. (2002). The start of the quiescent period of cortisol remains phase locked to the melatonin onset despite circadian phase alterations in humans working the night schedule. *Neuroscience Letters*, *318*, 89–92.

Weintraub, H. (1993). The MyoD family and myogenesis: Redundancy, networks, and thresholds. *Cell*, *75*, 1241–1244.

Welsh, D. K., Yoo, S. H., Liu, A. C., Takahashi, J. S., & Kay, S. A. (2004). Bioluminescence imaging of individual fibroblasts reveals persistent, independently phased circadian rhythms of clock gene expression. *Current Biology*, *14*, 2289–2295.

Wever, R. A. (1980). Phase shifts of human circadian rhythms due to shifts of artificial Zeitgebers. *Chronobiologia*, *7*, 303–327.

Whitmore, D., Foulkes, N. S., & Sassone-Corsi, P. (2000). Light acts directly on organs and cells in culture to set the vertebrate circadian clock. *Nature*, *404*, 87–91.

Wickland, C. R., & Turek, F. W. (1991). Phase-shifting effects of acute increases in activity on circadian locomotor rhythms in hamsters. *American Journal of Physiology*, *261*, R1109–R1117.

Wright, J. E., Vogel, J. A., Sampson, J. B., Knapik, J. J., Patton, J. F., & Daniels, W. L. (1983). Effects of travel across time zones (jet-lag) on exercise capacity and performance. *Aviation and Space Environmental Medicine*, *54*, 132–137.

Wu, C. C., Chiao, C. W., Hsiao, G., Chen, A., & Yen, M. H. (2001). Melatonin prevents endotoxin-induced circulatory failure in rats. *Journal of Pinealogy Research*, *30*, 147–156.

Wyszynska-Koko, J., Pierzchala, M., Flisikowski, K., Kamyczek, M., Rozycki, M., & Kuryl, J. (2006). Polymorphisms in coding and regulatory regions of the porcine MYF6 and MYOG genes and expression of the MYF6 gene in m. longissimus dorsi versus productive traits in pigs. *Journal of Applied Genetics*, *47*, 131–138.

Yagita, K., Tamanini, F., van Der Horst, G. T., & Okamura, H. (2001). Molecular mechanisms of the biological clock in cultured fibroblasts. *Science*, *292*, 278–281.

Yamamoto, T., Nakahata, Y., Soma, H., Akashi, M., Mamine, T., & Takumi, T. (2004). Transcriptional oscillation of canonical clock genes in mouse peripheral tissues. *BMC Molecular Biology*, *5*, 18.

Yamazaki, S., Numano, R., Abe, M., Hida, A., Takahashi, R. et al. (2000). Resetting central and peripheral circadian oscillators in transgenic rats. *Science*, *288*, 682–685.

Yoshida, N., Yoshida, S., Koishi, K., Masuda, K., & Nabeshima, Y. (1998). Cell heterogeneity upon myogenic differentiation: down-regulation of MyoD and Myf-5 generates 'reserve cells'. *Journal of Cellular Science*, *111*(Pt. 6), 769–779.

Zambon, A. C., McDearmon, E. L., Salomonis, N., Vranizan, K. M., Johansen, K. L. et al. (2003). Time- and exercise-dependent gene regulation in human skeletal muscle. *Genome Biology*, *4*, R61.

Zhang, X., Dube, T. J., & Esser, K. A. (2009). Working around the clock: Circadian rhythms and skeletal muscle. *Journal of Applied Physiology*, *107*, 1647–1654.

Zheng, B., Albrecht, U., Kaasik, K., Sage, M., Lu, W. et al. (2001). Nonredundant roles of the mPer1 and mPer2 genes in the mammalian circadian clock. *Cell*, *105*, 683–694.

Zhou, Z., & Bornemann, A. (2001). MRF4 protein expression in regenerating rat muscle. *Journal of Muscle Research and Cell Motility*, *22*, 311–316.

19 Mitochondrial genome: Clues about the evolution of extant equids and genomic diversity of horse breeds

Cynthia C. Steiner, Kateryna D. Makova, and Oliver A. Ryder

Introduction

Selection for domestication and for superior performance in a variety of activities under human influence has contributed to molding the genetic makeup of modern horse. This is dramatically evident for many breeds and, notably, the Thoroughbred racehorse. Accompanying this process of domestication and selection is a loss of genetic variation and the emergence of founder effect; In domestic horses, the number of male founders (stallions) is thought to be quite small, as indicated by the lack of variation in the Y-chromosome (Wallner et al., 2003).

The oldest pedigree record for domestic horses, the studbook of the Thoroughbred, documents the genetic bottleneck associated with founder effect. The 3 foundation stallions and approximately 30 foundation mares represent the ancestral gene pool of this breed of horses under domestication. The relative reduction in genetic variation in Thoroughbreds and other horse breeds is also demonstrable by comparison of levels of heterozygosity, now readily measured using techniques of DNA analysis and genomics.

Selection for performance is a fundamental aspect of management of domestic horses whereby understanding the genetic basis of breed's adaptations is of fundamental interest in the study of equine biology, and has value in the marketplace. Recent data regarding selection for genes involved in mitochondrial function – encoded by nuclear genes and the mitochondrial genome itself – in horses and other model species provide a perspective on research directions that can be anticipated to shed light on metabolic adaptations that affect performance in horses. The same loci are also candidates for elucidating the basis for human variation in exercise physiology and are potentially applicable to understanding the etiology of some metabolic disorders.

With the release of the assembled genome of a Thoroughbred horse and the accompanying effort at discovering single nucleotide polymorphisms (SNPs; Wade et al., 2009), a new era in genetic evaluation of horses is at hand. While detailed studies of sequence variation across the entire mitochondrial genome have not been undertaken on a sufficiently large population of individuals encompassing the diversity of breeds and environments in which domestic horses reside, emerging technologies are expected to quickly erase this deficit. The unique aspects of selection for strength, speed, endurance, and behavior in domestic horses suggest that expanded studies utilizing genomics technologies that explore the interaction of the mitochondrial genome with its host can provide new insights relevant to horse breeding, exercise physiology, and adaptive evolution.

Equine Genomics, First Edition. Edited by Bhanu P. Chowdhary.

In the present chapter, we discuss the influence of mitochondrial genetics, especially variation in the mitochondrial genome of horses framed in an evolutionary context. We address the evolution and diversity of genetic variation in the mitochondrial genome of domestic horses considering their wild ancestors and living relatives. We also provide insights on the genetic pathways involved in some of the unique morphological and physiological adaptations that characterize horse breeds.

Evolution of the Family Equidae

Although a remarkable radiation of horses occurred during the Miocene according to the fossil record (MacFadden, 2005), there is only one extant equid genus, *Equus*, which is comprised of eight species of horses, asses, and zebras that arose in the late Pliocene of North America between 2 and 4 million years ago (Mya; Lindsay et al., 1980; Hulbert, 1996). Current diversity is represented by the domestic horse, *E. caballus*, Przewalski's wild horse, *E. przewalskii*, kiang, *E. kiang*, Asiatic wild ass, *E. hemionus*, African wild ass, *E. asinus*, mountain zebra, *E. zebra*, Burchell's zebra, *E. burchelli*, and Grevy's zebra, *E. grevyi*.

The evolution of equids has been extensively studied, primarily because of their large fossil record (Bernor et al., 2010). Nevertheless, controversy regarding phylogenetic relationships among some of the extant taxa reflects inconsistencies of morphological and molecular studies. In the past 30 years, morphological surveys that include the analysis of dental, cranial, and postcranial traits (Eisenmann, 1979; Bennet, 1980; Harris & Porter, 1980) failed to establish a solid phylogenetic framework for extant equid evolution, resulting in dissimilar phylogenies from independent datasets. The molecular era provided a new tool for resolving morphological phylogenetic discrepancies by incorporating more characters in the analyses (e.g., nucleotides) and novel techniques (e.g., restriction enzyme maps, DNA sequencing). Among the first molecular studies were the analyses of protein sequences (Clegg, 1974; Kaminski, 1979), radioimmunological assays (Lowenstein & Ryder, 1985), and genetic variation of the mitochondrial DNA (George & Ryder 1986).

Mitochondrial DNA (mtDNA) quickly appeared to be a very informative source of genetic variation, due to its relatively rapid rate of evolution. In particular, the survey by George and Ryder (1986) analyzed mtDNA restriction-endonuclease maps of *E. asinus*, *E. burchelli*, *E. grevyi*, *E. hemionus*, and *E. zebra* and found that there were at least three major clades in modern *Equus*: the zebras, the wild asses, and the true horses. Later, Ishida et al. (1995) investigated the mitochondrial D loop region, which supported the interpretations of George and Ryder (1986). Another subsequent mtDNA study by Oakenfull et al. (2000) analyzed a large sample of equid species using two mtDNA genes (D loop region and *12s rRNA*), resulting in more details about equid phylogenetic relationships, such as the early split of caballines (domestic and Przewalski's Horses) and non-caballines (asses and zebras), and the basal position of Asiatic asses relative to African wild ass and zebras.

Additional molecular markers found in nuclear DNA were later explored in parallel with mtDNA genes to refine questions on the evolution of equids. For example, Oakenfull and Clegg (1998) were the first to investigate DNA sequences of nuclear genes (α and θ globins), also finding a marked genetic divergence between the zebras, wild asses, and the true horses. Wallner et al. (2003) analyzed Y chromosome sequences for multiple loci in all species of equids, identifying five fixed nucleotide differences between *E. caballus* and *E. przewalskii*. More recently, Krüger et al. (2005) used microsatellite data to estimate phylogenetic relationship in seven equid species, supporting the grouping of African and Asiatic asses.

In general, all molecular studies suggest that domestic and Przewalski's wild horses diverged early from all other species of *Equus*. The relationship between zebras and asses, however, is not clearly resolved. In particular, the phylogenetic positions of African (*E. asinus*) and Asiatic asses (*E. kiang* and *E. hemionus*) are highly controversial. For example, studies using postcranial morphology, mtDNA restriction fragment length polymorphism (RFLPs), protein electrophoresis, and mitochondrial and nuclear sequence data position the African wild ass as sister taxon to zebras (Kaminski, 1979; Harris & Porter, 1980; George & Ryder, 1986; Oakenfull & Clegg, 1998; Oakenfull et al., 2000). In contrast, cranial and dental morphology data (Eisenmann, 1979; Bennett, 1980) group the Asiatic and African wild asses as sister group to zebras. All previous surveys provide relatively poor to no levels of node support.

Thus, a persistent problem in the current molecular-based phylogenies of equids has been failure to include all recognized extant *Equus* species and a sufficient number of genetic markers (e.g., nuclear genes), which should provide stronger insights into *Equus* species relationships. Most studies investigated only a few mitochondrial or nuclear DNA genetic markers, inferring phylogenies with rather poorly supported nodes where the bias caused by genetic introgression or incomplete lineage sorting cannot be assessed. Recent efforts have been made toward a complete sampling of species, the incorporation of multiple genetic markers from different cellular compartments (mitochondrial and nuclear), and the application of more sophisticated phylogenetic inference algorithms (e.g., Bayesian) for supporting and testing phylogenetic hypotheses (Figure 19.1; Steiner & Ryder, 2011; Steiner et al., 2012).

For instance, the Bayesian analysis of the mtDNA data combined partial sequences from three regions: D loop, *12S rRNA*, and *Cytb* (2074 bp total; Figure 19.1a) from Genbank and recently acquired sequences (Steiner & Ryder, 2011). The analysis of the mitochondrial data set resulted

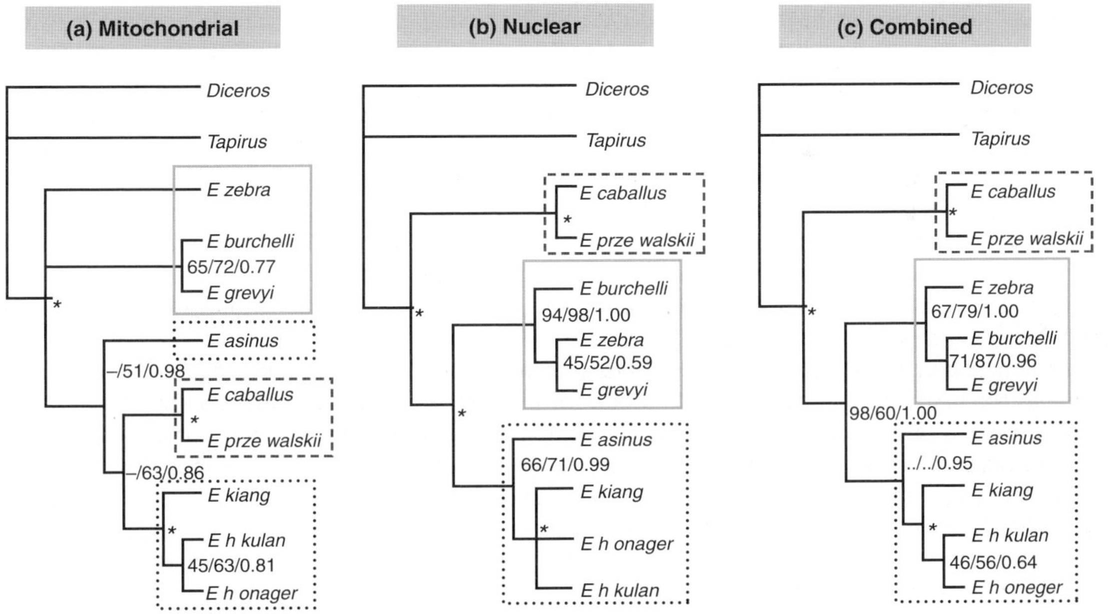

Figure 19.1 Comparison of Bayesian topologies obtained from (a) mitochondrial (GTR + Γ + I), (b) nuclear (GTR + Γ + I), and (c) the combined (two genetic partitions with GTR + Γ + I) data sets of equids. Maximum parsimony (MP) bootstrap, maximum likelihood (ML) bootstrap, and Bayesian posterior probabilities (PP) values are indicated for all nodes. A star indicates the highest support for all three approaches ($BP_{MP} = BP_{ML} = 100$, $PP = 1.00$). Boxes represent different equid groups including caballines (dashed dark grey lines), zebras (light grey line), and asses (dotted black lines).

in a lack of phylogenetic resolution and poor node support by both bootstrap (BP) and posterior probability (PP) values. The mitochondrial tree is characterized by a polytomy composed of the mountain zebra, a cluster of Burchell's and Grevy's zebras, and a third cluster formed by asses and caballines. In the latter, the African wild ass is basal relative to two clades, the caballines and Asiatic asses (kiang and the Asiatic wild ass). Well-supported nodes include the cluster of caballines, Asiatic asses, and the monophyly of Equidae (BP = 100, PP = 1.00).

The Bayesian analysis of the nuclear DNA comprised a large data set consisting of partial sequences of 20 genes (15791 bp) from coding and non-coding areas of the genome (Figure 19.1b; Steiner et al., 2012). The nuclear phylogenetic tree reveals a well-resolved and relatively well-supported topology for most nodes, identifying three groups of equids as proposed before: caballines, zebras, and asses (African and Asiatic). Specifically, the deepest node in the tree corresponds to the split between caballine and non-caballine equids. Within the asses group, the African wild ass appears basal and sister to Asiatic asses, while within zebras, the Burchell's zebra (*E. burchelli*) diverged first relative to mountain (*E. zebra*) and Grevy's (*E. grevyi*) zebras.

A more extensive analysis concatenated mitochondrial and nuclear genes into a single dataset (23 genes, 17865 bp; Figure 19.1c). Combining genes makes sense, as analyses of data consisting of a single or small number of genes have a high probability of supporting conflicting phylogenies. In contrast, analyses of concatenated genes have yielded fully resolved species trees with maximum support (Rokas et al., 2003). The combined Bayesian analysis showed similar results relative to the nuclear dataset, separating caballines from asses and zebras. However, in the zebra cluster, mountain zebra appears basal and sister to Burchell's (*E. burchelli*) and Grevy's (*E. grevyi*) zebras. The monophyly of all species of asses is not well supported by MP and ML approaches, but the relationship between kiang and wild Asiatic asses is resolved.

Lower nodal support in the combined data likely reflects phylogenetic incongruence between the mitochondrial and nuclear datasets. These discrepancies are likely due to differences in gene properties, including base composition and substitutions rate, or even substitution saturation in the case of mitochondrial genes (Brown et al., 1979; Galtier & Gouy, 1995; Perna & Kocher, 1995). In addition, and perhaps more important in the evolution of species, are the biological processes that occurred during speciation, such as incomplete genetic lineage sorting, genetic introgression, or hybridization among species (Rokas & Carroll, 2006; Hallström & Janke, 2008). For example, before closely related species become completely genetically isolated, they can hybridize with each other. This poses a problem for phylogenetic analysis, because the species history can be obscured by introgression or the incorporation of genes from one species into the gene pool of another.

In conclusion, these phylogenetic analyses seem to highlight the value of nuclear data as a complementary source of genetic information for inferring equid relationships relative to mtDNA, supporting (1) the taxonomic division of caballines and non-caballines initially proposed by morphological data (Forsten, 1992) and other molecular studies (Lowenstein & Ryder, 1985; George & Ryder, 1986; Oakenfull & Clegg, 1998; Krüger et al., 2005); and (2) the phylogenetic position of the African wild ass (*E. asinus*) as sister taxon to Asiatic asses as previously predicted by cranial characters (Bennet, 1980), immunological protein assays (Lowestein & Ryder 1985), Y chromosome sequences, and microsatellite data (Wallner et al., 2003; Krüger et al., 2005). This comprehensive phylogenetic framework provides a new outlook on the evolution of extant equids, allowing other issues related to the evolution of extinct equid species to be addressed, including the history of horse domestication, the origin of horse breeds, and their morphological and physiological adaptations.

Przewalski's Horse and Horse Domestication

Przewalski's Horse (*Equus przewalskii*) is an endangered species that is considered to be the only true "wild" extant horse species. Przewalski's Horse is the closest living relative to domestic horse (*Equus caballus*) and can be used to infer the history of horse domestication. Compared with domestic horse (2N = 64), Przewalski's Horse (2N = 66) has an extra chromosome pair due to one Robertsonian translocation: either fission of domestic horse chromosome 5 or fusion of Przewalski's Horse chromosomes 23 and 24 (Benirschke et al., 1965; Bowling & Ruvinsky, 2000; Myka et al., 2003; Yang et al., 2003; Ahrens & Stranszinger, 2005). In comparison, the chromosomal differences between domestic horses, asses, and zebras include numerous translocations, fusions, and inversions (Yang et al., 2004, Trifonov et al., 2008). The Przewalski's horse is known to have the highest diploid chromosome number among horse species (Ryder et al., 1978).

Although this chromosomal difference genetically separates Przewalski's horse from domestic horse, the two have been known to interbreed to produce fertile offspring (Short et al., 1974). In contrast, hybridizations between domestic horses and other related species such as donkeys usually result in viable but infertile offspring due to multiple chromosomal rearrangements that separate the species (Zong & Fan, 1989; Yang et al., 2004). Skull measurements often overlap between the Przewalski's and domestic horses (Forsten, 1987; Eisenmann & Baylac 2000), but other skeletal features highlight differences in the morphology between the two horses (Sasaki et al., 1999).

Przewalski's Horses probably once lived in a broader range throughout Europe and Asia, including the Dzungarian Basin, an area that is now part of Mongolia, Kazakhstan, and the Xinjiang-Uygur Autonomous Region of China (Ryder, 1993). By the mid-1960s, Przewalski's Horse had become virtually extinct in the wild due to human activity; however, it had subsequently been bred in captivity and reintroduced into its natural habitat in parts of Mongolia and China (Ryder & Wedemeyer, 1982; Ryder, 1993; Bouman & Bouman 1994). The present-day Przewalski's Horse population, consisting of more than 2,000 animals, originated from a mere 13 animals gathered at the turn of the nineteenth century, one of which was a domestic horse and the other a domestic/Przewalski's Horse hybrid (Oakenfull & Ryder, 1998). Notably, the founders included only four Przewalski's Horse females (Volf et al., 1991; Oakenfull & Ryder 1998). Later, additional introgression from domestic horse mares via their hybrid offspring occurred, but was not revealed until genetic studies were performed (Bowling et al., 2003).

The genetic relationship between Przewalski's and domestic horses has been subject to a heated debate, and several questions remain unanswered: Is Przewalski's Horse the direct ancestor of domestic horse? Or has Przewalski's Horse branched out from the ancestral population that included the domestic horse genetic pool?

Previous studies have used different DNA markers to determine the genetic relationship between Przewalski's and domestic horse and have resulted in contradictory results. Whereas some protein, microsatellite, and Y chromosome analyses have supported phylogenetic separation of the two taxa (Bowling et al., 2003; Wallner et al., 2003), the latest investigations of autosomal (Lau et al., 2009; Wade et al., 2009) and X chromosomal DNA (Lau et al., 2009) have not. For instance, our recent phylogenetic analysis of several autosomal and X chromosomal introns placed Przewalski's Horses within the domestic horse clade (Lau et al., 2009). Similarly, phylogenetic separation between Przewalski's versus domestic horses was not observed after typing ~1,000 autosomal SNPs as part of the analysis of the horse genome (Wade et al., 2009). This led to the speculation that either *E. przewalskii* is very recently derived from *E. caballus*, or that the two horses intermixed considerably following divergence from a common ancestor (Wade et al., 2009). Because Przewalski's Horses are the only truly wild horses existing today, they have alternatively been hypothesized to be the direct ancestors of domestic horses (Ryder et al., 1994).

Possible explanations for these findings are gene flow from domestic horse into the Przewalski's Horse population, different effective population sizes of males and females, or an unequal ratio of female to male founders (Wallner et al., 2003). Only relatively short genomic regions (or a small number of sites) have been analyzed in the reports mentioned above, inflating statistical fluctuations. Moreover, it is also possible that these discrepancies in results are due to differences in sampling depending on whether the Przewalski's Horses studied belong to the same mitochondrial lineage.

Studies of mtDNA of Przewalski's Horse were limited to sequencing the D loop region (Ishida et al., 1995; Oakenfull & Ryder, 1998; Oakenfull et al., 2000). There are four maternal lineages in the current Przewalski's Horse population (Ryder, 1994); however, sequencing of the mitochondrial D loop region showed only two mtDNA haplotypes in these lineages (Oakenfull & Ryder 1998). Moreover, other studies of mitochondrial DNA indicated that domestic horse and Przewalski's Horse do not form separate clades (Ishida et al., 1995, Oakenfull & Ryder, 1998).

Recently, using the massively parallel sequencing technology, we determined the sequence of complete mtDNA genomes in all four surviving matrilines of Przewalski's Horse and compared it to the completely sequenced domestic horse mtDNA genomes that are currently publicly available for six horse breeds (a Thoroughbred, Debao pony, Tibetan breed from Naqu, Tibetan breed from Deqin, Jeju breed, and Tibetan breed from Zhongdian), using the mitochondrial genome of the Somali wild ass as an outgroup (Goto et al., 2011). Three mitochondrial haplotypes were discovered in Przewlaski's horse – two similar ones, haplotypes I/II, and one substantially divergent from the other two, haplotype III. Haplotypes I/II versus III did not cluster together on a phylogenetic tree, rejecting the monophyly of Przewalski's Horse maternal lineages, and were estimated to split 0.117–0.186 million years ago, significantly preceding horse domestication, thought to occur only about 6,000 years ago (Goto et al., 2011; Outram et al., 2009).

The divergence of Przewalski's Horse mtDNA haplotypes I/II and III poses an interesting question. The drastically different haplotypes could represent genetic polymorphism that existed in the ancestral population of Przewalski's and domestic horses. As horse domestication is a relatively recent event, Przewalski's Horses could have retained such ancestral polymorphism in their current population. In this case, sequencing mtDNA from ancient horse remains is expected to reveal haplotypes similar to the ones found here for Przewalski's Horses. Second, one or both divergent mtDNA haplotypes could have been introduced from domestic to Przewalski's Horses via interbreeding. In particular, two closely related haplotypes are very similar in sequence and cluster together with domestic horse haplotype(s) (Goto et al., 2011), and thus might have been acquired by Przewalski's Horse through genetic introgression. Additional sequencing of mtDNA from modern domestic horses from Eurasia might identify mtDNA haplotypes similar in sequence to one or several Przewalski's Horse haplotypes. A combination of these two scenarios (e.g., one divergent haplotype acquired via introgression while another one inherited from the ancestral horse population) is also possible.

Furthermore, as a part of the same study (Goto et al., 2011), we sequenced partial nuclear genomes of the four Przewalski's Horses and analyzed these data together with the publicly available genomic sequences of the Thoroughbred domestic horse (Wade et al., 2009). In the phylogeny based on X chromosomal sequences, Przewalski's and domestic horse lineages were intermixed, while in that built from autosomal sequences, Przewalski's Horses lineages were monophyletic. The results indicate that Przewalski's Horses have ancient polyphyletic origins and are not the direct progenitors of domestic horses. The data also suggest that Przewalski's and domestic horse lineages diverged at least 0.117 million years ago, but since then have retained ancestral genetic polymorphism and/or experienced gene flow.

History of Horse Breeds

The different domestic breeds of horse are originally derived from a variety of wild populations distributed from Europe to Middle East. Breeds established prior to 1500 AD exhibit a pattern of geographic distribution and morphological stability that is the result of conservative breeding based on the "native broodmare" (Bennett & Hoffmann, 1999). In the last century, many horse populations have suffered severe decline throughout the world, when horses lost importance as instruments of transportation of mechanical power (Aberle & Distl, 2004).

The study of the genetic structure of different horse breeds has been essential in understanding the development of today's breeds, and in establishing conservation programs for preserving the genetic variability still present in horse populations. The history of horse breeds and domestication has been investigated largely from mitochondrial and Y chromosome sequences, leading to conclusions about a single patriline (Lindgren et al., 2004) but numerous matrilines (Vila et al., 2001) incorporated into the genetic pool of domestic horses. This is consistent with the breeding plans of most horse breeds, in which one male is bred with many different females (McDonnell, 2005).

The sequencing and genetic analyses of mtDNA sequences, specifically the hypervariable region of the D loop, have been extensively used. Multiple studies have addressed sequence variation among maternal lineages of horse breeds, including the Lipizzan horses (Kavar et al., 2002), Mongolian (Li et al., 2008), German heavy draught (Aberle et al., 2007), Iberian (Royo et al., 2005), Thoroughbred (Hill et al., 2002; Bower et al., 2011), Italian breeds (Cozzi et al., 2004), and New World horses (Luis et al., 2006). In general, these studies document a diversity of hypervariable region haplotypes suggesting mixed origins of horse breeds with high maternal haplotype variation, even within individual breeds. Accordingly, the origin of single breeds is rather obscure, although multiple domestication events are suggested (Vila et al., 2001; Jansen et al., 2002).

New approaches have also been implemented in horse population analyses facilitated by the development of genomic tools for the domestic horse as model organism. Specifically, SNP genotyping has characterized within- and across-breed genetic structure by analyzing patterns of linkage disequilibrium (LD; Wade et al., 2009). These analyses revealed a moderate within-breed LD, with similar values shared among the majority of breeds (see also Chapter 7). Latter implicates that horses do not appear to have undergone a tight domestication bottleneck as observed for example in cows. In addition to that, major haplotypes are frequently shared among diverse populations.

Adaptation in Horse Breeds

Domestic horses extensively vary in their morphological and physiological features. For instance, the warmblood horses are lightweight, quick, and more excitable animals, mostly used for races and leisure sports. The Arabian breed is possibly the best representative of this group. Coldblooded horses, in contrast, are typically tall, heavy animals, such as the Shire, the Clydesdale, or the Friesian breeds. They have been predominantly developed to become very strong and resistant horses, being subsequently used in agriculture and traction-type tasks like goods transportation.

Relatively few molecular studies have addressed the genetic bases of such characteristic phenotypes among horse breeds and most of the studies to date have analyzed mitochondrial genes.

Mitochondrial DNA is a candidate genome for this type of adaptive studies because of its vital role in energy metabolism through oxidative phosphorylation in eukaryotic cells (Fonseca et al., 2008). In addition, amino acid changes in mtDNA have shown to improve aerobic capacity and adaptation to thermal environments in some organisms (Mishmar et al., 2003; Jobson et al., 2004; Dalziel et al., 2006). Comparative studies in domestic horses by Xu et al. (2007) and Ning et al. (2010) investigated adaptation to high altitude in Tibetan and Chinese horses using whole mtDNA genomes and the candidate mitochondrial gene *ND6*, respectively. Both studies produced evidence for adaptive evolution in the *ND6* gene in plateau horses, suggesting the potential influence of oxygen deficits and cold temperatures in animals living in high-altitude environments.

The genetic basis of morphological traits in horses has also been studied using genome-wide approaches. Gu et al. (2009) for example analyzed candidate genes involved in functional adaptations contributing to elite athletic phenotypes in Thoroughbred horses. They used microsatellite-based linkage mapping for identifying potential genes under selection. Among the athletic-performance candidate genes identified were mitochondrial proteins such as *ACCS1*, *MTFR1*, and *PDK4*. These results emphasize the importance of energy production genes in horse adaptations, and highlight the relevance of functional co-adaptation studies between mitochondrial and nuclear-encoded proteins for understanding how species maintain optimal metabolic fitness (see Blier et al., 2001; Grossman et al., 2004; Shen et al., 2010).

Future surveys including whole mitochondrial genomes and genome-wide association studies appear as the most appropriate strategies for dissecting the genetic basis of morphological and physiological traits in horse breeds. The precipitous decline in genome sequencing costs and the advances in bioinformatics and comparative genomics, in combination with data on domestic horse pedigrees and phenotypes will make it possible. Genomic approaches will surely lead to the identification of mutations in genes related to morphology, metabolism, and diseases susceptibility in horses.

References

Aberle, K., & Distl, O. (2004). Domestication of the horse: Results based on microsatellite and mitochondrial DNA markers. *Archiv Tierzucht, Dummerstorf, 47*(6), 517–535.

Aberle, K., Hamann, H., Drogemuller, C., & Distl, O. (2007). Phylogenetic relationships of German heavy draught horse breeds inferred from mitochondrial DNA D-loop variation. *Journal of Animal Breeding and Genetics, 124*, 94–100.

Ahrens, E., & Stranzinger, G. (2005). Comparative chromosomal studies of E. caballus (ECA) and E. przewalskii (EPR) in a female F1 hybrid. *Journal of Animal Breeding and Genetics = Zeitschrift fur Tierzuchtung und Zuchtungsbiologie, 122*(Suppl. 1), 97–102.

Benirschke, K., Malouf, N., Low, R. J., & Heck, H. (1965). Chromosome Complement: Differences between Equus Caballus and Equus Przewalskii, Poliakoff. *Science, 148*, 382–383.

Bennett, D. (1980). Stripes do not a zebra make, Part I: A cladistic analysis of Equus. *Systematic Zoology, 29*(3), 272–287.

Bennett, D., & Hoffmann, R. (1999). Equus caballus. *Mammalian Species, 628*, 1–14.

Bernor, R., Armour-Chelu, M., Gilbert, H., Kaiser, T., & Schulz, E. (2010). Equidae. In L. Werdelin & B. Sanders (eds.), *Cenozoic mammals of Africa* (pp. 685–721). Berkeley: University of California Press.

Blier, P., Dufresne, F., & Burton, R. (2001). Natural selection and the evolution of mtDNA-encoded peptides: Evidence for intergenomic co-adaptation. *Trends in Genetics, 17*(7), 400–406.

Bouman, I., & Bouman, J. (1994). The history of the Przewalski's horse. In L. Boyd & K. A. Houpt (eds.), *Przewalski's Horse: The History and Biology of an Endangered Species* (pp. 5–38). Albany: State University of New York Press.

Bower, M., Campana, M., Whitten, M., Edwards, C., Jones, H. et al. (2011). The cosmopolitan maternal heritage of the Thoroughbred racehorse breed shows a significant contribution from British and Irish native mares. *Biology Letters, 7*, 316–320.

Bowling, A., & Ruvinsky, A. (eds.) (2000). *Genetic aspects of domestication, breeds, and their origins*. Oxon: CABI.

Bowling, A. T., Zimmermann, W., Ryder, O., Penado, C., Peto, S. et al. (2003). Genetic variation in Przewalski's horses, with special focus on the last wild caught mare, 231 Orlitza III. *Cytogenetic and Genome Research 102*(1–4), 226–234.

Brown, W., George, M., & Wilson, A. (1979). Rapid evolution of animal mitochondrial DNA. *Proceedings of the National Academy of Sciences USA*, *76*(4), 1967–1971.

Clegg, J. (1974). Horse hemoglobin polymorphism. *Annals of the New York Academy of Sciences*, *241*, 61–69.

Cozzi, M., Strillacci, M. G., Valiati, P., Bighignoli, B., Cancedda, M., & Zanotti, M. (2004). Mitochondrial D-loop sequence variation among Italian horse breeds. *Genetics Selection Evolution*, *36*, 663–672.

Dalziel, A., Moyes, C., Fredriksson, E., & Lougheed, S. (2006). Molecular evolution of cytochrome c oxidase in high-performance fish (teleostei: Scombroidei). *Journal of Molecular Evolution*, *62*(3), 319–331.

Eisenmann, V. (1979). Caracteres evolutifs et phylogeny du genre Equus (Mammalia, Perissodactyla). *Comptes Rendus de l'Academie des Sciences, Paris, ser D*, *288*, 497–500.

Eisenmann, V., & Baylac, M. (2000). Extant and fossil Equus (Mammalia, Perissodactyla) skulls: A morphometric definition of the subgenus Equus. *Zoologica Scripta*, *29*(2), 89–100.

Fonseca, R., Johnson, W., O'Brien, S., Ramos, M., & Antunes, A. (2008). The adaptive evolution of the mammalian mitochondrial genome. *BMC Genomics*, *9*, 119.

Forsten, A. (1987). A comment on equid craniology. *Mammalia*, *51*(3), 455–467.

Forsten, A. (1992). Mitochondrial-DNA time-table and the evolution of Equus: Comparison of molecular and paleontological evidence. *Annales Zoologici Fennici*, *28*, 301–309.

Galtier, N., & Gouy, M. (1995). Inferring phylogenies from DNA sequences of unequal base compositions. *Proceedings of the National Academy of Science USA*, *92*, 11317–11321.

George, J. M., & Ryder, O. (1986). Mitochondrial DNA evolution in the genus Equus. *Molecular Biology Evolution*, *3*(6), 535–546.

Goto, H., Ryder, O. A., Fisher, A. R., Schultz, B., Kosakovsky Pond, S. A. et al. (2011). A massively parallel sequencing approach uncovers ancient origins and high genetic variability of endangered Przewalski's horses. *Genome and Biology of Evolution*, *3*, 1096–1106.

Grossman, L., Wildman, D., Schmidt, T., & Goodman, M. (2004). Accelerated evolution of the electron transport chain in anthropoid primates. *Trends in Genetics*, *20*(11), 578–585.

Gu, J., Park, S., Katz, L., Sulimova, G., MacHugh, D., & Hill, E. (2009). A genome scan for positive selection in Thoroughbred horses. *PloS One*, *4*(6), e5767.

Hallström, B., & Janke, A. (2008). Resolution among major placental mammal interordinal relationships with genome data imply that speciation influenced their earliest radiations. *BMC Evolutionary Biology*, *8*(162), 162–174.

Harris, A., & Porter, L. (1980). Late Pleistocene horses of Dry Cave, Eddy County, New Mexico. *Journal of Mammalogy*, *61*, 46–65.

Hill, E., Bradley, D., Al-Barody, M., Ertugrul, O., Splan, R. et al. (2002). History and integrity of Thoroughbred dam lines revealed in equine mtDNA variation. *Animal Genetics*, *33*, 287–294.

Hulbert, R. J. (1996). The ancestry of the horse. In S. Olsen (ed.), *Horses through time* (pp. 11–34). Boulder, CO: Roberts Rinchart.

Ishida, N., Oyunsuren, T., Mashima, S., Mukoyama, H., & Saitou, N. (1995). Mitochondrial DNA sequences of various species of the genus Equus with special reference to the phylogenetic relationship between Przewalskii's wild horse and domestic horse. *Journal of Molecular Evolution*, *41*, 180–188.

Jansen, T., Forster, P., Levine, M. A., Oelke, H., Hurles, M. et al. (2002). Mitochondrial DNA and the origins of the domestic horse. *Proceedings of the National Academy of Science USA*, *99*(16), 10905–10910.

Jobson, R., Nielsen, R., Laakkonen, L., Wikstrom, M., & Albert, V. (2004). Adaptive evolution of cytochrome c oxidase: Infrastructure for a carnivorous plant radiation. *Proceedings of the National Academy of Science USA*, *101*(52), 18064–18068.

Kaminski, M. (1979). The biochemical evolution of the horse. *Comparative Biochemistry and Physiology*, *63B*, 175–178.

Kavar, T., Brem, G., Habe, F., Solkner, J., & Dovc, P. (2002). History of Lipizzan horse maternal lines as revealed by mtDNA analysis. *Genetics Selection Evolution*, *34*, 635–648.

Krüger, K., Gaillard, C., Stranzinger, G., & Rieder, S. (2005). Phylogenetic analysis and species allocation of individual equids using microsatellite data. *Journal of Animal Breeding and Genetics*, *122*(1), 78–86.

Lau, A. N., Peng, L., Goto, H., Chemnick, L., Ryder, O. A., & Makova, K. D. (2009). Horse domestication and conservation genetics of Przewalski's horse inferred from sex chromosomal and autosomal sequences. *Molecular Biology and Evolution*, *26*(1), 199–208.

Li, J., Shi, Y., Fan, C., & Manglai, D. (2008). mtDNA diversity and origin of Chinese Mongolian horses. *Asian-Australasian Journal of Animal Sciences*, *21*(12), 1696–1702.

Lindgren, G., Backstrom, N., Swinburne, J., Hellborg, L., Heinarsson, A. et al. (2004). Limited number of patrilines in horse domestication. *Nature Genetics*, *36*, 335–336.

Lindsay, E., Opdyke, N., & Johnson, N. (1980). Pliocene dispersal of the horse Equus and late Cenozoic mammalian dispersal events. *Nature*, *287*, 135–138.

Lowenstein, J., & Ryder, O. (1985). Immunological systematics of the extinct quagga (Equidae). *Experientia*, *41*, 1192–1193.

Luis, C., Bastos-Silveira, C., Cothran, E., & Oom, M. (2006). Iberian origins of new world horse breeds. *Journal of Heredity*, *97*(2), 107–113.

MacFadden, B. (2005). Fossil horses – evidence for evolution. *Science*, *307*, 1728–1730.

McDonnell, S. (2005). *Understanding your horse's behavior*. Lexington, KY: Blood-Horse Publications.

Mishmar, D., Ruiz-Pesini, E., Golik, P., Macaulay, V., Clark, A. et al. (2003). Natural selection shaped regional mtDNA variation in humans. *Proceedings of the National Academy of Science USA*, *100*(1), 171–176.

Myka, J. L., Lear, T. L., Houck, M. L., Ryder, O. A., & Bailey, E. (2003). FISH analysis comparing genome organization in the domestic horse (Equus caballus) to that of the Mongolian wild horse (E. przewalskii). *Cytogenetic and Genome Research*, *102*(1–4), 222–225.

Ning, T., Xiao, H., Li, J., Hua, S., & Zhang, Y. (2010). Adaptive evolution of the mitochondrial ND6 gene in the domestic horse. *Genetics and Molecular Research*, *9*(1), 144–150.

Oakenfull, E., & Clegg, J. (1998). Phylogenetic relationships within the Genus Equus and the Evolution of a and b globin genes. *Journal of Molecular Evolution*, *47*, 772–783.

Oakenfull, E., Lim, H., & Ryder, O. (2000). A survey of equid mitochondrial DNA: Implications for the evolution, genetic diversity and conservation of Equus. *Conservation Genetics*, *1*, 341–355.

Oakenfull, E. A., & Ryder, O. A. (1998). Mitochondrial control region and 12S rRNA variation in Przewalski's horse (Equus przewalskii). *Animal Genetics*, *29*(6), 456–459.

Outram, A. K., Stear, N. A., Bendrey, R., Olsen, S., Kasparov, A. et al. (2009). The earliest horse harnessing and milking. *Science*, *323*(5919), 1332–1335.

Perna, N., & Kocher, T. (1995). Patterns of nucleotide composition at fourfold degenerate sites of animal mitochondrial genomes. *Journal of Molecular Evolution*, *41*, 353–358.

Rokas, A., & Carroll, S. (2006). Bushes in the tree of life. *PLoS Biology 4*, e352.

Rokas, A., Williams, B., King, N., & Carroll, S. (2003). Genome-scale approaches to resolving incongruence in molecular phylogenies. *Nature*, *425*, 798–804.

Royo, L., Alvarez, I., Beja-Pereira, A., Molina, A., Fernandez, I. et al. (2005). The origins of Iberian horses assessed via mitochondrial DNA. *Journal of Heredity*, *96*(6), 663–669.

Ryder, O. (1994). Genetic studies of Przewalski's horses and their impact on conservation. In L. Boyd & K. A. Houpt (eds.), *Przewalski's horse: The history and biology of an endangered species* (pp. 75–92). Albany: The State University of New York Press.

Ryder, O. A. (1993). Przewalski's horse: Prospects for reintroduction into the wild. *Conservation Biology*, *7*(1), 13–19.

Ryder, O. A., Epel, N. C., & Benirschke, K. (1978). Chromosome banding studies of the Equidae. *Cytogenetics and Cell Genetics*, *20*(1–6), 332–350.

Ryder, O. A., & Wedemeyer, E. A. (1982). A cooperative breeding programme for the Mongolian wild horse Equus przewalskii in the United States. *Biological Conservation*, *22*(4), 259–271.

Sasaki, M., Endo, H., Yamagiwa, D., Yamamoto, M., Arishima, K., & Hayashi, Y. (1999). Morphological character of the shoulder and leg skeleton in Przewalski's horse (Equus przewalskii). *Annals of Anatomy*, *181*(4), 403–407.

Shen, Y.-Y., Liang, L., Zhu, Z.-H., Zhou, W.-P., Irwin, D., & Zhang, Y.-P. (2010). Adaptive evolution of energy metaboslim genes and the origin of flight in bats. *Proceedings of the National Academy of Science USA*, *107*(19), 8666–8671.

Short, R. V., Chandley, A. C., Jones, R. C., & Allen, W. R. (1974). Meiosis in interspecific equine hybrids. *Cytogenetic and Genome Research*, *13*(5), 465–478.

Steiner, C. C., Houck, M. L., & Ryder, O. A. (2011). Species identification and chromosome variation of captive two-toed sloths. *Zoo Biology 30*, 623–635.

Steiner, C. C., Mitelberg, A., Tursi, R., & Ryder, O. A. (2012). Molecular phylogeny of extant equids and effects of ancestral polymorphism in resolving species-level phylogenies. *Molecular Phylogenetics and Evolution 65*, 573–581.

Trifonov, V. A., Stanyon, R., Nesterenko, A. I., Fu, B., Perelman, P. L. et al. (2008). Multidirectional cross-species painting illuminates the history of karyotypic evolution in Perissodactyla. *Chromosome Research*, *16*, 89–107.

Vila, C., Leonard, J., Gotherstrom, A., Marklund, S., Sandberg, K. et al. (2001). Widespread origins of domestic horse lineages. *Science*, *291*, 474–477.

Volf, J., Kus, E., & Prokopovaá, L. (1991). *General studbook of the Przewalski horse*. Prague: Zoological Garden.

Wade, C. M., Giulotto, E., Sigurdsson, S., Zoli, M., Gnerre, S. et al. (2009). Genome sequence, comparative analysis, and population genetics of the domestic horse. *Science*, *326*(5954), 865–867.

Wallner, B., Brem, G., Muller, M., & Achmann, R. (2003). Fixed nucleotide differences on the Y chromosome indicate clear divergence between Equus przewalskii and Equus caballus. *Animal Genetics*, *34*, 453–456.

Xu, S., Luosang, J., Hua, S., He, J., Ciren, A. et al. (2007). High altitude adaptation and phylogenetic analysis of Tibetan horse based on the mitochondrial genome. *Journal of Genetics and Genomics*, *34*(8), 720–729.

Yang, F., Fu, B., O'Brien, P. C., Nie, W., Ryder, O. A., & Ferguson-Smith, M. A. (2004). Refined genome-wide comparative map of the domestic horse, donkey and human based on cross-species chromosome painting: insight into the occasional fertility of mules. *Chromosome Research*, *12*(1), 65–76.

Yang, F., Fu, B., O'Brien, P. C., Robinson, T. J., Ryder, O. A., & Ferguson-Smith, M. A. (2003). Karyotypic relationships of horses and zebras: Results of cross-species chromosome painting. *Cytogenetic and Genome Research*, *102*(1–4), 235–243.

Zong, E., & Fan, G. (1989). The variety of sterility and gradual progression to fertility in hybrids of the horse and donkey. *Heredity*, *62*(Pt. 3), 393–406.

Index

Acrosome – impaired acrosome reaction, 205
Adaptation in horse breeds (genetic underpinning), 318
Albinism, 163
Alopecia (Linear/Areata), 166
Arabian fading syndrome (AFS), 164–165

Banding techniques, 2–4
  C-banding, 3
  G-banding, 3, 50
  NOR-banding, 3, 51
  Q-banding, 3, 50
  R-banding, 3, 50
  T-banding, 3, 50

Cerebellar abiotrophy, 220–221
Chromosome aberrations, 4
  Autosome, 4
  Gametes and embryos, 201–202
  Mosaics/chimeras, 4–5
  Numerical, 4
  Structural, 4
  X chromosome, 4
  Y chromosome, 74, 83–84
Cervical vertebral malformation and malarticulation, 224–225
Chromosome number - horse, 1, 50
Chromosome size, 51–52
Chromosome/cytogenetic analysis (services), 248–250
Chromosomes and fertility, 200–201
Circadian clock – mammalian, 286–288
Circadian desynchrony, 299–301
Circadian-immune interaction, 297–299
Clock genes – peripheral, 290–292
Coat color genetics, 143–151
  Base colors (Black, Chestnut, Bay, Seal Brown), 143–144
  Dilutions (Cream, Pearl, Champagne, Dun and Lavender Foal), 144–146, 164
  White spotting and depigmentation (Frame, Tobiano, Sabino, Dominant
  White, Leopard Complex, Gray, Roan and White face & legs), 146–151, 163
Comparative map/genomics, 62, 64, 106–107
Cryptorchidism, 204–205
Curly coat syndrome, 165–166
Cytogenetic map, 54–56
Cytogenetic resources, 251

Degenerative myeloencephalopathy, 221–223
Dermal hypersensitivity – insect bite, 159–160
Dermatitis – pastern chronic, 160–162
Domestication of equids (mitochondrial basis), 315–317

Equine breeds – genomic attributes, 108–109
Equine genome sequence, 103–111
  functional elements, 126–127
  genes and numbers, 108
  genome assembly features, 105
  repetitive elements, 106
  special centromeres, 107
Equine SNP genotyping array (Beadchip)
  first generation, 113–115
  second generation, 122
  Use among Perrisodactyls, 120–122
  Use for genomewide mapping, 115–116
  Use for mapping across breeds, 118–119
  Use for mapping complex traits, 117–118
  Use for mapping simple traits within breeds, 116–117
  Use for population genetic analysis, 119–120
Evolution of *Equidae* – genetic basis (mitochondria), 312

Flowsorted and microdissected chromosomes, 56
Fluorescence in situ hybridization, 54–58
Forensic analysis, 243–244

Equine Genomics, First Edition. Edited by Bhanu P. Chowdhary.

Functional genomics, 125
  Biological understanding, 135–137
  Gene networks, 134

Genetic linkage maps, 12–46
  ECA1-ECAX maps, 17–44
  IEGMW map, 16
  IHRFP map, 14
  Newmarket map, 14
  Uppsala map, 14
Genetic testing
  diseases, 244–246
  phenotypic traits, 246–248
Glycogen branching enzyme deficiency (GBED), 173–174

Horse breeds history delineation (underlying genetics), 317
Hyperkalemic Periodic Paralysis (HYPP), 172
Hypotrichosis (follicular dysplasia), 165

Idiogram, 2
Immunodeficiency syndrome (foal), 165

Junctional epidermolysis bullosa, 157–159
Juvenile idiopathic epilepsy, 228–229

Karyotype, 2, 51, 53
Keratosis – linear, 162

Laminitis 255–262
  Functional genomics, 255–260
  Therapy trends, 261–262
  Transcriptional profiling, 260–261
Laryngeal neuropathy – recurrent, 233
Lavender foal syndrome, 164, 219–220

Major Histocompatibility Complex (MHC), 93–94
Malignant hyperthermia (MH), 172–173
Melanoma – Gray coat color, 163
Mitochondrial genome, 5, 311–319
Molecular map of equine MHC, 94–95
Monogeneic trait mapping using linkage map, 45
mtDNA and diversity, 6
mtDNA and phylogenetics, 6
Myoclonus – inherited, 227–228

Narcolepsy, 231–223
Navicular disease, 194
Nuclear genome, 1

Occipitoatlantoaxial malformation, 229–231
Optimum racing distance – GWAS (in Thoroughbreds), 275

Osteochondrosis (OC), 187–188
  Gene expression and candidate genes, 192–194
  Heretibility and genetic correlation, 188–191
  Quantative trait loci for OC, 191–192

Parentage analysis, 241–243
Performance – athletic, 265–278
  Circadian regulation of, 292–297
  Functional genomics and proteomics, 276
  Genetic variation impacting (in Thoroughbreds), 271–273
  Genomic regions under selection (in Thoroughbreds), 269–271
  Global gene expression variation, 277
  Heritability of, 266
  Mitochondrial genes impacting (in Thoroughbreds), 267–269
  Nuclear genes impacting, 267
Photic headshaking, 233–234
Physical maps, 49
Polymorphic genetic markers, 12–13
Polysaccharide storage myopathy – type I (PSSM), 174–178
Polysaccharide storage myopathy – type II (PSSM), 178–179
Proteome, 133–134
Pseudoautosomal region (PAR), 82–83

Radiation hybrid (RH) mapping, 58–62
Recurrent exertional rhabdomyolosis, 179–181
Regional dermal asthenia (HERDA), 156–157
Reproduction and fertility, 199

Sensory deafness (American Paint horse), 226–227
Sex reversal, 5, 84–85, 201
Shivers, 223–224
SNP map, 108
"Speed" gene, 273–275
Stallion fertility – genetics of, 202–205
Synteny mapping, 58

Trasncriptome, 128
  mRNA transcriptome, 128–133
  non-coding RNA transcriptome, 133
  sperm transcriptome, 207–210
  testis transcriptome, 206–207

White foal syndrome – lethal, 164, 217–218

Y chromosome (ECAY), 73–101
Y chromosome gene catalog, 77–80
Y chromosome map, 78, 80–82

Zoo-FISH map, 62, 63